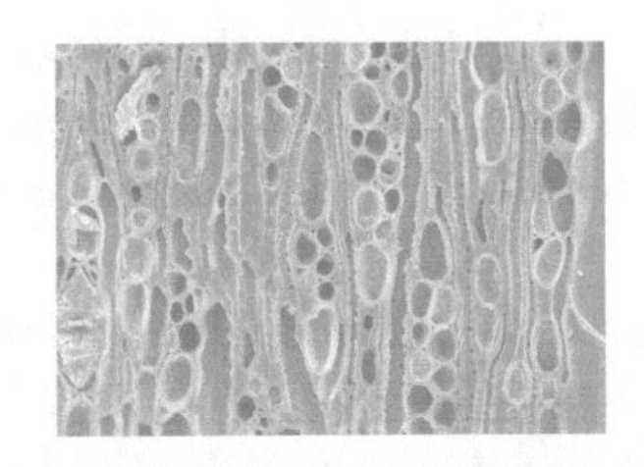

Archaeology of African Plant Use

Publications of the
Institute of Archaeology, University College London

Series Editor: Ruth Whitehouse
Director of the Institute: Stephen Shennan
Founding Series Editor: Peter J. Ucko

The Institute of Archaeology of University College London is one of the oldest, largest, and most prestigious archaeology research facilities in the world. Its extensive publications program includes the best theory, research, pedagogy, and reference materials in archaeology and cognate disciplines, through publishing exemplary work of scholars worldwide. Through its publications, the Institute brings together key areas of theoretical and substantive knowledge, improves archaeological practice, and brings archaeological findings to the general public, researchers, and practitioners. It also publishes staff research projects, site and survey reports, and conference proceedings. The publications program, formerly developed in-house or in conjunction with UCL Press, is now produced in partnership with Left Coast Press, Inc. The Institute can be accessed online at www.ucl.ac.uk/archaeology.

Recent Titles

Helen Dawson, *Mediterranean Voyages*
Chris J. Stevens, Sam Nixon, Mary Anne Murray, and Dorian Q Fuller (Eds.), *Archaeology of African Plant Use*
Andrew Bevan and Mark Lake (Eds.), *Computational Approaches to Archaeological Spaces*
Sue Colledge, James Conolly, Keith Dobney, Katie Manning, and Stephen Shennan (Eds.), *The Origins and Spread of Domestic Animals in Southwest Asia and Europe*
Julia Shaw, *Buddhist Landscapes of Central India*
Ralph Haeussler, *Becoming Roman?*
Ethan E. Cochrane and Andrew Gardner, *Evolutionary and Interpretive Archaeologies*
Andrew Bevan and David Wengrow (Eds.), *Cultures of Commodity Branding*
Peter Jordan (Ed.), *Landscape and Culture in Northern Eurasia*
Peter Jordan and Marek Zvelebil (Eds.), *Ceramics before Farming*
Marcos Martinón-Torres and Thilo Rehren (Eds.), *Archaeology, History, and Science*
Miriam Davis, *Dame Kathleen Kenyon*
Elizabeth Pye (Ed.), *The Power of Touch*
Russell McDougall and Iain Davidson (Eds.), *The Roth Family, Anthropology, and Colonial Administration*
Eleni Asouti and Dorian Q Fuller, *Trees and Woodlands of South India*
Tony Waldron, *Paleoepidemiology*
Janet Picton, Stephen Quirke, and Paul C. Roberts (Eds.), *Living Images*
Timothy Clack and Marcus Brittain (Eds.), *Archaeology and the Media*
Sue Colledge and James Conolly (Eds.), *The Origins and Spread of Domestic Plants in Southwest Asia and Europe*
Gustavo Politis, *Nukak*
Sue Hamilton, Ruth Whitehouse, and Katherine I. Wright (Eds.) *Archaeology and Women*

Information on older titles in this series can be obtained from the Left Coast Press, Inc. website www.LCoastPress.com

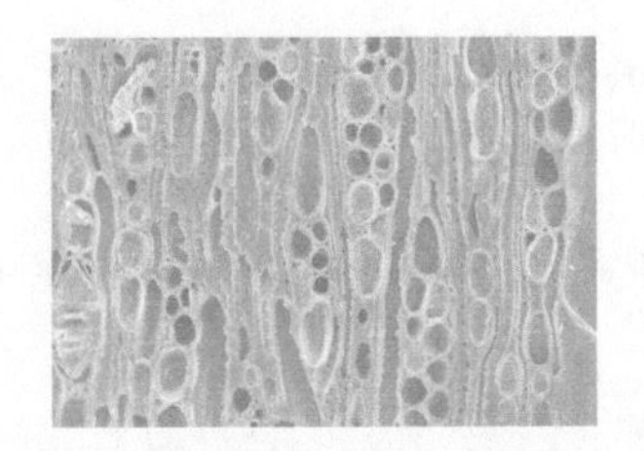

Archaeology of African Plant Use

EDITORS

CHRIS J. STEVENS
SAM NIXON
MARY ANNE MURRAY
DORIAN Q FULLER

LONDON AND NEW YORK

First published 2014 by Left Coast Press, Inc.

Published 2016 by Routledge
2 Park Square, Milton Park, Abingdon, Oxon OX14 4RN
711 Third Avenue, New York, NY 10017, USA

Routledge is an imprint of the Taylor & Francis Group, an informa business

Library of Congress Cataloging-in-Publication Data:
Archaeology of African plant use / Chris J. Stevens, Sam Nixon, Mary Anne Murray, Dorian Q Fuller, editors.
pages cm.— (Publications of the Institute of Archaeology, University College London) Includes bibliographical references and index.
ISBN 978-1-61132-974-2 (hardback : alk. paper)
ISBN 978-1-61132-976-6 (institutional eBook)
ISBN 978-1-61132-756-4 (consumer eBook)
1. Plant remains (Archaeology)—Africa. 2. Agriculture, Prehistoric—Africa. 3. Antiquities, Prehistoric—Africa. 4. Excavations (Archaeology)—Africa. 5. Africa—Antiquities. I. Stevens, Chris J., editor of compilation. II. Nixon, Sam (Archaeologist), editor of compilation. III. Murray, Mary Anne, editor of compilation. IV. Fuller, Dorian Q, editor of compilation. V. Fuller, Dorian Q African archaeobotany expanding. VI. Series: Publications of the Institute of Archaeology, University College London. DT13.A68 2013
561.196—dc23

2013022299

ISBN-13: 978-1-61132-974-2 (hbk)
ISBN-13: 978-1-61132-975-9 (pbk)

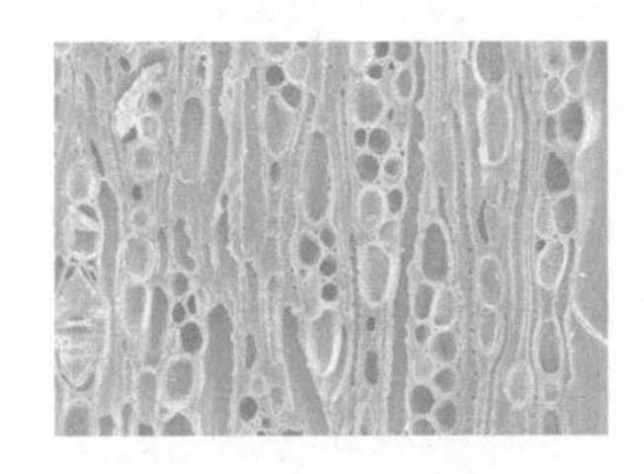

Contents

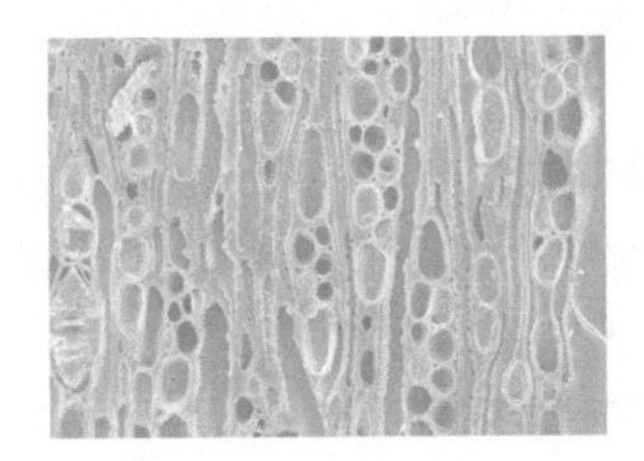

List of Illustrations

Figures

Tables

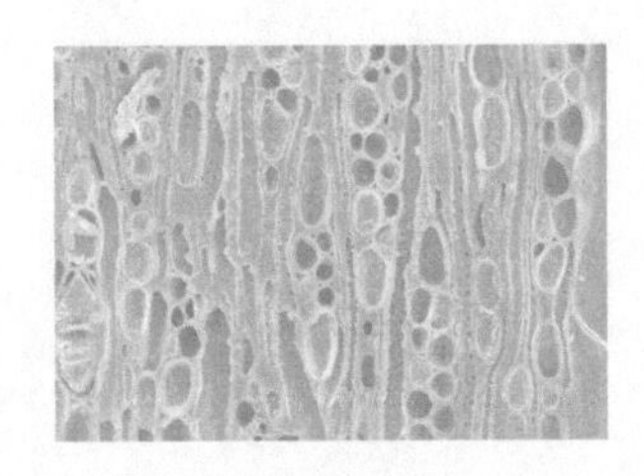

Acknowledgements

We dedicate this book to two past directors of the Institute of Archaeology, who did much to promote the development of archaeology and archaeobotany in Africa: David Harris, who inspired us to pursue past plant uses through archaeobotanical research in all regions, including Africa, and the late Peter Ucko, who encouraged us and supported us to host the African Archaeobotany conference.

We first must thank the many contributors to this volume, for their scholarship and insights. To those who were quick to get their chapters in, we are grateful for your patience with what has been a protracted editorial process.

When we held the International Workgroup for African Archaeobotany, we received support from the British Academy, which supported some scholars from various parts of Africa to attend the conference, including from Egypt, Nigeria, and Uganda. The conference also benefited from a picnic lunch, laboratory tour, and tea party at the Royal Botanical Gardens, Kew, and we must thank Mark Nesbitt, who made our visit there en masse smooth and fruitful. We also received support from our institution, numerous students who kept the conference running, and our past director Peter Ucko, whose energetic encouragement is deeply missed. We also wish to acknowledge our colleagues who chaired the sessions at the conference and helped to ensure beneficial discussions, including Katharina Neumann, René Cappers, and Marijke van der Veen.

The following illustrations were re-rendered by Chris Stevens: Figures 4.1, 8.1, 9.1, 9.2, 9.3, 12.1, 16.2, 18.3–18.8, 21.1, 21.3, 21.4, 21.7, 21.13, 21.16, 21.17, 21.18, 21.20, 21.21. In addition, Figure 5.5 was adapted from Table 5.2 created for Cartwright and associates. Figure 16.1 was created by Sam Nixon.

During the final editing process of this book, Chris Stevens has been a research fellow supported by the European Research Council grant to Dorian Fuller on "Comparative Pathways to Agriculture" (ComPAg, no. 323842). The research collected in this book enriches our understanding of pathways to agriculture in Africa.

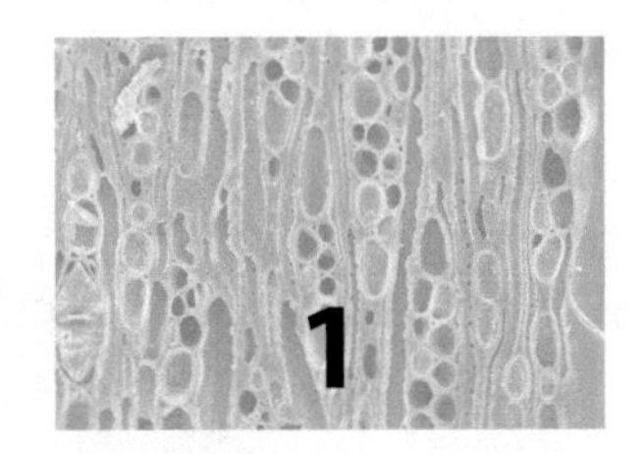

African Archaeobotany Expanding

An Editorial

Dorian Q Fuller, Sam Nixon, Chris J. Stevens, and Mary Anne Murray

Africa is a large continent, some 30.34 million square kilometres, dwarfed only by Asia (Times Atlas 2011). It stretches from the subtropics of the southern hemisphere (>34°S) to the subtropics of the north (>36°N). The Mediterranean environments of North Africa have their ecological parallel in parts of the southern Cape flora. Bordering these zones are deserts, the vast Sahara in the north, and the Namib and Kalahari deserts of the south. Between these deserts are diverse tropical environments of sub-Saharan Africa; savannahs, dry tropical woodlands, and moist rainforests. Although these environments boast the longest history of human occupation, as the cradle of hominid evolution during the Pliocene and Pleistocene they have received far less attention in terms of the documentation of past human plant use and archaeobotanical research. There are some 50 countries in Africa in addition to nearby Madagascar and the Canary Islands of Spain; however, half of these countries can speak of no archaeobotanical evidence at all. In this book, some 11 countries are represented by specific studies, which in itself reflects the slow, yet uneven advance of archaeobotanical research in Africa.

This book emerges from a meeting of the International Workshop for African Archaeobotany (IWAA) held in London in July 2006. This meeting of the IWAA has been held every three years since 1994, and all the conferences have produced edited volumes. Taken together these publications can be seen as representative of the general progress in the accumulation of African archaeobotanical knowledge (Figure 1.1). Since the 1994 conference's volume was published the following year, with a focus on the remains of Nabta Playa and some from Libya (Barakat 1995; Hather 1995; Mitka and Wasylikowa 1995; Wasylikowa et al. 1995; van der Veen 1995), the range of countries represented in the IWAA conferences has expanded, and correspondingly the number of studies has also expanded slightly for each country.

Nevertheless, there has been a marked dominance of studies on Egyptian sites. Although Egypt is unambiguously part of the African continent, for many archaeological specialists on Egypt (or Egyptologists) and for many Africanist archaeologists with a sub-Saharan focus, Egypt is regarded as a realm quite separate. Thus if we hold Egypt aside, we must admit that Africa remains a poorly studied continent by archaeobotanists. Although this lack of research is partly an inevitable result of political and logistical hurdles to working in many parts of Africa, there is still much scope for promoting the expansion of archaeobotany in Africa.

Figure 1.2 demonstrates just how many countries are untouched in archaeobotanical terms. For many countries where research has been conducted, the studies often cover just a few sites, and thus countries have been ranked in terms of whether they have been only minimally

Archaeology of African Plant Use by Chris J. Stevens, Sam Nixon, Mary Anne Murray, and Dorian Q Fuller, Eds., 17–24

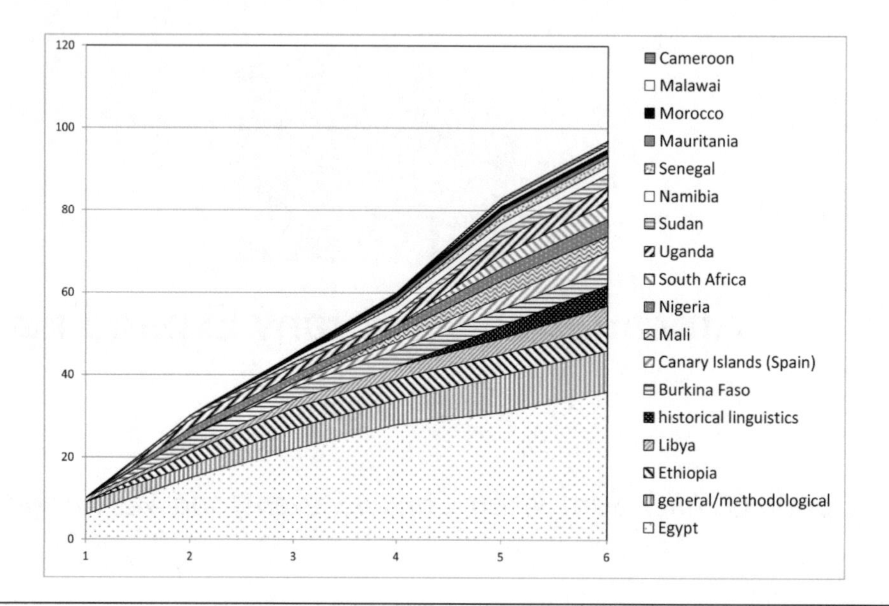

Figure 1.1 Graph of cumulative published contributions arising from the International Workshop for African Archaeobotany, subdivided by country, based on the published conference volumes. 1. *Acta Palaeobotanica* 35 1995; 2. van der Veen 1999; 3. Neumann, Butler, and Kahlheber 2003; 4. Cappers 2007; 5. this volume; 6. Fahmy, Kahlheber, and D'Andrea 2011.

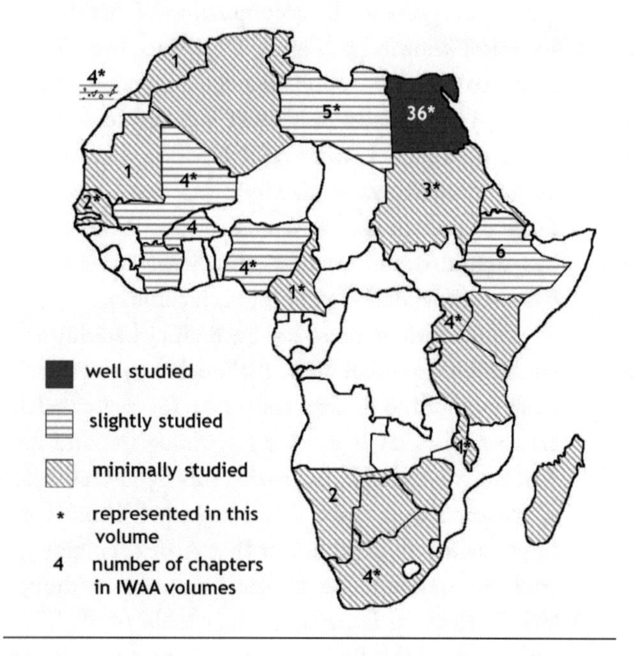

Figure 1.2 Map of the general distribution of archaeobotanical studies in Africa. Countries are shaded on the basis of how well-studied they are in three grades. Numbers of published articles in IWAA conference volumes are indicated on countries, as are countries studied in the present volume.

studied, slightly studied or, in the case of Egypt, well studied. Of those countries that have had some archaeobotanical work, just over half are represented in this volume. Geographically and chronologically expanding African archaeobotany and redressing the continuing bias toward Egypt should be priorities for the development of this field.

Toward the Study of African Agricultural Origins

Plant domestication is always one of the main research topics in archaeobotany. The origins of agriculture were one of the watersheds in studies of human economic history, the evolution of human social systems, and human modification of the environment (Bellwood 2005; Diamond 1997; Fuller, Willcox, and Allaby 2011). Archaeobotanical evidence provides a direct window on plant domestication processes (Fuller 2007; Purugganan and Fuller 2011), which can be regarded as complementing genetic evidence (Purugganan and Fuller 2009; Fuller and Allaby 2009).

Africa was first put on the map in terms of agricultural origins by Vavilov (1950), who defined Ethiopia as one of his centres of origin of cultivated plants. Vavilov and his colleagues had studied only crops in Ethiopia, and although it has many indigenous crop species, Ethiopia is also a region that has fostered significant diversity in crops introduced from Southwest Asia, particularly cereals such as emmer wheat and barley, which have always attracted much attention in agricultural origins research. In 1959 Murdock developed a hypothesis of a specifically western

African centre of domestication, from which many of the more widespread crops in Africa might have derived (Murdock 1959, pp. 64–77; see also the updated discussions of Harlan 1992; Harris 1976).

Although Murdock's deduction came in the context of reviewing ethnographic evidence for geographical patterns in various cultural behaviours, the 1960s also saw a significant development of botanical studies, especially by Harlan and colleagues, of modern landraces and wild populations of African crops, with particular studies focused on the three major African cereals—sorghum, pearl millet, and finger millet (Brunken, De Wet, and Harlan 1977; Harlan, De Wet, and Stemler 1976; Hilu and De Wet 1976; Hilu, De Wet, and Harlan 1979). Parallel botanical research from the French tradition included the early explorations of Chevalier (1932), especially of Saharan plants, and Porteres (1976) in western Africa. Such research encouraged the beginnings of the collection of archaeobotanical evidence in Africa, although flotation and the development of archaeobotanical laboratories with an African interest would develop only over subsequent decades. The foundation of what was known was drawn together in an edited volume on African plant domestication (Harlan, De Wet, and Stemler 1976). This baseline was expanded on by the proceedings of the IWAA conference published in the 1990s and the 21st century.

African Archaeobotany and the IWAA

The first IWAA conference held in 1994 in Krakow attempted to define some African archaeobotanical research problems. It was very much developed from the point of view of specialists on European archaeobotany, who were confronted with the rather different range of taxa encountered in African sites. Even though these were sites in the north, mainly Egypt, the early Holocene material, notably that of Nabta Playa, was dominated by tropical savannah grasses and could be discussed in relation to sorghum domestication (Wasylikowa et al. 1995).

This first small conference provided a focused breakaway meeting from the International Work Group for Palaeoethnobotany (IWGP), which had for more than two decades been a European-focused gathering for archaeobotanists. The IWAA was more than just a conference of presentations; it offered a chance for researchers to bring material with them and compare it at the microscope during a laboratory session. This model of a conference with open laboratory time remains a powerful platform for the advance of archaeobotanical research.

The second conference followed three years later, 1997, and was held in Leicester (van der Veen 1999). While this meeting also included updated reports on Nabta Playa and other early Holocene sites of Egypt's Western Desert, it had a wider geographical and chronological interest, including wood charcoal studies as well as West/Central African studies relating to such plant domesticates as pearl millet and rice, which were absent from Saharan Neolithic contexts such as Nabta Playa. Additionally, there were some papers on Roman-era long-distance trade in plants, inspired by the remarkably preserved material at Berenike on the Red Sea coast (Cappers 1999, 2003). This conference also saw the first African-focused discussions of crop-processing and the role of ethnoarchaeology in making sense of archaeological formation processes—and how these processes for African crops and cultural traditions might differ from those of the Mediterranean or European world (D'Andrea et al. 1999). One notable point of discussion in the conference was sorghum domestication, which raised issues about whether and why sorghum was such a 'late domesticate' in Africa—late compared to the early Holocene start of cultivation known from Southwest Asia or tropical America, and late compared to available finds in India. This critical discussion of sorghum problematised the finds of African crops outside Africa (Rowley-Conway, Deakin, and Shaw 1999), such as in India, and also raised questions about whether all crops were domesticated in the same way or whether crops such as sorghum were in some way different in their rate and pathway to domestication from better studied crops, such as wheat and barley (Haaland 1999).

The third conference, held in Frankfurt in 2000, continued to expand the geographical range covered in the proceedings as well as the methodological range (Neumann, Butler, and Kahlheber 2003). Noticeable for the first time was the involvement in the conference of historical linguists, who provided an alternative methodology for reconstructing past African agriculture much less patchy than that of archaeobotanical approaches, even if some inference of chronology and botanical identity are regarded as more problematic. This conference also included systematic reviews of African crops beyond Africa (De Moulins, Phillips, and Durrani 2003; Fuller 2003; Tengberg 2003), which demonstrated two key points, one methodological and the other historical. Methodologically, it is clear that much early work often requires critical appraisal, especially when it was carried out before the establishment of the research community per se, because identification criteria and dating associations can sometimes be wrong (a point already illustrated by Hilu, De Wet, and Harlan 1979). In terms of historical inference, however, it remained likely that finds in India are still somewhat earlier than those of the same crops in Africa, implying, given the range of the wild progenitors, that evidence for early domesticated pearl millet, sorghum, finger millet, and cowpea is still yet to be found in Africa. Rather than suggesting Early Holocene agricultural origins, however, these data pointed toward the mid-Holocene, 4,000–5,000 years ago, leading to the recognition that we needed to understand 'Africa behind-hand' (Neumann 2003). Most important, African archaeobotany has clearly called into question the often inferred synchronicity of agricultural origins in the early Holocene (Cohen 2009; Diamond 2002).

Since that conference African archaeobotany can be regarded as an established research focus, with the meetings held in Groningen in 2003 (Cappers 2007), London in 2006 (this volume), Cairo in 2009 (Fahmy, Kahlheber, and D'Andrea 2011), and Vienna in 2012 each contributing further to geographical, methodological, and chronological growth of the subject area. These meetings have promoted an eclectic view of archaeobotany, attracting papers in historical linguistics, comparative ethnography, and ethnoarchaeology, as well as varied archaeobotanical datasets, from phytoliths and wood charcoal to carpology (the study of seeds and related remains). The published volumes from these conferences serve as an index of the field's growth, as reference points in the history of African archaeobotany, and as repositories of new data and models of Africa's agricultural prehistory.

Although this book does not include chapters derived from all the oral presentations at the conference it does cover a representative majority. It also includes additional chapters that enhance the breadth of archaeobotanical topics. One of our aims has been to move beyond the traditional foci of agricultural origins and ancient Egyptian archaeobotany, although these are clearly important topics. We have therefore sought to highlight the importance of the growth of Palaeolithic archaeobotany in its broadest sense to include both empirical studies of Pleistocene plant remains and the theorization of the role of plant foods in human evolution and potential lines of enquiry into this topic. Additionally, we emphasise the importance of agricultural change and archaeobotanical research in the study of complex societies in Africa. Therefore, these two themes constitute two quarters of this book, which is organized around four areas of research.

STRUCTURE OF THE BOOK

ARCHAEOBOTANIES OF HOMINIDS AND THE PALAEOLITHIC

The study of early humans in Africa is interesting not only for the history of the continent but also, because Africa is the origin point of humans, on a world scale. Although hominid archaeology in Africa is a huge sphere of study, archaeobotanical study of hominids has never matched the pace of faunal research on hunting and scavenging and stone tool studies of human technological capability and cognition. Although plant foods are considered to be a central part of early hominin diet—and plant food processing, especially by cooking, a potential central technology of the human ecological niche (Carmody, Weintraub, and Wrangham 2011; Wollstonecroft 2011; Wollstonecroft et al. 2012; Wrangham et al. 1999)—archaeobotanical studies of Palaeolithic material are rare. Although this rarity has partly to do with the generally much poorer preservation of plant remains in Pleistocene sites, there remains a range of ways in which this problem can tackled—for example, by analogy and model-building, as well as by seeking more robust plant micro-remains (such as phytoliths). Africa has provided some of earliest examples of systematic collection of Pleistocene plant remains, admittedly of the very late Palaeolithic, at Wadi Kubbaniya, as well as drawing attention to those plant foods missing from the archaeological record (Hillman 1989; Hillman, Madeyska, and Hather 1989). The Palaeolithic record despite its challenges offers the potential for recovering plant remains and furnishing new understandings of plant use as our species evolved in a non-agricultural world.

This section of the book starts with Chapter 2, Haslam's fascinating new synthesis of plant use by the great apes, in the form of nut/tuber use. This study provides an essential background and lead-in to studies of hominid/Palaeolithic plant use, being parallel to ethnoarchaeology but in this case called 'ethoarchaeology', from ethology, the study of animal behaviour. As human evolution blurs the boundaries between human and nonhuman, the archaeobotany of human origins can strive to trace plant use through the comparison of chimpanzee plant processing 'cultures' (Lycett, Collard, and McGrew 2009; Whiten et al. 1999) and those of stone-tool and fire-using hominins.

Building on the analogical and model-building approach, Chapter 3 presents Hillman and Wollstonecroft's consideration of how the evolution of various food-processing techniques have been primary factors in increasing dietary diversity, possibly influencing evolutionary outcomes, such as brain size. Complementing the idea of early food use, research can also look at how archaeology can recover evidence of the environments and habitats of early humans in Africa.

This section then moves to more empirical studies of Pleistocene plant remains, which provide crucial evidence of the changing environments exploited by early humans. In Chapter 4, Sievers (née Scott) discusses Middle-Stone-age vegetation in KwaZulu-Natal in South Africa, providing a glimpse into past vegetation and showing the potential of seed remains to begin to reconstruct deep-time human environments in Africa. Although many archaeobotanists start with the assumption that seed remains are indicative of human food habits, she develops an alternative argument for seeds deriving from plants brought onto site for other uses, such as bedding. Likewise, in Chapter 5, Cartwright, Parkington, and Cowling show how charcoal remains can allow these deep-time environments to be reconstructed in considerable detail.

A West African Neolithic Revolution

The topic of agricultural origins is a mainstay of archaeobotanical research, and the best documented plant domestications in Africa are those of West Africa. This fact confirms Murdock's hypothesis that western Africa was an independent centre of plant domestication. This region also then had its own history of agricultural dispersal, which crossed diverse environmental zones from the Sahel through the savannahs and dry woodlands to the tropical rainforests. Here we include a number of contributions that fill in important details of this broader regional transition to farming.

We start this section with the best archaeologically documented domesticate in West Africa, pearl millet (*Pennisetum glaucum*). In Chapter 6, Manning and Fuller discuss a recently acquired early date for pearl millet, from the Tilemsi Valley. Here the evidence for domesticated pearl millet by the end of the third millennium B.C.E. points to earlier cultivation than hitherto known, which can perhaps be associated with the arrival of cattle-herding practices in the region. Nevertheless, early Holocene ceramic-using foragers have recently been recognized in the region—for example, at Ounjougou (Ozainne et al. 2009). Their presence means that there is a high potential for documenting cultural adaptations and subsistence changes through the Holocene that lead up to agriculture, which can in turn help disentangle the role of local adaptations and its influences from those of immigrants fleeing the drying Sahara in the middle Holocene.

In Chapter 7, Eichhorn and Neumann report on the wood charcoal sequence from Ounjougou, which provides a direct reflection of human exploitation of, and persistence within, changing environments over much of the Holocene in central Mali.

Moving farther west, in Chapter 8, Murray and Déme report early evidence from Senegal, which may relate to the establishment of agriculture there. It is unclear if agriculture in Senegal was established as early as the second millennium B.C.E., but the evidence from Walaldé clarified that agriculture focused on pearl millet was established by early in the first millennium B.C.E. Of potential interest is the absence of fonio (*Digitaria exilis*), since the later first millennium C.E. site of Cubalel produced significant quantities of grains of *Digitaria* sp. (Murray, Fuller, and Cappeza 2007). This and other small-grained cereals of western Africa remain poorly documented, and their precise origins in time and space are still elusive.

Subsequent chapters in this section relate to the establishment of cereal agriculture farther south in the wetter tropics. In Chapter 9, according to the palynological evidence Orijemie and Sowunmi show how pollen evidence provides a clear indication of the spread of agriculture and its effects on the forests in coastal Nigeria around the 'Dahomey gap', the break between the wet rainforests of west and central Africa. Of particular significance is the palynological argument that arboriculture based on oil palms (*Elaeis guineensis*) became a major part of a managed landscape.

In Chapter 10, Kahlheber, Höhn, and Neumann trace the arrival of West African grain agriculture, focused on pearl millet and legumes, in the wet rainforest zone of central Africa in Cameroon. Here they have found evidence for Iron Age millet cultivation in a region that generally has been regarded as too wet for this crop of Sahelian origin. By combining seed evidence with wood charcoal and phytolith analysis this experiment in wet tropical pearl millet cultivation can be situated within its local environmental context. This study provides clear evidence that the savannah agricultural package from West Africa initially thrust into the rainforest as farmers migrated southward. Subsequently, in this region grain crops were largely displaced by better adapted forest vegeculture systems. This situation raises questions about and the relationship between tuber and cereal agriculture in the African tropics and the dynamics that fuelled the early expansion of Bantu language speakers.

In Chapter 11, Bostoen uses linguistic methods to triangulate the origins of agriculture known to Bantu speakers over much of central, southern, and eastern Africa in the tuber crops and managed trees of tropical West Africa.

AGRICULTURE AND PLANT USE IN COMPLEX SOCIETIES

This section comprises chapters covering a range of archaeobotanical approaches to plant use and agriculture in complex societies. As one might expect, it includes studies of ancient Egypt, both the Middle Kingdom period at Saqarra and the New Kingdom period at Amarna. In Chapter 12, Fahmy, Kawai, and Yoshimura present evidence from cult chambers, and in Chapter 13, Stevens and Clapham examine the evidence from small households and their kitchen gardens.

The remarkable preservation in Egypt offer glimpses of categories of plant use rarely seen elsewhere, categories of plant use that must regarded as 'missing plants' in most world regions—spices such as coriander and vegetables such as carrot, both documented from the New Kingdom. Of interest in the Middle Kingdom material is evidence for date palms; this may be the period when systematic cultivation of dates, involving artificial pollination, began in Egypt.

Moving beyond Egypt we include studies from Nubia, Uganda, Malawi, and the Canary Islands. In Chapter 14, Fuller considers agricultural change and state collapse in Nubia, arguing that agricultural innovation brought in during the Meroitic empire that increased agricultural productivity contributed to the downfall and fragmentation of that kingdom by creating new opportunities through local wealth establishment by cotton cultivation and by increased population through double-cropping, including with sorghum.

In Chapter 15, Reid and Ashley explore the evidence for past intensive agricultural production on the islands of Lake Victoria, which supported a denser settlement pattern in that region than found in recent times.

In Chapter 16, Heijen reports the first archaeobotanical results from Malawi, associated with a large and relatively recent site, which among other things marks the presence of European-introduced maize as an addition to traditional sorghum-based subsistence.

In another island context, in Chapter 17, Morales, Rodríguez, and Marrero consider the likelihood that agriculture was abandoned during the history of the Canary Islands. Although Mediterranean agriculture came fairly late to those islands, it did not persist indefinitely once it was introduced.

ADVANCING THE ARCHAEOLOGY OF AFRICAN PLANT USE

In the final section we look at how archaeobotany has advanced in Africa and how related approaches can shed complementary light on the past of plant use. In Chapter 18, Pelling reports patterns from a large trans-Saharan archaeobotanical database, exploring how new patterns emerge as more data is analysed using multivariant methods.

In Chapter 19, Antonites and Raath Antonites sketch a critical history of archaeobotany in Southern Africa that highlights how much more could be done to promote such studies in the region.

In Chapter 20, Ehret provides a detailed example of how historical linguistics can build models for the origins and spread of agriculture. This chapter complements Chapter 11 in that it focuses on another region and other language groups, including Nilo-Saharan, Cushitic, and Omotic. Ehret outlines some methodological issues in building linguistic models for prehistory. For archaeobotanists, these models are a challenge, because they span both large geographical areas and time spans, for which, by and large, we lack any archaeobotanical record. This situation should encourage archaeobotanical sampling to test, tweak, and engage with this other line of evidence on the human past.

In Chapter 21, Blench provides us with another aspect of present variation, which relates to and can provide some models for the past through the consideration of modern ethnographic traditions of agricultural tool use. He reviews geographically, historically, and archaeologically some 21 categories of tools, including many subtypes. For some tools he also draws together comparative linguistic data. As do the other historical linguistic studies, this comparative material culture review poses a challenge for archaeologists to try to recover more physical evidence for the development and spread of these technologies.

The final two chapters consider plant impressions, not just as evidence for the plants themselves but also of how these plants relate to other technologies. In Chapter 22, through a comparison of an Irish case study and several African examples, McClatchie and Fuller consider how carbonized macro-remains often reveal a picture different from that seen in ceramic crop and plant impressions. Shifts in the nature of plant impressions are suggested to be driven by changes in how, when, and where pottery is produced, as much as by changes in plant use itself.

Finally, in Chapter 23, Iles presents a study of plant impressions in iron slag. Although these may be identifiable only to family level, they nevertheless show varying patterns in plant use that can be related to the wider vegetational environments in which iron-smelting was practiced. Both Chapters 22 and 23 point to the need for integrated or collaborative studies of plant remains alongside the artefact technologies that they have been incorporated in.

We hope that the range of approaches to African plant use, archaeobotany *sensu lato*, may inspire a wider international research community to be collaborative, comparative, and active.

REFERENCES

Barakat, H. N. (1995) Charcoals from Neolithic site at Nabta Playa (E-75-6), Egypt. *Acta Palaeobotanica* 35 (1), 163–66.

Bellwood, P. (2005) *First Farmers.* Oxford: Blackwell.

Brunken, J., J., M. J. De Wet, and J. R. Harlan (1977) The morphology and domestication of pearl millet. *Economic Botany 31,* 163–74.

Cappers, R. T. J. (1999) Trade and subsistence at the Roman Port of Berenike, Red Sea Coast, Egypt. In M. van der Veen (Ed.), *The Exploitation of Plant Resources in Ancient Africa,* pp. 185–98. London: Kluwer/Plenum.

———. (2003) Exotic imports of the Roman Empire: An exploratory study of potential vegetal products from Asia. In K. Neumann, A. Butler, and S. Kahlheber (Eds.), *Food, Fuel and Fields. Progress in African Archaeobotany, Africa Praehistorica* 15, 197–206. Köln: Heinrich-Barth-Institut.

———. (Ed.) (2007) *Fields of Change: Progress in African Archaeobotany,* Grongingen Archaeological Studies 5. Groningen: Barkhuis Publishing.

Carmody, R. N., G. S. Weintraub, and R. W. Wrangham (2011) Energetic consequences of thermal and nonthermal food processing. *Proceedings of the National Academy of Sciences 108* (48), 19199–203.

Chevalier, A. (1932) Les production végetales du Sahara. *Revuew de Botanique Appliqué et d'Agriculture Tropicale 12,* 669–924.

Cohen, M. (2009) Introduction: Rethinking the origins of Agriculture. *Current Anthropology* 50 (5), 591–95.

D'Andrea, A. C., D. E. Lyons, M. Haile, and A. Butler (1999) Ethnoarchaeological approaches to the study of prehistoric agriculture in the highlands of Ethiopia. In M. van der Veen (Ed.), *The Exploitation of Plant Resources in Ancient Africa,* pp. 101–22. London: Kluwer Academic/ Plenum.

De Moulins, D., C. Phillips, and N. Durrani (2003) The archaeobotanical record of Yemen and the question of Afro-Asian contacts. In K. Neumann, A. Butler, and S. Kahlheber (Eds.), *Food, Fuel and Fields, Progress in African Archaeobotany. Africa Praehistorica* 15, pp. 213–28. Köln: Heinrich-Barth-Institut.

Diamond, J. (1997) *Guns, Germs and Steel: The Fates of Human Societies.* New York: W.W. Norton.

———. (2002) Evolution, consequences and future of plant and animal domestication. *Nature 418,* 700–07.

Fahmy, A. G., S. Kahlheber, and A. C. D'Andrea (Eds.) (2011) *Windows on the African Past. Current Approaches to African Archaeobotany.* Reports in African Archaeology 3. Frankfurt: Africa Magna Verlag.

Fuller, D. Q (2003) African crops in prehistoric South Asia: A critical review. In K. Neumann, A. Butler, and S. Kahlheber (Eds.), *Food, Fuel and Fields. Progress in Africa Archaeobotany, Africa Praehistorica* 15, pp. 239–71. Köln: Heinrich-Barth-Institut.

———. (2007) Contrasting Patterns in Crop Domestication and Domestication Rates: Recent Archaeobotanical Insights from the Old World. *Annals of Botany 100* (5), 903–24.

Fuller, D. Q, and R. Allaby (2009) Seed dispersal and crop domestication: Shattering, germination and seasonality in evolution under cultivation. In L. Ostergaard (Ed.), *Fruit Development and Seed Dispersal, Annual Plant Reviews* 38, pp. 238–95. Oxford: Wiley-Blackwell.

Fuller, D. Q, G. Willcox, and R. G. Allaby (2011) Cultivation and domestication had multiple origins: Arguments against the core area hypothesis for the origins of agriculture in the Near East. *World Archaeology 43* (4), 628–52.

Haaland, R. (1999) The puzzle of the late emergence of domesticated sorghum in the Nile Valley. In C. Gosden and J. Hather (Eds.), *The Prehistory of Food. Appetites for Change,* pp. 397–418. London: Routledge.

Harlan, J. R. (1992) Indigenous African agriculture. In C. W. Cowan and P. J. Watson (Eds.), *The Origins of Agriculture,* pp. 59–70. Washington, D.C.: Smithsonian Institution Press.

Harlan, J. R., J. M. J. De Wet, and A. B. Stemler (Eds.) (1976) *Origins of African Plant Domestication.* The Hague: Mouton.

Harris, D. R. (1976) Traditional systems of plant food production and the origins of agriculture in West Africa. In J. R. Harlan, J. M. J. De Wet, and A. B. Stemler (Eds.), *Origins of African Plant Domestication,* pp. 311–56. The Hague: Mouton.

Hather, J. (1995) Parenchymatous tissues from the Early Neolithic site E-75-6 at Nabta Playa, Western Desert, South Egypt, preliminary report. *Acta Palaeobotanica* 35 (1), 157–62.

Hillman, G. C. (1989) Late Palaeolithic plant foods from Wadi Kubbaniya in Upper Egypt: Dietary diversity, infant weaning, and seasonality in a riverine environment. In D. R. Harris and G. C. Hillman (Eds.), *Foraging and Farming,* One World Archaeology 13. pp. 207–39. London: Unwin Hyman.

Hillman, G. C., E. Madeyska, and J. G. Hather (1989) Wild plant foods and diet of Late Palaeolithic Wadi Kubbaniya: The evidence from charred remains. In F. Wendorf, R. Schild, and A. Close (Eds.), *The Prehistory of Wadi Kubbaniya Vol. 2: Stratigraphy, Palaeoeconomy and Environment,* pp. 162–242. Dallas: Southern Methodist University Press.

Hilu, K. W., and J. M. J. De Wet (1976) Domestication of *Eleusine coracana. Economic Botany* 30, 199–208.

Hilu, K. W., J. M. J. De Wet, and J. R. Harlan (1979) Archaeobotanical studies of *Eleusine coracana* ssp. *coracana* (finger millet). *American Journal of Botany 66,* 330–33.

Lycett, S. J., M. Collard, and W. C. McGrew (2009) Cladistic analyses of behavioural variation in wild *Pan troglodytes:* Exploring the chimpanzee culture hypothesis. *Journal of Human Evolution 57,* 337–39.

Mitka, J., and K. Wasylikowa (1995) Numerical analysis of charred seeds and fruits from an 8000 years old site at Nabta Playa, Western Desert, South Egypt. *Acta Palaeobotanica* 35 (1), 175–84.

Murdock, G. P. (1959) *Africa: Its Peoples and Their Culture History.* New York: McGraw-Hill.

Murray, M. A., D. Q Fuller, and C. Cappeza. (2007) Crop production on the Senegal River in the early First Millennium AD: Peliminary archaeobotanical results from Cubalel. In R. Cappers (Ed.), *Fields of Change, Progress in African Archaeobotany.* Proceedings of the 4th International Workshop for African Archaeobotany, pp. 63–70. Groningen: Barkhuis Publishing.

Neumann, K. (2003) The late emergence of agriculture in sub-Saharan Africa: Archaeobotanical evidence and ecological considerations. In K. Neumann, A. Butler, and S. Kahlheber (Eds.), *Food, Fuel and Fields: Progress in African Archaeobotany,* Africa Praehistorica 15, pp. 71–92. Köln: Heinrich-Barth-Institut.

Neumann, K., A. Butler, and S. Kahlheber (Eds.) (2003) *Food, Fuel and Fields: Progress in African Archaeobotany, Africa Praehistorica* 15. Köln: Heinrich-Barth-Institut.

Ozainne, S., L. Lespez, Y. Le Drezen, B. Eichhorn, K. Neumann, and E. Huysecom (2009) Developing a chronology integrating archaeological and environmental data from different contexts: The Late Holocene Sequence of Ounjougou (Mali). *Radiocarbon 51,* 457–70.

Porteres, R. (1976) African cereals: Eleusine, Fonio, Black Fonio, Teff, Brachiaria, Pasapalum, Pennisetum, and African Rice. In J. R. Harlan, J. M. J. De Wet, and A. B. Stemler (Eds.), *Origins of African Plant Domestication,* pp. 409–52. The Hague: Mouton.

Purugganan, M., and D. Q Fuller (2009) The nature of selection during plant domestication. *Nature 457,* 843–48.

Purugganan, M., and D. Q Fuller (2011) Archaeological data reveal slow rates of evolution during plant domestication. *Evolution* 65 (1), 171–83.

Rowley-Conway, P., W. Deakin, and C. H. Shaw (1999) Ancient DNA from Sorghum: The evidence from Qasr Ibrim, Egyptian Nubia. In M. van der Veen (Ed.), *The Exploitation of Plant Resources in Ancient Africa,* pp. 55–62. London: Kluwer/Plenum.

Tengberg, M. (2003) Archaeobotany in the Oman peninsula and the role of Eastern Arabia in the spread of African crops. In K. Neumann, A. Butler, and S. Kahlheber (Eds.), *Food, Fuel and Fields: Progress in African Archaeobotany, Africa Praehistorica* 15, pp. 229–38. Köln: Heinrich-Barth-Institute.

Times Atlas (2011) *The Times Comprehensive Atlas of the World.* London: Times Books.

van der Veen, M. (1995). Ancient agriculture in Libya: A review. *Acta Palaeobotanica 35,* 85–98.

———. (Ed.) (1999) *The Exploitation of Plant Resources in Ancient Africa.* London: Kluwer/Plenum.

Vavilov, N. I. (1950) *The Origins, Variation, Immunity and Breeding of Cultivated Plants.* Translated by K. Starr Chester. *Chronica Botanica* 13. Waltham, MA: Chronica Botanica.

Wasylikowa, K., R. Schild, F. Wendorf, H. Krolik, L. Kubiak-Martens, and J. R. Harlan (1995) Archaeobotany of the early Neolithic site E-75-6 at Nabta Playa, Western Desert, South Egypt (preliminary results). *Acta Palaeobotanica* 35 (1), 133–55.

Whiten, A., J. Goodall, W. C. McGrew, T. Nishida, V. Reynolds, Y. Sugiyama, C. E. G. Tutin, R. W. Wrangham, and C. Boesch (1999) Cultures in Chimpanzees. *Nature 399,* 682–85.

Wollstonecroft, M. M. (2011) Investigating the role of food processing in human evolution: A niche construction approach. *Archaeological and Anthropological Sciences 3,* 141–50.

Wollstonecroft, M. M., P. R. Ellis, G. C. Hillman, D. Q Fuller, and P. J. Butterworth (2012) A calorie is not necessarily a calorie: Technical choice, nutrient bioaccessibility, and interspecies differences in plants. *Proceedings of the National Academy of Sciences 109* (17), E991.

Wrangham R., J. H. Jones, G. Laden, D. Pilbeam, and N. L. Conklin-Brittain (1999) The raw and the stolen: Cooking and the ecology of human origins. *Current Anthropology 40* (5), 567–94.

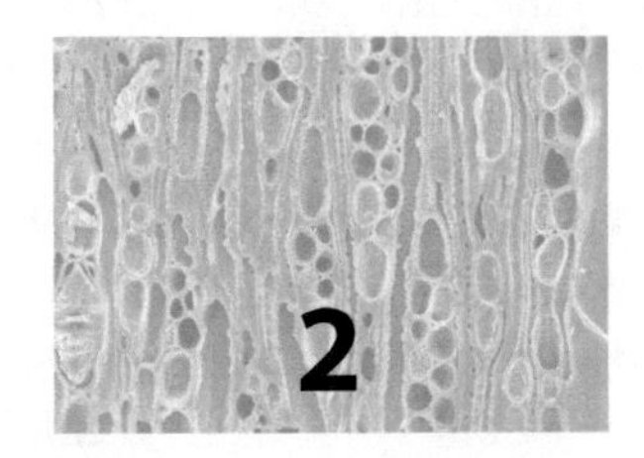

Primate Archaeobotany

The Potential for Revealing Nonhuman Primate Plant Use in the African Archaeological Record

Michael Haslam

Recognition is growing that valuable comparative data on human evolution may be gained through an examination of the archaeological remains left by nonhuman species, including the closely related and technology-proficient chimpanzee (*Pan troglodytes*) and extending to more distant members of the human family tree (Haslam 2012; Haslam et al. 2009). Because much of the justification for extending archaeology to nonhuman primates comes from plant processing and plant-based technologies of species such as chimpanzees and capuchin monkeys (Mercader et al. 2007; Mercader, Panger, and Boesch 2002; Ottoni and Izar 2008; Visalberghi et al. 2007), it is therefore necessary to explore the extent to which the activities of such animals can be detected and interpreted in the archaeobotanical record. This perspective is important not just for understanding the intraspecies chronological development of nonhuman primate plant use but also because it bears on discussions of factors such as cooking and nonlithic tool use in human evolution. Because Africa was the primary centre for the evolution of both humanity and our closest primate relatives, African archaeobotanists have a central role to play in exploring primate archaeobotany.

Archaeology of African Plant Use by Chris J. Stevens, Sam Nixon, Mary Anne Murray, and Dorian Q Fuller, Eds., 25–35

Primate Ethnobotany and Archaeological Potentials

Archaeobotanical studies cover a number of overlapping categories, including diet, domestication, construction, fuel, plant-processing technologies, plant-tools, and nonsubsistence exploitation of plant properties (for example, medicines, poisons). It would be impossible in a short chapter to review all such aspects of primate-plant interactions across Africa. Instead, here I focus on components of plant processing and exploitation (subsistence and technology) by the closest genetic relatives to humans, the chimpanzees, as a means of introducing the notion of primate archaeobotany. Possible extensions of these themes to other nonhuman primates are noted only where appropriate and are by no means systematic or exhaustive. Note that the term *ethobotany* is used here, in contrast to *ethnobotany*, to distinguish extant nonhuman plant exploitation.

Chimpanzee Ethobotany: Subsistence

Chimpanzees occupy a broader range of environments than many primates do, including lowland and montane forests, savannah, and woodlands; however, as would be expected from forest-dwelling common ancestry (Milton 1993), most primate species live today in tropical forests, and fruits, leaves, flowers, and other vegetation make up the majority of the typical primate diet. Of the great apes, gorillas and orangutans consume only a very small nonplant component (chiefly invertebrates), and, although chimpanzees eat a higher percentage of animal matter thanks to practices such as termite-fishing and the hunting of colobus monkeys, this amount is estimated to contribute at most around 5–10% of the total diet (Milton 1999, 2003; Tappen and Wrangham 2000). Evidence for bonobo (*Pan paniscus*) faunivory is rare (McGrew et al. 2007).

Chimpanzee diets, like those of the bonobo and gorilla, are preferentially composed of ripe tree-fruits (Laden and Wrangham 2005; Milton 2003). In contrast to the large gorillas and orangutans, however, chimpanzees also target nutrient-rich plant foods that require behavioural adaptations to access them. These include hard-shelled nut species that must be cracked using tools, palm pith pounded using a frond to produce an edible pulp, and underground storage organs (USOs) reached using digging sticks (Hernandez-Aguilar, Moore, and Pickering 2007; McGrew 1992; Yamakoshi and Sugiyama 1995). A non-exhaustive, selected list of plant taxa exploited for subsistence by chimpanzees is summarised in Table 2.1, providing an initial target list for archaeobotanical study of these apes. Foods processed using technology (for instance, *Strychnos* spp., *Coula edulis*, *Treculia africana*) have been emphasised, as these may leave processing residues (for example, Koops, McGrew, and Matsuzawa 2010) detectable on archaeologically recovered tools. More extensive species lists are provided in the primatological literature (for instance, McGrew, Baldwin, and Tutin 1988; Tweheyo, Lye, and Weladji 2004).

Recent emphasis on the role of 'fallback foods', typically low nutrient or difficult-to-access foods relied on at times of seasonal or other shortage, as drivers of evolutionary change (Laden and Wrangham 2005) has led to the suggestion that chimpanzee behavioural innovations such as tool use and fission-fusion foraging parties evolved to allow them to continue to subsist on high-quality foods year-round (Chapman, White, and Wrangham 1994; Lambert 2007). Tool use among New World capuchin monkeys (*Sapajus libidinosus*), including digging for tubers and cracking nuts with hammers and anvils (Ottoni and Izar 2008; Visalberghi et al. 2009), may represent an independently evolved approach similar to that of the chimpanzee (Lambert 2007), although food scarcity appears not to be a strong driver of nut-cracking among these monkeys (Spagnoletti et al. 2012). An alternative strategy, seen in gorillas, for example, involves anatomical (for instance, dental, digestive, and body-size) adaptations that permit processing of lower nutritional density but abundant foods such as mature leaves and bark (Harcourt and Stewart 2007). Clearly, *Homo sapiens* tend to the behavioural more than anatomical end of this adaptive continuum.

The potential for identifying nonhuman primate tools in the archaeobotanical record is discussed further later in the chapter, but first we consider the kinds of evidence that the chimpanzee and other primate diets may leave behind. The most direct evidence is found in fecal deposits, including seeds and chitinous insect exoskeletons, and plant remains integrated into dental calculus. There exists a substantial comparative record of fecal contents for many extant primate species (Behie, Pavelka, and Chapman 2010; Bradley et al. 2007; Moreno-Black 1978; Tutin and Fernandez 1993) collected over several decades, but high turnover of organic material within forested environments acts against long-term preservation. Drier areas inhabited by chimpanzees, such as Assirik (McGrew et al. 2003) and Fongoli (Pruetz and Bertolani 2007) in Senegal, may provide better opportunities for organic longevity. Perhaps the best opportunity for recognising past chimpanzee occupation via fecal material comes from build-up through repeated use of nesting sites (Sept 1998), since chimpanzees tend to defecate upon arising (McGrew et al. 2003). Dental calculus is receiving increasing attention as a degradation-resistant archaeobotanical trap, especially for plant microfossils (Cummings and Magennis 1997; Henry and Piperno 2008), with starches extracted and identified from the calculus of two recently deceased chimpanzees from Kibale, Uganda (Hardy et al. 2009). Claims have also been made for the recovery of diet-related phytoliths from the teeth of an extinct Asian ape, *Gigantopithecus* (Ciochon, Piperno, and Thompson 1990). The limits of this approach are as yet unknown, but the possibility cannot be ruled out that primate and ancestral human dietary niches and food acquisition strategies, including tool use where consumed foods are otherwise inaccessible to the studied species, may be identified dating back even millions of years.

A less direct form of evidence, but one that is potentially very important, is starch residues on artefacts. Starch residues have played an important role in the development of the emerging field of primate archaeology. It was the differentiation via palaeoamylogy (ancient starch analysis; Haslam 2004) of plants not consumed by humans that allowed for confidence in the attribution of 4,300-year-old stone tools to chimpanzee nut-cracking (Mercader et al. 2007). Excavated sites in Taï National Park, Côte d'Ivoire, contained both unquestionably human-manufactured flaked stone artefacts along with percussion-damaged stones, and the identification of starch residues from three

Table 2.1 Selected Plant Taxa Exploited by Chimpanzees (*Pan troglodytes*) for Subsistence, Maintenance, and Technological purposes.[†] (Note that variations exist between different chimpanzee subspecies, habitats, and cultural traditions and that this table is non-exhaustive in coverage.)

Taxa	Plant Part	Use	Area(s)
Ficus spp.	Fruit	Food	Kahuzi-Biega National Park, Democratic Republic of the Congo; Mt. Assirik, Senegal; Budongo Forest Reserve, Uganda; Mahale Mountains National Park, Tanzania; Ugalla, Tanzania
Myrianthus holstii *Rubus* spp.	Fruit	Food	Kahuzi-Biega National Park, Democratic Republic of the Congo
Coula edulis *Panda oleosa* *Parinari excelsa* *Sacoglottis gabonensis*	Nut	Food	Taï National Park, Côte de'Ivoire; Sapo National Park, Liberia
Detarium senegalense	Nut	Food	Taï National Park, Côte de'Ivoire; Tiwai Island, Sierra Leone
Dacryodes gabonensis *Trichoscypha arborea*	Fruit	Food	Taï National Park, Côte de'Ivoire
Elaeis guineensis	Nut	Food	Bossou, Guinea
Pseudospondias microcarpa	Fruit	Food	Kibale Forest, Uganda; Mt Assirik, Senegal
Uvariopsis congensis	Fruit	Food	Kibale Forest, Uganda
Strychnos spp.	Fruit	Food	Gombe National Park, Tanzania; Ugalla, Tanzania; Mt. Assirik, Senegal; Taï National Park, Côte de'Ivoire; Yealé, Côte de'Ivoire
Conopharyngia sp.	Fruit	Food	Gombe National Park, Tanzania; Kibale Forest, Uganda
Diplorhynchus condylocarpon	Seeds	Food	Gombe National Park, Tanzania
Balsamocitrus sp.	Fruit	Food	Budongo Forest Reserve, Uganda
Broussonetia papyrifera	Leaves, Fruit	Food	Budongo Forest Reserve, Uganda
Cynometra alexandri *Pterygota mildbraedii*	Fruit	Food	Virunga National Park, Democratic Republic of the Congo
Harungana madagascariensis	Fruit	Food	Mahale Mountains National Park, Tanzania
Adansonia digitata *Saba senegalensis,* *Cola cordifolia* *Lannea* spp. *Grewia lasiodiscus* *Hexalobus monopetalus* *Tamarindus indica* *Zizyphus* sp.	Fruit	Food	Mt. Assirik, Senegal
Pterocarpus erinaceus	Bark	Food	Mt. Assirik, Senegal
Detarium microcarpum	Nut	Food	Gashaka Gumti National Park, Nigeria
Treculia africana	Fruit	Food	Nimba Mountains, Guinea; Taï National Park, Côte de'Ivoire
Grewia spp. *Canthium hispidum*	Fruit	Food	Ugalla, Tanzania
Dolichos kilimandscharicus *Tacca leontopetaloides* *Raphionacme welwitschii* *Brachystegia bussei* *Smilax* sp. *Fadogia quarrei* *Costus macranthus*	Underground storage organs (USOs)	Food	Ugalla, Tanzania
Unspecified	*	Extractive foraging	All chimpanzee sites

(*Continued*)

Table 2.1 Continued

Taxa	Plant Part	Use	Area(s)
Thomandersia hensii	Sticks	Puncture termite nests	Nouabalé-Ndoki National Park, Republic of the Congo
Sarcophrynium spp. *Megaphrynium* sp.	Stalks	Termite-fishing, ant-dipping	Nouabalé-Ndoki National Park, Republic of the Congo
Ataenidia conferta *Haumania danckelmaniana*	Stalks	Termite-fishing	Nouabalé-Ndoki National Park, Republic of the Congo
Grewia lasiodiscus *Cissus* sp. *Pericopsis laxiflora* *Landolphia heudelotii* *Oxytenanthera abyssinica* *Pterocarpus erinaceous*	Twigs, leaf-stalks, vines, bark	Termite-fishing	Mt. Assirik, Senegal
Ficus vallis-choudae *Saba comorensis* *Landolphia owariensis* *Grewia flavescens*	Leaves	Water-sponge	Mahale Mountains National Park, Tanzania
Ficus thonningii	Fruit	Water-sponge	Mahale Mountains National Park, Tanzania
Elaeis guineensis	Palm frond	Pounding palm apex	Bossou, Guinea
Hybophrynium braunianum	Leaves	Water-sponge	Bossou, Guinea
Aspilia spp.	Leaves	Medicinal	Gombe and Mahale Mountains National Parks, Tanzania
Vernonia amygdalina	Shoots	Medicinal	Mahale Mountains National Park, Tanzania
Brachystegia bussei *Julbernardia globiflora* *Combretum molle* *Pterocarpus tinctorius*	Branches	Nesting	Ugalla, Tanzania
Crysophyllum albidum	Branches	Nesting	Virunga National Park, Democratic Republic of the Congo; Kibale Forest, Uganda
Cynometra alexandri	Branches	Nesting	Virunga National Park, Democratic Republic of the Congo
Garcinia sp.	Branch	Honey extraction	Loango National Park, Gabon
Thomandersia laurifolia *Milletia* sp.	Sticks	Puncture termite nest	Dzanga-Sangha region, Central African Republic
Haumania danckelmaniana *Sarcophrynium* spp. *Dalhousiea africana*	Sticks	Termite-fishing	Dzanga-Sangha region, Central African Republic
Ataenidia conferta *Sarcophrynium* spp.	Petiole; Sticks	Termite-fishing	Dja Biosphere Reserve, Cameroon; Dzanga-Sangha region, Central African Republic
Alchornea floribunda *Tabernaemontana crassa*	Sticks	Puncture termite nests	Dja Biosphere Reserve, Cameroon
Megaphrynium macrostachyum	Sticks	Termite-fishing	Dja Biosphere Reserve, Cameroon

*Many reports do not present data on plant taxa used for activities such as ant-dipping, termite-fishing, and honey-gathering; the tools are typically described only as sticks, twigs, branches, leaves, leaf mid-ribs, and so on. In some cases chimpanzees may not be selective in the taxa used for these activities, but modification to shape or trim the tool may still be practiced (for example, Boesch, Head, and Robbins 2009; Fowler and Sommer 2007; Sanz and Morgan 2007). A similar lack of botanical identification applies in many instances to plant parts used as weapons (for example, Pruetz and Bertolani 2007) or digging sticks (Hernandez-Aguilar, Moore, and Pickering 2007) and in social displays, and for a variety of self-directed maintenance and stimulation activities (McGrew 1992).

†Data compiled from (Anderson, Williams, and Carter 1983; Boesch and Boesch-Achermann 2000; Boesch, Head, and Robbins 2009; Deblauwe et al. 2006; Fay and Carroll 1994; Fowler and Sommer 2007; Hernandez-Aguilar 2009; Hernandez-Aguilar, Moore, and Pickering 2007; Huffman and Seifu 1989; Isabirye-Basuta 1988; Koops, McGrew, and Matsuzawa 2010; Marchant and McGrew 2005; Matsusaka et al. 2006; McBeath and McGrew 1982; McGrew 1992, 2004; McGrew, Baldwin, and Tutin 1988; McGrew et al. 1999; Nishida 1989; Sanz and Morgan 2007; Sept 1992; Sugiyama 1997; Tonooka 2001; Tweheyo, Lye, and Weladji 2004; Whiten et al. 2001; Whitesides 1985; Wrangham and Nishida 1983; Yamagiwa and Basabose 2009).

nut species exploited exclusively by chimpanzees (*Parinari excelsa*, *Panda oleosa*, and *Detarium senegalense*) provided the strongest evidence in favour of assigning these tools to chimpanzees. To date this study remains the only dated and confirmed nonhuman ape archaeological site, but the potential for further such work and resultant insights into the time-span and evolutionary role of ape (including human) technology are significant.

Starchy foods are preferentially targeted by technology-using nonhuman primates, including nut-cracking by chimpanzees and capuchins and more rarely digging for USOs by chimpanzees, capuchins, and, anecdotally, also by baboons (Hernandez-Aguilar, Moore, and Pickering 2007; Mannu and Ottoni 2009; Marlowe and Berbesque 2009). Plant residues, including starches and fibrous material from nut processing, are routinely embedded and retained on stone and wooden hammers and anvils during use (for example, Barton 2007). Plant use-residues can be differentiated even on wooden digging utensils, as demonstrated by study of ethnographic Australian digging sticks (Nugent 2006). Guidance as to sampling locations for primate archaeobotanical residues is provided by distinct wear traces left by pounding activities, such as cupule formation in sandstone anvils by capuchins (Visalberghi et al. 2007) and impact pitting on chimpanzee hammer stones (Haslam et al. 2009).

Starches also play a prominent role in discussions of the evolution of cooking and other behavioural adaptations by humans. Cooking improves foods for consumption by breaking down physical barriers, altering molecular structure, and reducing toxin load (Wrangham 2007; Wrangham et al. 1999), which in turn make more items edible and may reduce energetic expenditure on digestion. A preference for cooked foods, including starchy tubers, has been shown in a pilot study among all great apes (Wobber, Hare, and Wrangham 2008), suggesting that once cooking began in the hominin line it could have been taken up relatively rapidly and had follow-on effects on the life history and social behaviour of our ancestors. Documenting the effects of primate plant processing activities on starch residues (including pounding and grinding) may provide an avenue for identifying the entry and spread of cooked foods into the hominin diet and for differentiating hominin from nonhuman primate processing activities. However, further comparative work is required to document the effects of mechanical damage (Babot 2003), and gelatinisation through heat and moisture (Crowther 2012; Henry, Hudson, and Piperno 2009), on starches from different species.

Durable material surface remains also act as guides to the location of buried primate archaeological deposits, with present-day and recently buried nut-cracking localities liberally strewn with nutshells (Mercader, Panger, and Boesch 2002; Visalberghi et al. 2007) and broken hammer fragments. Under favourable conditions shells may survive for a considerable time, as evidenced by remains found with pitted stones at the Early-Middle Pleistocene hominin site of Gesher Benot Ya'aqov in Israel (Goren-Inbar et al. 2002). Of relevance here is that not all chimpanzee groups crack nuts, although the hypothesised restriction of this activity to sites west of the N'Zo-Sassandra River in Côte d'Ivoire has recently been shown to be incorrect (Morgan and Abwe 2006). Nevertheless, there remain chimpanzee groups with access to the relevant nut species and percussion materials that do not use these resources (Boesch, Head, and Robbins 2009; Whiten et al. 2001). Whether this absence results from local loss of technological proficiency, lack of innovative ability, or other explanations is currently unknown (Wrangham 2006).

Chimpanzee Ethobotany: Technology

Vegetation is the primary raw material for nonhuman primate tools (McGrew 1992), although the relative roles of stones and plants vary between taxa (Mannu and Ottoni 2009). Greater recognition by archaeologists of the varied forms of primate plant technology is necessary to identify nonhuman contributions to the archaeobotanical record and aids both primatological and human evolution studies by building a record of technological development for extant species outside our own ancestral lineage. Chimpanzees were the first nonhuman species for which intensive observations of tool use were made, and some field sites (for example, Gombe and Mahale M in Tanzania) have mutlidecadal data (McGrew 2004), making this species arguably the most appropriate for discussing nonhuman plant-tool use.

The most comprehensive study of variation in extant chimpanzee behaviour (Whiten et al. 2001) identified 65 behaviour patterns that were present at multiple sites. These included such actions as probing, pounding, dipping, and clasping, although not all involved the use of external objects. It's important to note that 57 of these behaviours involve plants in some fashion, and 47 involve detached pieces of vegetation such as sticks and stems used to fish for insects, leaves used to collect water by sponging and wipe surfaces, and wooden hammers used to crack nuts (Whiten et al. 2001). Note also that 10 of the 19 'very likely cultural variants' identified for extant orangutan populations (van Schaik et al. 2003) likewise involve detached plant tools. Table 2.1 summarises selected plant taxa used as tools by chimpanzees.

Modification and even standardisation of chimpanzee plant implements occurs for activities such as ant-fishing and honey extraction (Boesch and Boesch 1990; Boesch, Head, and Robbins 2009; McGrew 2004), and together with the location of discarded tools near extraction sites this purposeful modification allows primatologists to readily identify such items even in the absence of the tool user (for instance, Fowler and Sommer 2007). The same recognition should be feasible for archaeobotanists encountering these tools in archaeological contexts, provided taphonomic factors permit survival and the analyst is aware of the characteristics of nonhuman primate tools. Residue studies may be possible when such implements are identified, especially on tools used to access starchy foods or gather persistent materials such as honey (Boesch, Head, and Robbins 2009; Sanz and Morgan 2009). Residues (in the form of prey-species hair) have been observed adhering to a wooden tool used in the manner of a spear by savannah chimpanzees to immobilise lesser bushbabies (*Galago senegalensis*) (Pruetz and Bertolani 2007), and if this tradition persists then retention of blood and other bodily components on a subset of such tools is likely. As in the case of vegetal probes, the hierarchical manufacturing process employed for these 'hunting' tools may assist with their identification in the archaeological record.

A key element of the chimpanzee ethobotanical record, noted previously, for example, in the context of nut-hammering behaviour, is the presence of cultural variation between different communities. In this instance 'cultural' is defined as 'sufficiently frequent at one or more sites to be consistent with social transmission, yet absent at one or more others where environmental explanations [for the absence] were rejected' (Whiten et al. 2001, pp. 148–82). In practice cultural variation means that the full range of material culture employed by any given chimpanzee group (past or present) cannot be predicted in advance, even where full ecological information is known. It also means that the chimpanzee archaeobotanical record in any one region is expected to show cultural drift over time, with tool-forms coming in and out of fashion within functional constraints. The pace at which this occurs is very unlikely to be anywhere near as rapid as that seen in modern humans, especially since chimpanzees appear to lack a strong 'ratchet effect' for cumulative cultural accumulation over generations (Tennie, Call, and Tomasello 2009). Nevertheless, any longitudinal data collected by archaeobotanists that bear on this issue would provide useful guiding parameters as to the rate of cultural change for early hominin technologies. It has been noted that primatology benefits significantly over archaeology in having living subjects to study (Haslam et al. 2009); however, from an evolutionary perspective this supposition is valid only to the extent that extant primate populations are seen as socially and technologically representative of past behaviours. Archaeobotanical study of chimpanzees, and other potentially cultural primates such as orangutans and capuchin monkeys, will help ascertain if the recognised fallacy of uncritically projecting modern human activities back into prehistory also holds for nonhuman primates (Haslam 2012).

A final category of botanical exploitation likely to have bearing on our views of human evolution is the construction of nests (Sept 1998). All great apes daily construct nests from bent and interwoven branches and leaves for resting and sleeping (Fruth and Hohmann 1994; van Schaik et al. 2003; Yamagiwa 2000). The precise function of arboreal chimpanzee nests is not well understood (McGrew et al. 2003), with hypotheses including thermoregulation or predator avoidance, but the spatial patterning and repeated use of ape nesting sites can broaden our perspective on early human site formation (Hernandez-Aguilar 2009; Sept 1992; Stewart, Piel, and McGrew 2011). Nesting sites act as foci for activity and potentially for accumulation of plant-food debris such as seeds and other feeding or self-maintenance debris, with the nests themselves surviving typically for a few weeks to several months or even years depending on the tree species used, the season of construction, and whether the branches used were broken during construction (McGrew et al. 2003). Nests are, of course, extremely unlikely to be observed archaeologically unless a fallen tree or ground nest is covered under anaerobic conditions; it is the spatial and temporal shaping of other behaviours that is of greater interest.

DEVELOPING PRIMATE ARCHAEOBOTANY

For all archaeobotanical studies there is a prime concern with taphonomy. Most ephemeral actions of chimpanzee and other primates of relevance to human evolution (by way of relatedness or technological aptitude) do not typically leave material traces that survive natural nutrient recycling processes. However, primate stone tools demonstrably are recoverable several thousand years after their use (Mercader et al. 2007; Mercader, Panger, and Boesch 2002), and microbotanical residues on these offer a rare opportunity to record nonhuman technological and subsistence behaviour in the distant past. Hard woods selected and/or modified for use as digging sticks and probes, and dense scatters of primate-targeted nut-shells, provide the most promising nonlithic avenues for recovery. These materials have the same likelihood of survival as artefacts and debris created by humans from the same materials, with the exception that charring is less likely for the nonhuman assemblage. Hominin Middle Pleistocene wooden artefacts and nut-cracking debris are known but rare (Goren-Inbar et al. 2002; Thieme 1997), and recognition of past chimpanzee or other primate behaviours even much more recent than this would still be valuable. Although taphonomy is not unproblematic, the importance of archaeobotanists analysing primate records to reconstruct evolutionary and ethological data lies in the fact that in most instances plant remains will be the only evidence of activity at primate sites. Whether this means that the time-depth of the identifiable primate archaeological record is limited to a few years in the absence of accompanying stone tools or, rarely, prey bones (Tappen and Wrangham 2000) is unknown at present.

Differentiating human records from those of nonhuman primates is likely to be unproblematic in many cases, given the human propensity for accumulating durable artefacts at most activity sites. Flaked stone artefacts have been the hominin calling card for 2.5 million years (Semaw et al. 1997), and no extant animal other than humans regularly creates them. Yet long-term sympatry of humans with other apes means that not all hominin sites are necessarily exclusively hominin (Mercader et al. 2007), especially since humans and chimpanzees have similarities in their preference for nonrandomly distributed resources such as ripe fruits. An alternate avenue for differentiating primate archaeological remains may rely on the fact that human use of tools and fire to modify foods after extraction are also unique to our lineage. For example, if nut starch residues on a stone or wooden anvil display evidence of systematic grinding and not just incidental pounding, then they may potentially be attributed to human agency, as the chimpanzee and capuchin aim is to open a nut, not crush it. The same applies to gelatinised starches, which require both moisture and heat to gelatinise and would not ordinarily result from natural forest fires. Furthermore, although subsistence plants exploited by humans and nonhuman primates do have some overlap, this is not an intractable problem. For example, Peters and O'Brien (1981) found that of 461 genera of food plants consumed in eastern and southern Africa by humans, chimpanzees, and baboons, 36% were consumed only by humans, 17% each were consumed solely by chimpanzees or baboons, and only 6% overlapped all three primates. At the time of that study the best chimpanzee-human comparison came from the Kasakati Basin/Mahale Mountains region of Tanzania, where of 161 genera 19% were consumed by humans only, 60% by chimpanzees only, and 21% by both (Peters and O'Brien 1981). Consideration of palaeobotanical (primarily palynological) analyses at the Pleistocene early hominin locales of Koobi Fora and Omo demonstrates that identification of leaf, shoot, and fruit food taxa exploited by modern humans and large primates is possible at such sites (Peters and O'Brien 1981).

Extraction of a chimpanzee or other primate botanical signature from background noise requires the same attention to primate plant exploitation practices as that traditionally afforded to humans, alongside ongoing study of botanical components of primate site creation and longevity.

Practical application of archaeobotanical methods to primate field sites does not require modifications of existing protocols or techniques. What is required are working collaborations with primatologists who have experience locating and working with their study species in its environment, in addition to extensive behavioural records. Reconstruction of site environments for past primate evolution is as important as that for hominins in determining causal from incidental ecological factors, and in this regard existing archaeobotanical approaches are appropriate. Even seemingly simple research into the longevity of a common activity, such as termite-fishing at the famous Gombe National Park, would be of great benefit to our understanding of primate traditions. As noted by McGrew (1992, p. 196), 'termite fishing may just as well have been invented in 1959, the year before Jane Goodall arrived, or a million years ago'. No living primate species is a direct model for early hominin behaviour or capabilities, but the more we examine the long-term behavioural trajectory of animals with genetically and anatomically similar constitutions to our own, the more chance we have of identifying truly unique aspects of human behaviour.

CONCLUSION

The notion of primate archaeobotany builds on current trends to integrate nonhuman primates more usefully into discussions of human evolution. It benefits from having living subjects to study, and an established primatological literature recording plant consumption and exploitation among a large number of species. Beyond diet, these studies have demonstrated the importance of plant-processing and vegetation tools to extant African chimpanzees, Southeast Asian orangutans and South American capuchin monkeys in particular, and further discoveries among other taxa (such as bonobos and gorillas) are likely. Developing a more explicitly archaeobotanical approach to nonhuman primate species also benefits the study of primate evolution in its own right, and aids conservation efforts by increasing our understanding of changes in primate-plant interactions over time. Importantly, recognising the existence of a nonhuman archaeobotanical record provides comparative data for archaeologists working in areas with extant tool-using primates, giving an alternate perspective on ways that intelligent species with similar physiological adaptations to our own make use of the botanical environment.

ACKNOWLEDGEMENTS

My thanks to Dorian Fuller for the invitation to contribute to this volume and to Alison Crowther, Adriana Hernandez-Aguilar, Susana Carvalho, William McGrew, and Elisabetta Visalberghi for helpful discussions. Funding was received from European Research Council Starting Grant #283959 (Primate Archaeology), awarded to M. Haslam.

REFERENCES

Anderson, J., E. Williamson, and J. Carter (1983) Chimpanzees of Sapo Forest, Liberia: Density, nests, tools and meat-eating. *Primates 24*, 594–601.

Babot, M. d. P. (2003) Starch grain damage as an indicator of food processing. In D. M. Hart and L. A. Wallis (Eds.), *Phytolith and Starch Research in the Australian-Pacific-Asian Regions: The State of the Art*, pp. 69–81. Canberra: Pandanus Books.

Barton, H. (2007) Starch residues on museum artefacts: Implications for determining tool use. *Journal of Archaeological Science 34* (10), 1752–62.

Behie, A., M. Pavelka, and C. A. Chapman (2010) Sources of variation in fecal cortisol levels in howler monkeys in Belize. *American Journal of Primatology 72*, 600–06.

Boesch, C., and H. Boesch (1990) Tool use and tool making in wild chimpanzees. *Folia Primatologica 54*, 86–99.

Boesch, C., and H. Boesch-Achermann (2000) *The Chimpanzees of the Tai Forest. Behavioural Ecology and Evolution*. Oxford: Oxford University Press.

Boesch, C., J. Head, and M. Robbins (2009) Complex tool sets for honey extraction among chimpanzees in Loango National Park, Gabon. *Journal of Human Evolution 56*, 560–69.

Bradley, B. J., M. Stiller, D. Doran-Sheehy, T. Harris, C. A. Chapman, L. Vigilant, and H. N. Poinar (2007) Plant DNA sequences from feces: Potential means for assessing diets of wild primates. *American Journal of Primatology 69*, 699–705.

Chapman, C. A., F. J. White, and R. Wrangham (1994) Party size in chimpanzees and bonobos: A reevaluation of theory based on two similarly forested sites. In R. Wrangham, W. C. McGrew, F. B. deWaal, and P. Heltne (Eds.), *Chimpanzee Cultures*, pp. 41–58. Cambridge: Harvard University Press.

Ciochon, R. L., D. R. Piperno, and R. G. Thompson (1990) Opal phytoliths found on the teeth of the extinct ape *Gigantopithecus blacki*: Implications for paleodietary studies. *Proceedings of the National Academy of Sciences 87*, 8120–24.

Crowther, A. (2012) The differential survival of native starch during cooking and implications for archaeological analyses: A review. *Archaeological and Anthropological Sciences 4*, 221–35.

Cummings, L. S., and A. Magennis (1997) A phytolith and starch record of food and grit in Mayan human tooth tartar. In A. Pinilla, J. Juan-Tresserras, and M. J. Machado (Eds.), *Primer Encuentro Europeo Sobre el Estudio de Fitolitos*, pp. 211–18. Madrid: Centro de Ciencias Medioambientales.

Deblauwe, I., P. Guislain, J. Dupain, and L. van Elsacker (2006) Use of a tool-set by *Pan troglodytes troglodytes* to obtain termites (Macrotermes) in the periphery of the Dja Biosphere Reserve, Southeast Cameroon. *American Journal of Primatology 68*, 1191–96.

Fay, J., and R. Carroll (1994) Chimpanzee tool use for honey and termite extraction in Central Africa. *American Journal of Primatology 34*, 309–17.

Fowler, A., and V. Sommer (2007) Subsistence technology of Nigerian chimpanzees. *International Journal of Primatology 28*, 997–1023.

Fruth, B., and G. Hohmann (1994) Comparative analyses of nest-building behavior in bonobos and chimpanzees. In R. Wrangham, W. C. McGrew, F. de Waal, and P. Heltne (Eds.), *Chimpanzee Cultures*, pp. 109–28. Cambridge, MA: Harvard University Press.

Goren-Inbar, N., G. Sharon, Y. Melamed, and M. Kislev (2002) Nuts, nut cracking, and pitted stones at Gesher Benot Ya'aqov, Israel. *Proceedings of the National Academy of Sciences 99*, 2455–60.

Harcourt, A. H., and K. J. Stewart (2007) Gorilla society: What we know and don't know. *Evolutionary Anthropology 16*, 147–58.

Hardy, K., T. Blakeney, L. Copeland, J. Kirkham, R. Wrangham, and M. Collins (2009) Starch granules, dental calculus and new perspectives on ancient diet. *Journal of Archaeological Science 36*, 248–55.

Haslam, M. (2004) The decomposition of starch grains in soils: Implications for archaeological residue analyses. *Journal of Archaeological Science 31*, 1715–34.

———. (2012) Towards a prehistory of primates. *Antiquity 86*, 299–315.

Haslam, M., A. Hernandez-Aguilar, V. Ling, S. Carvalho, I. de la Torre, A. DeStefano, A. Du, B. L. Hardy, J. Harris, L. Marchant, T. Matsuzawa, W. McGrew, J. Mercarder, R. Mora, M. Petraglia, H. Roche, E. Visalberghi, and R. Warren (2009) Primate archaeology. *Nature 460*, 339–44.

Henry, A., H. Hudson, and D. R. Piperno (2009) Changes in starch grain morphologies from cooking. *Journal of Archaeological Science 36* (3), 915–22.

Henry, A., and D. R. Piperno (2008) Using plant microfossils from dental calculus to recover human diet: A case study from Tell al-Raqā'i, Syria. *Journal of Archaeological Science 35* (7), 1943–50.

Hernandez-Aguilar, A. (2009) Chimpanzee nest distribution and site reuse in a dry habitat: Implications for early hominin ranging. *Journal of Human Evolution 57*, 350–64.

Hernandez-Aguilar, A., J. Moore, and T. Pickering (2007) Savanna chimpanzees use tools to harvest the underground storage organs of plants. *Proceedings of the National Academy of Sciences 104*, 19210–13.

Huffman, M., and M. Seifu (1989) Observations on the illness and consumption of a possibly medicinal plant *Vernonia amygdalina* (Del.), by a wild chimpanzee in the Mahale Mountains National Park, Tanzania. *Primates 30*, 51–63.

Isabirye-Basuta, G. (1988) Food competition among individuals in a free-ranging chimpanzee community in Kibale Forest, Uganda. *Behaviour 105*, 135–47.

Koops, K., W. C. McGrew, and T. Matsuzawa (2010) Do chimpanzees (*Pan troglodytes*) use cleavers and anvils to fracture *Treculia africana* fruits? Preliminary data on a new form of percussive technology. *Primates 51*, 175–78.

Laden, G., and R. Wrangham (2005) The rise of the hominids as an adaptive shift in fallback foods: Plant underground storage organs (USOs) and australopith origins. *Journal of Human Evolution 49*, 482–98.

Lambert, J. E. (2007) Seasonality, fallback strategies and natural selection: A chimpanzee and cercopithecoid model for interpreting the evolution of hominin diet. In P. S. Ungar (Ed.), *Evolution of the Human Diet*, pp. 324–43. Oxford: Oxford University Press.

Mannu, M., and E. Ottoni (2009) The enhanced tool-kit of two groups of wild bearded capuchin monkeys in the Caatinga: Tool making, associative use, and secondary tools. *American Journal of Primatology 71*, 242–51.

Marchant, L. F., and W. C. McGrew (2005) Percussive technology: Chimpanzee baobab smashing and the evolutionary modeling of hominid knapping. In V. Roux and B. Bril (Eds.), *Stone Knapping: The Necessary Conditions for a Uniquely Hominin Behaviour*, pp. 341–50. Cambridge: McDonald Institute for Archaeological Research.

Marlowe, F. W., and J. C. Berbesque (2009) Tubers as fallback foods and their impact on Hadza hunter-gatherers. *American Journal of Physical Anthropology 140*, 751–58.

Matsusaka, T., H. Nishie, M. Shimada, N. Kutsukake, K. Zamma, M. Nakamura, and T. Nishida (2006) Tool-use for drinking water by immature chimpanzees of Mahale: Prevalence of an unessential behavior. *Primates 47*, 113–22.

McBeath, N., and W. C. McGrew (1982) Tools used by wild chimpanzees to obtain termites at Mt Assirik, Senegal: The influence of habitat. *Journal of Human Evolution 11*, 65–72.

McGrew, W. C. (1992) *Chimpanzee Material Culture: Implications for Human Evolution*. Cambridge: Cambridge University Press.

———. (2004) *The Cultured Chimpanzee: Reflections on Cultural Primatology*. Cambridge: Cambridge University Press.

McGrew, W. C., P. J. Baldwin, L. F. Marchant, J. Pruetz, S. E. Scott, and C. Tutin (2003) Ethoarchaeology and elementary technology of unhabituated wild chimpanzees at Assirik, Senegal, West Africa. *PaleoAnthropology 2003.05.02*, pp. 1–20.

McGrew, W. C., P. Baldwin, and C. Tutin (1988) Diet of wild chimpanzees (*Pan troglodytes verus*) at Mt. Assirik, Senegal: I. Composition. *American Journal of Primatology 16*, 213–26.

McGrew, W. C., L. F. Marchant, M. M. Beuerlein, D. Vrancken, B. Fruth, and G. Hohmann (2007) Prospects for bonobo insectivory: Lui Kotal, Democratic Republic of Congo. *International Journal of Primatology 28*, 1237–52.

McGrew, W. C., L. F. Marchant, R. Wrangham, and H. Klein (1999) Manual laterality in anvil use: Wild chimpanzees cracking *Strychnos* fruits. *Laterality 4* (1), 79–87.

Mercader, J., H. Barton, J. Gillespie, J. Harris, S. Kuhn, R. T. Tyler, and C. Boesch (2007) 4,300-year-old chimpanzee sites and the origins of percussive stone technology. *Proceedings of the National Academy of Sciences 104* (9), 3043–48.

Mercader, J., M. Panger, and C. Boesch (2002) Excavation of a chimpanzee stone tool site in the African rainforest. *Science 296*, 1452–55.

Milton, K. (1993) Diet and primate evolution. *Scientific American* August, 86–93.

Milton, K. (1999) Nutritional characteristics of wild primate foods: Do the diets of our closest living relatives have lessons for us? *Nutrition 15*, 488–98.

———. (2003) The critical role played by animal source foods in human (*Homo*) evolution. *Journal of Nutrition Supplement*, 3886S–92S.

Moreno-Black, G. (1978) The use of scat samples in primate diet analysis. *Primates 19*, 215–21.

Morgan, B. J., and E. E. Abwe (2006) Chimpanzees use stone hammers in Cameroon. *Current Biology 16*, R632–33.

Nishida, T. (1989) A note on the chimpanzee ecology of the Ugalla Area, Tanzania. *Primates 30*, 129–38.

Nugent, S. J. (2006). Applying use-wear and residue analysis to digging sticks. *Memoirs of the Queensland Museum, Cultural Heritage Series 4* (1), 89–95.

Ottoni, E., and P. Izar (2008). Capuchin monkey tool use: Overview and implications. *Evolutionary Anthropology 17*, 171–78.

Peters, C., and E. O'Brien (1981) The early hominid plant-food niche: Insights from an analysis of plant exploitation by *Homo*, *Pan*, and *Papio* in eastern and southern Africa. *Current Anthropology 22*, 127–40.

Pruetz, J., and P. Bertolani (2007) Savanna chimpanzees, *Pan troglodytes verus*, hunt with tools. *Current Biology 17*, 412–17.

Sanz, C., and D. Morgan (2007) Chimpanzee tool technology in the Goualougo Triangle, Republic of Congo. *Journal of Human Evolution 52*, 420–33.

———. (2009) Flexible and persistent tool-using strategies in honey-gathering by wild chimpanzees. *International Journal of Primatology 30*, 411–27.

Semaw, S., P. Renne, J. W. K. Harris, C. S. Feibel, R. L. Bernor, N. Fesseha, and K. Mowbray (1997) 2.5-million-year-old stone tools from Gona, Ethiopia. *Nature 385*, 333–36.

Sept, J. (1992) Was there no place like home? A new perspective on early hominid archaeological sites from the mapping of chimpanzee nests. *Current Anthropology 33*, 187–207.

———. (1998) Shadows on a changing landscape: Comparing nesting patterns of hominids and chimpanzees since their last common ancestor. *American Journal of Primatology 46*, 85–101.

Spagnoletti, N., E. Visalberghi, M. Verderane, E. Ottoni, P. Izar, and D. Fragaszy (2012) Stone tool use in wild bearded capuchin monkeys, *Cebus libidinosus*. Is it a strategy to overcome food scarcity? *Animal Behaviour 83*, 1285–94.

Stewart, F., A. Piel, and W. C. McGrew (2011) Living archaeology: Artefacts of specific nest site fidelity in wild chimpanzees. *Journal of Human Evolution 61*, 388–95.

Sugiyama, Y. (1997) Social tradition and the use of tool-composites by wild chimpanzees. *Evolutionary Anthropology 6*, 23–27.

Tappen, M., and R. Wrangham (2000) Recognizing nominoid-modified bones: The taphonomy of colobus bones partially digested by free-ranging chimpanzees in the Kibale Forest, Uganda. *American Journal of Physical Anthropology 113*, 217–34.

Tennie, C., J. Call, and M. Tomasello (2009) Ratcheting up the ratchet: On the evolution of cumulative culture. *Philosophical Transactions of the Royal Society of London Series B 364*, 2405–15.

Thieme, H. (1997) Lower Palaeolithic hunting spears from Germany. *Nature 385*, 807–10.

Tonooka, R. (2001) Leaf-folding behavior for drinking water by wild chimpanzees (*Pan troglodytes verus*) at Bossou, Guinea. *Animal Cognition 4*, 325–34.

Tutin, C., and M. Fernandez (1993) Composition of the diet of chimpanzees and comparisons with that of sympatric lowland Gorillas in the Lope Reserve, Gabon. *American Journal of Primatology 30*, 195–211.

Tweheyo, M., K. Lye, and R. Weladji (2004) Chimpanzee diet and habitat selection in the Budongo Forest Reserve, Uganda. *Forest Ecology and Management 188*, 267–78.

van Schaik, C., M. Ancrenaz, G. Borgen, B. Galdikas, C. Knott, I. Singleton, A. Suzuki, S. Utami, and M. Merrill (2003). Orangutan cultures and the evolution of material culture. *Science 299*, 102–05.

Visalberghi, E., E. Addessi, V. Truppa, N. Spagnoletti, E. Ottoni, P. Izar, and D. Fragaszy (2009) Selection of effective stone tools by wild bearded Capuchin monkeys. *Current Biology 19*, 1–5.

Visalberghi, E., D. Fragaszy, E. Ottoni, P. Izar, M. de Oliveira, and F. Andrade (2007) Characteristics of hammer stones and anvils used by wild bearded capuchin monkeys (*Cebus libidinosus*) to crack open palm nuts. *American Journal of Physical Anthropology 132*, 426–44.

Whiten, A., J. Goodall, W. C. McGrew, T. Nishida, V. Reynlds, Y. Sugiyama, C. Tutin, R. Wrangham, and C. Boesch (2001) Charting cultural variation in chimpanzees. *Behaviour 138*, 1481–16.

Whitesides, G. (1985) Nut cracking by wild chimpanzees in Sierra Leone, West Africa. *Primates 26*, 91–94.

Wobber, V., B. Hare, and R. Wrangham (2008) Great apes prefer cooked food. *Journal of Human Evolution 55*, 340–48.

Wrangham, R. (2006) Chimpanzees: The culture-zone concept becomes untidy. *Current Biology 16*, R634–35.

———. (2007) The cooking enigma. In *The Evolution of the Human Diet*. In P. S. Ungar (Ed.), pp. 308–23. Oxford. Oxford University Press.

Wrangham, R., J. H. Jones, G. Laden, D. Pilbeam, and N. Conklin-Brittain (1999) The raw and the stolen: Cooking and the ecology of human origins. *Current Anthropology 40* (5), 567–94.

Wrangham, R., and T. Nishida (1983) *Aspilia* spp. leaves: A puzzle in the feeding behavior of wild chimpanzees. *Primates 24*, 276–82.

Yamagiwa, J. (2000) Factors influencing the formation of ground nests by eastern lowland gorillas in Kahuzi-Biega National Park: Some evolutionary implications of nesting behaviour. *Journal of Human Evolution 40*, 99–109.

Yamagiwa, J., and A. K. Basabose (2009) Fallback foods and dietary partitioning among *Pan* and *Gorilla*. *American Journal of Physical Anthropology 140*, 739–50.

Yamakoshi, G., and Y. Sugiyama (1995) Pestle-pounding behavior of wild chimpanzees at Bossou, Guinea: A newly observed tool-using behavior. *Primates 36*, 489–500.

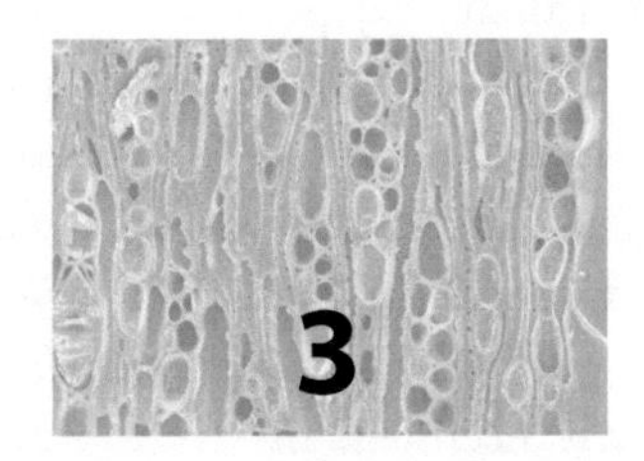

3

Dietary Diversity

Our Species-Specific Dietary Adaptation

Gordon Hillman and Michèle Wollstonecroft

> One of the hallmarks of our species is the variety of foods we can eat. Thus our species-specific dietary adaptation seems to involve the consumption of many foods . . . the key to understanding our unusual dietary breadth rests in the study of the evolution of the human diet. (Ungar and Teaford 2002, p. 5)

We embarked on our hominin evolution ca. 7 million years ago (MYA) (Brunet et al. 2005; Zollikofer et al. 2005) equipped with a set of basic nutrient requirements and features of digestive physiology already shaped by 30 million years of earlier evolution as anthropoid primates (Hohmann 2009; Milton 1999). These inherited features 'set limitations on the types of foods they could successfully exploit' and, in the hominoid line, required that 'dietary quality be kept high for a physically active and highly social lifestyle' (Milton 2002, p. 113). Given that these general adaptations operated for many millions of years, it is not surprising that we inherited from our anthropoid primate ancestors an instinct for 'sensory specific satiety', which favours the consumption of a variety of foods or food classes (dietary diversity) and precludes the overconsumption of individual foods (Johns 1999, p. 30). More specifically, 'humans are well adapted for lean meat, fish, insects and highly diverse plant foods without being dependent on any particular proportions of plants versus meat' (Lindeberg 2009, p. 43).

Other features of the human diet can be traced to our hominoid ancestors, and, because we share our hominoid ancestors with other great apes, most of the recent attempts to model hominoid and early human diets have been based on the dietary behaviour of chimpanzees and other great apes (for example, Conklin-Brittain, Wrangham, and Smith 2002; Hohmann 2009; Milton 1993, 1999, 2000; Rodman 2002; Snodgrass, Leonard, and Robertson 2009; Wrangham et al. 1999). Such studies have shown that there are numerous similarities in the dietary strategies of both humans and great apes, including an array of combinations of dietary diversity, dietary flexibility (the use of similar edible resources interchangeably), dietary selectivity, high quality in terms of concentrated energy and nutrients, and a general avoidance of toxins, as well as food sharing and rudimentary tools used to access otherwise embedded foods (Conklin-Brittain, Wrangham, and Hunt 1998; Hohmann 2009; Wrangham et al. 1999).

Nevertheless, the human diet is significantly richer than that of the great apes in terms of the wide range of terrestrial and aquatic plants and animals and birds that we eat, and the forms in which we eat them. Major increases in dietary breadth occurred at two critical points in our evolution: the hominin/ape split and the emergence of *Homo* (Sponheimer and Dufour 2009). Dietary change and shifts in body composition are considered to be the two most important preconditions for

Archaeology of African Plant Use by Chris J. Stevens, Sam Nixon, Mary Anne Murray, and Dorian Q Fuller, Eds., 37–49

the split with the apes and the later extraordinary expansion of hominin brain size (cm^3) (Aiello and Wheeler 1995); over the past 3.9 million years the *Homo* brain increased to three times the size of that of our earliest australopithecine ancestors (see Snodgrass, Leonard, and Robertson 2009, Table 2.2).

Because of evidence of evolutionary advancements in the craniodental toolkit, it has been inferred that a significant step toward greater dietary heterogeneity occurred when our ancestors began targeting more concentrated sources of energy and protein somewhere between 2 and 4.5 MYA (or possibly substantially earlier if the recently recovered bipedal hominoid, *Sahelanthropus tchadensis*, dating to 7 MYA, were to be placed on the hominin side of the great ape/hominin split; see Brunet et al. 2005; Zollikofer et al. 2005).

The effects of climate change, the thinning of forests and the broadening array of available food classes on the emergence of terrestriality, the development of bipedalsim, increased brain-size, and the eventual emergence of language have been explored in detail by Aiello (1996), Aiello and Dunbar (1993), Leonard and Robertson (1994, 1997), Milton (1993, 1999, 2000), and many others. Thus we will not discuss them here except to note that each of these developments had a significant impact on palaeodietary variety, even more on dietary flexibility, but above all on the burgeoning need for more concentrated sources of energy.

Shifts toward greater dietary diversity appear to have served as both a cause and a consequence of human evolutionary history. Before the emergence of any hominins at c. 7 MYA, our ancestral hominoids appear to have been limited to consuming plants that could be gathered without mechanical aids, such as digging sticks or stone blades, and eaten without prior treatment with processing equipment and/or fire and without the formidable dentition and associated mandibular musculature that eventually evolved in early hominins.

These restrictions clearly constrained the range of potential caloric staples to which hominoids had access and limited them to their erstwhile primary energy source of tender, young, proteinaceous foliage and sugar-rich fruits. Significantly, such restrictions precluded access to more concentrated forms of energy, such as the carbohydrates found in underground storage organs (USOs), nuts, and seeds that could not be collected without mechanical aids. Thus these hominoids' diets were significantly higher in protein, dietary fibre, and most vitamins and minerals—and significantly lower in carbohydrates—than that of modern humans (Milton 2000, 2002).

The complex factors that influenced the expanding dietary needs of our hominin ancestors continue to be debated. The traditional view was that a great-ape-style diet (that is, one in which protein, energy, and other nutrients were obtained primarily by consuming young leaves and shoots [see Conklin-Brittain, Wrangham, and Smith 2002]) continued with only minor changes among our australopithecine ancestors, who, together with the 'earliest *Homo*' (*H. rudolfensis* and *H. habilis*) were regarded as having had ape-like life histories on the basis of brain size, dental eruption schedule, patterns of epiphysical closure, body weight, and stature (O'Connell, Hawkes, and Blurton Jones 2002, citing Smith and Tomkins 1995). However, Ungar and Teaford (2002, p. 5) argue that the 'evidence suggests that we began to evolve our unique ability to assimilate a broad spectrum of foods more than four million years ago, shortly after our evolutionary split with chimpanzees', and, 'thickened dental enamel and jaws, and larger, flatter teeth . . . increased the range of foods that our earliest ancestor could process'. More specifically, Ungar and Teaford (2002) note that these craniodental changes are already apparent as one moves from *Ardipithicus ramidus* to *Australopithecus africanus*, two of the earliest hominins. Blumenschine (1992) similarly cites archaeological evidence of a seemingly more routine incorporation of concentrated sources of energy, from proteins and fats and from foods such as bone marrow and meat that were (mainly) scavenged from the corpses of vertebrates.

A model of more specific dietary developments at this stage has been proposed by Conklin-Brittain, Wrangham, and Smith (2002). They argue that evolution of the early australopithecines from chimp-like ancestors involved the progressive reduction of dietary fibre by (1) eating low-fibre underground storage organs (USOs) as a source of nonstructural carbohydrates in place of their less carbohydrate-rich, higher-fibre, stem-pith foods; (2) eating meaty animal foods in place of leaf protein of many of the fibrous leafy foods; and (3) processing the more fibrous foods before consuming them. However, experiments by Hillman (unpublished) with large-scale harvesting and processing of wild USOs from terrestrial plants, using the simplest possible tools (or none at all), suggest that such activities by australopithecines would have been limited. And, although the edible roots of wetland plants of the sort proposed by Conklin-Brittain and colleagues (2002) would have presented an attractive target, such foods could probably have been used on only a relatively small scale, owing to technological limitations.

We do not know when our ancestors mastered the ability to use rocks and stones to access these types of embedded resources. The earliest archaeological evidence—namely, stone tools and cut marks on bone from the Ethiopian sites of Gona and Bouri—have been dated to ca. 2.5 MYA (Lee-Thorp and Sponheimer 2006). Because modern chimps and bonobos regularly crack nuts with hammer-stones and stone anvils, which they are apparently prepared to carry from considerable distances (Goren-Inbar et al. 2002), it is likely that such skills were already in use among our common ancestors before the ape/hominin split. However, recent trace element and carbon isotope studies by Sponheimer and Dufour (2009) suggest that the most important distinctions between the diets of apes and the earliest hominins probably reflected ecology rather than technology. They interpret the biogeochemical data as indicating that australopithecines and early *Homo* routinely consumed C_4 savannah foods, particularly the underground storage organs (USOs) of C_4 grasses and C_4 aquatics of the sedge family (Cyperaceae) and other monocotyledons with edible USOs, and/or animals that consumed C_4 savannah plants, directly or indirectly (see Sponheimer and Dufour 2009). Their argument is supported by recent analyses of fossilised tooth enamel from a range of Plio-Pleistocene hominins species from archaeological sites in central, eastern, and southern Africa (Cerling et al. 2013b; Sponheimer et al. 2013; Wynn et al. 2013). Despite interspecies differences in the hominin specimens examined, as well as variations between their habitats, these analyses confirm a general trend toward the consumption of C_4 as well as CAM foods (that is, succulents that are more typical of deserts). Neither C_4 nor CAM plants are eaten in significant amounts by most other primates or other hominoids (exceptions being extinct *Theropithecus* spp. and some extant baboons; see Cerling et al. 2013a). These results indicate decreases in the intake of foods from woodland trees and shrubs and increases in the intake of savannah grassland resources—that is, C_4/CAM plants and/or animals that consumed C_4/CAM plants. Significantly, these carbon isotope studies also confirm a continuous expansion of dietary diversity, with hominins consuming foods from a wide range of C_3 as well as C_4/CAM sources.

A tuberous Cyperaceae that is frequently mentioned in the human evolution literature (for example, Copeland 2007; Lee-Thorp et al. 2012; Peters and Vogel 2005) and that we regard as a possible hominin food source, but only after our ancestors had mastered control of fire, is *Cyperus rotundus* (purple nut-grass). This C_4 sedge grows in dry to damp conditions in a range of soil types (Townsend and Guest 1985). It is widespread in the temperate and tropical zones of both hemispheres, and Copeland (2007) included it in her model for plants that likely populated the Olduvai lowermost Bed II palaeolandscape. Most authors dismiss *C. rotundus* tubers from their lists of potential C_4 Hominin foods owing to the unpleasant, bitter taste, woody texture, and the toxicity of the mature specimens. However, we would argue that purple nut-grass tubers offer great food potential because detoxification requires only rudimentary processing techniques and are relatively easy to uproot and are a sustainable resource owing to their positive response to harvesting. Intensive, routine harvesting stimulates the sprouting of new tubers, prevents the build-up of old, woody tubers, and creates space for new growth.

Archaeobotanical evidence of the consumption of *C. rotundus* tubers by modern human populations has been identified by Hillman and colleagues (1989) at three Upper Palaeolithic sites at Wadi Kubbaniya, on the Nile River in northern Egypt, near Aswad. *C. rotundus* tubers dominated the archaeobotanical assemblages from these sites. The tubers were found mixed with fuel, charred faeces, and foods, including dom palm fruit, and in association with grindstones, mortars, pestles, and hearths (Hillman, Madeyska, and Hather 1989, p. 191). The likely use of these tools for plant processing was confirmed by the results of a Pyrolysis Mass Spectrometry analyses on several of the groundstone tools that had plant residues on their working surfaces (Jones 1989).

Hillman and associates (1989) inferred that Palaeolithic people had been processing the *C. rotundus* tubers with a sequence of grinding, possibly leaching, and roasting to both detoxify and soften them before eating them. His experiments with modern *C. rotundus* (Hillman, Madeyska, and Hather 1989) confirm that in the earliest stages of growth the tubers can be consumed after one merely (1) roasts and then (2) rubs them to remove the outer layers. Once the tubers are fully mature, a more complex sequence is necessary: roasting, grinding or pounding, leaching, and cooking; the tubers can then be eaten as gruel, mush, or griddle cakes.

Biological productivity (tuber density) and harvesting returns were measured and observations were made on the effects of intensive harvesting on the quality and yield of wild stands (Hillman 1989; Hillman, Madeyska, and Hather 1989). It was found that it was possible to harvest several days' carbohydrate requirements in a matter of hours owing to the shallow depth of the tubers and the density of specimens per m^2. Harvesting returns averaged

260 g per m^2; tuber density was observed to vary significantly between patches, ranging from several hundred tubers per (37 g) m^2 to <21,000 (3.3 kg) per m^2, possibly because of differences in the moisture level of the substrate.

FOOD PROCESSING AND DIETARY DIVERSITY

Overall, radical advances in the craniodental toolkit, which escalated with the emergence of *Homo* (and currently provide the best archaeological evidence for increased dietary breadth; see Sponheimer and Dufour 2009) must have been strongly selectively advantaged by radical improvements in the diet. Archaeologists continue to debate what those improvements were and whether they involved increasing amounts of meat or energy-rich plant foods or both, but there is general agreement that they involved increasing dietary breadth. One argument is that the incorporation of meat, which contains adequate amounts and types of protein, essential fatty acids, and vitamins and minerals necessary for normal metabolic performance, facilitated a broadening of the range of plant foods ingested, which became the primary energy source (Milton 1999; see also Sponheimer and Dufour 2009). Lindeberg (2009, p. 52) concludes that current evidence indicates the relative amount of meat versus plant foods has varied considerably throughout hominin evolution. Johns (1999) points out that increasing brain and body size, and the consumption of a widening range and greater proportions of animal foods, were major selective pressures favouring the ingestion of more antioxidants.

Perhaps the most obvious advantage of dietary diversity is that it increases the probability of an individual obtaining the full range of essential macronutrients (lipids, proteins, and carbohydrates) and micronutrients (minerals, vitamins) for adequate diet quality and energy availability (Foote et al. 2004; Lindeberg 2009; Ruel 2003; Sponheimer and Dufour 2009). It also increases the likelihood of an individual obtaining the exogenous antioxidants needed to counterbalance oxidative stress, a problem that escalated in the genus *Homo* as a consequence of the increasing energy demands of encephalization and locomotor adaptations (Johns 1990, 1999). Additionally, dietary diversity clearly minimizes the effect of any particular toxin or digestibility-reducing compound ingested at any one time (Clutton-Brock and Harvey 1977; Johns 1990; Stahl 1989).

We believe that a major catalyst that has facilitated *Homo* dietary diversity is food processing. Our definition of food processing encompasses the use of pulverizing, grinding, grating, using wet and dry thermal treatments, soaking, malting, fermenting, and dehydrating—and/or combinations of these techniques and/or others—for the purposes of (1) modifying otherwise inedible or toxic plant and animal tissue into edible and/or safer foods or beverages; (2) modifying a single plant or animal organ into a variety of food forms, each with a different taste and/or texture; (3) preparing compound foods such as soups, stews, gruels, breads, cakes, and beverages by mixing plant and/or animal tissue (making resources go farther); (4) preserving foods for storage. Our investigations into food processing are part of a relatively new and growing avenue of archaeological research that is based on arguments, first proposed by Ann Stahl (1984, 1989; and to some degree Yen 1975, 1980) and recently taken up by Carmody and Wrangham (2009), Hillman (2004), Jones (2009), Lyons (2007), Lyons and D'Andrea (2003), Munro and Bar-Oz (2005), Samuel (1996), Speth (2004), Wandsnider (1997), Wollstonecroft and associates (2008), and Wrangham and associates (1999), among others, who regard food processing as a critical variable in human diet, health and evolution, ethnic identity, and social and economic change (intensification).

The relationships between cuisine, diet, nutrition, and food processing are multidimensional and complex (see Aworh 2008; Ellis et al. 2004; Parada and Aguilera 2007; Schlemmer et al. 2001; Stahl et al. 2002; Verhagen et al. 2001; Waldron et al. 1997; Walingo 2009; Wills et al. 1998). Dietary choice and food are strongly linked with ethnic/group identity, and, as explained by Lyons (2007, p. 347), 'identity is constructed not just through the foods that we consume, but how we prepare them as cuisine'; thus 'technology is not secondary to food in cuisine'.

Among the more significant implications of the available data, particularly in studies of present-day rural Africa, where food security is low and malnutrition is widespread, is that traditional food processing techniques do indeed promote food variety, palatability, safety, and stability; furthermore, studies show that improved nutrition can best be obtained by upgrading the existing methods or fortifying the food with specific ingredients, rather than replacing the types of processing methods or foods (Aworh 2008; Blandino et al. 2003). Moreover, these studies demonstrate that species selection and technical choice are intrinsically linked, because the functional properties of edible plants can promote or discourage their consumption, delimit the forms in which they can be eaten, and promote or inhibit the release of nutrients. In the present context the term *functional properties* of a species and/or plant part refers to the genetically determined response of a plant tissue to different processing techniques and conditions—for example, particle reduction, wet and dry thermal processes, malting, fermentation, and soaking. Depending on the species/plant part, some processes work better than others; even the sequence in which different processes are applied makes a difference to palatability, taste, texture, and nutritional potential (Lyons and D'Andrea 2003; Wandsnider 1997; Wills et al. 1998; Wollstonecroft et al. 2008, 2012).

Furthermore, research shows that food processing promotes not only greater food variety, palatability, safety, stability and nutrient quality but also enhances the bioaccessibility of macro-nutrients (protein, fats, carbohydrates, fibre), micronutrients (minerals and vitamins), and antioxidants (Johns 1999; and see the contributors in Pfannhauser, Fenwick, and Khokhar 2001). *Bioaccessibility* is defined as the fraction of a nutrient that is released from a food matrix during processing and/or consumption and its potential availability for absorption in the gastrointestinal tract (Parada and Aguilera 2007; Stahl et al. 2002). Bioaccessibility is an important factor in *bioavailability*, which is measured as the rate and proportion of a food that is absorbed by the digestive system to become metabolically active (Ellis et al. 2004; Verhagen et al. 2001). In the raw form, many types of plant tissue—for example, almond and carrot—pass along the gastrointestinal tract without releasing their nutrients, because much of their tissue remains intact and the cell walls unbroken (Ellis et al. 2004; Holden 1990; Stahl et al. 2002). Food processing, such as grinding, fermenting, or heating, can help promote nutrient bioaccessibility by transforming the chemistry and microstructure of plant tissue by disrupting the cell walls, changing nutrient-matrix complexes, and/or transforming tissue substances into more active molecular structures (Parada and Aguilera 2007).

Ultimately, the data from these studies show that the functional properties of the plant and animal parts, and how amenable they are to processing with the available technology, may be as great an influence on human selection decisions as the energy and protein rank of that species (Lyons and D'Andrea 2003; Wollstonecroft et al. 2008). Thus a number of archaeologists (for instance, Carmody, Weintraub, and Wrangham 2011; Carmody and Wrangham 2009; Fuller and Rowlands 2009; Hillman 2004; Lyons and D'Andrea 2003; Munro and Bar-Oz 2005; Stahl 1984, 1989; Speth 2004; Wandsnider 1997; Wollstonecroft 2011; Wollstonecroft et al. 2012; Wrangham et al. 1999) have begun to consider how food processing—including its origins, technical innovations, geographic spread, and regional distinctions—is of significance in human evolution. Certainly rudimentary innovations, some of them thought to date from at least early *Homo*, carry implications for changes in dietary selection and ingestive behaviours and for physical, mental and ecological repercussions, such as shifts in foraging movements, new foraging skills and knowledge and methods of transmitting them from one generation to the next, changes in brain and body sizes, and longevity and disease-prevention abilities, particularly for infant health and survival.

This expanding archaeological literature has brought to light numerous interpretative problems with assumptions about human diet and its evolution—problems that are partly the result of the paucity of archaeological evidence (particularly plant foods) and limitations of our archaeological methods. But a major cause of interpretive problems is the lack of understanding about the multidimensionality and complexity of diet—for example, optimization and diet breadth models that are based on ranking foods according to protein or energy (protein or Kcal/hr obtained, less protein or Kcal/hr expended) (for instance, Barlow and Heck 2002; Broughton 1999; Winterhalder and Goland 1997) but that overlook the necessity of dietary diversity and the non-energy, nonprotein critical nutrients necessary for human diet and health as well as the functional properties of the foods in question.

Another significant interpretative problem in the archaeological literature is the lack of consensus about the terms that we use to discuss palaeodiet, a problem that has resulted in significant confusion in the use of terms such as *dietary diversity*, *dietary breadth*, and *dietary quality*, with the last term sometimes referring to a diet of decreased fibre or increased protein or energy concentrations. (For a summary of the problems with the use of the term *dietary quality* in the archaeological literature, see Sponheimer and Dufour 2009.)

As a step toward addressing this problem, the remaining paragraphs explore ways to define and measure dietary diversity and summarise the reasons why food processing is an essential component of human dietary diversity.

DIETARY DIVERSITY, DIETARY FLEXIBILITY, AND FOOD QUALITY

In the archaeological literature, dietary diversity (diet breadth, variety) is loosely used to indicate the range of individual foods consumed by the group(s) under study. Within the nutritional and economic literature, *dietary diversity* (DD) is a specialized term that refers to counts of major food groups or individual foods over a reference period of typically 1–3 days. DD is linked to *dietary quality* (DQ), a term that nutritionists use to describe the degree to which a diet provides an individual with the nutrients and energy that he or she needs for good health, growth, and disease prevention; however, unlike DD, DQ has no official definition within the nutritional literature (Ruel 2003, p. 3912S). DD is considered an essential component in, and therefore a reliable measure of, whether a diet provides adequate nutrients. DD is also used to address economic questions concerning food security in developing nations, because it is associated with food access, per capita caloric acquisition, and economic status (Hoddinott and Yohannes 2002).

There is no global consensus among food scientists about how to measure DD and DQ. Many different combinations of food counts and food group classification systems have been used in conjunction with different reference periods, scoring systems, and cut-off points between low and high diversity (Ruel 2003). At the root of this problem is the fact that the nutrient adequacy of an individual's diet is influenced by a multitude of ecological, economic, pathological, and ethnic factors; therefore, measures that are useful in some contexts are problematic in others (Drewnowski et al. 1996; Ruel 2003). Moreover, there are significant variations in the amount of nutrients that an edible plant contributes to the diet, which are highly influenced by the conditions in which the plant was grown, harvested, transported, and stored and according to in what quantities and how often it was eaten, as well as the bioaccessibility of the nutrients, which again is highly influenced by food processing, the inherent functional properties of the species/plant part as well as the form in which it is eaten (for example, immature/mature, fresh/dried/cooked, hot/cold, whole/grated/sliced, fermented/not fermented) and what other foods it is typically eaten with (Wills et al. 1998).

In most cases, counts of food groups are considered more meaningful measures of DD than are counts of individual foods, because individual foods are often based on the same or similar ingredients (Swindale and Bilinsky 2006; but see the section on Food Processing, Dietary Diversity, and Dietary Flexibility: African Examples regarding how the processing of a single plant into different products can promote nutrient diversity). For example, Bruneton (1975) reported that Moroccan High Atlas Berber peoples consume the cereals millet (*Pennisetum glaucum*), barley (*Hordeum vulgare*), wheats (*Triticum aestivum/durum*), and/or maize (*Zea mays*) at each of their three daily meals but in different forms at each meal: boiled (as gruel), fragmented and steamed as couscous, and baked into breads.

DD studies vary in terms of how many food groupings are employed and the types of additional, associated questions that are asked. Depending on the research questions and culture under study, as few as four and as many as 20 or more food groupings have been reported; some studies further include questions about portion size, number of servings, and within-food group counts—for example, the number and types of vegetables and/or fruit consumed (Ruel 2003, p. 3924S). The range and types of food groups depend on the research questions and the ethnic and ecological context of the population being studied. The Food and Agriculture Organization of the United Nations (FAO), for example, recommends 12 categories for investigating food security: cereals, tubers, vegetables, fruits, meat, fish, eggs, legumes, milk and milk products, fats and oils, sugars and sweets, and beverages. But for investigating nutritional status, the FAO recommended increasing the number of food groups to create categories that highlight specific critical nutrients, such as Vitamin A-rich fruit and vegetables and iron-rich organ meats (FAO 2008).

Addressing Archaeological Questions Concerning Dietary Diversity

Comparing the levels of dietary diversity of our ancient ancestors is even more problematic than comparing those of modern populations, given that we do not know what our ancestors actually ate for much of our evolution. Nevertheless, at the same time we do have access to a wealth of information on the diets of our closest relatives, the great apes, and other species that may have a similar diet to our ancestors—for example, baboons (see Lee-Thorp and Sponheimer 2006; and Sponheimer and Dufour 2009)—as well as much ethnographic information about the eating habits across our species, in addition to the, albeit patchy, archaeological data. As a basis for meaningfully compiling and assessing what these sources can tell us about the evolution of the human diet, we propose three interconnected dietary diversity measures: individual dietary range (IDR), groupings of dietary range (GDR), and dietary flexibility.

Individual Dietary Range (IDR)

Individual dietary range, as applied here, is the total number of individual foods consumed by any one individual. It is like DD in that it involves counts of unique species eaten over a reference period, but differs from DD because the score per species depends on how many different types of organs of that species are consumed; that is, plant foods may include underground storage organs, nuts, fleshy fruit, seeds, leaves, shoots, flower-buds, and mature pollen. Two other factors influence IDR: (1) the mean taxonomic distance separating the species consumed and (2) the amount of the food that is eaten and how often it is eaten. IDR can be calculated by simply counting the number of different species eaten within a specific time frame (day, month, year), with the score per food species being adjusted for the number of different types of organs or tissue of that species consumed (for example, root, stem leaf, fruit, muscle, liver, brain, bone grease) or the number of life-forms (for instance, immature shoots/mature stems, larvae/pupae/adults). Many plants produce a wide range of edible organs, such as cattail or bulrush (*Typha latifolia*), which produces four edible organs, including the rhizomes, immature flowering heads, mature pollen, and 'Cossack asparagus' (the core of flowering aerial shoots).

As we discussed previously, DD counts are normally calculated over 1–3 days. For archaeological purposes (IDR), however, annual totals are more useful, because they offer the advantage of allowing easy comparisons of ethnographic and archaeological data. But presented without details of seasonal patterns of availability and consumption, annual totals can give an exaggerated impression of IDR. Therefore, to obtain more meaningful comparisons, we recommend recording not only the number of species consumed but also the total number of different genera, families, and higher orders represented (as a crude indicator of taxonomic 'spread') and their patterns of consumption around the full cycle of seasons, making note of which foods serve as staples in each season.

GROUPINGS OF DIETARY RANGE (GDR)

Groupings of dietary range are proposed as a method for addressing questions about diet, nutrition, and food processing of ancient peoples. Like DD groupings, GDR classifications depend on specific questions and the geographic, ecological, and cultural contexts of the prehistoric people being studied, as well as the time frame. For example, we suggest that GDR investigations into dietary diversity and nutrition classify foods according to their principal contributions to the diet—such as Vitamin A, Vitamin C, antioxidants. Alternatively, GDR investigations into possible relationships between food selection and technical choice will entail comparisons between sets of groupings, the first set comprising the preferred edible plants, classified according to their most common processing methods (single or multiple stage), the second set comprising available and potentially edible plants that were not selected, which are classified according to how they are known to be processed in societies that do consume them. Another potential application of GDR studies is within-group classifications to explore long-term trends in diet by detecting points in which foods have been added or dropped from the diet. The paucity of archaeological data can be addressed with modelling based on ethnographic studies at one end of the spectrum and great ape studies at the other. It is also clearly important here to take account of taphonomic processes affecting the likelihood that particular foods will survive in identifiable forms.

DIETARY FLEXIBILITY

Dietary flexibility is the ability to alternate between different foods or, in physiological terms, the ability to obtain the necessary nutrients from any one of a range of very different combinations of food plants, animals, and fungi. In this chapter we apply this term to populations of humans or hominoids, but it can apply equally to an individual. Ungar, Grine, and Teaford (2006, p. 210) suggest that high levels of dietary flexibility date back to the earliest *Homo's* biological and cultural adaptations for a more versatile and flexible diet and the absence of a need for 'keystone foods', which are specific foods essential for survival and reproduction. Aspects of dietary flexibility are visible throughout the hominin lineage; for example, Lindeberg (2009, p. 52) observed that the proportion of meat versus plant foods appears to have varied considerably throughout our evolution.

Our species demonstrated extreme dietary flexibility when it first spread from Africa into almost every ecosystem available around the planet. In so doing, the different populations subsequently adapted to totally different dietary packages available from environments ranging from tropical rain forests to ice caps, and from deserts to coniferous taiga. Within each of these regions, distinct food customs and food-processing traditions (cuisine) developed (see Fuller and Rowlands 2009; Haaland 2007; Lyons 2007; Lyons and D'Andrea 2003).

FOOD PROCESSING, DIETARY DIVERSITY, AND DIETARY FLEXIBILITY: AFRICAN EXAMPLES

For present purposes, the most useful examples of food processing in promoting dietary diversity and dietary flexibility are nutritional reports from rural Africa. As part of a global effort to address the problems of low food security and malnutrition in rural Africa, traditional African processing methods, and their role in diet and nutrition, have been extensively investigated. These studies are of particular relevance here because most food processing techniques used in rural Africa are non-mechanised, often with tools and techniques that have long histories of use in continental Africa, for example, grinding, which Mercader (2009) recently identified at a Middle Stone Age site in Mozambique, and fermentation, which is thought to date to at least Predynastic times in Egypt (ca. 3,800 B.C.E.) and the Early Dynastic period in Mesopotamia (ca. 2nd millennium B.C.E.) (Haaland 2007; Samuel 1996; Wengrow 2010).

In the next few paragraphs we briefly summarise some of the advantages of fermentation, a processing method that is of critical importance in sub-Saharan Africa, where fermented USO and cereal foods constitute a high proportion of the diet and are particularly important weaning foods (Odunfa 1988; Osungbaro 2009; Oyewole 2002). In fact, Osungbaro (2009) reports that fermented cereals provide as much as 77% of the caloric consumption in that region. Odunfa (1988, p. 259) listed more than 30 fermented food products consumed in Africa, which he classified into five categories:

1. Fermented starchy roots, for example, cassava (*Manihot esculenta* Crantz) and yam (*Dioscoreae* spp.);
2. fermented cereals such as maize, teff (*Eragrostis tef*), sorghum (*Sorghum bicolor*), and one of the millets (for instance, finger millet *Eleusine coracana*);
3. alcoholic beverages such as palm wine and the sorghum beer *pito*, which is popular in Nigeria and Ghana;
4. fermented vegetable proteins, particularly legumes such as the West African locust bean (*Parkia biglobosa*);
5. fermented animal proteins, for example, milk and fish products.

Fermentation is a complex procedure involving parallel or sequential interactions of mixed cultures of yeasts, bacteria, and fungi. Four main fermentation processes are used worldwide: alcoholic, lactic acid, acetic acid, and alkali fermentation. Natural lactic acid bacteria, which are the most common fermentation agents for cereal and milk products, are the most widely used in Africa (Blandino et al. 2003). Fermentation improves food texture, taste and aroma, nutrient bioaccessibility, and shelf-life, thus preventing loss of food quality and quantity, (Obadina, Oyewole, and Odusami 2009; Odunfa 1988). Odunfa (1988, p. 264), for example, reported that, owing to the production of organic acids during fermentation, *ogi* (a fermented cereal product) can be kept for up to 14 days or more by periodic decanting and replacing the supernatant water. Blandino and associates (2003, p. 530) further observed that lactic acid fermentation improves food shelf life because the bacteria reduce the pH to below 4.0, thus creating a food matrix that is inhospitable for most microorganisms.

Fermentation promotes nutrient increases as well as enhancing nutrient bioavailability. Fermented legumes, for example, are higher in B vitamins than legumes that have simply been cooked; fermentation also increases the protein concentrations in some legumes—for example, fermented African locust-bean (*Parkia biglobosa*) (Odunfa 1988). Among the ways that fermentation improves bioavailability of nutrients in plant foods is that it reduces toxic and other antinutrient substances—such as phytic acid, tannins, and polyphenols in cereals—substances that otherwise limit the digestion and absorption of protein, vitamins, and minerals, particularly iron (Blandino et al. 2003; Odunfa 1988). Blandino and associates (2003) explain that, *depending on the form of food*, fermentation can facilitate optimum pH conditions for enzymatic degradation of phytates, which in turn promotes substantial increases in bioavailabilty of soluble iron, zinc, and calcium as well as protein and amino acids. They further explain that the bioaccessibility of iron increases, because lactic acid fermentation reduces tannin levels. Although it is also true that fermentation sometimes causes nutrient losses, such as a reduction of amino acids and starch, the gains are more often greater than the losses, because fermentation typically transforms foods into safer, better tasting, and more stable products. Also, protein-poor or low-quality foods can be fortified with legume flours or fermented with enriched starter cultures, a necessity for the poorer African households, where protein is scarce and the main meal of the day comprises exclusively USO or cereal products, and in cases where the food raw materials are low in quality (macronutrient and/or micronutrient poor) (see Achi 2005 and Osho 2003).

Fermentation facilitates dietary diversity in a number of ways. It permits people to incorporate otherwise inedible foods into their diet—for example, the African locust-bean, an indigenous legume that is highly toxic in the raw form (Odunfa 1988). Fermentation also provides a means of transforming individual raw foods into a variety of edible products, each with a unique taste, texture, and nutrient composition and nutrient bioaccessibility. Fermentation also promotes dietary flexibility, a point that is exemplified by the African adoption of cassava and maize, which are USO and seed foods introduced from the Americas in the 16th century. Both these products have become staple foods throughout Africa. The fact that these introduced foods were so readily accepted by Africans is undoubtedly because people were able to transform them into foodstuff with acceptable tastes, textures, and shelf-life expectancies, using existing indigenous food fermentation methods.

West African fermented cassava foods provide particularly good examples of the ways that food processing promotes dietary diversity. Approximately a dozen fermented foods and beverages are made from cassava in Africa, including beer, porridge, bread, pudding, dumplings, and sweets. Cassava is high in carbohydrates and dietary fibre and also contains several important macronutrients and micronutrients, including vitamin C and zinc, but is low in protein (cf. Osho 2003).

Traditional cassava processing involves peeling, particle reduction by cutting, grating and/or pulverising, fermentation, sieving, and dehydration. The sequence of these activities depends on the desired final product (Figure 3.1). As well as providing a means of creating a variety of food products from cassava, traditional processing detoxifies the tubers, which contain two cyanogenic glucosides known as *linamarin* and *lotaustralin* (Obadina, Oyewole, and Odusami 2009; Oyewole 2002).

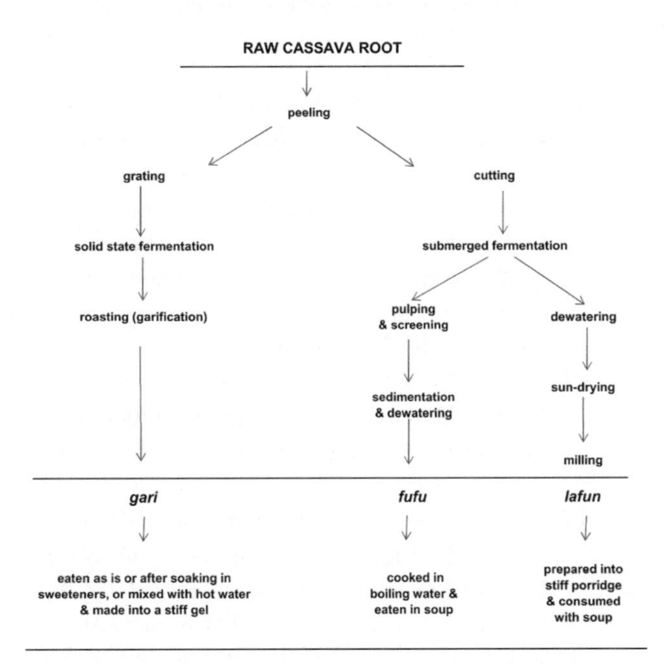

Figure 3.1 Diagram of processing pathways for preparation of three alternative foods derived from cassava (*Manihot esculenta* Crantz) in western Africa (Kuboye, 1985; Obadina, Oyewole, and Odusami 2009; Oyewole 2002).

Among the more popular West African cassava foods are *fufu*, a dumpling that is eaten in soups; *lafun*, a porridge made from fermented cassava flour; and *gari*, a granular fermented food that can be eaten raw or cooked as a savory or a sweet (Oyewole 2002, p. 8). The initial stages of preparing *fufu* and *lafun* are similar: the tubers are peeled, cut into pieces, and subjected to submerged fermentation (Obadina, Oyewole, and Odusami 2009; Oyewole 2002), but well after the earliest evidence for the controlled use of fire, (for example, Gesher Benot Ya'aqov, Israel, Goren-Inbar, et al. 2002). Following fermentation, *fufu* is processed into a pasty mash by pulping and screening to remove the fibres, followed by sedimentation and dehydration, and finally boiling. *Lafun* has a coarser texture than *fufu*, because the plant fibres are not removed. Subsequent to fermentation it is dewatered, sun-dried, and milled into a flour, which is boiled into a porridge. Cassava that is intended for *gari* is peeled, grated, and fermented in the solid state (rather than submerged) and then pulverised and roasted.

These examples of cassava exemplify some of the ways that our species has learned to use food processing to increase dietary diversity, to facilitate dietary flexibility, and to promote nutrient bioaccessibility.

CONCLUSION

Shifts toward greater dietary diversity appear to have served as both a cause and a consequence of human evolutionary history. The dietary breadth enjoyed and required by humans today is an outcome of an ever-expanding food processing tradition that possibly began when our Lower Palaeolithic *Homo* ancestors initiated the routine use of rocks to get at otherwise unavailable, embedded edible parts of plants and animals. It has been argued that the dietary requirements that shaped the human genome were established by 150,000–200,000 years ago, before the emergence of modern humans (see Lindeberg 2009, p. 49), but well after the earliest evidence for the controlled use of fire, (for example, Gesher Benot Ya'aqov, Israel, Goren-Inbar, et al. 2002). Nevertheless, humans continued to develop new ways to incorporate new foods and to upgrade the nutritional benefits of those foods, as observed with the African example of cassava processing, which may have implications about more recent changes in human dietary requirements. However, for many populations, after eons of expansion of dietary diversity the inception of agriculture—starting a mere 13,000 years ago—has culminated in the drastic narrowing of dietary diversity, with dire effects on human health. Whether or not food processing does, or can, counter this narrowing of diversity has yet to be established.

REFERENCES

Achi, O. K. (2005) The potential for upgrading traditional fermented foods through biotechnology. *Journal of Biotechnology 4*, 375–80.

Aiello, L. C. (1996) Terrestriality, bipedalism and the origin of language. *Proceedings of the British Academy 88*, 269–89.

Aiello, L. C., and R. I. M. Dunbar (1993) Neocortex size, group size, and the evolution of language. *Current Anthropology 34*, 184–93.

Aiello, L. C., and P. Wheeler (1995) The expensive tissue hypothesis: The brain and the digestive system in human primate evolution. *Current Anthropology 36*, 199–221.

Aworh, O. C. (2008) The role of traditional food processing in national development: The West African experience. In G. L. Robertson and J. R. Lupien (Eds.), *Using Food Science and Technology to Improve Nutrition and Promote National Development*, Chapter 3. International Union of Food Science and Technology, www.iufost.org/publications/books/documents/Revd.pdf. Accessed July 28, 2012.

Barlow, K. R., and M. Heck (2002) More on acorn eating during the Natufian: Expected patterning in diet and the archaeological record of subsistence. In S. L. R. Mason and J. G. Hather (Eds.), *Hunter-Gatherer Archaeobotany: Perspectives from the Northern Temperate Zone*, pp. 128–45. London: Institute of Archaeology, UCL.

Blandino, A., M. E. Al-Aseeri, S. S. Pandiella, D. Cantero, and C. Webb (2003) Cereal-based fermented foods and beverages. *Food Research International 36*, 527–43.

Blumenschine, R. J. (1992) Hominid carnivory and foraging strategies and the socio-economic function of early archaeological sites. In A. Whiten and E. N. Widdowson (Eds.), *Foraging Strategies and Natural Diets of Monkeys, Apes and Humans*, pp. 51–61. Oxford: Oxford University Press.

Broughton, J. M. (1999) *Resource Depression and Intensification during the Late Holocene, San Francisco Bay: Evidence from the Emeryville Shellmound*. Berkeley and Los Angeles: University of California.

Brunet, M., F. Guy, D. Pilbeam, D. Lieberman, A. Likius, H. T. Mackaye, M. S. Ponce de Leon, C. P. E. Zollikofer, and P. Vignaud (2005) New material of the earliest hominin from the Upper Miocene of Chad. *Nature 434*, 752–55.

Bruneton, A. (1975) Bread in the region of the Moroccan High Atlas: A chain of daily technical operations in order to provide daily nourishment. In M. L. Arnott (Ed.), *Gastronomy: The Anthropology of Food and Food Habits*, pp. 275–85. The Hague: Mouton Publishers.

Carmody, R. N., G. S. Weintraub, R. W. Wrangham (2011) Energetic consequences of thermal and nonthermal food processing. *Proceedings of the National Academy of Sciences 108*, 19199–203.

Carmody, R. N., and R. W. Wrangham (2009) The energetic significance of cooking. *Journal of Human Evolution 57*, 379–91.

Cerling, T. E., K. L. Critz, N. G. Jablonski, and M. G. Leakey (2013a) Diet of Theropithecus from 4 to 1 Ma in Kenya. *Proceedings of the National Academy of Sciences of the United States*, published online June 3, 2013, doi:10.1073/pnas.1222571110.

Cerling, T. E., F. K. Manthi, E. N. Mbua, L. N. Leakey, M. G. Leakey, R. E. Leakey, F. H. Brown, F. E. Grine, J. A. Hart, R. H. Prince Kaleme, K. T. Uno, and B. A. Wood (2013b) Stable isotope-based diet reconstructions of Turkana Basin Hominins. *Proceedings of the National Academy of Sciences of the United States*, published online June 3, 2013, doi:10.1073/pnas.1222568110.

Clutton-Brock, T. H., and P. H. Harvy (1977) Primate ecology and social organization. *Journal of Zoology 183*, 1–39.

Conklin-Brittain, N. L., R. W. Wrangham, and K. D. Hunt (1998) Dietary response of chimpanzees and cercopithicines to seasonal variation in fruit abundance. II, Macronutrients. *International Journal of Primatology 19*, 971–98.

Conklin-Brittain, N. L., R. W. Wrangham, and C. C. Smith (2002) A two-stage model of increased dietary quality in early hominind evolution: The role of fiber. In P. S. Ungar and M. F. Teaford (Eds.), *Human Diet: Its Origin and Evolution*, pp. 61–76. Westport, CT: Bergin and Garvey.

Copeland, S. R. (2007) Vegetation and plant food reconstruction of lowermost Bed II, Olduvai Gorge, using modern analogs. *Journal of Human Evolution 53*, 146–75.

Drewnowski, A. S., Ahlstrom Henderson, A. B. Shore, C. Fischler, P. Prezosi, and S. Hercberg (1996) Diet quality and dietary diversity in France: Implications for the French paradox. *Journal of the American Dietetic Association 96*, 663–69.

Ellis, P. R., C. W. C. Kendall, Y. Ren, C. Parker, J. F. Pacy, K. W. Waldron, and D. J. A. Jenkins (2004) Role of cell walls in the bioaccessibility of lipids in almond seeds. *American Journal of Clinical Nutrition 80*, 604–13.

FAO (2008) *Guidelines for Measuring Household and Individual Dietary Diversity, Version 4. Prepared by the Nutrition and Consumer Protection Division of the Food and Agriculture Organization of the United Nations*. Rome: FAO.

Foote, J., S. Murphy, L. Wilkens, P. Basiotis, and A. Carlson (2004) Dietary variety increases the probability of nutrient adequacy among adults. *Journal of Nutrition 134*, 1779–85.

Fuller, D. Q, and M. Rowlands (2009) Towards a long-term macro-geography of cultural substances: Food and sacrifice tradition in East, West and South Asia. *Chinese Review of Anthropology 12*, 1–37.

Goren-Inbar, N., G. Sharon, Y. Melamed, and M. Kislev (2002) Nuts, nut-cracking and pitted stones at Gesher Benot Ya'aqov, Israel. *Proceedings of the National Academy of Sciences 99*, 2455–60.

Haaland, R. (2007) Porridge and pot, bread and oven: Food ways and symbolism in Africa and the Near East from the Neolithic to the Present. *Cambridge Archaeological Journal 17*, 165–82.

Hillman, G. C. (1989) Late Palaeolithic plant foods from Wadi Kubbaniya in Upper Egypt: Dietary diversity, infant weaning, and seasonality in a riverine environment. In D. R. Harris and G. C. Hillman (Eds.), *Foraging and Farming*, One World Archaeology 13. pp. 207–39. London: Unwin Hyman.

———. (2004) *The Rise and Fall of Dietary Diversity*. Paper presented at the 2004 meeting of the Society of Economic Botany, Canterbury.

Hillman, G. C., E. Madeyska, and J. G. Hather (1989) Wild plant foods and diet of Late Palaeolithic Wadi Kubbaniya: The evidence from charred remains. In F. Wendorf, R. Schild, and A. Close (Eds.), *The Prehistory of Wadi Kubbaniya*, Vol. 2: *Stratigraphy, Palaeoeconomy and Environment*, pp. 162–242. Dallas: Southern Methodist University Press.

Hoddinott, J., and Y. Yohannes (2002) *Dietary Diversity as a Household Security Index: Technical Appendix*. Food and Nutritional Technical Assistance Project, Academy for Educational Development. Washington, D.C.: FANTA.

Hohmann, G. (2009) The diets of nonhuman primates: Frugivory, food processing and food sharing. Evolution of human diets. In J. J. Hublin and M. P. Richards (Eds.), *The Evolution of Hominin Diets: Integrative Approaches to the Study of Palaeolithic Subsistence*, pp. 1–14. Dordrecht: Springer.

Holden, T. G. (1990) *Taphonomic and Methodological Problems Reconstructing Diet from Ancient Human Gut and Faecal Remains*. Ph.D. Dissertation, Institute of Archaeology, University of London.

Johns, T. (1990) *With Bitter Herbs They Shall Eat It: Chemical Ecology and the Origin of Human Diet and Medicine*. Tucson: University of Arizona Press.

———. (1999) The chemical ecology of human ingestive behaviours. *Annual Review of Anthropology 28*, 27–50.

Jones, C. E. R. (1989) Archaeochemistry: Facts or fancy. In F. Wendorf, R. Schild, and A. Close (Eds.), *The Prehistory of Wadi Kubbaniya*, Vol. 2: *Stratigraphy, Paleoeconomy, and Environment*, pp. 162–242. Dallas: Southern Methodist University Press.

Jones, M. (2009) Moving north: Archaeobotanical evidence for plant diet in Middle and Upper Palaeolithic Europe. In J. Humblin and M. P. Richards (Eds.), *The Evolution of Hominin Diets: Integrating Approaches to the Study of Palaeolithic Subsistence*, pp. 171–80. Dordrecht: Springer.

Kuboye, A. O. (1985) Traditional fermented foods and beverages of Nigeria. In L. Prage (Ed.), *Proceedings of the IFA/UNU Workshop on the Development of Indigenous Fermented Foods and Food Technology in Africa*, Douala, Cameroon, pp. 224–36. Stockholm: International Foundation for Science. Cited in Latunde-Dada, G. O. (1997) Fermented foods and cottage industries in Nigeria. *Food and Nutrition Bulletin of the United Nations University Press* in collaboration with the United Nations ACC Subcommittee on Nutrition, Vol. 18, http://archive.unu.edu/unupress/food/V184e/begin.htm.

Lee-Thorp, J., A. Likius, H. T. Mackaye, P. Vignaud, M. Sponheimer, and M. Brunet (2012) Isotopic evidence for an early shift to C4 resources by Pliocene hominins in Chad. *Proceedings of the National Academy of Sciences of the United States 109*, 20369–372.

Lee-Thorp, J. A., and M. Sponheimer (2006) Contributions of biogeochemistry to understanding early hominin ecology. *Yearbook of Physical Anthropology 49*, 131–48.

Leonard, W. R., and M. L. Robertson (1994) Evolutionary perspectives on human nutrition: The influence of brain and body size on diet and metabolism. *American Journal of Human Biology 6*, 77–88.

———. (1997) Comparative primate energetics and hominid evolution. *American Journal of Physiological Anthropology 102*, 265–81.

Lindeberg, S. (2009) Modern human physiology with respect to evolutionary adaptations that relate to diet in the past. In J. J. Hublin and M. P. Richards (Eds.), *The Evolution of Hominin Diets: Integrating Approaches to the Study of Paleolithic Subsistence*, pp. 43–57. Dordrecht: Springer.

Lyons, D. (2007) Integrating African cuisines: Rural cuisine and identity in Tigray, highland Ethiopia. *Journal of Social Archaeology 7*, 346–71.

Lyons, D., and A. C. D'Andrea (2003) Griddles, ovens and agricultural origins: An ethnoarchaeological study of bread baking in Highland Ethiopia. *American Anthropologist 105*, 515–30.

Mercader, J. (2009) Mozambican grass seed consumption during the Middle Stone Age. *Science 376*, 1680–83.

Milton, K. (1993) Diet and primate evolution. *Scientific American 269*, 86–93.

———. (1999) Nutritional characteristics of wild primate foods: Do the diets of our closest living relatives have lessons for us? *Nutrition 15*, 488–98.

Milton, K. (2000) Back to basics: Why foods of wild primates have relevance for modern human health. *Nutrition 16*, 480–83.

———. (2002a) Hunter-gatherer diets: Wild foods signal relief from diseases of affluence. In P. S. Ungar and M. F. Teaford (Eds.), *Human Diet: Its Origin and Evolution*, pp. 111–22. Westport, CT: Bergin and Garvey.

Milton, K. (2002b) Back to basics: Why foods of wild primates have relevance for modern human health. *Nutrition* 16, 480–83.

Munro, N. D., and G. Bar-Oz (2005) Gazelle bone fat processing in the Levantine Epipalaeolithic. *Journal of Archaeological Science* 32, 223–39.

Obadina, A. O., O. B. Oyewole, and A. O. Odusami (2009) Microbiological safety and quality assessment of some fermented cassava products (*lafun, fufu, gari*). *Scientific Research and Essay* 4 (5), 432–35.

O'Connell, J. F., K. Hawkes, and N. Blurton Jones (2002) Meat-eating, grandmothering and the evolution of early human diets. In P. S. Ungar and M. F. Teaford (Eds.), *Human Diet: Its Origin and Evolution*, pp. 49–60. Westport, CT: Bergin and Garvey.

Odunfa, S. A. (1988) African fermented foods: From art to science. *MIRCEN Journal* 4, 259–73.

Osho, S. M. (2003) The processing and acceptability of a fortified cassava-based product (gari) with soybean. *Nutrition and Food Science* 33, 278–83.

Osungbaro, T. O. (2009) Physical and nutritive properties of fermented cereal foods. *African Journal of Food Science* 3, 24–27.

Oyewole, O. B. (2002) *The Powers at the Roots: Food and Its Microbial Allies*. Paper presented at the 15th Inaugural Lecture. University of Agriculture, Abeokuta, Nigeria, October 9, 2002.

Parada, J., and J. M. Aguilera (2007) Food microstructure affects the bioavailability of several nutrients. *Journal of Food Science* 72 (2), R21–32.

Peters, C. R., and, J. C. Vogel (2005) Africa's wild C4 plant foods and possible early hominid diets. *Journal of Human Evolution* 48, 219–36.

Pfannhauser, W., G. R. Fenwick, and S. Khokhar (Eds.) (2001) *Biologically Active Phytochemicals in Food: Analysis, Metabolism, Bioavailability and Function*. Cambridge: Royal Society of Chemists.

Ruel, M. T. (2003) Operationalizing dietary diversity: A review of measurement issues and research priorities. *Journal of Nutrition* 133, 3911S–26S.

Rodman, P. S. (2002) Plants of the apes: Is there a hominoid model for the origin of the hominid diet? In P. S. Ungar and M. F. Teaford (Eds.), *Human Diet: Its Origin and Evolution*, pp. 77–109. Westport, CT: Bergin and Garvey.

Samuel, D. (1996) Investigation of ancient Egyptian baking and brewing methods by correlative microscopy. *Science* 273, 488–90.

Schlemmer, R., E. Razzazi, H. W. Hulan, F. Bauer, and W. Luf (2001) The effect of the addition of different spices on the development of oxysterols in fried meat. In W. Pfannhauser, G. R. Fenwick, and S. Khokhar (Eds.), *Biologically Active Phytochemicals in Food: Analysis, Metabolism, Bioavailability and Function*, pp. 471–73. Cambridge: Royal Society of Chemists.

Smith, B. H., and R. L. Thompkins (1995) Towards a life history of the Hominidae. *Annual Review of Anthropology* 24, 257–79.

Snodgrass, J. J., W. R. Leonard, and M. L. Robertson (2009) The energetics of encephalization in early hominids. In J. J. Hublin and M. P. Richards (Eds.), *The Evolution of Hominin Diets: Integrating Approaches to the Study of Paleolithic Subsistence*, pp. 15–30. Dordrecht: Springer.

Speth, J. (2004) *The Emergence of Bone Boiling: Why It Matters*. Paper presented at the 69th Annual Meeting of the Society for American Archaeology, April 3, 2004, Montreal.

Sponheimer, M., Z. Alemseged, T. E. Cerling, F. E. Grine, W. H. Kimbel, M. G. Leakey, J. A. Lee-Thorp, F. K. Manthi, K. E. Reed, B. A. Wood, and J. G. Wynn (2013) Isotopic evidence of early hominin diets. *Publication of the National Academy of Science*, published June 3, 2013, doi:10.1073/pnas.1222579110.

Sponheimer, M., and D. L. Dufour (2009) Increased dietary breadth in early *Hominin* evolution: Revisiting arguments and evidence with a focus on biological contributions. In J. Humblin and M. P. Richards (Eds.), *The Evolution of Hominin Diets: Integrating Approaches to the Study of Palaeolithic Subsistence*, pp. 229–40. Dordrecht: Springer.

Stahl, A. B. (1984) Hominid dietary selection before fire. *Current Anthropology* 25, 151–68.

———. (1989) Plant-food processing: Implications for dietary quality. In D. R. Harris and G. C. Hillman (Eds.), *Foraging and Farming: The Evolution of Plant Exploitation*, pp. 171–96. London: Unwin Hyman.

Stahl, W., H. van den Berg, J. Arthur, A. Bast, J. Dainty, R. M. Faulks, C. Gartner, G. Haenen, P. Hollman, B. Holst, F. J. Kelly, M. C. Polidori, C. Rice-Evans, S. Southon, T. van Vliet, J. Vina-Ribes, G. Williamson, and S. B. Astley (2002) Bioavailability and metabolism. *Molecular Aspects of Medicine* 23, 39–100.

Swindale, A., and P. Bilinsky (2006) *Household Dietary Diversity Score (HDDS) for Measurement of Household Food Access: Indicator Guide* (v.2). Washington, D.C.: Food and Nutrition Technical Assistance Project, Academy for Educational Development, 2006.

Townsend, C. C., and E. Guest (1985) *Flora of Iraq, Volume 8: Monocotyledons*. Baghdad: Ministry of Agriculture and Agrarian Reform.

Ungar, P. S., F. E. Grine, and M. F. Teaford (2006) Diet in early Homo: A review of the evidence and a new model of adaptive versatility. *Annual Review of Anthropology* 35, 209–29.

Ungar P. S., and M. F. Teaford (2002) Perspectives on the evolution of human diet. In P. S. Ungar and M. F. Teaford (Eds.), *Human Diet: Its Origins and Evolution*, pp. 1–6. Westport, CT: Greenwood Publishing.

Verhagen, H., S. Coolen, G. Duchateau, J. Mathot, and T. Mulder (2001) Bioanalysis and biomarkers: The tool and the goal. In W. Pfannhauser, G. R. Fenwick, and S. Khokhar (Eds.), *Biologically Active Phytochemicals in Food: Analysis, Metabolism, Bioavailability and Function*, pp. 125–30. Cambridge: Royal Society of Chemists.

Waldron, K. W., A. C. Smith, A. J. Parr, A. Ng, and M. L. Parker (1997) New approaches to understanding and controlling cell separation in relation to fruit and vegetable texture. *Trends in Food Science and Technology* 8, 213–21.

Walingo, M. K. (2009) Indigenous food processing methods that improve zinc absorption and bioavailability of plant diets consumed by the Kenyan population. *African Journal of Food, Nutrition and Development* 9, 523–35.

Wandsnider, L. (1997) The roasted and the boiled: Food composition and heat treatment with special emphasis on pit-hearth cooking. *Journal of Anthropological Archaeology* 16, 1–48.

Wengrow, D. (2010) *What Makes Civilization? The Ancient Near East and the Future of the West.* Oxford: Oxford University Press.

Wills, R., B. McGlasson, D. Graham, and D. Joyce (1998) *Post-Harvest: An Introduction to the Physiology and Handling of Fruit, Vegetables and Ornamentals.* Sydney: University of New South Wales Press.

Winterhalder, B., and C. Goland (1997) An evolutionary ecology perspective on diet choice, risk, and plant domestication. In K. J. Gremillion (Ed.), *People, Plants and Landscapes: Studies in Palaeoethnobotany*, pp. 123–60. Tuscaloosa: University of Alabama Press.

Wollstonecroft, M. (2011) Investigating the role of food processing in human evolution: A niche construction approach. *Journal of Archaeological and Anthropological Sciences* 3, 141–50.

Wollstonecroft, M., P. R. Ellis, G. C. Hillman, D. Q Fuller, and P. J. Butterworth (2012) A calorie is not necessarily a calorie: Technical choice, nutrient bioaccessibility, and interspecies differences of edible plants. *Proceedings of the National Academy of Sciences* 109, published online: www.pnas.org/content/109/17/E991.extract.

Wollstonecroft, M., P. R. Ellis, G. C. Hillman, D. Q Fuller (2008) Advancements in plant food processing in the Near Eastern Epipalaeolithic and implications for improved edibility and nutrient bioaccessibility: An experimental assessment of sea club-rush (*Bolboschoenus maritimus* [L.] Palla). *Vegetation History and Archaeobotany* 17 (Suppl. 1), S19–27.

Wrangham, R., J. H. Jones, G. Laden, D. Pilbeam, and N. L. Congklin-Brittain (1999) The raw and the stolen: Cooking and the ecology of human origins. *Current Anthropology* 40, 567–94.

Wynn, J. G., M. Sponheimer, W. H. Kimbel, Z. Alemseged, K, Reed, Z. K. Bedaso, and J. N. Wilson (2013) Diet of *Australopithecus afarensis* from the Pliocene Hadar Formation, Ethiopia. *Publication of the National Academy of Science*, published online June 3, 2013, doi:10.1073/pnas.1222559110.

Yen, D. E. (1975) Indigenous food processing in Oceania. In M. L. Arnott (Ed.), *Gastronomy, the Anthropology of Food Habits*, pp. 147–68. The Hague: Mouton Publishers.

———. (1980) Food crops. In R. B. Ward and A. Proctor (Eds.), *South Pacific Agriculture: Choices and Constraints*, South Pacific Agricultural Survey 1979, pp. 197–234. Canberra: Asian Development Bank and the Australian National University Press.

Zollikofer, C. P. E., M. S. Ponce de Leon, D. Lieberman, F. Guy, D. Pilbeam, A. Likius, H. T. Mackaye, P. Vignaud, and M. Brunet (2005) Virtual cranial reconstruction of *Sahelanthropus tchadensis*. *Nature* 434, 755–59.

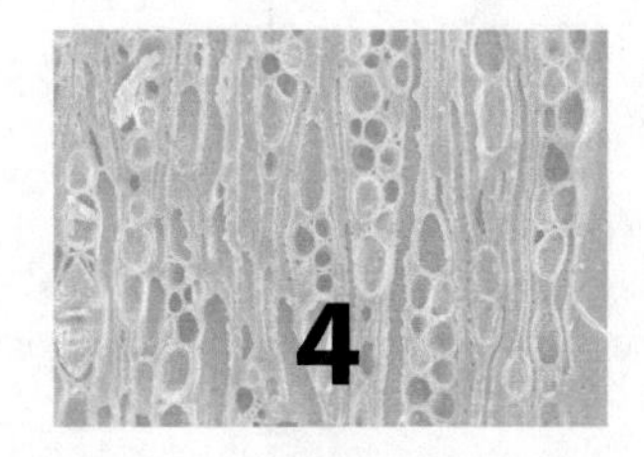

4

Seeds at Sibudu

A Glimpse of Middle Stone Age Vegetation at Sibudu Cave, KwaZulu-Natal, South Africa

Christine Sievers

Plant fruiting structures from archaeological deposits at Sibudu Cave, KwaZulu-Natal (KZN), South Africa, have been analysed to identify the vegetation history of the area. Wadley (2004) studied the material from the Middle Stone Age (MSA) layers dating from around 38 ka (thousand years ago) to more than 72 ka, and Scott (2005) studied the material from the Iron Age layers. This chapter summarises further MSA seed analysis and discusses the comparison of evergreen and deciduous taxa, suggested by Wadley (2004), as an approach to identify vegetation change in an assemblage that otherwise lacks clear indicators of vegetation change and contains elements of evergreen forest, forest margin, riverine, and more open vegetation, throughout the MSA sequence.

Background

Geographical, Climatic and Vegetation Setting

Sibudu Cave overlooks the uThongathi River about 12 km from where it flows into the Indian Ocean, on the east coast of South Africa (Figure 4.1). The approximately 55 m × 18 m shelter was formed by erosion of Natal Group Sandstone and on the precipitous cliffs around the shelter and in the steep valleys nearby, inaccessible dense evergreen forest has been able to escape conversion into farmland. Although categorized evergreen, the forest also has deciduous and semi-deciduous species, particularly along the forest margins. Similarly, along the perennial uThongathi River, there is evergreen and deciduous vegetation. It is likely that before the extensive replacement of natural vegetation by sugarcane and subsistence farmers, the vegetation in the area consisted of Typical Coast-belt Forest, as defined by Acocks (1988). This vegetation type comprises forest and thornveld, which consists of various stages in the succession between forest and grassveld, but only rarely open grassy savannah. Annual rainfall is in the vicinity of 1,000 mm per annum. Summers are wet and hot with mean temperatures of 22–25°C; winters are mild (15–17°C) and drier than the summers (Grant and Thomas 1998), although there is seldom a month without rain (Scott 2005). The shelter is well protected from the elements despite its southwesterly aspect.

Both the absolute and relative values of moisture availability, insolation, wind, and temperature have a major effect on plant growth. The average mean daily temperatures for the hottest and coolest months are February, 24.5°C, and July, 16.5°C, and the diurnal range is small (Moll 1976). No frost occurs. The absolute maximum temperature is high, but its effect on the vegetation is dampened by high midday humidity in the summer months. Prevalent winds are northeast and southwest. Fronts from the southwest, often preceded by hot northwest Berg winds, are the primary source of precipitation. The Berg wind conditions,

Archaeology of African Plant Use by Chris J. Stevens, Sam Nixon, Mary Anne Murray, and Dorian Q Fuller, Eds., 51–58

Figure 4.1 Geographical location of Sibudu Cave.

although occasional, are severe and act as a limiting factor for sensitive plants, because relative humidity is low during Berg wind conditions; low humidity associated with high temperatures can cause extreme stress to plants.

THE EXCAVATIONS AND DATING

Professor Lyn Wadley, director of the ACACIA (Ancient Culture and Cognition in Africa) project at the University of the Witwatersrand, excavated at Sibudu from 1998 to 2011 and uncovered an area of 22 square m, approximately 3 m deep in the 2 m × 2 m trial trench and various depths in adjacent parts of the grid. Excavations are continuing under the directorship of Professor Nicolas Conard of Tübingen University, Germany.

The shelter contains a long sequence of well-preserved MSA deposits that are directly overlain by Iron Age deposits that date to the last 1,000 years (Wadley and Jacobs 2006; Jacobs et al. 2008a, b). The sediments are largely anthropogenically derived (Pickering 2006) and contain much ashy material (Schiegl and Conard 2006), even beyond the numerous hearths. The hearths are generally irregular-circular patches and never surrounded by, or built with, stones (Wadley and Jacobs 2006). Optically Stimulated Luminescence (OSL) (Jacobs et al. 2008a, b) of quartz grains has provided ages for layers that are combined according to the cultural sequence determined by lithic analysis—namely, Howiesons Poort ~65–62.5 ka; post-Howiesons Poort ~58 ka; late MSA, ~48 ka; and final MSA, ~38 ka. Occupation is not continuous, and there are hiatuses of approximately 10 ka between the last three periods of occupation.

ENVIRONMENTAL INVESTIGATIONS

The good preservation of organic material has provided a large corpus of material for environmental interpretations. The sediments have been subjected to geomagnetic susceptibility and geological and mineralogical investigations. Studies of excavated bone, pollen, phytoliths and charcoal, and proxy geological data from farther afield add complementary strands of evidence for a composite view of past vegetation in the area (Clark and Plug 2008; Wadley 2006, 2013).

Interpretations of past vegetation were the focus of research of the plant fruiting structures, rather than plant/people interrelationships, because of the difficulty of identifying various possible cultural and/or natural taphonomic routes leading to the accumulation of the plant fruiting structures (hereafter referred to collectively as seeds). Although it is possible that the seeds were blown into, or fell into, the shelter from vegetation growing on the cliff above, it is also probable that many seeds were brought in by people, birds, and small mammals, both deliberately and inadvertently. The seeds are likely to have been carbonized in the anaerobic conditions in the characteristically fine sediments beneath the ubiquitous hearths in the shelter (Sievers and Wadley 2008). Studies to determine the various agents leading to the accumulation and movement of seeds across and within the deposits are continuing (Sievers 2011).

METHODS

Each stratigraphical layer was removed separately in 50 cm × 50 cm quadrants within 1 m squares. The deposits were initially dry-sieved through a 2-mm sieve and, since 2003, through a 1-mm sieve. All seeds from the sieves were collected and analysed. The seeds were identified by comparison to modern reference material at the University of the Witwatersrand, and at the South African National Biodiversity Institute (SANBI) in Pretoria.

Uncarbonized remains were noted, but not included in the analysis. Carbonized seeds that could not be identified to family, genus, or species were divided into types on the basis of shape, size, and any other distinctive features, such as breakage patterns. The parts of the fruiting structures recovered differed according to taxa but were specific to individual taxa, which were variously identified by their nutlets, seeds, seed coats, or stones. Stones were defined as the hardened, woody endocarps of drupes or generally single-seeded,

Table 4.1 Sibudu Cave MSA Seed Frequencies (Sievers 2006) Grouped According to OSL Age Clusters (Jacobs et al. 2008a, b) ('x' indicates the presence of a taxon represented by a fragment, or fragments, less than half the size of the seed)

	Family	~64–62 ka	~58 ka	~48 ka	~38 ka	Totals
Acacia sp.	Fabaceae	–	1	1	–	2
Asparagus sp.	Asparagaceae	1	67	22	4	94
Calodendrum capense	Rutaceae	–	x	x	x	x
Indet. seed	Celastraceae	–	1	–	–	1
Chrysophyllum viridifolium	Sapotaceae	–	–	2	–	2
Clerodendrum glabrum	Lamiaceae	–	1	1	–	2
Cordia cf. *caffra*	Boraginaceae	–	1	2	1	4
Cussonia sp.	Araliaceae	–	–	1	–	1
Indet.	Cyperaceae	24	44	504	–	572
Cyphostemma sp.	Vitaceae	–	–	4	–	4
Diospyros sp.	Ebenaceae	–	–	1	–	1
Ehretia rigida	Boraginaceae	–	–	4	1	5
Euclea spp.	Ebenaceae	–	3	4	–	7
Grewia sp.	Malvaceae	x	13	1	–	14
Harpephyllum caffrum	Anacardiaceae	–	–	11	–	11
Lantana cf. *rugosa*	Verbenaceae	1	–	24	–	25
Pappea capensis	Sapindaceae	–	–	1	–	1
Pavetta spp.	Rubiaceae	–	3	108	–	111
Indet. seed	Poaceae	–	–	1	–	1
Podocarpus falcatus	Podocarpaceae	–	x	–	–	x
Protorhus longifolia	Anacardiaceae	–	3	6	–	9
Psychotria cf. *capensis*	Rubiaceae	–	–	2	–	2
Rapanea melanophloeos	Myrsinaceae	–	–	1	–	1
Rhoicissus cf. *digitata*	Vitaceae	–	2	–	–	2
Searsia sp.	Anacardiaceae	–	2	8	–	10
Sideroxylon inerme	Sapotaceae	1	1	16	1	19
Strelitzia sp.	Strelitziaceae	–	–	2	–	2
Indet. seed	Rubiaceae	–	–	2	–	2
Vepris lanceolata	Rutaceae	–	–	6	–	6
Ziziphus mucronata	Rhamncaeae	x	1	20	3	24
Totals		27	143	755	10	935

fleshy fruits. Frequencies were based on whole structures or those constituting more than half of the structure.

To increase the sample size, remains from stratigraphically adjacent layers were combined (Sievers 2006; Table 4.1) according to OSL-determined age clusters and cultural designations based on the lithic sequence. To further increase the sample size in a comparison between deciduous and evergreen taxa, data from charcoal studies (Allott 2006) were included (Table 4.2).

RESULTS

Thirty taxa were identified to family, genus, or species from charred seeds (Table 4.1), and the more common taxa have been illustrated (Wintjes and Sievers 2006). Fruiting

Table 4.2 Distribution of Evergreen and Deciduous Taxa in the Sibudu Cave Combined Charcoal (Allott 2006) and Seed Assemblages (Sievers 2006) (grouped according to the cultural sequence based on lithic designations and OSL age clusters; Jacobs et al. 2008a, b). The number in brackets ( ), according to the column in which it occurs, indicates the number of deciduous or evergreen taxa exclusive to the layer; underlining indicates the names of these exclusive taxa; and **bold** lettering records the frequency of evergreen or deciduous taxa in the layer. For example, 6 evergreen taxa occur in the final MSA, and 2 of these, cf. *Phoenix reclinata* and cf. *Xylotheca kraussiana*, occur only in the final MSA.

	Evergreen taxa	Deciduous taxa
final MSA ~38 ka	**6** (2) *Brachylaena* sp. *Erica* sp. *Mystroxylon* cf. *aethiopicum* cf. *Phoenix reclinata* *Sideroxylon inerme* cf. *Xylotheca kraussiana*	**8** *Acacia* spp. *Calodendrum capense* *Clerodendrum glabrum* *Cordia* cf. *caffra* *Ehretia rigida* *Kirkia* sp. *Ziziphus mucronata*
late MSA ~48 ka	**21** (6) *Buxus* sp. *Chrysophyllum viridifolium* *Cunonia capensis* *Curtisia dentata* *Cyphostemma* sp. *Diospyros* sp. *Drypetes* sp. *Erica caffra* *Erica* sp. *Euclea* sp. *Harpephyllum caffrum* *Mystroxylon* cf. *aethiopicum* *Nuxia* sp. *Podocarpus* sp. *Protorhus longifolia* *Psychotria* cf. *capensis* *Rapanea melanophloeos* *Rhus* sp. *Sideroxylon inerme* *Strelitzia* sp. *Vepris lanceolata*	**22** (8) *Acacia* spp. cf. *Afzelia* sp. *Albizia* spp. *Bridelia* sp. cf. *Burkea africana* *Calodendrum capense* *Celtis* sp. *Clerodendrum glabrum* *Cordia* cf. *caffra* *Ehretia rigida* cf. *Erythrina* sp. *Grewia* sp. cf. *Heteromorpha arborescens* *Kirkia* sp. *Lantana* cf. *rugosa* *Macaranga* cf. *capensis* *Pappea capensis* *Spirostachys/Sapium* *Ximenia* sp. *Ziziphus mucronata*
post-Howiesons Poort ~58 ka	**24** (7) *Brachylaena* cf. *discolor* *Buxus* sp. *Cryptocarya* sp. *Cunonia capensis* *Deinbollia* sp. *Diospyros* sp. *Erica* sp. *Erica caffra* *Euclea* sp. *Leucosidea sericea* *Manilkara* sp. *Morella* cf. *pilulifera* *Mystroxylon* cf. *aethiopicum* *Nuxia* sp. *Podocarpus falcatus* *Podocarpus* spp. *Protorhus longifolia* *Rapanea melanophloeos* *Rhoicissus* cf. *digitata* *Rhus* sp. *Sideroxylon inerme* *Syzygium* sp. *Vepris* sp.	**14** (1) *Acacia* spp. *Bridelia* sp. *Calodendrum capense* *Celtis africana* *Celtis* sp. *Clerodendrum glabrum* *Cordia* cf. *caffra* *Grewia* sp. *Ochna* sp. *Ptaeroxylon obliquum* *Spirostachys/Sapium* *Ximenia* sp. *Ziziphus mucronata*
Howiesons Poort ~62–65 ka	**8** *Brachylaena* sp. *Buxus* sp. *Curtisia dentata* *Drypetes* sp. *Mystroxylon* cf. *aethiopicum* *Podocarpus* spp. *Sideroxylon inerme*	**7** *Grewia* sp. *Kirkia* sp. *Lantana* cf. *rugosa* *Ochna* sp. *Ptaeroxylon obliquum* *Spirostachys/Sapium* *Ziziphus mucronata*

structures that could not be identified to family and/or genus were assigned type numbers. Type 5 comprised 45 specimens and Type 13, 10 specimens. The remaining 30 types were represented by one or occasionally two or more seeds and constituted about 4% of the total assemblage.

Cyperaceae dominated throughout the assemblage, and carbonized nutlets of the sedge *Cladium mariscus* have been identified using Scanning Electron Microscopy (Sievers and Muasya 2011). The other carbonized seeds are listed in Table 4.1. The nomenclature follows Germishuizen and Meyer (2003). The taxa comprise woody canopy and subcanopy trees, shrubs, and lianas. Except for *Podocarpus*, these taxa are all found in the various habitats near Sibudu or are likely to be found in similar habitats. *Podocarpus falcatus* is characteristic of moist inland afromontane forests, but it is also found along the coast adjacent to swamps at Richards Bay, north of Sibudu.

A comparison of deciduous and evergreen taxa, as identified in the combined seed (Sievers 2006) and charcoal (Allott 2006) assemblages (Table 4.2), indicates a marked increase in the proportion of deciduous taxa at about ~48 ka as opposed to ~58 ka. At ~48 ka the number of taxa that occur in the deposit for the first time is relatively similar for both deciduous and evergreen taxa, but there is a proportional increase in the number of deciduous taxa—that is, at ~58 ka evergreen taxa clearly dominated, whereas at ~48 ka evergreen and deciduous taxa appear to occur in similar numbers.

Discussion

The Cyperaceae in the Sibudu assemblage, specifically *Cladium mariscus*, which occurs throughout the sequence after 65 ka, indicates the presence of standing, open water. The sedges were probably harvested and brought to the site from the nearby uThongathi River and indicate that that the river was perennial throughout the period when deposits were forming in the shelter.

Some of the Sibudu assemblage taxa that are categorized as forest taxa are found only in forests, for example,

Chrysophyllum viridifolium, but many of the so called forest taxa are not exclusive to forests. For instance, *Harpephyllum caffrum*, which like *C. viridifolium* is also represented in the ~48 ka layers, is a forest tree (Pooley 1993) but is successfully grown as a street tree in coastal areas and in inland regions where it tolerates dry, cold conditions. *Ziziphus mucronata* is found throughout the Sibudu MSA, and, although huge specimens of *Z. mucronata* are found in forests along the KZN coast, the species has a very wide distribution in wooded grassland and even in semidesert, where it is found in alluvial river beds (Coates Palgrave 2002). Many southern African species are found in diverse habitats and tolerate a variety of temperature and moisture regimes, and because of this fact communities of species, rather than individual species, are generally used in the definition of vegetation types. The vegetation types are defined by various parameters, such as the co-occurrence of species and the relative abundances of these species (Acocks 1988). However, interpretations drawn from the co-occurrence and the abundances of taxa in the Sibudu assemblage must be cautious because of the bias introduced by variations in deposition, preservation, recovery, and the basis of frequency determination for each taxon; although intra-taxa comparisons of frequencies may be valid, inter-taxa frequency comparisons are meaningless, because the frequencies represent various fruiting structures of various sizes from taxa with widely divergent fruit and seed production capabilities.

The archaeobotanical interpretations are further complicated by the mosaic of habitats in the Sibudu area. A major influence on the vegetation is exerted by the diverse topography that leads to variations in insolation (that is, the degree of solar radiation), temperature, wind, shade tolerance, and moisture availability. The steep south-facing valleys are subjected to less insolation than the north-facing slopes, which as a result are much hotter and drier. Along the forest margins, plants are also subjected to the desiccating effects of more sunlight, and vegetation along the river banks may be sustained by moisture seeping down from distant catchment areas, allowing for the presence of plants that may not otherwise survive in the area. Moisture availability peaks in the river bed where various hydrophytes and helophytes are able to thrive.

Plants have developed various strategies to tolerate stress, and one of the adaptations is to lose leaves under adverse conditions. Deciduous plants are those that shed all their leaves during a certain season; in the summer rainfall areas of southern Africa this occurs during the cold, dry winters. The seasonal variation is more marked on a gradient westward from the Indian Ocean into the colder, drier interior, where the coastal forest vegetation is replaced by savannah in which deciduous species are more prominent. Savannah refers to a vegetation type made up of a herbaceous layer dominated by Poaceae (grasses) and an upper layer of woody plants that can vary from widely spaced to a 75% canopy cover (Smit 1999). Wadley (2004) argued for an analogy between the pattern of increased deciduous species on a westward gradient from coastal forest to savannah and an increase in deciduous taxa in the Sibudu seed assemblage—namely, that the latter could indicate an increase in savannah (bushveld) type vegetation in the Sibudu area. The attractiveness of Wadley's argument was that it avoided dependence on relative abundance within taxa, because it was based on the range of taxa present. However, biases that affect abundance within taxa may also affect taxa diversity and the degree to which the archaeobotanical evidence reflects that of the prevailing vegetation needs discussion.

A basic assumption underlying the validity of any interpretations is that present day ecological observations are likely to reflect those in the MSA. It is possible that modern analogues do not exist for some vegetation types (Allott 2004; Wadley 2013). Furthermore, there is no guarantee that the range of taxa in the deposit is necessarily an accurate representation of the proportion of evergreen and deciduous species in the surrounding vegetation, even though bias introduced by preservation and recovery would presumably apply equally to evergreen and deciduous species. Further untestable assumptions are that deciduous and evergreen trees are equally likely to have fruit, which is produced just as regularly and profusely; that human agents are as likely to harvest fruits from deciduous trees as from evergreen trees, or indeed collect wood for fuel; and that the fruits are equally likely to become charred and be deposited in the cave. Another complication is that some plants may act as deciduous, semideciduous, or evergreen, depending on environmental conditions. Therefore, in the comparison of deciduous and evergreen taxa (Table 4.2), I have recorded the taxon according to their most characteristic habit and have omitted any taxa whose behaviour could not be identified securely. For example, *Pappea capensis* is recorded as deciduous, because it is generally considered to be deciduous (Johnson and Johnson 1997; Van Wyk and Van Wyk 1997), or at least primarily deciduous, although it may be evergreen in higher rainfall areas and along rivers (Grant and Thomas 2000).

The results indicate a proportional increase in deciduous taxa at ~48 ka, as opposed to ~58 ka. It is difficult to be certain what this apparent increase means in terms of the vegetation and in terms of the causative factors. It could indicate a decrease in forest cover and a corresponding increase in savannah grassland. Savannah elements were present at ~58 ka, and it is possible that at ~48 ka the savannah woodland had become increasingly open. The reasons for this vegetation change remain obscure, because it could be the result of any combination of relative or absolute variations in climatic factors such as moisture availability, insolation, wind, and temperature. I turn to evidence from relevant multidisciplinary data for possible clues and corroboration of the changing vegetation trends.

Charcoal analyses (Allott 2006) indicate that at ~48 ka there were fewer evergreen forest components and more taxa common in the drier northern regions of South Africa (such as *Kirkia* sp. and cf. *Burkea africana*) than at ~58 ka. Although the patterns may be a result of environmental change, a change in wood selection, charcoal fragmentation, or sample bias, an increase in *Acacia* spp. and other Fabaceae taxa also suggest that conditions were drier and that the vegetation was perhaps more open at ~48 ka than during the ~58 ka occupations of the shelter.

In a palynological study of 15 samples from Sibudu (Renault and Bamford 2006), Poaceae (grasses) pollen was dominant. This is not surprising since, with the exception of grasses, most indigenous plants in southern Africa are insect-pollinated rather than wind-pollinated. Poor preservation of pollen in all the Sibudu samples resulted in totals below the predetermined minimum pollen count deemed reliable for any palaeoenvironmental interpretations. Thus the pollen counts cannot be relied on to demonstrate the relative abundances of grasses versus shrub or tree cover at ~48 ka and ~58 ka and cannot confirm a possible increase in grass cover at ~48 ka. Moreover, grass phytoliths in the deposit are present probably as a result of the use of grass for tinder (Schiegl et al. 2004), and their frequencies are thus unlikely to be a reflection of the relative grass cover in the area. Sedge phytoliths corroborate the evidence from sedge pollens (Renault and Bamford 2006) and sedge nutlets. The phytoliths indicate combustion of plant matter but could possibly also be related to microbial degradation of the fresh plants (Schiegl and Conard 2006). The sedge's presence indicates of the proximity of the river, but, because of the extent of the catchment area, moisture in the river is not necessarily a reflection of local precipitation and humidity regimes.

Fluctuations in micromammal species potentially could provide an indication of changes in the proportions of vegetation types, but except for layers in the ~38 ka and ~48 ka age clusters, the sample is too sparse for environmental interpretations (Glenny 2006). The micromammalian analysis does, however, bear out a complex mosaic of vegetation dominated by open savannah grassland at ~48 ka. Macrofaunal analyses indicate large equids and large to very large bovids at both ~48 ka and ~58 ka, and there is an increase in NSIP (number of identified specimens) of large grazers, such as the extinct Cape Horse, Giant Hartebeest, and buffalo at ~48 ka (Plug 2004). The increase in large grazers could be agency related or reflect an increase in grazers as a result of increasing grass cover in a savannah environment. Forest and woodland species are still present (Wells 2006), but there is the first confirmed presence of giraffe and eland at ~48 ka (Plug 2004). Eland are grazers and giraffe are browsers in open wooded environments. Giraffe indicate dry savannah (Plug 2004), and the dry condition is relevant to the argument of decreasing moisture availability with respect to the warming trend to be discussed.

The relationship between the relative grass and woody taxa coverage in savannahs is a dynamic relationship influenced by a number of factors, such as moisture availability, shade tolerance, fire, and the presence of herbivores. Proxy evidence that informs on one of these variables—namely, moisture availability—is Herries's (2006) magnetic susceptibility data, which indicates an oscillating but gradual warming trend from ~58 ka onward. Even without a change in precipitation, an increase in temperature could increase evapo-transpiration and have a negative effect on moisture availability. The resulting harsher conditions could lead to an increase in deciduous species.

Recently, the GIS-based coexistence approach (CAGIS) has been applied to analysis of the combined seed and charcoal assemblages at Sibudu (Bruch, Sievers, and Wadley 2012). The mean minimum temperature of coldest winter month is the only parameter where the coexistence interval values do not overlap with modern values; that is, it was clearly colder at 65–62.5 ka, ~58 ka, and 48 ka. The lowest mean minimum temperature of the coldest winter month occurs around ~58 ka and was followed by a gradual warming. Trends with respect to precipitation and temperature both annually and in the hottest and coldest quarters are not as conclusive. All precipitation parameters that were analysed do not show values significantly different from modern ones. More precision and firmer interpretations will be possible with larger samples and tighter tolerance ranges for taxa.

CONCLUSIONS

When the results of the seed analysis are assessed in relation to other relevant evidence it is clear that the different strands of environmental evidence complement each other in the reconstruction of a complex mosaic of evergreen forest, riverine vegetation, and deciduous woodland with varying proportions of grass cover, with the nearby uThongathi River supplying a constant source of moisture throughout the time that the deposits were forming in the shelter. Given the range of ecological tolerances of the taxa present, the variety of habitats in the Sibudu area and the relatively low taxa richness for each age cluster, one is not surprised that the seed identifications alone are insufficient to provide clear indications of a succession of vegetation types during the Sibudu MSA occupations. However, a comparison of the frequencies of evergreen and deciduous taxa, incorporating both seed (Sievers 2006) and charcoal data (Allott 2006) to increase sample size, provides a method of identifying vegetation change. Amalgamation of data from adjacent layers is necessary to boost sample sizes for the comparison, and the conclusions show general trends that disguise possible smaller scale variations within the different layers of the various age clusters. Reservations concerning the interpretive validity of a comparison between deciduous and evergreen species over time are mitigated by other evidence supporting the trend indicated by the seed and charcoal data, suggesting increasingly open savannah vegetation in the Sibudu area at ~48 ka as opposed to ~58 ka. The increase of deciduous taxa may be related to increased evapo-transpiration caused by the gradually warming trend suggested by magnetic susceptibility studies (Herries 2006) and coexistence analysis (Bruch, Sievers, and Wadley 2012).

ACKNOWLEDGEMENTS

I thank Professor Lyn Wadley for most generous financial and other support for this analysis, as well as for the funding to attend the 5th International Workshop for African Archaeobotany (IWAA). Participation in the workshop was an invaluable experience, and sharing of expertise in the practical sessions led to the identification of the Cyperaceae nutlets, thus demonstrating the value of the workshop. I heartily thank the organisers of the 5th IWAA, the other archaeobotanists who so generously shared their knowledge. For valuable support and advice gratefully appreciated, I give Lucy Allott many thanks. I am also grateful to Marinda Koekemoer of SANBI for access to comparative material.

REFERENCES

Acocks, J. P. H. (1988) Veld types of South Africa (3rd ed.). *Memoirs of the Botanical Survey of South Africa 57*, 1–146.

Allott, L. F. (2004) Changing environments in Oxygen Isotope Stage 3: Reconstructions using archaeological charcoal from Sibudu Cave. *South Africa Journal of* Science *100*, 79–185.

———. (2006) Archaeological charcoal as a window on palaeovegetation and wood-use during the Middle Stone Age at Sibudu Cave. *Southern African Humanities 18* (1), 173–201.

Bruch, A. A., C. Sievers, and L. Wadley (2012) Quantification of climate and vegetation from southern African Middle Stone Age sites: An application using Late Pleistocene plant material from Sibudu, South Africa. *Quaternary Science Reviews 45*, 7–17.

Clark, J. L., and I. Plug (2008) Animal exploitation strategies during the South African Middle Stone Age: Howiesons Poort and post-Howiesons Poort fauna from Sibudu Cave. *Journal of Human Evolution 54*, 886–98.

Coates Palgrave, M. (2002) *Trees of Southern Africa* (3rd ed.). Cape Town: Struik.

Germishuizen, G., and N. L. Meyer (Eds.) (2003) Plants of southern Africa: An annotated checklist. *Strelitzia 14*, 1–1231.

Glenny, W. (2006) Report on the micromammal assemblage analysis from Sibudu Cave. *Southern African Humanities 18* (1), 279–88.

Grant, R., and V. Thomas (1998) *Sappi Tree Spotting KwaZulu-Natal*. Johannesburg: Jacana.

———. (2000) *Sappi Tree Spotting Bushveld*. Johannesburg: Jacana.

Herries, A. I. R. (2006) Archaeomagnetic evidence for climate change at Sibudu Cave. *Southern African Humanities 18* (1), 131–47.

Jacobs, Z., R. G. Roberts, R. F. Galbraith, H. J. Deacon, R. Grün, A. Mackay, P. J. Mitchell, R. Vogelsang, and L. Wadley (2008a) Ages for the Middle Stone Age of southern Africa: Implications for human behavior and dispersal. *Science 322*, 733–35.

Jacobs, Z., A. G. Wintle, G. A. T. Duller, R. G. Roberts, and L. Wadley (2008b) New ages for the post-Howiesons Poort, late and final Middle Stone Age at Sibudu Cave, South Africa. *Journal of Archaeological Science 35*, 1790–807.

Johnson, D., and S. Johnson (1993) *Gardening with Indigenous Trees and Shrubs*. Halfway House: Southern Book Publishers.

Moll, E. J. (1976) The vegetation of the three rivers region, Natal. *Natal Town and Regional Planning Reports 33*, 1–102.

Pickering, R. (2006) Regional geology, setting and sedimentology of Sibudu Cave. *Southern African Humanities 18* (1), 123–29.

Plug, I. (2004) Resource exploitation: Animal use during the Middle Stone Age at Sibudu Cave, KwaZulu-Natal. *South African Journal of Science 100*, 151–58.

Pooley, E. (1993) *The Complete Field Guide to Trees of Natal, Zululand and the Transkei*. Durban: Natal Flora Publications Trust.

Renaut, R., and M. K. Bamford (2006) Results of preliminary palynological analysis at Sibudu Cave. *Southern African Humanities 18* (1), 235–40.

Schiegl, S., and N. J. Conard (2006) The Middle Stone Age sediments at Sibudu: Results from FTIR spectroscopy and microscopic analyses. *Southern African Humanities 18* (1), 149–72.

Schiegl, S., P. Stockhammer, C. Scott [Sievers], and L. Wadley (2004) A mineralogical and phytolith study of the Middle Stone Age hearths in Sibudu Cave, KwaZulu-Natal, South Africa. *South African Journal of Science 100*, 185–94.

Scott [Sievers], C. (2005) *Analysis and Interpretation of Archaeobotanical Remains from Sibudu Cave, KwaZulu-Natal*. Unpublished M.Sc. dissertation, University of the Witwatersrand.

Sievers, C. (2006) Seeds from the Middle Stone Age layers at Sibudu Cave. *Southern African Humanities 18* (1), 203–22.

———. (2011). Sedges from Sibudu, South Africa: Evidence for their use. In A. G. Fahmy, S. Kahlheber, and A. C. D'Andrea (Eds.), *Windows on the African Past: Current Approaches to African Archaeobotany. Proceedings of the 6th International Workshop on African Archaeobotany*, June 13–15, 2009, Helwan University, Cairo, Egypt. *Reports in African Archaeobotany 3*, 225–41.

Sievers, C., and A. M. Muasya (2011) Identification of the sedge *Cladium mariscus* subsp. *jamaicense* and its possible use in the Middle Stone Age at Sibudu, KwaZulu-Natal. *Southern African Humanities 23*, 77–86.

Sievers, C., and L. Wadley (2008) Going underground: Experimental carbonization of fruiting structures under hearths. *Journal of Archaeological Science 35*: 209–17.

Smit, N. (1999) *Guide to the Acacias of South Africa*. Pretoria: Briza.

Van Wyk, B., and P. Van Wyk (1997) *Field Guide to Trees of Southern Africa*. Cape Town: Struik.

Wadley, L. (2004) Vegetation changes between 61,500 and 26,000 years ago: The evidence from seeds in Sibudu Cave, KwaZulu-Natal. *South African Journal of Science 100*, 167–74.

———. (2006). Partners in grime: Results of multi-disciplinary archaeology at Sibudu Cave. *Southern African Humanities 18* (1), 315–41.

———. (2013) MIS4 and MIS3 occupations at Sibudu, KwaZulu-Natal, South African. *South African Archaeological Society Bulletin 68*: 41–51.

Wadley, L., and Z. Jacobs (2006) Sibudu Cave: Background to the excavations, stratigraphy and dating. *Southern African Humanities 18* (1), 1–26.

Wells, C. R. (2006) A sample integrity analysis of faunal remains from the RSp layer at Sibudu Cave. *Southern African Humanities 18* (1), 261–77.

Wintjes, J., and C. Sievers (2006) Seeing Sibudu seeds: an illustrated text of the more frequent Middle Stone Age stones, nuts and seeds. *Southern African Humanities 18* (1), 223–33.

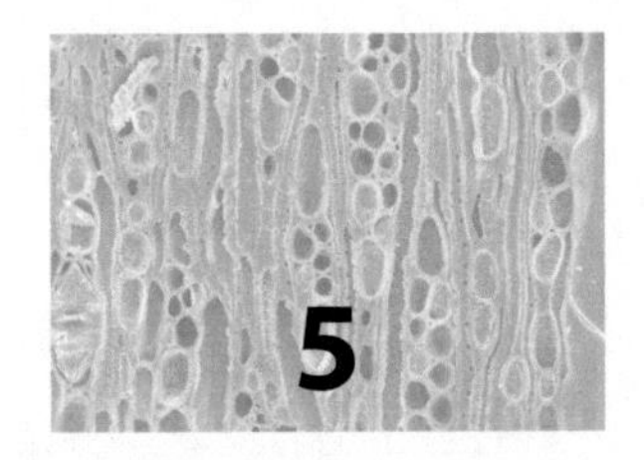

Understanding Late and Terminal Pleistocene Vegetation Change in the Western Cape, South Africa

The Wood Charcoal Evidence from Elands Bay Cave

Caroline Cartwright, John Parkington, and Richard Cowling

Elands Bay Cave (hereafter EBC) is located in the Baboon Point headland in the Western Cape about 200 km north of Cape Town, South Africa (Figure 5.1). It was chosen for excavation in the 1970s as a site likely to provide bioarchaeological evidence of coastal-inland seasonal mobility (Parkington 1972; 1976; Parkington and Poggenpoel 1971). The excavated evidence revealed that EBC was visited episodically by people over some tens of thousands of years. Its deposits show evidence for brief and possibly seasonal visits to EBC, rather than a continuous sequence of permanent occupation. They are a fragmented reflection of change in the local terrestrial and marine environments during parts of the Holocene, much of the terminal Pleistocene, and the last glacial maximum. EBC contains one or more periods of occupation that are beyond the range of the conventional radiocarbon dating technique. The principal deposits comprise a series of shell middens overlying Pleistocene loams that accumulated during the last glacial period when sea level was much lower and the coastline up to 35 km to the west. Figure 5.2 shows the present-day proximity of the shoreline, viewed from EBC.

This chapter consists of a brief summary of the research carried out on the analysis, interpretation, and ecological and environmental reconstruction of the charcoal assemblages from EBC (Cartwright and Parkington 1997; Cowling et al. 1999; Parkington et al. 2000). Research continues at the present day, not only on additional charcoal samples from EBC but also on an ever-increasing corpus of archaeobotanical material from Diepkloof Rock Shelter, a site of considerable significance and interest for comparison with EBC.

Methodology

When EBC was excavated in the 1970s charcoal fragments were sampled only for radiocarbon dating from securely stratified contexts, usually in hearths. Additionally, many bulk samples were taken, including residues from both the 12-mm and 3-mm mesh sieves, as well as unprocessed sediments. Both of these categories yielded the charcoal used for the EBC published research. This charcoal is reliably associated and contemporary with excavated materials from the same depositional envelopes.

Collection of comparative reference samples essential for the identification of the archaeological charcoal commenced in the immediate vicinity of the site and expanded to other locations in the southwestern, northwestern, and southern Cape. Woody taxa were collected and fully documented from a range of substrata and topographical locations in order to assess variability introduced by microenvironments. This assessment included characterisation of extant vegetational communities in their

Archaeology of African Plant Use by Chris J. Stevens, Sam Nixon, Mary Anne Murray, and Dorian Q Fuller, Eds., 59–72

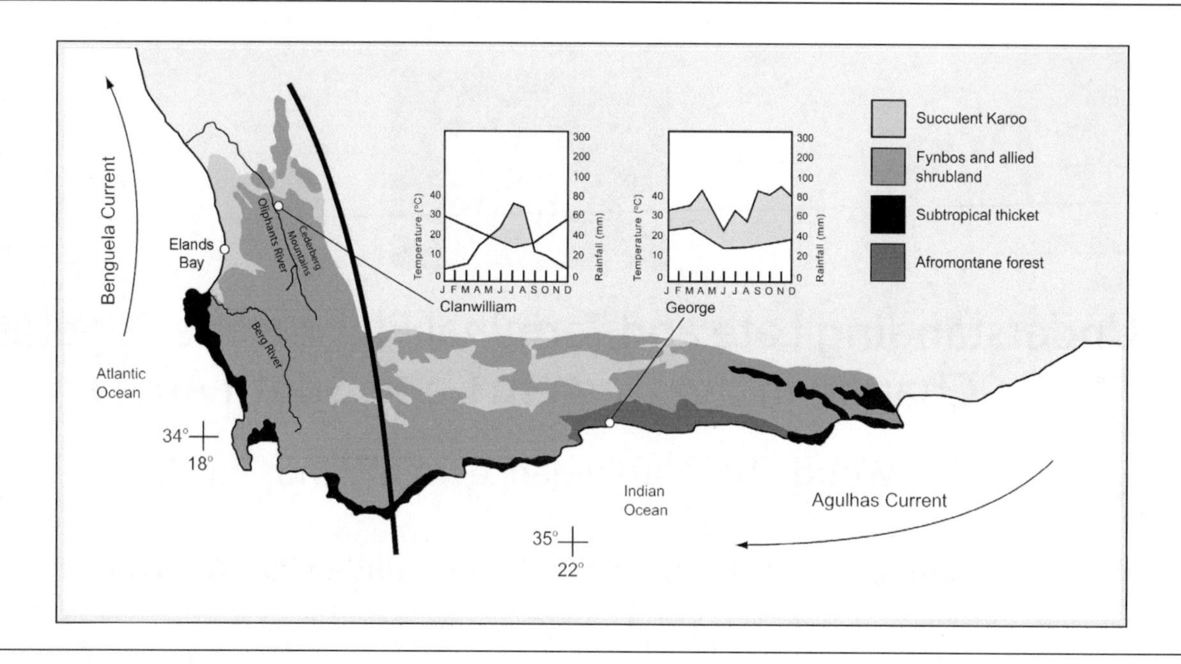

Figure 5.1 Map showing the location of Elands Bay Cave in relation to the major vegetation types of southwestern Africa. The bold line divides the western winter-rainfall area from the eastern area with its spring and autumn maximum rainfall (after Cowling et al. 1999).

Figure 5.2 View from the mouth of Elands Bay Cave toward the sea.

climatic and habitat envelopes covering steep gradients in temperature and rainfall regimes in the region of the Berg and Olifants Rivers and inland to the lower slopes of the Cederberg Mountains. It has resulted in a broad set of available environmental correlates vital to the interpretation of the EBC assemblages (Cowling et al. 1999). The 470 woody taxa collected not only characterise the region today but also mirror the vegetational changes seen in the archaeological charcoal assemblages.

Charcoal extracted from bulk soil samples was subsampled to avoid any bias toward selection of larger-sized pieces. Over 6,700 fragments of charcoal were identified in the initial analysis (Cartwright and Parkington 1997). Each piece was fractured to show transverse, radial, and tangential longitudinal (TS, RLS, TLS) surfaces for examination by standard techniques of optical and scanning electron microscopy (SEM). Following IAWA protocols (for example, Wheeler, Baas, and Gasson 1989; Wheeler et al. 1986) the diagnostic anatomical characteristics permitted identification to taxon. Subsequent study of the EBC charcoal fragments has been carried out using variable pressure (VP) and field emission (FE) SEM (Figures 5.3 and 5.4, respectively). FE-SEM has revealed fine details of the anatomical structure of the charcoal in a particularly clear manner (Figure 5.4). In this example, the TLS of *Cassine peragua*, the image shows distinct 'edge effects' that highlight cell walls, particularly of the rays, while greatly enhancing the visibility and structure of intercellular pitting. Although the charcoal fragments from EBC are generally in good condition, FE-SEM ultra-resolution and increased range of magnification is particularly useful for charcoal older than 20,000 years.

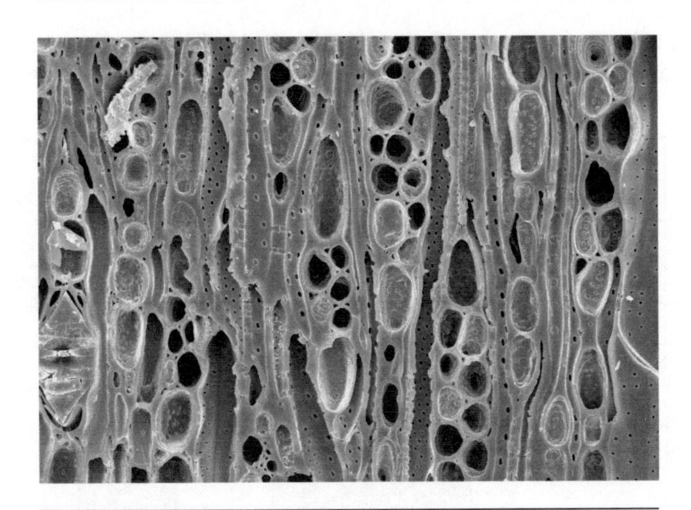

Figure 5.3 VP-SEM (variable pressure scanning electron microscope) photomicrograph of the transverse surface of *Cassine peragua* charcoal.

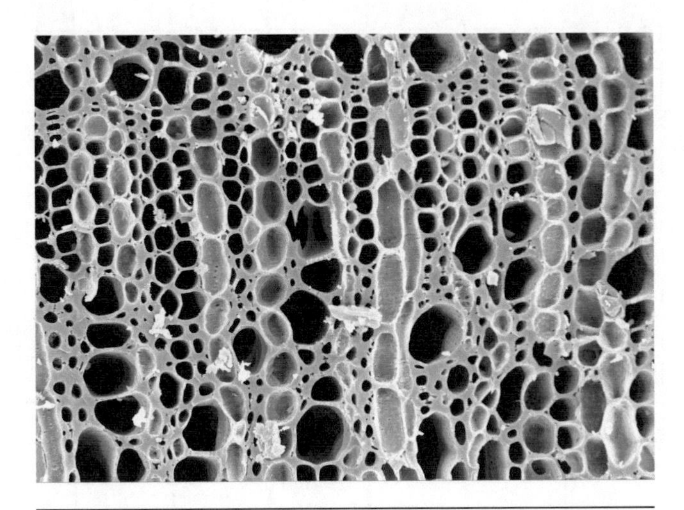

Figure 5.4 FE-SEM (field emission scanning electron microscope) photomicrograph of the tangential longitudinal surface of *Cassine peragua* charcoal.

Taxonomic nomenclature reflects the state-of-play at 2010, but revisions continue to occur.

RESULTS AND DISCUSSION

Charcoal fragments were extremely common in the EBC deposits, both as hearth concentrations and dispersed across the domestic areas. As with other bioarchaeological evidence from EBC, the plant remains reflected the environment through time, although seldom in a simple or direct manner. Much of the charcoal can be related to human selection of local woody resources for intentional fuelwood burning, but wood can also be accidentally caught up in a fire. Furthermore, wooden artefacts or structural material may be thrown on to the fire when no longer useful. In some instances, preferred fuelwood might be entirely reduced to ash. Nevertheless, the charcoal fragments must reflect the local environment, albeit in a complex way, requiring careful interpretation.

The EBC charcoal identifications suggest a series of community associations that clearly express remarkable patterning through time and include xeric thicket, asteraceous shrubland, mesic thicket, proteoid fynbos, riverine woodland, and Afromontane forest elements (Cartwright and Parkington 1997; Cowling et al. 1999). Marked changes in diversity have been recognised. For example, the Holocene samples usually comprise only 5 to 7 identified taxa, whereas this number rises consistently in the terminal Pleistocene to 11 and is greatest before and around the last glacial maximum with 12 or more identified taxa. It is clear that, although the taxonomic diversity is partly a measure of human selectivity, this pattern is likely to reflect an underlying shift in the diversity of available woody species in the EBC landscape.

There are a number of considerations, however, relating to a detailed interpretation and archaeobotanical reconstruction, which need to be taken into account. Many of the species in the EBC charcoal assemblages are extremely long-lived plants with wide distributions in southern Africa. These are mainly multistemmed shrubs and trees that resprout after fire or harvesting. They also have extreme phenotypic plasticity, often growing as prostrate shrubs or tall trees, depending on moisture availability and exposure to wind or fire. They might persist for many thousands of years after conditions optimal for their growth and regeneration have disappeared. For example, with regard to contemporary populations in the EBC area of *Pterocelastrus tricuspidatus* and *C. peragua*, both have their optimum distribution in mesic forest and may be persistors from a wetter epoch. Their presence as elements for palaeoenvironmental reconstruction within the assemblage, therefore, must be treated with caution. Conversely, fynbos taxa are ideal palaeoenvironmental indicators because they are short-lived and need to regenerate repeatedly in order to persist in a location. As this regeneration stage is highly sensitive to climatic conditions, fynbos populations are very vulnerable to environmental change (Cowling 1992; Cowling et al. 1999).

Examining the charcoal sequence in more detail, one notices an obvious and remarkable change in assemblage composition through time (Tables 5.1 and 5.2;

Table 5.1a Wood Charcoal from Elands Bay Cave in Deposits Dating from 320 to 10,500 Years BP, Showing Acronyms of Depositional Units Sampled, Weight of Samples, and Approximate Age (filled cell = taxon present; empty cell = taxon absent)

Depositional Unit	NKOM	MRSB	BUTH	RETS	JOFR	SOY1	BARH	RADS	SHAK(1)	SHAK(2)
sample size (n)	500	100	100	200	104	150	100	200	150	150
mass (g)	52.32	10.74	9.82	21.64	10.24	12.12	11.32	18.63	16.23	14.84
age (years before present)	320	500	1400	3290	3780	3780	3940	4160	4370	4370
Xeric thicket taxa										
Euclea racemosa	■									
Rhus cf. *undulata*	■	■	■							
Ruschia maxima	■	■	■				■	■		
Zygophyllum morgsana	■	■				■	■		■	
Lycium sp.		■			■			■		■
Ruschia sp.			■		■	■	■	■	■	■
Asteraceous shrubland										
Eriocephalus cf. *aromaticus*			■		■					
Aspalathus sp.				■		■				■
Passerina glomerata					■					
Chrysanthemoides sp.						■				
Euryops cf. *speciosissimus*			■		■					■
Hymenolepis cf. *parviflora*				■						
Salvia africana-lutea				■						
Phylica sp.					■	■				
Mesic thicket taxa										
Heeria argentea								■	■	■
Maytenus cf. *oleoides*										
Cassine cf. *peragua*										
Diospyros cf. *glabra*										
Dodonaea angustifolia										
Hartogiella schinoides										
Tarchonanthus camphoratus										
Maytenus cf. *heterophylla*										
Olea europaea subsp. *africana*										
Ficus cf. *cordata*										
Colpoon compressum										
Proteoid fynbos taxa										
Protea cf. *glabra*										
Leucadendron cf. *pubescens*										
Euryops sp.										
Cliffortia sp.										
Erica cf. *verecunda*										
Erica cf. *caffra*										
Protea cf. *nitida*										

Table 5.1a Continued

Depositional Unit	NKOM	MRSB	BUTH	RETS	JOFR	SOY1	BARH	RADS	SHAK(1)	SHAK(2)
General thicket taxa										
Rhus sp.		■	■	■	■	■	■			■
Euclea sp.		■		■		■	■	■	■	
Pterocelastrus tricuspidatus	■		■	■					■	■
Afromontane forest taxa										
Kiggelaria africana										
Podocarpus cf. *elongatus*										
Celtis africana										
Grewia cf. *occidentalis*										
Passerina sp.										
Myrsine africana										
Maytenus sp.										
Halleria cf. *lucida*										
Hartogiella sp.										
Chionanathus foveolatus										
Riverine woodland taxa										
Ficus cf. *sur*										
Salix cf. *mucronata*										

Table 5.1b Wood Charcoal from Elands Bay Cave in Deposits Dating from 320 to 10,500 Years BP, Showing Acronyms of Depositional Units Sampled, Weight of Samples, and Approximate Age (filled cell = taxon present; empty cell = taxon absent)

MARO(1)	MARO(2)	WIRO	BURO	PWBO(1)	PWBO(2)	PWBO(3)	GNOM	BSBP	Depositional Unit
150	100	100	32	100	100	150	104	100	sample size (n)
10.14	11.63	10.77	3.3	10.66	12.14	16.44	10.25	12.53	mass (g)
7910	7910		8860				9510	9600	age (years before present)
									Xeric thicket taxa
									Euclea racemosa
									Rhus cf. *undulata*
									Ruschia maxima
									Zygophyllum morgsana
							■		*Lycium* sp.
■		■							*Ruschia* sp.
									Asteraceous shrubland
									Eriocephalus cf. *aromaticus*
									Aspalathus sp.
									Passerina glomerata
									Chrysanthemoides sp.
■	■	■	■	■	■	■			*Euryops* cf. *speciosissimus*
				■		■		■	*Hymenolepis* cf. *parviflora*
					■				*Salvia africana-lutea*
									Phylica sp.

(Continued)

Table 5.1b Continued

MARO(1)	MARO(2)	WIRO	BURO	PWBO(1)	PWBO(2)	PWBO(3)	GNOM	BSBP	Depositional Unit
									Mesic thicket taxa
									Heeria argentea
									Maytenus cf. *oleoides*
									Cassine cf. *peragua*
									Diospyros cf. *glabra*
									Dodonaea angustifolia
									Hartogiella schinoides
									Tarchonanthus camphoratus
									Maytenus cf. *heterophylla*
									Olea europaea subsp. *africana*
									Ficus cf. *cordata*
									Colpoon compressum
									Proteoid fynbos taxa
									Protea cf. *glabra*
									Leucadendron cf. *pubescens*
									Euryops sp.
									Cliffortia sp.
									Erica cf. *verecunda*
									Erica cf. *caffra*
									Protea cf. *nitida*
									General thicket taxa
									Rhus sp.
									Euclea sp.
									Pterocelastrus tricuspidatus
									Afromontane forest taxa
									Kiggelaria africana
									Podocarpus cf. *elongatus*
									Celtis africana
									Grewia cf. *occidentalis*
									Passerina sp.
									Myrsine africana
									Maytenus sp.
									Halleria cf. *lucida*
									Hartogiella sp.
									Chionanathus foveolatus
									Riverine woodland taxa
									Ficus cf. *sur*
									Salix cf. *mucronata*

Table 5.2a Wood Charcoal from Elands Bay Cave in Deposits Dating from 10,500 to More Than 40,000 Years Ago, Showing Acronyms of Depositional Units Sampled, Weight of Samples, and Approximate Age (filled cell = taxon present; empty cell = taxon absent)

Depositional Unit	CRAY(1)	CRAY(2)	FOAM	SMOK	ASHE	GBAN	DSO2	OBS1	OBS2(1)	0BS2(2)
sample size (n)	200	400	163	100	100	500	200	70	100	200
mass (g)	26.85	47.64	18.54	15.33	13.95	62.84	26.65	10.23	13.43	28.25
age (years before present)	9950	9950	10460	10560	10660	10700	12450	12450		
Xeric thicket taxa										
Euclea racemosa										
Rhus cf. *undulata*										
Ruschia maxima										
Zygophyllum morgsana										
Lycium sp.										
Ruschia sp.										
Asteraceous shrubland										
Eriocephalus cf. *aromaticus*										
Aspalathus sp.										
Passerina glomerata										
Chrysanthemoides sp.										
Euryops cf. *speciosissimus*										
Hymenolepis cf. *parviflora*	■	■	■							
Salvia africana-lutea			■			■				
Phylica sp.	■	■							■	
Mesic thicket taxa										
Heeria argentea		■								
Maytenus cf. *oleoides*										
Cassine cf. *peragua*	■	■					■			■
Diospyros cf. *glabra*	■	■	■	■	■	■	■	■	■	
Dodonaea angustifolia	■	■	■	■	■	■		■	■	
Hartogiella schinoides		■					■			
Tarchonanthus camphoratus	■	■								
Maytenus cf. *heterophylla*	■	■	■	■	■	■	■	■	■	■
Olea europaea subsp. *africana*			■	■	■	■	■	■	■	■
Ficus cf. *cordata*				■	■	■				
Colpoon compressum					■		■			■
Proteoid fynbos taxa										
Protea cf. *glabra*							■	■	■	■
Leucadendron cf. *pubescens*								■	■	■
Euryops sp.			■				■			■
Cliffortia sp.					■				■	
Erica cf. *verecunda*										■
Erica cf. *caffra*										
Protea cf. *nitida*										

(Continued)

Table 5.2a Continued

Depositional Unit	CRAY(1)	CRAY(2)	FOAM	SMOK	ASHE	GBAN	DSO2	OBS1	OBS2(1)	0BS2(2)
General thicket taxa										
Rhus sp.	■	■	■	■		■	■	■	■	■
Euclea sp.	■	■	■	■	■	■	■	■	■	■
Pterocelastrus tricuspidatus					■		■			■
Afromontane forest taxa										
Kiggelaria africana										
Podocarpus cf. *elongatus*										
Celtis africana										
Grewia cf. *occidentalis*										
Passerina sp.							■			
Myrsine africana										
Maytenus sp.										
Halleria cf. *lucida*										
Hartogiella sp.										
Chionanathus foveolatus										
Riverine woodland taxa										
Ficus cf. *sur*										
Salix cf. *mucronata*										

Table 5.2b Wood Charcoal from Elands Bay Cave in Deposits Dating from 10,500 to More Than 40,000 Years Ago, showing Acronyms of Depositional Units Sampled, Weight of Samples, and Approximate Age (filled cell = taxon present; empty cell = taxon absent)

SOSE(1)	SOSE(2)	MOS1	SPIN	OAKO	LIME	SPAL	TAPT	PATT	NORT	Depositional Unit
120	100	100	30	200	200	150	600	340	200	sample size (n)
13.53	11.73	12.24	6.1	32.15	28.44	18.54	72.97	45.64	27.16	mass (g)
13260	13260	13600	17800	20500			>40000	>40000	>40000	age (years before present)
										Xeric thicket taxa
										Euclea racemosa
										Rhus cf. *undulata*
										Ruschia maxima
										Zygophyllum morgsana
										Lycium sp.
										Ruschia sp.
										Asteraceous shrubland
										Eriocephalus cf. *aromaticus*
										Aspalathus sp.
										Passerina glomerata
										Chrysanthemoides sp.
										Euryops cf. *speciosissimus*
										Hymenolepis cf. *parviflora*
										Salvia africana-lutea
										Phylica sp.

Table 5.2b Continued

SOSE(1)	SOSE(2)	MOS1	SPIN	OAKO	LIME	SPAL	TAPT	PATT	NORT	Depositional Unit
										Mesic thicket taxa
										Heeria argentea
										Maytenus cf. *oleoides*
										Cassine cf. *peragua*
										Diospyros cf. *glabra*
										Dodonaea angustifolia
										Hartogiella schinoides
										Tarchonanthus camphoratus
										Maytenus cf. *heterophylla*
										Olea europaea subsp. *africana*
										Ficus cf. *cordata*
										Colpoon compressum
										Proteoid fynbos taxa
										Protea cf. *glabra*
										Leucadendron cf. *pubescens*
										Euryops sp.
										Cliffortia sp.
										Erica cf. *verecunda*
										Erica cf. *caffra*
										Protea cf. *nitida*
										General thicket taxa
										Rhus sp.
										Euclea sp.
										Pterocelastrus tricuspidatus
										Afromontane forest taxa
										Kiggelaria africana
										Podocarpus cf. *elongatus*
										Celtis africana
										Grewia cf. *occidentalis*
										Passerina sp.
										Myrsine africana
										Maytenus sp.
										Halleria cf. *lucida*
										Hartogiella sp.
										Chionanathus foveolatus
										Riverine woodland taxa
										Ficus cf. *sur*
										Salix cf. *mucronata*

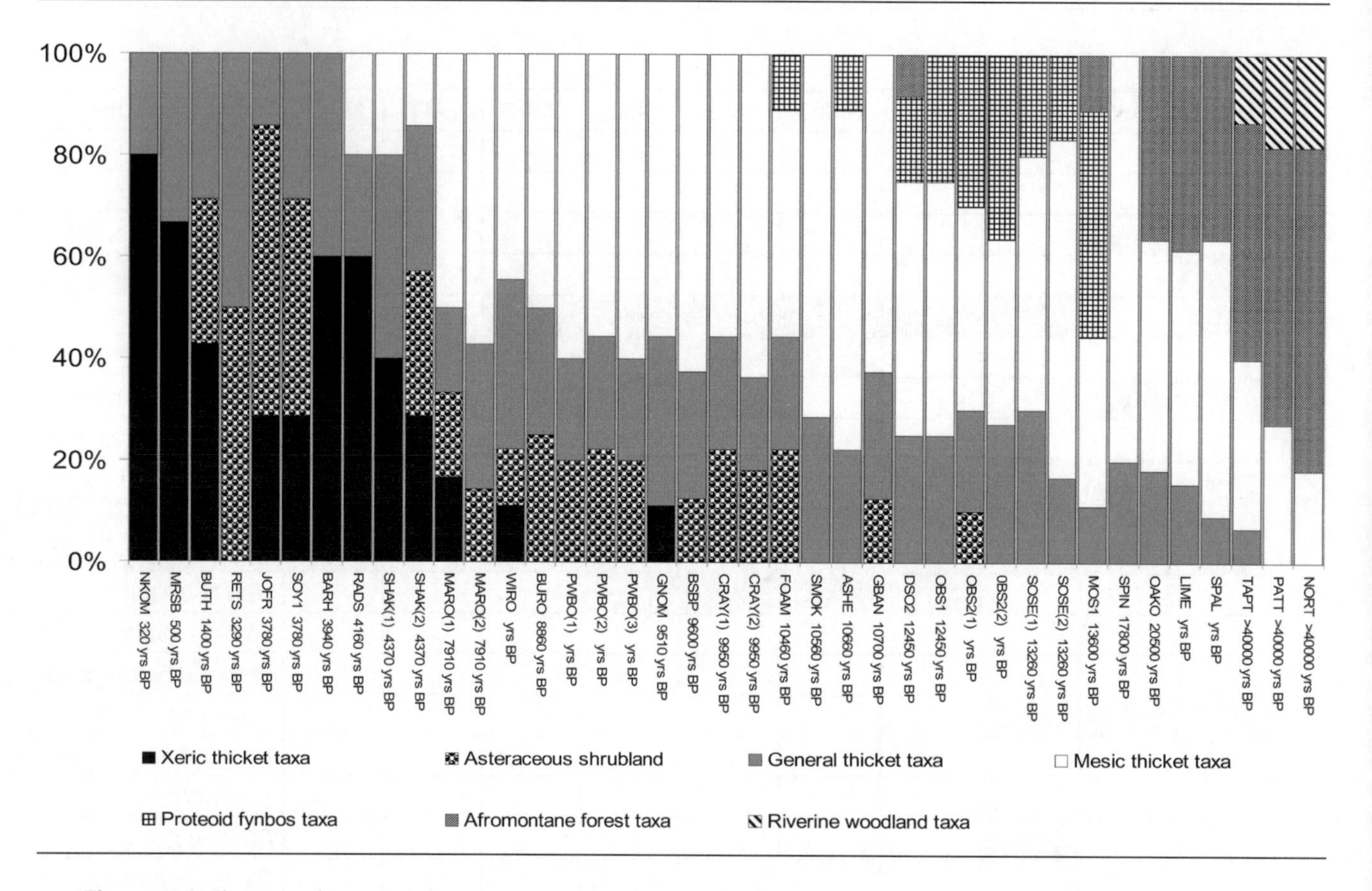

Figure 5.5 Showing changes through time in the representation of the seven community associations as seen from the taxon identified from wood charcoal in each depositional unit. The deposits are ordered in age from youngest to oldest from left to right. Each bar shows the proportion of identified species (by presence) representative of each community association in each unit as given in Tables 5.1 and 5.2.

Figure 5.6 Xeric thicket vegetation at Elands Bay.

Figure 5.5). The most recent charcoal reflects fuelwood collected from xeric thicket (strandveld) vegetation as found outside the cave today (Figures 5.2 and 5.6), including *Euclear acemosa, Rhus undulata, Ruschia maxima,* and *Zygophyllum morgsana*. Asteraceous shrubland elements (Figure 5.7) in the form of *Eriocephalus* sp., *Aspalathus* sp., *Passerina glomerata, Euryops speciosissimus, Chrysanthemoides* sp., *Hymenolepis parvifolia, Salvia africanalutea,* and *Phylica* sp. were common between 3,000 and 4,300 years ago. This evidence for relative stasis in

Figure 5.7 Asteraceous shrubland vegetation near Elands Bay.

Figure 5.8 Thicket vegetation near Elands Bay.

vegetation composition in the EBC area over the past 4,000 years comes from the asteraceous shrubland and thicket elements (Figure 5.8) associated with the Table Mountain Group-derived soils and older sands on the plateau of the Baboon Point headland. Vegetation with a similar composition exists along the semi-arid coast of the eastern fynbos biome with bimodal rainfall (Figure 5.1). Species such as *P. tricuspidatus* and *E. racemosa* are common trees in mesic forests throughout the fynbos biome and beyond. In particular, *R. undulata* is a typical component of contemporary xeric thickets in the northwestern coastal forelands of the fynbos biome.

Before about 4,000 years ago EBC people were collecting increasing quantities of wood from species indicative of more mesic thickets than those currently growing in the area. Mesic thicket elements, including *Heeria argentea, Maytenus oleoides, C. peragua, Diospyros glabra,* and *Dodonaea angustifolia* are notably present at the Pleistocene/

Holocene boundary between about 8,000 and 10,000 years ago. The absence of strandveld taxa could be attributed to a more distant coastline as a result of a lower sea level. The decline in asteraceous shrubland elements recorded for this period is less explicable. These thickets show a gradual compositional change indicating more mesic conditions with increasing sample age. Some of these species, such as *M. oleoides*, *D. glabra*, and *C. peragua*, can still be found in an isolated thicket a few km south of EBC, but modern analogues are most common on the lower slopes of the inland mountain ranges, which have much higher rainfall (about 400–600 mm/yr) than the EBC area does today.

Between 12,400 and 13,600 years ago there is a distinctive and diagnostic phase in the charcoal record in which EBC residents collected the wood of taxa indicative of proteoid fynbos (*Protea glabra*, *Leucadendron pubescens*, and *Erica* spp.). Today, *P. glabra* grows under conditions considerably moister than the contemporary EBC environment, and *L. pubescens* is widely distributed in the northwestern part of the fynbos biome under higher rainfall conditions. Furthermore, Ericaceae are largely confined to the moist, upper slopes of the inland (Cederberg) mountains or grow along the banks of perennial streams and rivers. Taken together, the presence of these taxa at this point in the EBC charcoal record strongly suggests relatively moist conditions, indicating that this period was wetter than the Holocene.

Before about 20,000 years ago, at and beyond the last glacial maximum, there is evidence for marked environmental change in the vicinity of EBC with taxa characteristic of Afromontane rainforest dominating the charcoal assemblages. Although *Kiggelaria africana*, *Podocarpus elongatus* (Figure 5.9) and *Myrsine africana* are common in patches of Afromontane thicket and scrub-forest in the inland mountains today, *Celtis africana*, *Grewia occidentalis* and *Chionanthus foveolatus* no longer grow in this region. These three have their distribution limits to the south, where they grow in mesic thicket and in Afromontane forest that is not subject to pronounced summer drought. *C. africana*, being a winter-deciduous tree, is unlikely to be tolerant of summer drought conditions.

In addition to the charcoal of Afromontane taxa, the three oldest deposits at EBC, lying beyond the range of conventional radiocarbon dating, have considerable amounts of *Ficussur* and *Salix mucronata* charcoal. *S. mucronata* is very common in woodland that fringes the Olifants River, an elevated valley inland of EBC. *F. sur*, however, is a distinctly tropical species that extends only into the extreme east of the fynbos biome, where it is also a component of tall, and subtropical, riverine forest. It seems highly likely that these taxa were part of a riverine community alongside the palaeoriver that today forms the Verlorenvlei.

Any evaluation of changes through time of the woody vegetation at EBC must take into account the specific location of the Western Cape at the southern tip of Africa. Figure 5.1 (after Cowling et al. 1999) sets out the main features: at 30 degrees south, the Cape is affected by westerly cyclonic lows in winter but bypassed in the summer, giving the local mountains and coastal plains a Mediterranean-type winter rainfall climate, unlike the rest of the subcontinent. The fynbos vegetation thrives on the nutrient-poor soils of the Cape Fold Belt mountains, contrasting strongly with the karroid semidesert and subtropical bushveld to the north and east. Offshore, heat is transferred from the warm Agulhas current to the cold Benguela current around the Cape. Thus any changes in atmospheric or oceanic systems would affect the climate and vegetation of the Cape, resulting in shifts in the distributions of plant and animal species. Using this model, the changes through time in the EBC charcoal record can be examined with more clarity.

Figure 5.9 Small *Podocarpus* near Elands Bay.

It has been suggested that the EBC charcoal assemblages provide a rare opportunity for palaeoenvironmental reconstruction in the fynbos biome. The comprehensive and well-dated set of samples span the critical period through the peak of the last glacial maximum at 18,000 to 20,000 years ago, thereby avoiding the limitations of other studies in the western fynbos biome (Parkington et al. 2000). The charcoal assemblages summarised in this chapter indicate that soil moisture conditions have deteriorated steadily over the past 40,000 years or more in the EBC area. Independent observations corroborating this trend come from faunal and pollen evidence from the same levels at EBC that were analysed for charcoal (Parkington et al. 2000). The pollen evidence is remarkably consistent with that of the charcoal in reflecting greater terminal Pleistocene moisture availability, despite the pollen obviously having been introduced into the cave by very different processes. It is astonishing that at about the time of the last glacial maximum, in the vicinity of EBC, there was a relatively diverse mixed subtropical/Afromontane thicket or forest as opposed to the situation seen today and for most of the Holocene, in which xeric thicket and asteraceous shrubland are dominant.

The inevitable conclusion is that soil moisture conditions were probably considerably higher during glacial than in Holocene times, although the relative roles of increased rainfall or reduced evapotranspiration to explain this pattern are not known in detail—hence the need to investigate water use efficiency using isotopes. It is clear, however, that the trends observed in the eastern part of the biome—that is, a drier glacial—are reversed for the west and therefore cannot be generalised for the entire biome. This observation is consistent with the hypothesis that the eastern part of the fynbos biome was deprived of its major source of moisture during glacial time—namely, moist air drawn across a warm Indian Ocean as a result of atmospheric instabilities during the equinoxes. Decreased moisture from weakened anticyclonic fronts combined with a colder Indian Ocean would leave the southern Cape mountain valleys drier. It would help to know if the cyclonic lows that currently bring rain to the southwestern Cape in winter were more extended through the year in the wetter glacial (Figure 5.1). Frontal systems emerging from a colder Atlantic would have been less moisture-laden but more vigorously 'squeezed' of their moisture by the more charged circulation of the terminal Pleistocene. Such fronts may not have crossed the north-south barrier of the western part of the Cape Fold Belt, leaving the eastern part in a rain shadow. Any contraction of the Benguela upwelling to a position north of Elands Bay would imply a reduced effect of southerly winds and a greater role for north westerlies. With greater cloud cover and lower temperatures, any increase in precipitation would have been much more effective. The sustained presence of *C. africana* charcoal in assemblages older than 20,000 years ago at EBC would then seem to be a clear indicator of such moisture availability throughout the year, although one also needs to consider whether it might just have been growing in the Verlorenvlei riparian zone.

There are obviously very dramatic differences in the composition of the charcoal assemblages at EBC 20,500, 13,500, and 10,500 years ago. All three are marked by the presence of many more arboreal taxa than, for example, at 3,500 years ago, implying a greater level of soil moisture. Looking more closely at the three patterns and their sequenced occurrence could provide the key to understanding the 'deglaciation' process along the Cape west coast. Taking into consideration that there is a pulse of occupation in the now arid interior Karoo between 13,500 and 9,500 years ago (Parkington et al. 2000), the underlying atmospheric shifts are probably of subcontinental significance and scale. In this context, it must again be noted that the critical taxa are the mesicfynbos group, including *Protea*, *Leucadendron*, and *Erica*, which are present only between 13,500 and 11,500 years ago. In contrast, *Celtis* and *Grewia* occur only around or before the last glacial maximum and prefer habitats not exposed to pronounced summer drought. One might speculate that the distinction between the two assemblages is the resumption of a largely winter rainfall system at about 14,000 years ago—a system with more effective moisture than today.

Conclusions

The EBC charcoal record, supported by the pollen and faunal evidence, shows remarkable changes in vegetation composition with time. The most recent charcoal reflects the xeric thicket (strandveld) vegetation outside the EBC today. Asteraceous shrubland elements were common between 3,000 and 4,300 years ago, and mesic thicket elements flourished at the Pleistocene/Holocene boundary between about 8,000 and 10,000 years ago. Between 12,400 and 13,600 years ago proteoid fynbos dominated, but at and beyond the last glacial maximum most of the woody taxa were from riverine woodland and Afromontane forest communities.

The evidence from the plant and animal assemblages at EBC during the last glacial maximum suggests that it was wetter, colder, cloudier, and more grassy than today.

It seems likely that the local vegetation was also far more diverse than at present, with a wider range of plant communities on the exposed coastal plain, rocky cliffs, and riverbanks. In conjunction with the evidence of the stone tools, the plants and animals chart the changing record of the intensity and nature of site use at EBC. All can clearly be interpreted as a set of human responses to the increasing aridity and seasonality of the early Holocene in this part of the Western Cape.

REFERENCES

Cartwright, C. R., and J. E. Parkington (1997) The wood charcoal assemblages from Elands Bay Cave, southwestern Cape: Principles, procedures and preliminary interpretation. *South African Archaeological Bulletin 52*, 59–72.

Cowling, R. M. (Ed.) (1992) *The Ecology of Fynbos: Nutrients, Fire and Diversity*. Cape Town: Oxford University Press.

Cowling, R. M., C. R. Cartwright, J. E. Parkington, and J. C. Allsopp (1999) Fossil wood charcoal assemblages from Elands Bay Cave, South Africa: Implications for Late Quaternary vegetation and climates in the winter-rainfall fynbos biome. *Journal of Biogeography 26*, 367–78.

Parkington, J. (1972) Seasonal mobility in the Later Stone Age. *African Studies 31*, 223–43.

———. (1976) Coastal settlement between the mouths of the Berg and Olifants rivers, Cape Province. *South African Archaeological Bulletin 31*, 127–40.

Parkington, J., C. R. Cartwright, R. M. Cowling, A. Baxter, and M. Meadows (2000)Palaeovegetation at the last glacial maximum in the Western Cape, South Africa: Wood charcoal and pollen evidence from Elands Bay Cave. *South African Journal of Science 96*, 543–46.

Parkington, J., and C. Poggenpoel (1971) Excavations at De Hangen, 1968. *South African Archaeological Bulletin 101/102*, 3–36.

Wheeler, E. A., P. Baas, and P. E. Gasson (Eds.) (1989) IAWA list of microscopic features for hardwood identification. *IAWA Bulletin 10*, 219–332.

Wheeler, E. A., R. G. Pearson, C. A. LaPasha, T. Zack, and W. Hatley (1986). *Computer-aided wood identification*. North Carolina Agriculture Research Service. Bulletin 474.

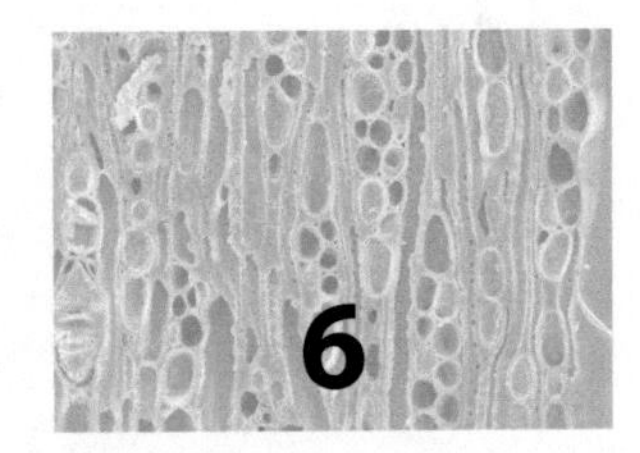

Early Millet Farmers in the Lower Tilemsi Valley, Northeastern Mali

Katie Manning and Dorian Q Fuller

The spread of agricultural and herding lifeways in Africa presents an intriguing alternative to the Near Eastern or European model in that the raising of livestock—in particular, cattle—precedes the appearance of fully domesticated crops by some 5,000 years. That cattle herding and fishing dominated the food-producing systems of Saharan Africa for much of the Middle to Late Holocene is evidence of an alternative developmental trajectory to the typical sedentary agricultural model (Garcea 2006; Holl 2005; MacDonald 1998; Sutton 1974, 1977). Instead, it is widely believed that the spread of domestic livestock and high mobility levels supported significant east-west migrations, reflected in a broadly shared Saharan material culture. The process of integrating domestic cereals into this long-standing food-production system remains poorly understood and yet poses a pivotal moment in the social and economic history of sub-Saharan Africa. This chapter examines recent evidence from the Lower Tilemsi Valley, presenting new insights into the timing and spread of domestic pearl millet.

The Lower Tilemsi Valley

The Tilemsi Valley is a critical region for the study of agricultural development in sub-Saharan Africa. During the Mid- to Late Holocene the Tilemsi supported an active hydrological system, linking the Saharan zones of West Africa, where the wild progenitors of pearl millet are reported (Fuller 2003; Harlan 1992; Tostain 1998), with the West African Sahel. Recent research on modern stands of domestic millet demonstrate regionalised enzyme groupings indicating southeastern Mauritania (Tostain 1998) and/or a stretch from northeastern Mali to Lake Chad (Fuller 2003; Oumar et al. 2008) as the centres of pearl millet domestication (Figure 6.1). It is plausible that both areas were a focus for early pearl millet cultivation, although both Tostain (1998) and Oumar and associates (2008) use cluster analyses interpreted as phylogenies to assert just a single point of origin in Mauritania and northeast Mali, respectively. However, the relatively slow process of domestication means phylogenies are expected to not reflect origins (starts of cultivation) but post-domestication expansion events (as per Allaby, Fuller, and Brown 2008). In either scenario, the northern terminus of the Tilemsi Valley matches well with these purported domestication centres. Furthermore, the valley would have provided a fertile and accessible corridor into sub-Saharan Africa at a time of increasing aridification and southward displacement of Saharan populations.

In 1972, Andrew Smith undertook test excavations at the sites of Karkarichinkat Nord (KN) and Karkarichinkat Sud (KS), reporting the presence of domesticated *Pennisetum* in pot sherd impressions from KS (Smith 1992, p. 74). All of the sherds with identified *Pennisetum*, however, were of surface provenience and were

Archaeology of African Plant Use by Chris J. Stevens, Sam Nixon, Mary Anne Murray, and Dorian Q Fuller, Eds., 73–81

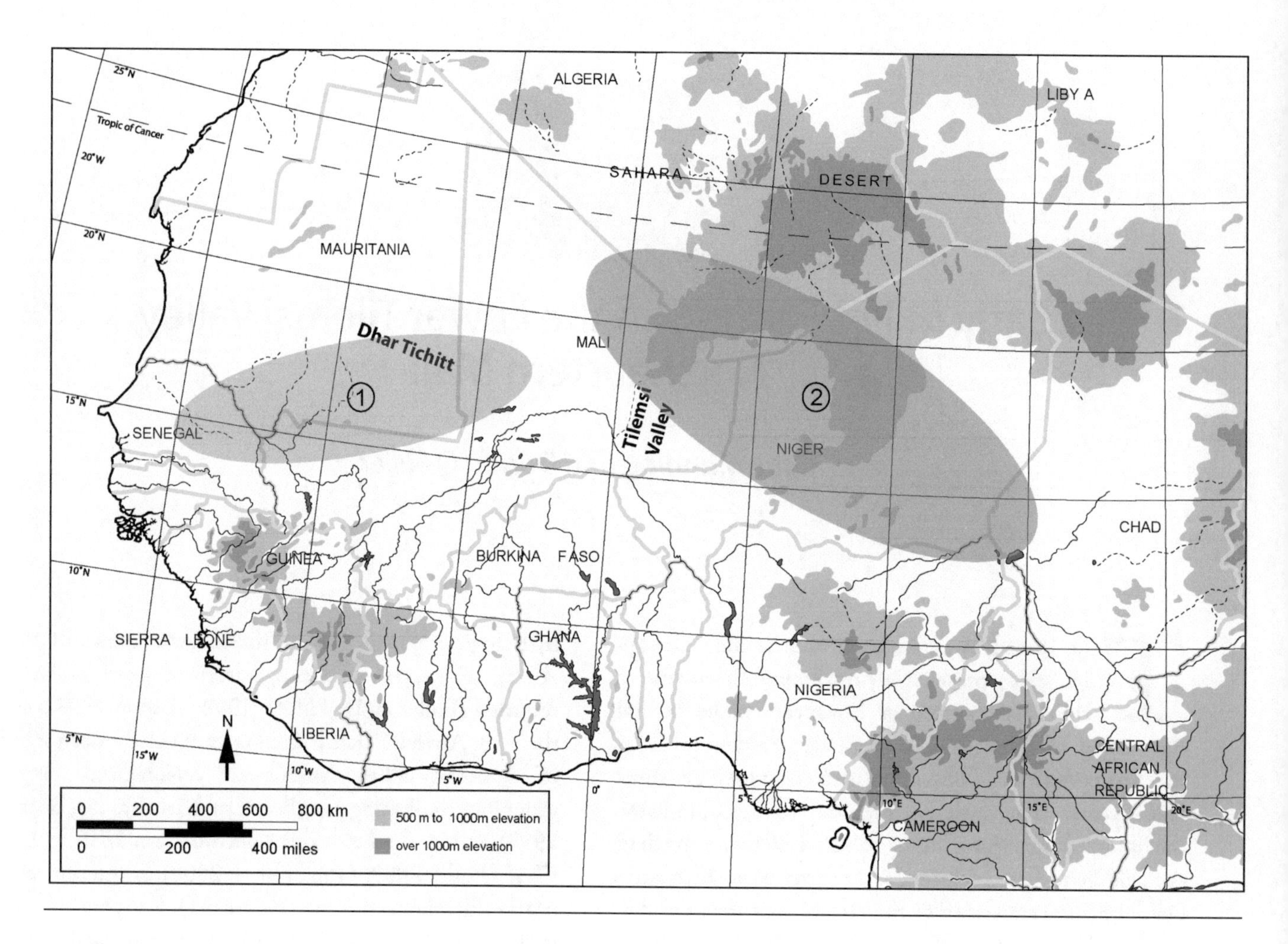

Figure 6.1 Proposed centres for pearl millet domestication.

dated only on the basis of stylistic similarity with excavated material to c. 2000 B.C.E. The absence of larger scale, systematic archaeobotanical sampling or direct dating has meant that, until now, these finds were hard to substantiate, contributing little to our understanding of the earliest history of this cereal (Neumann 2005, p. 259).

THE LOWER TILEMSI VALLEY PROJECT

In 2005 the Lower Tilemsi Valley Project was initiated by one of the authors (KM) in collaboration with the Direction National du Patrimoine Culturelle in Mali. The aims of this project were to refine the archaeological chronology of the Lower Tilemsi region and test prevailing assumptions about the social and economic organisation of the Late Stone Age (LSA) occupants. Specific objectives were formulated in response to hypotheses presented by Gaussen and Gaussen (1988) and Smith (1974a)—namely, that two culturally distinct populations co-inhabited the lower portion of the Tilemsi Valley (Gaussen and Gaussen 1988) and that at least one of these (denoted as the 'Facies K') practised small-scale millet cultivation on a seasonal basis (Smith 1974a, b; 1975).

Two seasons of fieldwork were undertaken in 2005–2006 and 2007. The first season was dedicated to excavations at the site of Karkarichinkat Nord (KN05), whereas deposits at the southern site (KS05) were found to be unconsolidated and stratigraphically compromised. As a result, in 2006 a survey was undertaken of the Karkarichinkat hinterland, and 86 multiperiod sites were mapped within a 20 × 20 km area. Five of the seemingly Neolithic sites (based on surface material) were chosen for test excavation in 2007 (Figure 6.2; Ebelelit-EB07, Tiboubija-TB07, Tin Alhar-TA07, Er Negf-EN07, and Jsmagamag-JS07-1 and JS07-2). The results of this work shed important light on intersite variability in economic and social practice in the Lower Tilemsi and provided the first reliable chronology for the region (Manning 2008a). Recent Bayesian modelling of

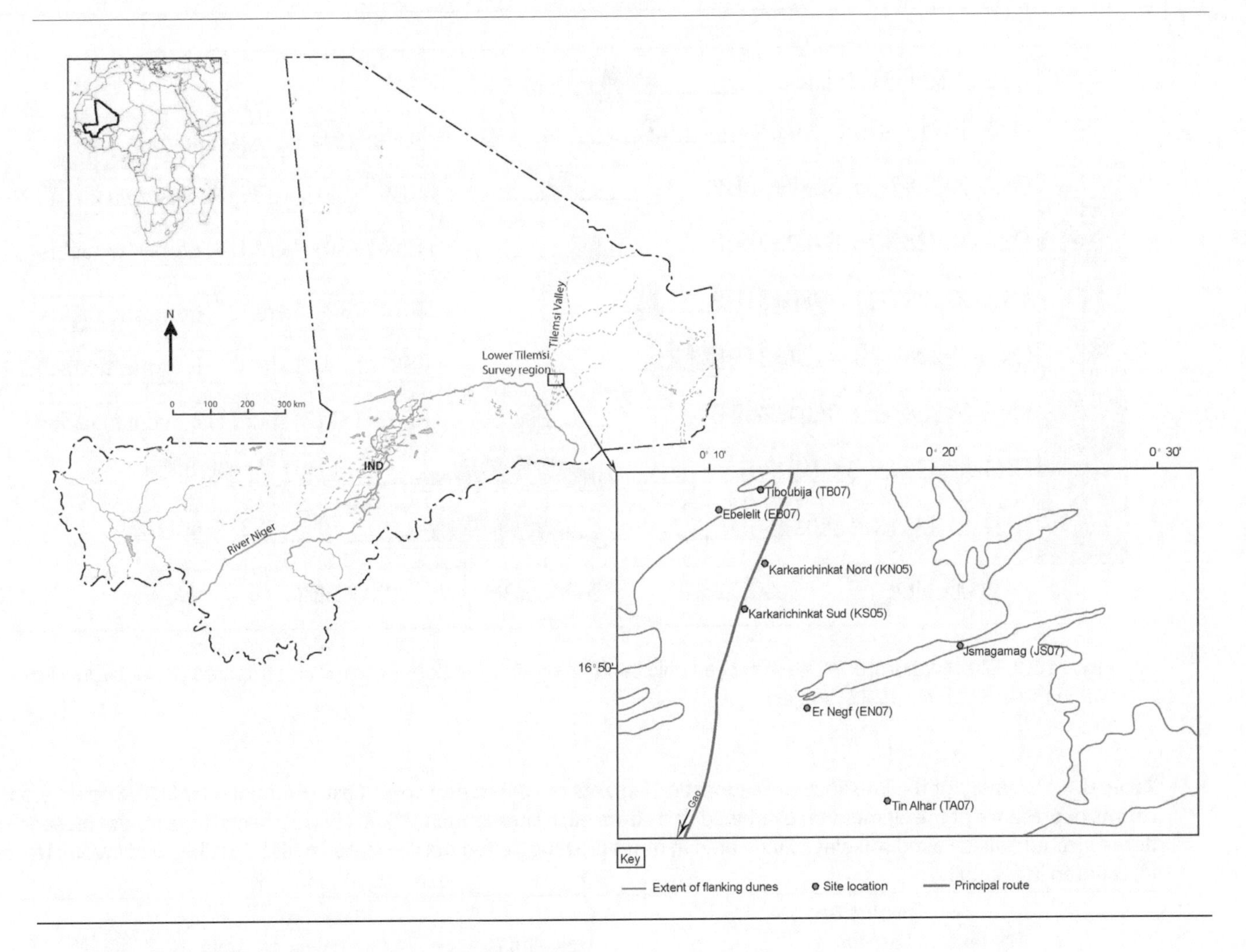

Figure 6.2 Area of the Lower Tilemsi Valley Project.

the 26 AMS dates and 5 OSL dates from these excavations indicate two distinct phases (Manning et al. 2011). Occupation at EN07, JS07-2, and EB07 begins after 2000 B.C.E. and appears to have lasted for around 100 years. In contrast, the modelled results from Karkarichinkat Nord suggest that occupation began around 2500 B.C.E. and finished between 57 and 100 years (68.2% probability) later. Neither TB07 nor TA07 had sufficient AMS dates for chronometric modelling. The two AMS samples from Tiboubija, however, yielded fairly early dates of 3808 ± 34 BP (OxA-18224), or 2434 to 2138 B.C.E. (95.5% probability), and 3958 ± 34 BP (OxA-18225), or 2572 to 2346 B.C.E. (95.5% probability), and may be more in line with the earlier phase of occupation c. 2500 B.C.E (Figure 6.3).

EVIDENCE FOR DOMESTICATED PEARL MILLET (*PENNISETUM GLAUCUM*)

Intensive palaeobotanical sampling was adopted in an attempt to retrieve macro plant remains of domestic pearl millet. A total of 36 soil samples, providing 110 litres of soil, were retrieved from KN05, and 22 soil samples totalling 166 litres of soil were retrieved from the 2007 test excavations. Despite the quantity of soil that was sieved, however, very few macro plant remains were recovered, *Celtis* sp. being the exception as a result of its high silica content.

In addition to sampling for macro plant remains on site, analysis of the pottery assemblage revealed that a large number of sherds were tempered with chaff. At EB07, EN07, and JS07-2, approximately one-third of all body and rim sherds contained chaff, and a subsample of these was chosen for analysis. In total, 27 mounted samples from 19 potsherds were examined by Scanning Electron Microscopy. Each image was assessed, and the diagnostic traits of each plant fragment were individually recorded.

Almost all of the examined potsherds contained impressions of pearl millet, with no other plant species. This result is to be expected if pearl millet was the only seed crop being repeatedly harvested, because weeds are rarely incorporated in the crop-processing by-products of

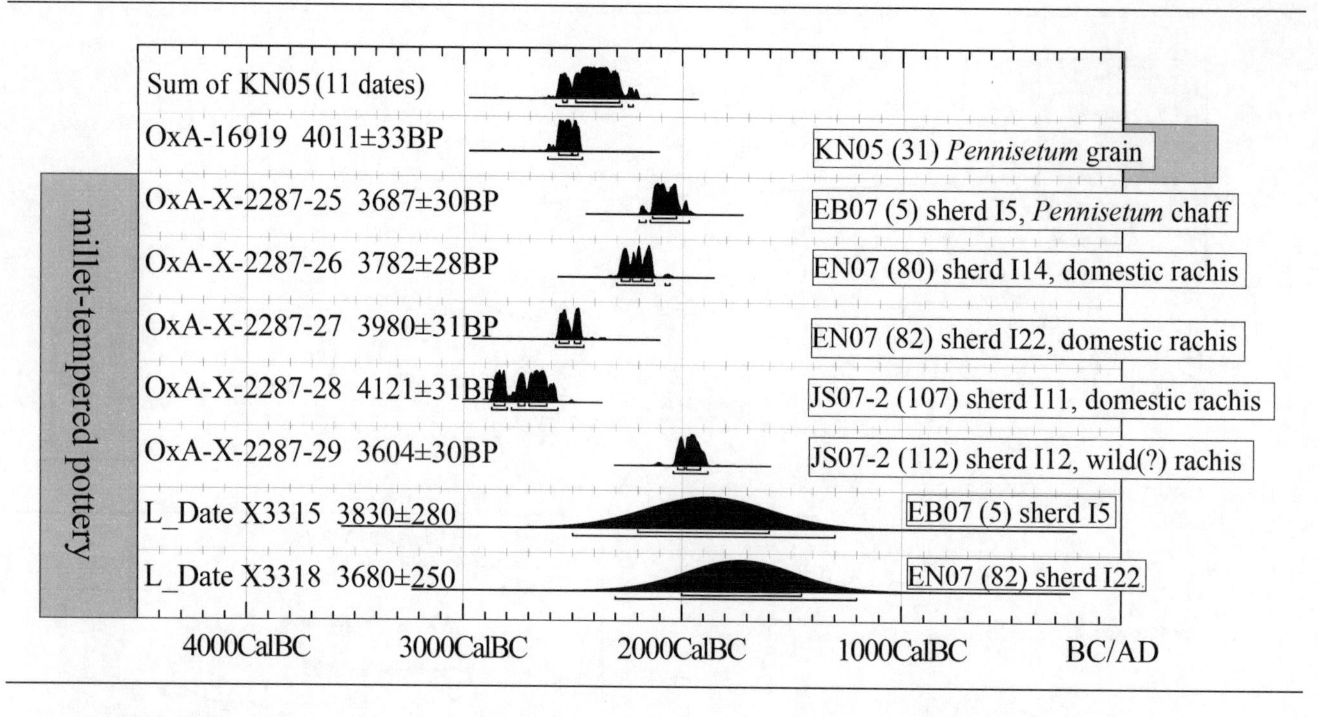

Figure 6.3 Direct radiocarbon dates on pearl millet and on organic fractions from millet- tempered pottery from sites within the Lower Tilemsi Valley Project.

Table 6.1 Summary of the Presence of Diagnostic Elements of *Pennisetum* sp. in Casts Examined by SEM. Shown are the counts of presence of the element on examined casts from each site/context (**). Rachis elements indicate the presence of domesticated millet. Paired spikelet (*) indicated in notes also suggested domesticated millet. Possible wild involucre base (#) is noted in the text.

Site Name	Context No.	Involucre Base w/ Bristles	Rachis**	Bristle	Unicellular Hairs	Spikelet	Lemma/ Palea Frag.	Grain (?)	Notes
KN05	1	0	0	1	0	0	1	0	
	77	1	0	3	0	2	2	0	
JS07-2	107	1	0	1	0	1	0	0	
	110	5	1**	3	1	2	1	0	paired spikelet (?)*
	112	3	1**	1	0	3	1	0	
EN07	80	5	1**	3	0	3	0	0	possible wild #;
	81	2	1**	4	2	1	3	1	paired spikelet*
	82	1	1**	4	2	1	2	0	paired spikelet*
EB07	3	2	6**	8	5	10	2	1	paired spikelet*
	5	0	0	2	1	1	0	0	
Total		24	11	27	10	23	11	2	
%		39%	18%	44%	16%	38%	18%	3%	

pearl millet on account of its dense spikes which are readily harvested alone (Reddy 1997).

Sixty-one pearl millet fragments were identified in the SEM and can be divided into six plant part categories (Table 6.1). These categories include the presence of involucre bases from which bristles arise; the presence of the rachis, which occurs below this bristle base (and indicates a domesticate); bristle fragments (and a subcategory of fragments with unicellular hairs on the bristle surface); spikelets (that is, husk, including lemma and/or palea);

and smaller husk fragments The most widespread, noted in 38% of the casts, were fragments of pearl millet bristles; of these more than a third (10 of 27) preserved evidence of rows of unicellular trichomes on their surface. This frequency of bristles is likely to be an underestimate, since many casts that showed only small bristle fragments were not subjected to SEM investigation, and this superficial examination of the sherds suggests that bristle fragments are nearly ubiquitous. Spikelet impressions were noted on 38%, but if smaller fragments of husk are included this number goes up to 54%. Involucre bases, with the base of the bristle clusters, were noted in 38% of the studied casts, but only one third of these (9 of 24) showed evidence for the base, with either a rachis (8 cases) or possible smooth scar of the wild type (1 case). Of two grain impressions, one was measurable (Figure 6.4A). Although the size of this grain (3 × 1.7 mm) is slightly larger than expected for wild-type millet, if the grain was wet when added to the clay its size may have expanded somewhat, and the metrics of this one grain impression do not provide good unambiguous evidence that grain size increase had occurred by this date. Flotation samples from Karkarichinkat Nord (KN05) also produced two grains, one of which was directly dated (Figure 6.3) and one of which was drawn and measured (1.84 × 1.11 × 1.02 mm: L × B × T). The latter fits within the wild size range, although its shape does suggest the club-shaped form of domesticated types (Figure 6.5). (For further discussion of grain size increase in pearl millet, see Manning et al. 2011 and comparisons to other crops in Fuller, Asouti, and Puruggnan 2012.)

The spikelet impressions were examined for evidence of paired spikelets. Wild *Pennisetum* normally has

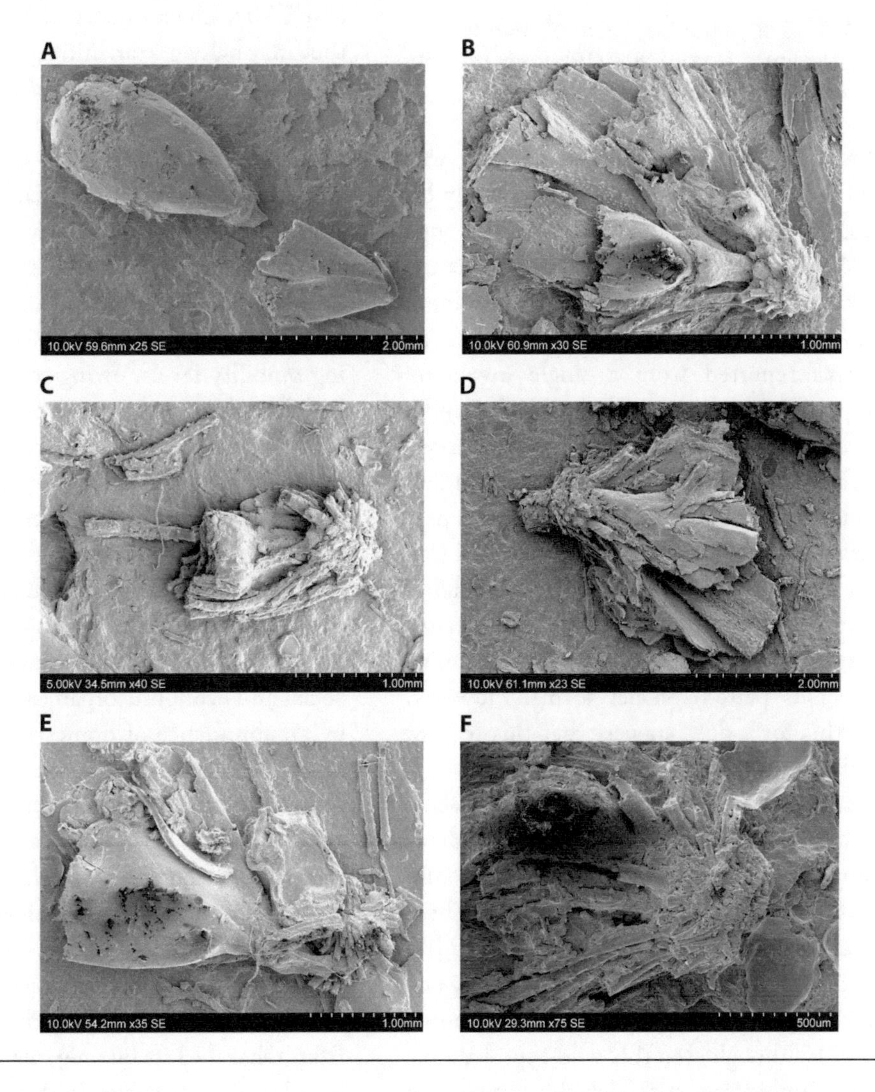

Figure 6.4 SEM images of grain and chaff impressions of *Pennisetum* within pottery. (A) Probable wild *Pennisetum* grain; (B, C) paired spikelets; (D, E) involucre base impressions with preserved rachis fragments indicating the stalked, nondehiscent morphotype of the domesticate; (F) possible wild type involucre base impressions.

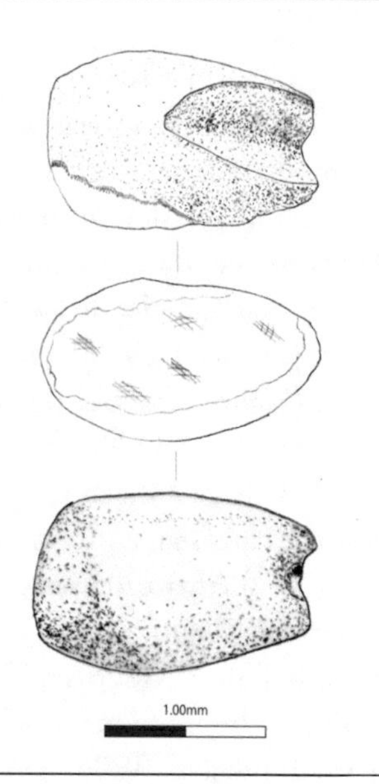

Figure 6.5 Drawings of carbonized grain of *Pennisetum* grain from Karkarichinkat Nord, KN05 (18); from top to bottom, dorsal, cross section, and ventral views.

a single grain in each bristly involucre, whereas domesticated forms often have multiples grains. The study by Godbole (1925) of domestic Indian pearl millet suggests that ~70% of involucres include two spikelets (each with a grain), whereas ~20% are single grained. The other ~10% include more than 2 grains, with as many as 9 grains reported from a single involucre. Archaeologically, early impressions of pearl millet indicate the presence of paired spikelets, such as those previously reported from Dhar Nema, Mauritania (see Fuller, Macdonald, and Vernet 2007). Four examples of a pair of such spikelets were noted in the Tilemsi material (for example, Figure 6.4B and 6.4C). The generally highly fragmentary nature of the chaff impressions in this material does not permit quantification of the proportion of single-spikelet versus paired-spikelet forms. However, the presence of this form does suggest that this feature had evolved in the cultivar.

Eight involucre base impressions with preserved rachis fragments were identified indicating the stalked, nondehiscient morphotype of the domesticate (Figure 6.4D and 6.4E), of which one was ambiguous (Figure 6.4B). These are distributed across three sites and six contexts (EB07 [3], JS07-2 [112] and (110); EN07 [80], [81], [82]). Only one possible wild type involucre base (Figure 6.4F) was noted from JS07-2 (107), although even this wild-type example remains somewhat ambiguous. Although this sample size is small it suggests the predominance of the domesticated form (that is, 89% of preserved rachis remains). Because of the evidence from other cereals that nonshattering evolved gradually, on the order of 1,000–2,000 years (Fuller 2007; Fuller et al. 2009), we would infer cultivation to have begun perhaps in the 4th millennium B.C.E. (Manning et al. 2011). This inferred earlier domestication would imply that when these sites were settled the incoming population brought with them an established economic suite, including domesticated millet and livestock.

DISCUSSION

The recent Bayesian modelling of the Lower Tilemsi Valley dates (Manning et al. 2011) suggests that the phasing for occupation of this region is more complex than originally presumed. Direct evidence for domesticated pearl millet is currently absent from KN05, TB07, and TA07. The sequence of dates for these sites suggests they may belong to an initial phase of occupation between 2500 and 2400 B.C.E. Evidence from KN05 also points to higher mobility levels. In particular, the stratigraphy at KN05 is characterised by thin horizons of alternating ashy and calcitic deposits relating to seasonal episodes of river inundation and occupation. Size reconstruction of the cattle from KN05 also reveals a broad osteometric range (Manning 2008b). Although this has previously been interpreted as a possible indication of declining mobility levels, owing to the likelihood that diverse breeding populations are more congenial to a sedentary-mixed economy, the presence of several distinct herding groups may equally be an indication of herd pooling at certain times of the year. Herd pooling, or joint herding, refers to the combining of distinct herds for communal grazing and is often associated with agro-pastoral groups as a means of enabling cropping work alongside stock keeping. Livestock clearly played an important role in the social and economic organisation at KN05, as evidenced by an abundance of domestic cattle, sheep, and goat. In addition, a fully articulated cow burial with associated wooden awning was found in the earliest occupation layers also in association with a large subterranean pit containing vast numbers of fish and cattle remains, which may be indicative of seasonal feasting (Manning 2011b).

Secure evidence for the exploitation of domesticated millet does not, therefore, occur until the end of the 3rd millennium B.C.E., predating the otherwise earliest remains from Dhar Tichitt by only one or two centuries. The earlier phase of occupation in the Lower Tilemsi Valley, however, raises a number of important questions about the initial stages of pearl millet domestication. In particular, the presence of fully domesticated pearl millet at EB07, EN07, and JS07-2 suggests the crop was introduced to this particular region in a fully domesticated form. Yet, there is little indication of a cultural hiatus between the initial phase of occupation and later agricultural activities. Although it has previously been suggested that the Lower Tilemsi was host to two contemporary but culturally distinct groups referred to as the Facies K and the Facies B (Gaussen and Gaussen 1988), we found no obvious deviations in either technological or stylistic aspects of the material culture from the excavated sites. There were indications of functional variation—namely, in the presence of lithic workshop sites such as TA07 and Lagraich, with evidence of specialized lithic production (Duhard 2003). Other subtle differences were noted in the faunal composition at different sites, although mapping of these variations suggest environmental influences rather than cultural ones (Manning 2011b). Similarly, there are differences in the decorative attributes of the pottery assemblages from some sites (Manning 2011a), but again these do not correlate with the selective evidence for domesticated pearl millet, nor do they parallel the disparity in site function or chronology. Instead, the underlying technical attributes of the region's pottery, as well as lithic assemblages and small finds, are highly standardised and indicate no obvious disparity within the material culture.

At our current state of evidence, we are left with two different hypotheses concerning the introduction of domestic pearl millet into the Lower Tilemsi Valley:

1. the valley underwent two separate waves of immigration, with only the second group possessing domestic pearl millet;
2. the valley underwent an initial period of occupation, by what appears to have been reconnaissance-like seasonal pastoral-fisherfolk, at sites such as Karkarichinkat Nord. Several centuries later the region was more intensively occupied by the same/related cultural group. This later full-scale occupation brought with it incipient specialisation and a developed agricultural component by at least 2000 B.C.E., implying an earlier origin for cereal domestication in Africa.

In light of the current evidence, the second scenario seems more likely. There is little indication in the region's material culture for either a technological or stylistic hiatus. Despite representing a key area in the later prehistory of sub-Saharan West Africa, the Tilemsi Valley has been only sporadically surveyed. The current study revealed 86 archaeological sites within a contained 20 × 20 km survey area. It is evident from these results that we have only scratched the surface in what this region has to offer. Furthermore, the detailed chronological modelling of AMS and OSL dates reveals a more complex picture of occupation than what is suggested when we consider only the independent dates. We believe it is highly likely that with more survey work, the chronological gaps in the occupation of the Tilemsi Valley will quickly be filled. Furthermore, the evidence for domestic pearl millet occurs in the earliest occupation levels at EB07, EN07, and JS07-2, where it appears to dominate the chaff component, suggesting it was introduced in fully domesticated form from outside the immediate area. Other clues point toward an earlier domestication event (Manning et al. 2011). Although rare, pot sherds with *Pennisetum* chaff temper were recovered from KN05 and JS07-1. Although no confirmed domestic traits were identified, the impressions appear to consist only of *Pennisetum* chaff, with no other plant species. Owing to the rare occurrence of chaff temper in this early phase of occupation, it is likely that the pottery was being produced and imported from outside the Tilemsi Valley, perhaps reflecting seasonal movement between the Tilemsi and millet-growing regions further afield. The two *Pennisetum* grains from Karkarichinkat Nord are further evidence that pearl millet was being exploited at this time, and, although from a limited sample, the shape is suggestive of a cultivar.

Conclusion

The spread of domestic pearl millet was rapid and widespread. Domestic finds have been dated in India to between 2000 and 1700 B.C.E. (Fuller 2003) and are found in northern Ghana at the site of Birimi around the same time (D'Andrea, Klee, and Casey 2001). So far, all the earliest finds of domestic pearl millet in Africa are closely associated with evidence for intensive pastoralism. In the Lower Tilemsi Valley and at Dhar Tichitt (Amblard 1984, 1996; Amblard and Pernès 1989; Holl 1985; MacDonald et al. 2003; MacDonald et al. 2009), the pastoral component in the economies of these regions reflects a broadly shared Saharan style. Continuity in the material culture of Holocene pastoral groups in Africa implies considerable interaction and mobility across both time and space. It is likely that the rapid rate of spread of domestic pearl millet was aided by the suitability of this grain for cultivation by mobile pastoralist societies owing to the crop's minimal water requirements and its ability to produce high yields over a relatively short growing season. Such cultural dynamics, however, tend to leave little archaeological trace, and it is therefore not surprising that earlier indications of the domestication process have yet to be found.

REFERENCES

Allaby, R. G., D. Q. Fuller, and T. A. Brown (2008) The genetic expectations of a protracted model for the origins of domesticated crops. *Proceedings of the National Academy of Sciences USA 105*, 13982–86.

Amblard, S. (1984) *Tichitt-Walata (R.I. Mauritanie): Civilisation et industrie lithique.* Paris: Éditions Recherche sur les civilisations.

———. (1996) Agricultural evidence and its interpretation on the Dhars Tichitt and Oualata, south-eastern Mauritania. In G. Pwiti and R. Soper (Eds.), *Aspects of African Archaeology. Papers from the 10th Congress of the Pan-African Association for Prehistory and Related Studies*, pp. 421–27. Harare: University of Zimbabwe Publications.

Amblard, S., and J. Pernès (1989). The identification of cultivated pearl millet (*Pennisetum*) amongst plant impressions on pottery from Oued Chebbi (Dhar Oualata, Mauritania). *African Archaeological Review 7*, 117–26.

D'Andrea, A. C., M. Klee, and J. Casey (2001). Archaeobotanical evidence for pearl millet (*Pennisetum glaucum*) in sub-Saharan West Africa. *Antiquity 75*, 341–48.

Duhard, J.-P. (2003) La découverte de Lagraich par Michel Gaussen. *Bulletin de la Société Préhistorique Ariège-Pyrénées (Préhistoire, Art et Société-Mélanges Jean Gaussen)* 58: 189–203.

Fuller, D. Q (2003) African crops in prehistoric South Asia: A critical review. In K. Neumann, S. Kahlheber, and E. A. Butler (Eds.), *Food, Fuel and Fields: Progress in African Archaeobotany*, pp. 239–71. Köln: Heinrich-Barth Institut.

———. (2007) Contrasting patterns in crop domestication and domestication rates: Recent archaeobotanical insights from the Old World. *Annals of Botany 100*, 903–24.

Fuller, D. Q, E. Asouti, and M. D. Purugganan (2012) Cultivation as slow evolutionary entanglement: Comparative data on rate and sequence of domestication. *Vegetation History and Archaeobotany 21*, 131–45.

Fuller, D. Q, K. Macdonald, and R. Vernet (2007) Early domesticated pearl millet in Dhar Nema (Mauritania): Evidence of crop processing waste as ceramic temper. In R. Cappers (Ed.), *Field of Change. Proceedings of the 4th International Workshop for African Archaeobotany*, pp. 71–76. Groningen: Barkhuis and Groningen University Library.

Fuller, D. Q, L. Qin, Y. Zheng, Z. Zhao, X. Chen, L. A. Hosoya, and G. Sun (2009) The domestication process and domestication rate in rice: Spikelet bases from the Lower Yangtze. *Science 323*, 1607–10.

Garcea, E. A. A. (2006) Semi-permanent foragers in semi-arid environments of North Africa. *World Archaeology 38*, 197–219.

Gaussen, J., and M. Gaussen (1988) *Le Tilemsi préhistorique et ses abords.* Paris: Centre National de la Recherche Scientifique.

Godbole, S. V. (1925) *Pennisetum Typhoideum. Studies on the Bajra Crop I: The Morphology of Pennisetum Typhoideum. Memoirs of the Department of Agriculture in India (Agricultural Research Institute, Pusa) 14*, 247–68.

Harlan, J. R. (1992) Indigenous African Agriculture. In C. W. Cowan and P. J. Watson (Eds.), *The Origins of Agriculture: An International Perspective*, pp. 59–70. Washington, D.C.: Smithsonian Institution Scholarly Press.

Holl, A. F. C. (1985) Subsistence patterns of the Dhar Tichitt Neolithic, Mauritania. *African Archaeological Review 3*, 151–62.

———. (2005) Holocene 'aquatic' adaptations in North Tropical Africa. In A. B. Stahl (Ed.), *African Archaeology: A Critical Introduction*, pp. 174–86. Oxford: Blackwell Publishing Ltd.

MacDonald, K. C. (1998) Before the Empire of Ghana: Pastoralism and the origins of cultural complexity in the Sahel. In I. G. Connah (Ed.), *Transformations in Africa: Essays on Africa's Later Past*, pp. 71–103. London: Leicester University Press.

MacDonald, K. C., R. Vernet, D. Q Fuller, and J. Woodhouse (2003) New light on the Tichitt tradition: A preliminary report on survey and excavation at Dhar Nema. In P. Mitchell, A. Haour, and J. Hobart (Eds.). *Researching Africa's Past: New Contributions from British Archaeologists*, pp. 73–80. Oxford: Oxford University School of Archaeology.

MacDonald, K. C., R. Vernet, M. Martinón Torres, and D. Q Fuller (2009) Dhar Néma: From early agriculture to metallurgy in southeastern Mauritania. *Azania: Archaeological Research in Africa 44* (1), 3–48.

Manning, K. (2008a) *Mobility, Climate Change and Cultural Development: A Revised View from the Lower Tilemsi Valley, Northeastern Mali.* Ph.D. thesis, University of Oxford.

———. (2008b) Mobility amongst LSA Sahelian pastoral groups: A view from the Lower Tilemsi Valley, Eastern Mali. *Archaeological Review from Cambridge 23* (2), 125–45.

———. (2011a) Potter communities and technological tradition in the Lower Tilemsi Valley, Mali. *Azania: Archaeological Research in Africa 46*, 1–87.

———. (2011b) The first herders of the West African Sahel: Inter-site comparative analysis of zooarchaeological data from the Lower Tilemsi Valley, Mali. In H. Jousse and J. Lesur (Eds.), *People and Animals in Africa: Recent Advances in Archaeozoology*, pp. 75–85. Frankfurt: Africa Magna Verlag.

Manning, K., R. Pelling, T. Higham, J.-L. Schwenniger, and D. Q Fuller (2011). 4500-year old domesticated pearl millet (*Pennisetum glaucum*) from the Tilemsi Valley, Mali: New insights into an alternative cereal domestication pathway. *Journal of Archaeological Science 38* (2), 312–22.

Neumann, K. (2005) The romance of farming: Plant cultivation and domestication in Africa. In A. Stahl (Ed.), *African Archaeology: A Critical Introduction*, pp. 249–75. Oxford: Blackwell Publishing Ltd.

Oumar, I., C. Marciac, J.-L. Pham, and Y. Vigouroux (2008) Phylogeny and origin of pearl millet (*Pennisetum glaucum* [L.] R. Br) as revealed by microsatellite loci. *Theoretical and Applied Genetics 117*, 489–97.

Reddy, S. N. (1997) If the threshing floor could talk: Integration of agriculture and pastoralism during the Late Harappan in Gujarat, India. *Journal of Anthropological Archaeology 16*, 162–87.

Smith, A. B. (1974a) *Adrar Bous and Karkarichinkat Examples of Post-Palaeolithic Human Adaptation in the Saharan and Sahel Zones of West Africa*. Ph.D. thesis, University of California.

———. (1974b) Preliminary report of excavations at Karkarichinkat Nord and Sud, Tilemsi Valley, Mali, spring 1972. *West African Journal of Archaeology 4*, 33–55.

———. (1975) A note on the flora and fauna from the post-Palaeolithic sites of Karkarichinkat Nord and Sud. *West African Journal of Archaeology 5*, 201–04.

———. (1992) *Pastoralism in Africa. Origins and Development Ecology*. London: Hurst and Company.

Sutton, J. E. G. (1974) The aquatic civilization of middle Africa. *Journal of African History 15*, 527–46.

———. (1977) The African aqualithic. *Antiquity 51*, 25–34.

Tostain, S. (1998) Le mil, une longue histoire: Hypotheses sur domestication et ses migrations. In M. Chastenet (Ed.), *Plantes et Paysages d'Afrique. Une Histoire a explorer*, pp. 461–90. Paris: Editions Karthala.

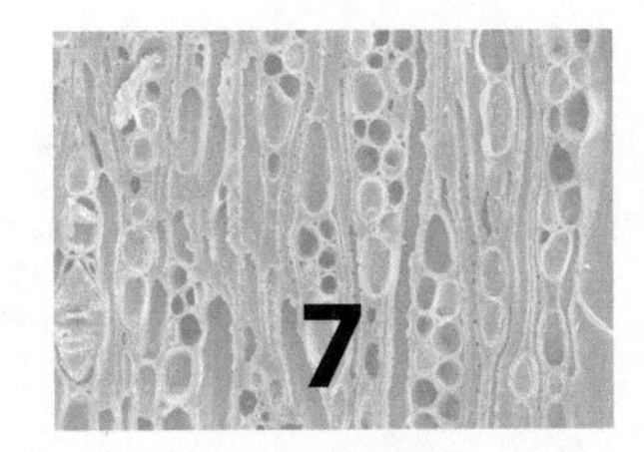

Holocene Vegetation Change and Land Use at Ounjougou, Mali

Barbara Eichhorn and Katharina Neumann

During the Holocene, West Africa witnessed important climatic fluctuations coupled with vegetation changes (Ballouche and Neumann 1995; DeMenocal et al. 2000; Lézine 1989; Lézine, Duplessy, and Cazet 2005; Maley 2004a, b; Marchant and Hooghiemstra 2004; Russell, Talbot, and Haskell 2003; Salzmann and Hoelzmann 2005; Salzmann and Waller 1998; Salzmann, Hoelzmann, and Morczinek 2002). Increasing human impact on the vegetation during this period can be assumed but can hardly be discriminated from climate-induced changes (summarised in Neumann, Hahn-Hadjali, and Salzmann 2004; Salzmann 2000; Salzmann and Waller 1998). Terrestrial proxy data for the vegetation history in the key areas of human occupation are still rare in West Africa, mostly due to the lack of sediments suitable for pollen preservation. Therefore, charcoal sequences become a valuable alternative tool to describe changes in woody vegetation through time (Neumann, Kahlheber, and Uebel 1998; Ballouche and Neumann 1995; Frank et al. 2001; Höhn 2002, 2005, 2007; Klee, Zach, and Neumann 2000) but may have a limited chronological resolution, and anthropogenic selection has to be considered (Neumann 1999).

We present here a chronological synthesis on the ecologically most significant charcoal taxa from Ounjougou, a site complex on the Dogon sandstone Plateau in Mali (Figure 7.1; Figure 7.2; Huysecom 2002). Preservation of Holocene plant micro- and macro-remains is excellent, comprising charred wood, fruits, seeds, leaf imprints and other vegetative parts, pollen, and phytoliths. Among archaeobotanical remains, charred wood is by far the most common, present in large amounts and sometimes forming distinct charcoal layers. The plant remains are only rarely associated with *in situ* buried archaeological sites (for example, the lowest layers of site 'Varves Ouest'). Most rather originate from natural deposition in colluvial and alluvial deposits; thus the influence of human selection is negligible. Although the river vegetation contributed considerably to the assemblages, remains from the surrounding savannahs and woodlands were also deposited in the sediments. The catchment area is estimated to cover 500 km^2 (Huysecom et al. 2006), providing a good representation of regional vegetation types.

Archaeology of African Plant Use by Chris J. Stevens, Sam Nixon, Mary Anne Murray, and Dorian Q Fuller, Eds., 83–96

Site Characteristics and Chronology

The complex at Ounjougou (14°24' N, 3°31' W) covers an area of 10 km^2 around the confluence of the Yamé River, a perennial tributary of the Niger with two seasonal side branches, the Menié-Menié and the Boumbangou (Figure 7.1; Figure 7.2). A series of gullies incise into a complex sequence of Quaternary aeolian, alluvial, and colluvial deposits, with archaeological sites sealed *in situ* (Huysecom et al. 2005; Mayor et al. 2005; Ozainne et al. 2009; Rasse et al. 2006). The stratigraphy has yielded archaeological material dating from the lower Palaeolithic to modern times (Huysecom et al. 2004b; Rasse et al. 2004; Robert et al. 2003). For the Holocene, at least five principal occupation phases can be discriminated (referred to here as Phase 1 to Phase 5) covered by a large number of

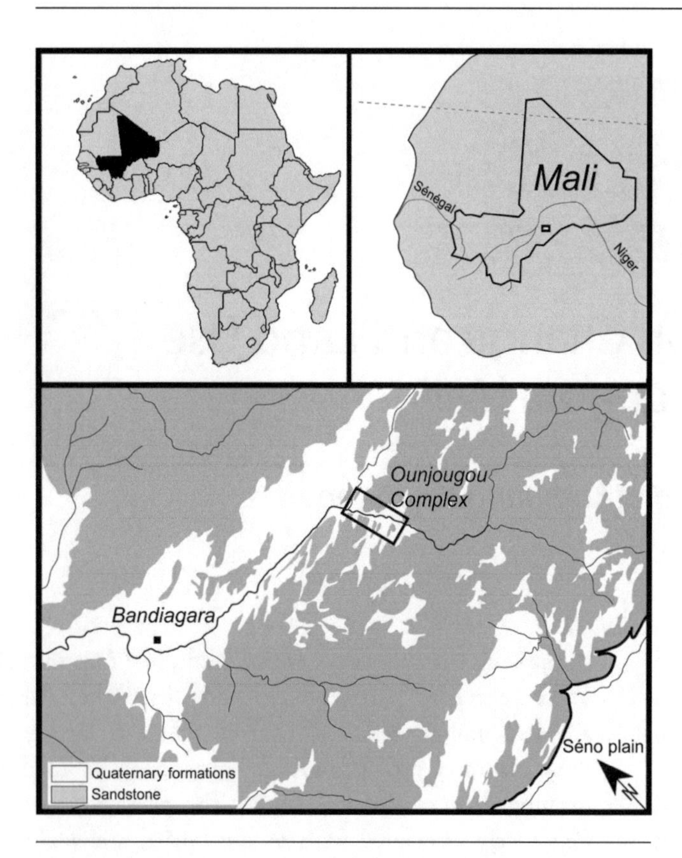

Figure 7.1 Map showing the location of the Ounjougou Complex (defined area in right-hand inset is shown in detail in Figure 7.2).

radiocarbon dates (Huysecom et al. 2004b; Ozainne et al. 2004; Ozainne et al. 2009), while the corresponding geomorphological sequence distinguishes four subphases for the Holocene (Rasse et al. 2006).

The Early Holocene is represented by the sites Ravin de la Mouche (Phase 1, ca. 9500–8500 B.C.E.), Ravin du Hibou and Damatoumou (Phase 2, 8000–ca. 7000 B.C.E.), reflecting the reoccupation of the area correlated with the climatic amelioration after the Pleistocene/Holocene transition. Most remarkable are the finds of the possibly oldest African ceramics at the site Ravin de la Mouche; their appearance probably correlated with the abundance of annual Panicoid grasses with edible grains, which have to be boiled before human consumption (Huysecom et al. 2009; Neumann et al. 2009). The Early Holocene sediments are mostly sandy, resulting from moderate to strong fluvial activity, and contain few plant remains.

For the Middle Holocene, a major sedimentary and archaeological hiatus between 7000 and 3500 B.C.E. has been assumed (Huysecom et al. 2004b). However, recent field surveys have identified botanical remains from archaeologically sterile sedimentary layers. The sites Coupe

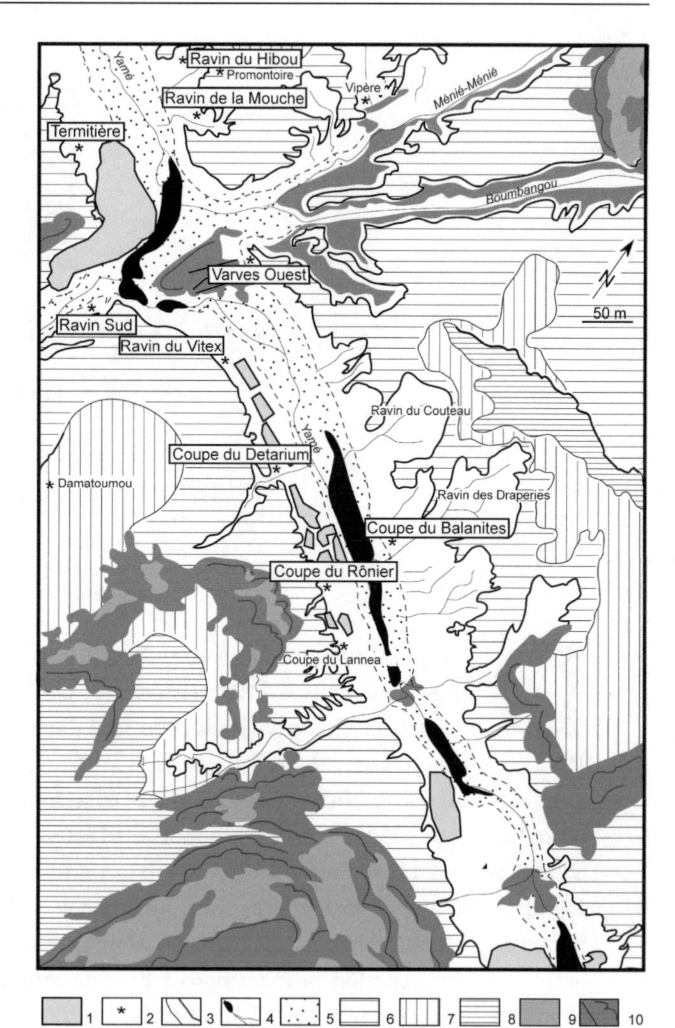

Figure 7.2 Geomorphology and location of sites at Ounjougou (as defined in Figure 7.1), based on the 1982 aerial photograph (modified from Rasse et al. 2006): (1) cultivated areas; (2) location of archaeological sites and sediment profiles; (3) terraces and glacis edges (<2 m, >2 m); (4) hydrological network and permanent lakes (in 1982); (5) modern alluvial sands; (6) secondary terrace-glacis; (7) early Holocene terrace (8th millennium B.C.E.); (8) main terrace-glacis; (9) eroded sectors of main glacis; (10) sandstone/fractures and scarps in sandstone.

du Detarium, Ravin du Vitex, and Coupe du Balanites are placed between 5300 and 4000 B.C.E. by radiocarbon dating (Rasse et al. 2006; Huysecom et al. 2007).

Archaeological evidence for the archaeological Phase 3 (ca. 3400–3000 B.C.E.) is sparse, but several sites (Ravin du Hibou, Ravin Sud) indicate a denser occupation during Phase 4 (3rd millennium B.C.E.). The archaeological material shows Saharan affinities probably reflecting migration and cultural contacts associated with increasing aridification. Phase 5 (1700–1000 B.C.E.) is mainly represented at Varves Ouest and characterised by a radical change in ceramic tradition, the presence of numerous grinding and mill stones, and the appearance of settlement structures (Huysecom et al. 2004b; Ozainne et al. 2004, 2009).

A major regional archaeological hiatus follows from ca. 1000 B.C.E. to 600 C.E. (Mayor et al. 2005). During the 1st millennium B.C.E. archaeological evidence is rare in some parts of West Africa, whereas in others large proto-urban communities appear and a diversified food production develops (Breunig and Neumann 2002a, b; Kahlheber and Neumann 2007; McIntosh 2006). Among the newly detected sites of the recent field campaigns, one site, Coupe du Rônier, dates to the middle of the 1st millennium B.C.E. (Huysecom et al. 2007).

Charcoal-rich deposits are preserved at Fiko, about 20 km to the West of Ounjougou. This site comprises several huge iron slag heaps (Huysecom et al. 2004a, 2005, 2006), beginning at least in the 13th and ending in the 19th century C.E. Biased by human fuel selection, the anthracological results are not directly comparable to those from Ounjougou, yet the site is of special importance, because its charcoal assemblages reflect the most recent development of the woody vegetation on the Dogon plateau.

General outlines of the Holocene regional vegetation history, based on palynological and preliminary anthracological data, were also available before this project began (Huysecom et al. 1999, 2000, 2001, 2002, 2004b; Ballouche 2002; Ballouche, Doutrelepont, and Huysecom 2003).

Modern Vegetation and Landuse

Ounjougou is currently located at the transition between the Sahelian and the Sudanian phytogeographic zones (Figure 7.3), with a rainy season of three and a half months and an average annual rainfall of 500–600 mm (Bandiagara, based on data before 1973; Coutard 1999; Nouaceur 2001). Ounjougou lies at the limit of the Sudanian and the Sahelian savannahs (Figure 7.3; Granier 2001). According to Jaeger and Winkoun (1962), the natural vegetation of 'favourable' sites on the Bandiagara plateau is a typical Sudanian savannah characterised by *Daniellia oliveri* in association with *Vitellaria parodoxa*, *Detarium* sp., *Parkia biglobosa*, *Terminalia macroptera*, *Khaya senegalensis*, *Vitex doniana*, *Prosopis africana*, *Entada africana*, and shrub species such as *Combretum micranthum*, *Ozoroa insignis*, and *Guiera senegalensis*. However, in reality, a cultural landscape covers the Bandiagara plateau and the adjacent Seno plain—that is, extensively cultivated park savannahs with few selected tree species. Pearl millet is the most important crop, and fallow cycles are usually short or the fields are permanently cultivated, with small savannah patches persisting away from settlements or on rocky slopes not suitable for agriculture. Whereas the sparse gallery forest of the Yamé River around Ounjougou is poor in species and dominated by *Syzygium guineense*, *Guiera senegalensis*, and the probably introduced *Andira inermis* (Arbonnier 2002; Aubréville 1950), floristically diverse ravine forests with Guinean and Sudano-Guinean species, such as *Aphania senegalensis*, *Cola laurifolia*, *Malacantha alnifolia*, *Manilkara multinervis*, or *Trema orientalis*, persist at the escarpment (Jaeger and Winkoun 1962; observations of the authors 2005).

Methods

The majority of the samples were collected by Hugues Doutrelepont (Royal Museum of Central Africa, Tervuren) during the 1998 to 2001 field seasons. Charcoal pieces of >2 mm, large fruits, and seeds were picked out by hand from sieves during the excavations. Fruits and seed samples were retrieved by dry sieving of sediment samples with 2 mm, 1 mm, and 0.5 mm mesh width and subsequent flotation in the Frankfurt laboratory. The charcoal fragments were split into the three diagnostic planes—transverse, tangential, and radial—and studied with an incident light microscope (range of magnifications: 50x–500x) and occasionally Scanning Electron Microscopy. Charcoal types were identified with the reference collection of modern African woods in the Frankfurt archaeobotanical laboratory, the Frankfurt DELTA/Intkey anatomical database, and wood anatomical atlases and anthracological catalogues (Eichhorn 2002; Höhn 2005; Neumann et al. 2001; Normand 1950, 1955, 1960). Identifications follow the IAWA standards (Wheeler, Baas, and Gasson 1989).

Results (Table 7.1)

Early Holocene: Ravin de la Mouche (Layer HA 3, ca. 9500–8500 B.C.E., Phase 1), Ravin du Hibou (8000–ca. 7000 B.C.E., Phase 2)

The charcoal assemblage of Ravin de la Mouche contains few taxa. The dominant species is *Syzygium guineense*, followed by *Prosopis africana* and *Anogeissus leiocarpus*, while other species occur in minor proportions (Table 7.1; Neumann et al. 2009). The 8th millennium B.C.E. is not well represented by charcoal, nevertheless, in comparison to Phase 1, a strong diversification of mainly Sudanian taxa is evident at Ravin du Hibou, including *Vitex* sp., represented by one fruit stone fragment. *Kigelia/Stereospermum* occurs in the charcoal assemblage of Phase 2 only and is virtually absent later. *Uapaca togoensis* is present from this phase until at least the 2nd millennium B.C.E. Charred bamboo

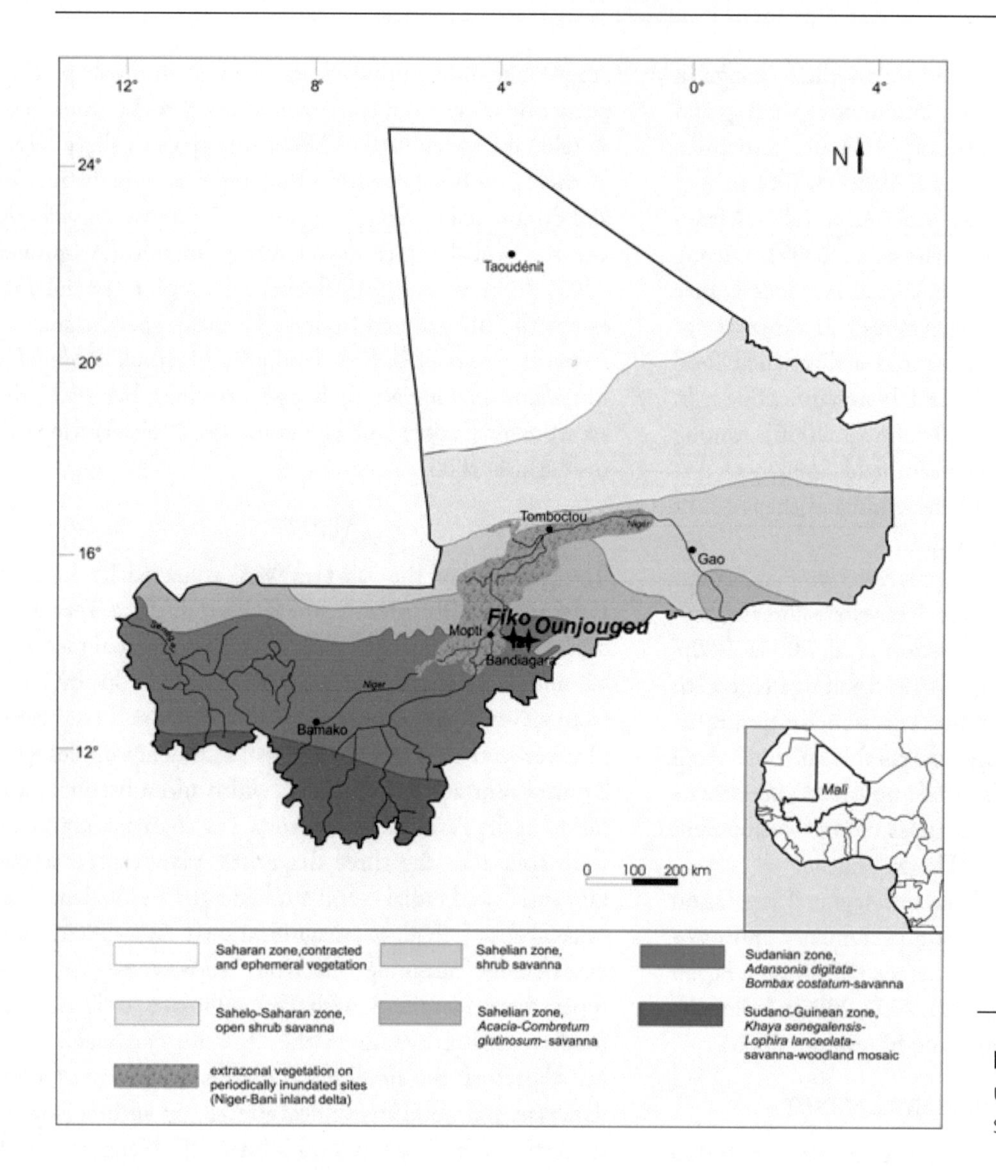

Figure 7.3 Major vegetation units of Mali and location of the sites Ounjougou and Fiko.

stalks are present, but only in small numbers. Additional taxa not mentioned in Table 7.1 are Rubiaceae type *Feretia*, *Annona senegalensis*, and *Pseudocedrela kotschyi*.

MIDDLE HOLOCENE: COUPE DU BALANITES (CA. 5300 B.C.E., NO ARCHAEOLOGICAL FINDS)

The Middle Holocene layers differ markedly from the preceding phases by virtue of a high proportion of thick carbonised monocot stalks, mainly originating from the bamboo species *Oxytenanthera abyssinica*. Charcoal of trees and shrubs is less numerous.

MIDDLE/LATE HOLOCENE TRANSITION: TERMITIÈRE (CA. 3400–3000 B.C.E., PHASE 3), RAVIN SUD AND RAVIN DU HIBOU (3RD MILLENNIUM B.C.E., PHASE 4)

Phases 3 and 4 are well represented by charcoal, the dominant taxa including those of gallery forest and well-drained soils. Several Sudano-Guinean and Sudanian savannah trees and shrubs appear during Phase 3, which were not present within earlier sediments, along with *Swartzia madagascariensis* type, *Flueggea virosa/Hymenocardia acida* (not listed in Table 7.1). In Phase 4, *Daniellia oliveri*, *Bombax costatum/Ceiba pentandra*, and *Cola* sp. appear for the first time at Ravin du Hibou.

LATE HOLOCENE: VARVES OUEST (1700–900 B.C.E., 5A: CA. 1700–1400 B.C.E., 5B: CA. 1400–1000 B.C.E.), COUPE DU RÔNIER (MID-1ST MILLENNIUM B.C.E.)

At Varves Ouest, in contrast to earlier phases, *Daniellia oliveri* (Figure 7.4, a–c) gains importance, whereas taxa with Sahelo-Sudanian affinities either appear for the first time or their relative abundance increases. The Guinean species *Uapaca togoensis* declines and finally disappears with the onset of Phase 5b. *Lophira lanceolata* (Figure 7.4, d–f),

Table 7.1 Synthesis of Semiquantitative Charcoal Analysis from Ounjougou (selected taxa)

Possible Habitat, Phytogeographical Classification		Age of Deposits	9500–8500 B.C.E.	8000–7000 B.C.E.	ca. 5300 B.C.E.	3400–3000 B.C.E.	3000–2000 B.C.E.	1700–1400 B.C.E.	1400–1000 B.C.E.	Early 1st mill. B.C.E.
		Archaeological Phase	Phase 1	Phase 2	–	Phase 3	Phase 4	Phase 5 a	Phase 5 b	–
Gallery Forest	Guinean	*Syzygium guineense*	XXXX	XXX	XX	XX	XX	XX	X	X
		Uapaca togoensis		X		XX	XX	X		
		Alchornea cordifolia	X			X	X	XX	XX	XX
	Sudano-Guinean	*Cola* sp.					X	X	X	X
		Khaya cf. *senegalensis*		X		XX	XX	XX	XX	XX
		Oxytenanthera abyssinica		X	XXXX	XXX	XXX	XXX	XX	XX
		Vitex sp.		X	X	X	X	XX	XX	X
Savannas and Partly Dry Forests		*Vitellaria/Manilkara* type			X	X	XX	XX	XX	X
		Parinari cf. *curatellifolia*				XX	X	X	X	X
		Lophira lanceolata				XX	XX	X	X	
		Daniellia oliveri					X	XX	XX	XX
		Kigelia/Stereospermum				X	X			
	Sudanian	*Anogeissus leiocarpus*	X				X	X	X	X
		Pterocarpus cf. *erinaceus*	X	X		X	X	X	X	X
Savannas and Fallows		*Diospyros mespiliformis*	X	X					X	X
		Prosopis africana	X	X	X	X	X	X	X	X
		Tamarindus indica				X	X	X	X	X
		Bombax/Ceiba					X	X	X	X
		Detarium cf. *microcarpum*		X	XX	XX	XX	X	X	X
		Terminalia sp.		X	X	X	XX	XX	X	X
		Lannea sp.						X	X	X
	Sahelo-Sudanian	*Combretum glutinosum*		X		X	X	X	X	X
		Combretum micranthum	X	X			X	XX	XX	XX
		Guiera senegalensis						X	X	X
		Bauhinia/Piliostigma type						X	X	X

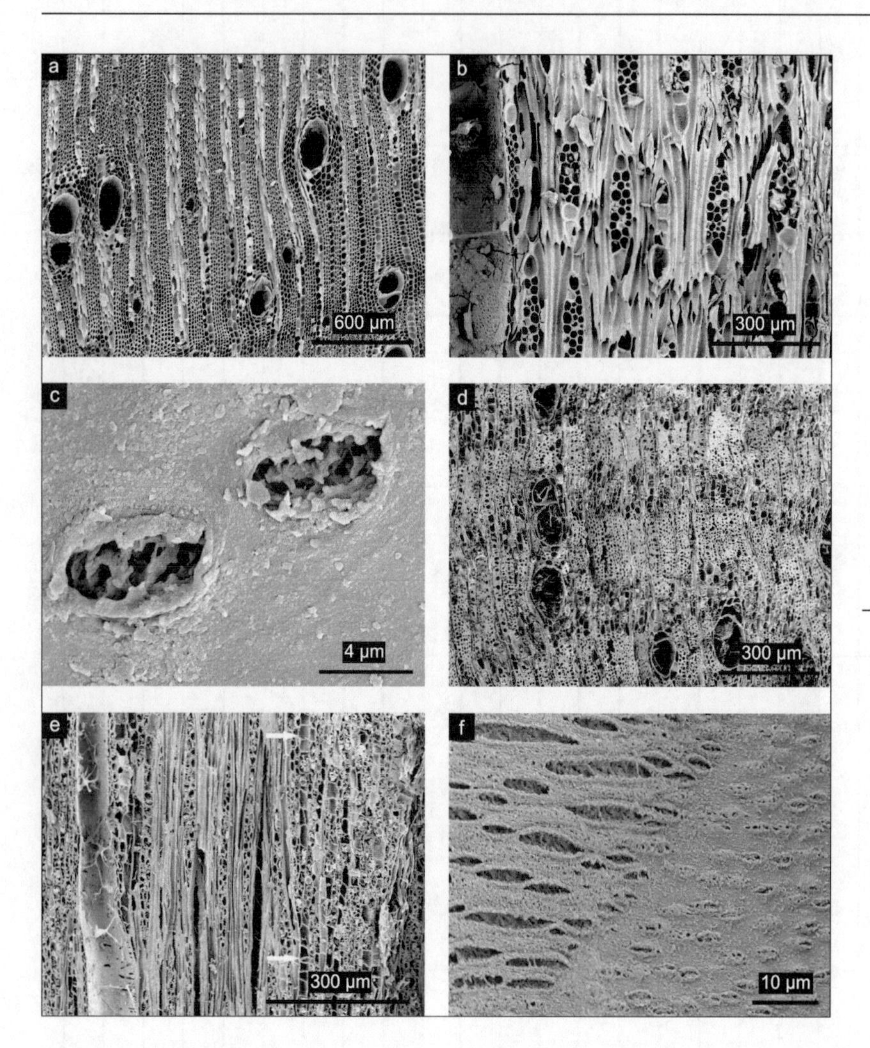

Figure 7.4 SEM photographs of wood charcoal from Ounjougou. (a-c) *Daniellia oliveri*, Fabaceae-Caesalpinioideae—(a) transverse view: paratracheal vasicentric parenchyma; (b) tangential view: heterocellular rays, storied structure; (c) tangential view: vestured intervessel pits; (d-f) *Lophira lanceolata*, Ochnaceae—(d) transverse view: banded parenchyma; (e) tangential view: arrows indicate multicelled axial parenchyma strands; (f) radial view: small vestured pits with coalescent apertures.

still present during Phase 5b, is absent in the uppermost sediments of Varves Ouest and at Coupe du Rônier. The preliminary results from Coupe du Rônier do not indicate major changes in comparison to the preceding phases at Varves Ouest.

FIKO (13TH–19TH CENTURIES C.E.)

At the iron reduction site of Fiko (Figure 7.3), *Terminalia* sp., *Pterocarpus* cf. *lucens*, *Prosopis africana*, and *Combretum glutinosum* dominate the charcoal assemblages (Figure 7.5). The sequence consists of three representative bulk samples of different ages, which show distinct changes in the abundance of the dominant taxa. During the course of occupation, *Terminalia* sp. and *P. africana* decline, with *C. glutinosum* and *P.* cf. *lucens* increasing significantly. The shea butter tree (*Vitellaria paradoxa*) is increasingly used, and Sudanian species are gradually replaced by Sahelian trees and shrubs, among them *Balanites aegyptiaca*. This last species was found neither at Ounjougou nor in samples related to early iron production at Fiko.

INTERPRETATION

EARLY HOLOCENE

The tree *Syzygium guineense*, dominant at Ravin de la Mouche and present throughout the Holocene, grows today in the gallery forest along the Yamé River. During the Early and Middle Holocene it was widespread in the present Sahelian and Sudanian zones. It has also been found in the Middle to Late Holocene site at Pentènga in southeast Burkina Faso (as Euphorbiaceae type 1 in Frank et al. 2001), and *Syzygium* pollen has been recorded from several Sahelian and Sudanian sites (Lézine 1987; Salzmann 2000; Salzmann and Waller 1998), where it often co-occurs with *Uapaca*. *Uapaca* is represented at Ounjougou by charcoal from 8000 B.C.E., but earlier by pollen (Ballouche, personal communication). *Syzygium guineense*, *Uapaca* and *Alchornea cordifolia* have Guinean affinities, but despite their wide ecological range, we assume that they were restricted to the gallery forest as extrazonal elements, because other Guinean species are absent.

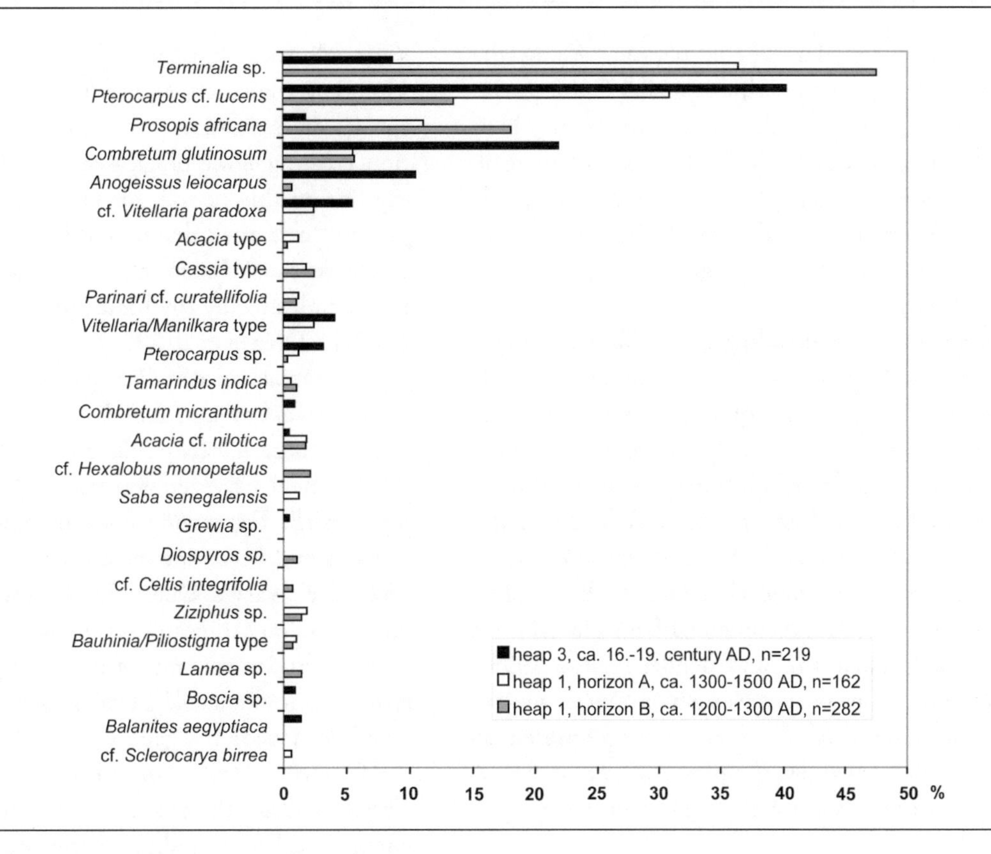

Figure 7.5 Charcoal diagram of the iron reduction site Fiko.

The remainder of the assemblage at Ravin de la Mouche comprises few Sudanian and Sahelo-Sudanian species. Although *Anogeissus leiocarpus, Pterocarpus erinaceus, Diospyros mespiliformis,* and *Prosopis africana* may occur in dry forest (Frank et al. 2001; Neumann and Müller-Haude 1999), the low species diversity indicates that they were part of the gallery forest, while a sparsely wooded savannah was growing on well-drained soils. This vegetation pattern is corroborated by palynological and phytolith data pointing to an open environment dominated by annual grasses and including xerophytic taxa—for example, Asteraceae and Chenopodiaceae (Rasse et al. 2006; Neumann et al. 2009).

The results seem to contradict the generally accepted view of an Early Holocene African humid period (for example, DeMenocal et al. 2000; Gasse 2000; Kuper and Kröpelin 2006; Lézine and Casanova 1989; Lézine, Duplessy, and Cazet 2005; Neumann 1989). However, it seems likely that the woody flora at Ounjougou was impoverished after a climatic deterioration during the Younger Dryas (13,400–12,100 cal. BP) and possibly the '11,400–11,200 cal. BP event' (Lézine, Duplessy, and Cazet 2005). The latter is almost contemporaneous with the charcoal deposits at Ravin de la Mouche and might have been responsible for the open character of the woody vegetation. At Lake Tilla in Nigeria, relatively cool Early Holocene conditions have been inferred from the high pollen percentages of the montane species *Olea hochstetteri* (Salzmann 2000; Salzmann, Hoelzmann, and Morczinek 2002).

Natural or anthropogenic fires seem rare at Ounjougou during this period (Huysecom et al. 2004b; Rasse et al. 2006), whereas at Lake Tilla regular fires promoted the open character of the woody vegetation throughout the Holocene. Bush fires occur mainly in Sudanian and Guinean long-grass savannah and woodland and only rarely in forests and Sahelian savannahs that have little combustible biomass (Ballouche and Dolidon 2005). An extension of forests at Ounjougou is contradicted by the high grass pollen values. The open nature of the grass layer and little human influence might explain the low fire frequency during this phase. Charred bamboo stalks, frequent during later phases, are absent from the assemblage of Ravin de la Mouche.

Around 8000 B.C.E., a Sudanian vegetation with some Sudano-Guinean elements became fully established. Frequently burnt savannahs or open woodlands ('forêts claires', C.C.T.A./C.S.A. 1956) were growing on the well-drained soils of the Dogon plateau. Along the river, a dense gallery forest with extrazonal Guinean elements persisted. Smaller patches of dry forests may have occurred, but unequivocal forest markers (see Frank et al. 2001) have not been found. A preponderance of savannahs during this phase was also concluded for Sahelian and Sudanian sites in Nigeria (Salzmann 2000; Salzmann and Waller 1998).

At Ounjougou, undemanding species like *Combretum glutinosum* probably grew on shallow soils of the sandstone plateau, whereas taxa requiring humid conditions were found either on the river banks or like *Kigelia/Stereospermum* and *Pseudocedrela kotschyi* on damp soils in the savannah (cf. Hahn 1996, pp. 61–64). Together the assemblage represents a mosaic of several vegetation types, typical of the Sudanian zone (Ballouche 2002; Hahn-Hadjali and Schmid 1999; Neumann, Hahn-Hadjali, and Salzmann 2004; Neumann and Müller-Haude 1999). The presence of Guinean species such as *Uapaca togoensis* and *Alchornea cordifolia*, which are no longer present at Ounjougou, confirm that precipitation during that part of the Early Holocene was distinctly higher than today.

MIDDLE HOLOCENE AND MIDDLE/LATE HOLOCENE TRANSITION

The charcoal data from Coupe du Balanites point to the persistence of a Sudanian mosaic during the Middle Holocene, with African bamboo along water courses and lake edges.

The Middle/Late Holocene transition is characterised by the appearance of some Sudano-Guinean savannah taxa. *Lophira lanceolata* is a characteristic species in the *Khaya senegalensis-Lophira lanceolata* savannahs (Granier 2001; Figure 7.3), currently restricted to the southernmost regions in Mali with higher rainfall. However, vegetation with dominant *Lophira* savannahs and *Isoberlinia* dry forests, as described by Huysecom and associates (2000) and Ballouche (2002), seems unlikely. The semiquantitative charcoal analysis shows that *Lophira lanceolata* is notsufficiently abundant to justify this interpretation, and *Isoberlinia* type charcoalis almost absent.

Lophira lanceolata and *Parinari curatellifolia* are fire-resistant trees (Ballouche 2002; Ballouche and Dolidon 2005; Burkill 1985–2000; Höhn 2005) and were probably promoted by increasing frequency of bushfires (Huysecom et al. 2004b, 2006). At Ounjougou, where they are no longer present, they grew on sites where water capacity was high enough to sustain Sudano-Guinean species but that were also regularly affected by fire, for example, at the outer margins of the gallery forest or on damp savannah soils. It seems that during the Middle/Late Holocene transition, regular bush fires initially favoured species diversification through the opening of the vegetation, creating new ecological niches. Other species appearing during this period, for example, *Tamarindus indica*, point to a marked dry season (Arbonnier 2002). A distinctly seasonal climate, although more humid than today, is also indicated by rhythmic sediments (Huysecom et al. 2007; Le Drézen et al. 2010).

The appearance of *Daniellia oliveri*, *Bombax costatum/Ceiba pentandra*, and *Entada africana* (not in Table 7.1) during Phase 4 can partly be interpreted as a gradual displacement of vegetation zones resulting from declining rainfall. The woody flora increasingly resembles the modern one, becoming most prominent from 1700 B.C.E. onward at Varves Ouest. In the savannahs and woodlands Sudano-Guinean taxa become rarer, being replaced first by Sudanian and eventually by Sahelo-Sudanian species. Simultaneously, taxa such as *Cola* (probably *C. laurifolia*) invade the gallery forest, replacing the retreating Guinean element *Uapaca togoensis*. Almost certainly climate was not the only factor responsible for vegetation change. Light-demanding species such as *Daniellia oliveri* (Arbonnier 2002; Aubréville 1950) were encouraged by fire and clearance related to denser settlement on the Dogon plateau,whereas others, such as *Bombax costatum*, with a fire-resistant thick bark (Ballouche and Dolidon 2005), may have benefited directly.

LATE HOLOCENE (1700–CA. 500 B.C.E.)

Palaeoecological data point to a pronounced shift toward more arid conditions in tropical Africa about 4000 years ago (Marchant and Hooghiemstra 2004). At Ounjougou, rhythmically laminated sediments indicate pronounced wet and dry seasons. At Coupe du Rônier, they are regularly separated by oxidation horizons with desiccation fissures in their upper parts (Huysecom et al. 2006, 2007; Le Drézen et al. 2010; Rasse et al. 2006), implying increasing aridity and at least partial desiccation of rivers and lakes during the dry season. This phenomenon, although less distinct, is already traceable at Varves Ouest (Ozainne et al. 2009). The high amounts and the distribution of microcharcoal in the laminated sediments point to an annual occurrence of anthropogenic bush fires during the dry season.

Charred caryopses of domesticated *Pennisetum glaucum* at Varves Ouest indicate the onset of agriculture in the region during the first half of the 2nd millennium B.C.E. Two grains were AMS-radiocarbon dated to 1882–1608 cal. BC/68.2% (3416±109 BP; Erl-9196, level 13) and

1493–1129 cal. BC/68.2% (3078±131 BP; Erl-9197, level 23; Ozainne et al. 2009). The dates are broadly contemporaneous with the other early domesticated *Pennisetum* finds in West Africa (Amblard 1996; D'Andrea and Casey 2002; D'Andrea, Klee, and Casey 2001; Kahlheber and Neumann 2007; MacDonald 1996; Manning et al. 2011; Neumann 2005; see Chapter 6 this volume).

The appearance of Sahelo-Sudanian woody taxa after 2000 B.C.E., the gradual retreat of Guinean species from the gallery forest and the disappearance of *Lophira lanceolata* after 1000 B.C.E. are unequivocally correlated with decreasing rainfall and increasing seasonality. However, clearance for fields and bush fires must have enhanced this effect. Shrubs with a high regeneration capacity, such as *Guiera senegalensis* and Bauhiniae, are abundant in modern fallows of the Sudanian zone (Neumann, Hahn-Hadjali, and Salzmann 2004), whereas *Lannea microcarpa* often occurs in park savannahs as a protected fruit tree (Boffa 1999).

The relative abundance of *Daniellia oliveri*, today a common fallow species in the southern Sudanian zone (Neumann, Hahn-Hadjali, and Salzmann 2004), increases at the beginning of Phase 5. However, this signal is not unambiguous, because *Daniellia oliveri* is also a characteristic species of gallery forests or at the feet of slopes with extraordinary water supply (Arbonnier 2002, p. 17; Küppers 1996, pp. 104–09; Wittig, Hahn-Hadjali, and Thiombiano 2000). Its growing importance may simply reflect that it penetrated the gallery forest as a consequence of diminishing rainfall.

Combretum micranthum is a fire-resistant species with a high regeneration capacity (Ballouche 2002), but in the Sudanian zone it is associated with poor, shallow soils (Arbonnier 2002; Aubréville 1950; Keay 1959). In the Sahelian zone of Iron Age Burkina Faso, Höhn, Kahlheber, and Hallier-von Czerniewicz (2004) interpret the increase of its charcoal, together with *Guiera senegalensis*, as an indicator of degraded soils due either to climatic factors or to agricultural activities. At Pentènga in southeastern Burkina Faso, the increase of *C. micranthum* charcoal begins with the onset of the Late Holocene and continues until the abandonment of the site (Frank et al. 2001). At Ounjougou, distinct hydrological changes might also explain the high numbers of *C. micranthum* charcoal, as the species often grows at the margins of temporary lakes and water courses possibly profiting from the annual desiccation of former perennial water bodies.

Despite aridification, the Sudanian mosaic persisted along with small fields. In addition to climatic change and fire, agriculture initially resulted in a diversification of habitats (cf. Frank et al. 2001), which might explain the high number of taxa in the charcoal samples. A species-rich gallery forest persisted, and even patches of dry forests might have existed, indicated by the rare occurrence of *Allophylus africanus* and lianas such as *Saba senegalensis* (not in Table 7.1). In spite of the apparent vegetation changes, the climate was probably still more humid than today. For the 1st millennium B.C.E., the anthracological data of the site Coupe du Rônier do not support the simplistic view of an extremely dry period (Mayor et al. 2005; Rasse et al. 2006).

IRON REDUCTION AT FIKO

The Fiko charcoal assemblage provides no evidence for a highly selective preference for one or a few species as seen for other metal smelting sites (cf. Neumann 1999; Neumann and Vogelsang 1996). On the contrary, a mixture of species is used, with the dominant ones yielding high-quality fuel woods. Kiethega (1997) also lists a large number of species used for iron production in Burkina Faso. With the exception of *Burkea africana*, as yet not identified at Fiko, there is a striking resemblance of taxa exploited in both regions. *Prosopis africana*, a common species at Fiko, is selectively protected by smiths in the Seno plain (Ballouche and Huysecom, personal communication) and is the preferred fuel of iron smiths in West Africa (Arbonnier 2002; Hahn 1996).

As selectivity is not pronounced, the charcoal assemblage of Fiko is suitable for inferences on the more recent vegetation history of the region. The assemblage represents typical woody vegetation in the modern Sahel-Sudanian transition zone of Central Mali (Figure 7.3; Touré, Grandtner, and Hiernaux 1994). With the exception of *Parinari* cf. *curatellifolia*, Sudano-Guinean or Guinean species are absent. It is thus evident that even at the beginning of the Fiko sequence climate was more arid than in the 2nd and 1st millennia B.C.E. Typical fallow and park savannah species are present from the beginning, pointing to an established agroforestry system. The subsequent changes in the assemblage are less unequivocal to interpret. Mayor and associates (2005) postulate an unstable climate with progressive aridification between 1000/1100 and 1800 C.E. This might explain the disappearance of *Parinari*, the appearance of Sahelo-Sudanian taxa such as *Boscia* sp. and *Balanites aegyptiaca*, and the growing importance of *Combretum glutinosum*. However, the latter is also a typical fallow species, and its increase together with the use of the typical park savannah tree *Vitellaria paradoxa* (Boffa 1999; Kahlheber 1999, 2004; Neumann 1999) might also indicate that larger areas were under cultivation. *Balanites aegyptiaca* is often seen as an indicator of anthropogenic influence in the Sudanian zone, invading from the Sahel with the arrival of Fulbe pastoralists, because the species is zoochorous (spread by animals) (Hahn 1996; Touré, Grandtner, and Hiernaux 1994). Furthermore, the direct consequences of wood exploitation for metal production should not be underestimated, with a decrease of the most preferred species leading to an increase of species with the ability to resprout after having been cut.

Conclusion

The charcoal data from Ounjougou reveal gradual but distinct changes in species composition and abundance during the Holocene within the frame of a Sudanian mosaic. The retreat of species with southern affinities and the appearance of taxa with more northern affinities in the second and at the turn to the 1st millennium B.C.E. are clear indicators of increasing aridity and seasonality. The growing influence of fire during the Holocene is well reflected in the anthracological spectra. Both factors are also unequivocally demonstrated by results of sedimentological analyses conducted at the same sites. However, from the 2nd millennium B.C.E. onward, the consequences of climatic deterioration and human impact on the woody vegetation are not easy to separate. It seems that after about 2000 B.C.E., the effect of decreasing rainfall was further amplified by increasing anthropogenic pressure. For the past millennium, the combined effects of climate change and anthropogenic influence are corroborated by the charcoal data of the iron-smelting site Fiko illustrating the continuous trend toward increasing aridity as well as direct consequences of fuel wood exploitation.

Acknowledgements

We are very much indebted in the German Research Foundation (DFG) for funding our research at Ounjougou. We thank all members of the project Peuplement humain et paléoenvironnement en Afrique de l'Ouest, under the leadership of Eric Huysecom, for their friendly cooperation. Special thanks are due Hugues Doutrelepont, who took most of the charcoal samples and Aziz Ballouche, Yann Le Drezen, Sylvain Ozainne, Sébastien Perret, and Michel Rasse. Figures 7.1 and 7.2 are modified from Rasse and associates (2006), with the kind permission of Michel Rasse. We are indebted in Stefanie Kahlheber for her help with the identification of domesticated *Pennisetum*, Monika Heckner and Manfred Ruppel for technical assistance, and Richard J. Byer for language editing. Thanks are also due the government of Mali for granting necessary visa and permits and the Mission culturelle de Bandiagara for the support of our work.

References

Amblard, S. (1996) Agricultural evidence and its interpretation on the Dhars Tichitt and Oualata, south-eastern Mauritania. In G. Pwiti and R. Soper (Eds.), *Aspects of African Archaeology. Papers from the 10th Congress of the Panafrican Association for Prehistory and Related Studies*, pp. 421–27. Harare: University of Zimbabwe Press.

Arbonnier, M. (2002) *Arbres, Arbustes et Lianes des Zones Sèches d'Afrique de l'Ouest*, Montpellier: CIRAD-MNHN.

Aubréville, A. (1950) *Flore Forestière Soudano-Guinéenne*. Paris: Société d'éditions géographiques, maritimes et coloniales.

Ballouche, A. (2002) Histoire des paysages végétaux et mémoire des sociétés dans les savanes ouest-africaines. *Historiens et Géographes 381*, 379–88.

Ballouche, A., and H. Dolidon (2005) Forêts claires et savanes ouest-africaines: Dynamiques et évolution de systèmes complexes à l'interface nature/société. In M. Taabni (Ed.), *La forêt: Enjeux comparés des formes d'appropriation, de gestion et d'exploitation dans les politiques environnementales et le contexte d'urbanisation généralisée* (Actes du colloque international à Poitiers), pp. 56–70. Poitiers: Mission de l'homme et de la Société (Université de Poitiers).

Ballouche, A., H. Doutrelepont, and E. Huysecom (2003) Données archéobotaniques et palynologiques préliminaires des dépôts holocènes du site d'Ounjougou (Mali). *Résumées Colloque Archéométrie 2003, GMPCA, Bordeaux*, 26.

Ballouche, A., and K. Neumann (1995) A new contribution to the Holocene vegetation history of the West African Sahel: Pollen from Oursi, Burkina Faso and charcoal from three sites in northeast Nigeria. *Vegetation History and Archaeobotany* 4, 31–39.

Boffa, J.-M. (1999) *Agroforestry Parklands in Sub-Saharan Africa*. Rome: FAO Conservation Guide 34.

Breunig, P., and K. Neumann (2002a) From Hunter-Gatherers to food producers: New archaeological and archaeobotanical evidence from the West African Sahel. In F. A. Hassan, (Ed.), *Droughts, Food and Culture*, pp. 123–55. New York: Academic Publishers.

———. (2002b)Continuity or discontinuity? The 1st millennium B.C. crisis in West African prehistory. In Jennerstrasse 8 (T. Lenssen-Erz et al.) (Eds.), *Tides of the Desert: Gezeiten der Wüste. Contributions to the Archaeology and Environmental History of Africa in Honour of Rudolph* Kuper (Africa Praehistorica 14), pp. 499–505. Köln: Heinrich-Barth-Institut.

Burkill, H. M. (1985–2000) *The Useful Plants of West Tropical Africa*. Vols. 1–5. Kew: Royal Botanic Gardens.

C.C.T.A./C.S.A. (1956) *Phytogéographie/Phytogeography (Yangambi 1956)*. C.C.T.A./C.S.A. Publications 53.

Coutard, S. (1999) Étude de l'environnement géologique des sites archéologiques holocènes d'Ounjougou (Plateau dogon, Mali). DEA thesis, Laboratoire de Sciences du Sol de l'INA Grignon.

D'Andrea, A. C., and J. Casey (2002) Pearl millet and Kintampo subsistence. *African Archaeological Review 19* (3), 147–73.

D'Andrea, A. C., M. Klee, and J. Casey (2001) Archaeobotanical evidence for pearl millet (*Pennisetum glaucum*) in sub-Saharan West Africa. *Antiquity 75*, 341–48.

DeMenocal, P., J. Ortiz, T. Guilderson, J. Adkins, M. Sarntheim, L. Baker, and M. Yarusinsky (2000) The abrupt onset and termination of the African humid period: Rapid climate responses to gradual insolation forcing. *Quaternary Science Reviews 19*, 347–61.

Eichhorn, B. (2002) *Anthrakologische Untersuchungen zur Vegetationsgeschichte des Kaokolandes, Nordwest-Namibia*. Ph.D. thesis, University of Cologne, http://kups.ub.unikoeln.de/volltexte/2004/1178/.

Frank, T., P. Breunig, P. Müller-Haude, K. Neumann, W. Van Neer, R. Vogelsang, and H. P. Wotzka (2001) The Chaîne de Gobnangou, SE Burkina Faso: Archaeological, archaeobotanical and geomorphological studies. *Beiträge zur allgemeinen und vergleichenden Archäologie 21*, 127–90.

Gasse, F. (2000) Hydrological changes in the African tropics since the Last Glacial Maximum. *Quaternary Science Reviews 19*, 189–211.

Granier, P. (2001) Biogéographie. In J. C. Arnaud (Ed.), *Atlas du Mali*, pp. 22–24. Paris: Jeune Afrique.

Hahn, H. P. (1996) Eisentechniken in Westafrika: Systeme der Ressourcennutzung und der Distribution. *Zentralblatt für Geologie und Paläontologie I 3/4*, 447–57.

Hahn, K. (1996) *Die Pflanzengesellschaften der Savannen im Südosten Burkina Fasos (Westafrika)*. Ph.D. thesis, Goethe Universität Frankfurt am Main.

Hahn-Hadjali, K., and S. Schmid (1999) Untersuchungen von Savannengesellschaften der Sudanzone Burkina Fasos (Westafrika) mit multitemporalen SPOT-Satellitendaten. *Die Erde 130*, 1–16.

Höhn, A. (2002) Vegetation changes in the Sahel of Burkina Faso (West Africa): Analysis of charcoal from the Iron Age sites Oursi and Oursi-village. In S. Thiébault, (Ed.), *Charcoal Analysis: Methodological Approaches, Palaeoecological Results and Wood Uses. Proceedings of the Second International Meeting of Anthracology, Paris, September 2000* (British Archaeological Reports International Series 1063), pp. 133–39. Oxford: Archaeopress.

———. (2005) Zur eisenzeitlichen Entwicklung der Kulturlandschaft im Sahel von Burkina Faso: Untersuchungen von archäologischen Holzkohlen. Ph.D. thesis, Goethe Universität Frankfurt am Main, http://publikationen.ub.uni-frankfurt.de/volltexte/2005/2253/.

———. (2007) Where did all the trees go? Changes of the woody vegetation in the Sahel of Burkina Faso during the last 2000 years. In R. Cappers (Ed.), *Fields of Change: Progress in African Archaeobotany* (Groningen Archaeological Studies 5), pp. 35–41. Groningen: Barkhuis Publishing.

Höhn, A., S. Kahlheber, and M. Hallier-von Czerniewicz (2004)Den frühen Bauern auf der Spur: Siedlungs- und Vegetationsgeschichte der Region Oursi (Burkina Faso).In K.-D. Albert, D. Löhr, and K. Neumann (Eds.), *Mensch und Natur in Westafrika*, pp. 221–55. Weinheim: Wiley/VCH.

Huysecom, E. (2002) Palaeoenvironment and human population in West Africa: An international research project in Mali. *Antiquity 76*, 335–36.

Huysecom, E., A. Ballouche, E. Boëda, L. Cappa, L. Cissé, A. Dembélé, A. Gallay, D. Konaté, A. Mayor, S. Ozainne, F. Raeli, M. Rasse, A. Robert, C. Robion, K. Sanogo, S. Soriano, O. Sow, and S. Stokes (2002) Cinquième campagne de recherches à Ounjougou (Mali). *SLSA Jahresbericht 2001*, 55–113.

Huysecom, E., A. Ballouche, L. Cissé, A. Gallay, D. Konaté, A. Mayor, K. Neumann, S. Ozainne, S. Perret, M. Rasse, A. Robert, C. Robion, K. Sanogo, V. Serneels, S. Soriano, and S. Stokes (2004a) Paléoenvironnement et peuplement humain en Afrique de l'ouest: Rapport de la sixième campagne de recherche à Ounjougou (Mali). *SLSA Jahresbericht 2003*, 27–68.

Huysecom, E., A. Ballouche, A. Gallay, N. Guindo, D. Keita, S. Kouti, Y. Le Drezen, A. Mayor, K. Neumann, S. Ozainne, S. Perret, M. Rasse, C. Robion-Brunner, K. Schaer, V. Serneels, S. Soriano, S. Stokes, and C. Tribolo (2005) La septième campagne de terrain à Ounjougou (Mali) et ses apports au programme interdisciplinaire. Paléoenvironnement et peuplement humain en Afrique de l'Ouest: *Résultats préliminaires. SLSA Jahresbericht 2004*, 57–142.

Huysecom, E., H. Beeckman, E. Boëda, H. Doutrelepont, N. Fedoroff, A. Mayor, F. Raeli, A. Robert, and S. Soriano (1999) Paléoenvironnement et peuplement humain en Afrique de l'Ouest. Rapport de la seconde mission de recherche (1998–1999) sur le gisement d'Ounjougou (Mali). *SLSA Jahresbericht 1998*, 153–204.

Huysecom, E., E. Boëda, K. Deforce, H. Doutrelepont, A. Downing, N. Fedoroff, A. Gallay, D. Konaté, A. Mayor, S. Ozainne, F. Raeli, A. Robert, S. Soriano, O. Sow, and S. Stokes (2001) Ounjougou (Mali) résultats préliminaires de la quatrième campagne de recherches.*SLSA Jahresbericht 2000*, 105–50.

Huysecom, E., E. Boëda, K. Deforce, H. Doutrelepont, A. Downing, N. Fedoroff, D. Konaté, A. Mayor, S. Ozainne, F. Raeli, A. Robert, E. Roche, O. Sow, and S. Stokes (2000) Ounjougou (Mali) troisième campagne de recherches dans les cadre du programme Paléoenvironnement et peuplement humain en Afrique de l'Ouest.*SLSA Jahresbericht 1999*, 97–149.

Huysecom, E., S. Ozainne, F. Raeli, A. Ballouche, M. Rasse, and S. Stokes (2004b) Ounjougou (Mali) A history of Holocene settlement at the southern edge of the Sahara. *Antiquity 78*, 579–93.

Huysecom, E., S. Ozainne, C. Robion-Brunner, A. Ballouche, N. Coulibaly, N. Guindo, D. Keïta, Y. Le Drezen, L. Lespez, A. Mayor, K. Neumann, B. Eichhorn, M. Rasse, C. Selleger, V. Serneels, S. Soriano, A. Terrier, B. D. Traoré, and C. Tribolo (2007) Peuplement humain et paléoenvironnement en Afrique de l'Ouest: Résultats de la neuvième année de recherches. *SLSA Jahresbericht 2006*, 41–122.

Huysecom, E., S. Ozainne, K. Schaer, A. Ballouche, R. Blench, D. Douyon, N. Guindo, D. Keita, Y. Le Drezen, K. Neumann, S. Perret, M. Rasse, C. Robion-Brunner, V. Serneels, S. Soriano, and C. Tribolo (2006) Peuplement humain et paléoenvironnement en Afrique de l' Ouest: Apports de la huitième année de recherches interdisciplinaires. *SLSA Jahresbericht 2005*, 79–160.

Huysecom, E., M. Rasse, L. Lespez, K. Neumann, A. Fahmy, A. Ballouche, S. Ozainne, M. Maggetti, C. Tribolo, and S. Soriano (2009) The emergence of pottery in Africa during the tenth millennium cal BC: New evidence from Ounjougou (Mali). *Antiquity 83*, 905–17.

Jaeger, P., and D. Winkoun (1962) Premier contact avec la flore et la végétation du plateau de Bandiagara. *Bulletin de l'I.F.A.N. 24* (A1), 68–111.

Kahlheber, K. (1999) Indications for agroforestry: Archaeobotanical remains of crops and woody plants from medieval Saouga, Burkina Faso. In M. van der Veen, (Ed.), *The Exploitation of Plant Resources in Ancient Africa*, pp. 89–100. New York: Kluwer Academic/Plenum.

Kahlheber, S. (2004) *Perlhirse und Baobab*. Ph. D. thesis, Goethe Universität Frankfurt am Main, http://publikationen.ub.uni-frankfurt.de/volltexte/2005/561/.

Kahlheber, S., and K. Neumann (2007) The development of plant cultivation in semi-arid West Africa. In T. P. Denham, J. Iriarte, and L. Vrydaghs (Eds.), *Rethinking Agriculture: Archaeological and Ethnoarchaeological Perspectives*, pp. 320–46. London: One World Archaeology/UCL Press.

Keay, R. W. J. (1959) *An Outline of Nigerian Vegetation*. Lagos: Federal Government Printer.

Kiethega, J. B. (1997) *La métallurgie lourde du fer au Burkina Faso*. Ph.D. thesis, University of Paris I.

Klee, M., B. Zach, and K. Neumann (2000) Four thousand years of plant exploitation in the Chad Basin of northeast Nigeria I: The archaeobotany of Kursakata.*Vegetation History and Archaeobotany 9*, 223–37.

Kuper, R., and S. Kröpelin (2006) Climate controlled Holocene occupation in the Sahara: Motor of Africa's evolution. *Science 313*, 803–07.

Küppers, K. (1996) *Die Vegetation der Chaîne de Gobnangou*. Ph.D. thesis, Goethe Universität Frankfurt am Main.

Le Drézen, Y., L. Lespez, M. Rasse, A. Garnier, S. Coutard, E. Huysecom, and A. Ballouche (2010) Hydrosedimentary records and Holocene environmental dynamics in the Yamé Valley (Mali, Sudano-Sahelian West Africa). *Comptes Rendus Géoscience 342* (3), 244–52.

Lézine, A.-M. (1987) *Paléoenvironnements végétaux d'Afrique nord tropicale depuis 12.000 BP*. Ph.D. thesis, University of Aix-Marseille II.

———. (1989) Late Quaternary vegetation and climate of the Sahel. *Quaternary Research 32*, 317–34.

Lézine, A.-M., and J. Casanova (1989) Pollen and hydrological evidence for the interpretation of past climates in tropical West Africa during the Holocene. *Quaternary Science Reviews 8*, 45–55.

Lézine, A.-M., J.-C. Duplessy, and J.-P. Cazet (2005) West-African monsoon variability during the last deglaciation and the Holocene: Evidence from fresh water algae, pollen and isotope data from core KW31, Gulf of Guinea. *Palaeogeography, Palaeoclimatology, Palaeoecology 219*, 225–37.

MacDonald, K. C. (1996) The Windé Koroji Complex: Evidence for the peopling of the eastern Inland Niger Delta (2100–500 BC). *Préhistoire, Anthropologie méditerranéenne 5*, 147–65.

Maley, J. (2004a) Le bassin du Tchad au Quaternaire récent: formations sédimentaires, paléoenvironnements et préhistoire: La question des paléotchads. In A.-M. Sémah and J. Renault-Miskovsky (Eds.), *L'Évolution de la végétation depuis deux millions d'années*, pp. 179–217. Paris: Artcom'/Errance.

———. (2004b) Les variations de la végétation et des paléoenvironnements du domaine forestier africain au cours de Quaternaire récent. In A.-M.Sémah and J. Renault-Miskovsky (Eds.), *L'Évolution de la végétation depuis deux millions d'années*, pp. 143–78. Paris: Artcom'/Errance.

Manning, K., R. Pelling, T. Higham, J. L. Schwenniger, and D.Q Fuller (2011) 4,500-year-old domesticated pearl millet (*Pennisetum glaucum*) from the Tilemsi Valley, Mali: New insights into an alternative cereal domestication pathway. *Journal of Archaeological Science* 38: 312–22.

Marchant, R., and H. Hooghiemstra (2004) Rapid environmental change in African and South American tropics around 4000 years before present: A review. *Earth-Science Reviews* 66, 217–60.

Mayor, A., E. Huysecom, A. Gallay, M. Rasse, and A. Ballouche (2005) Population dynamics and paleoclimate over the past 3000 years in the Dogon country, Mali. *Journal of Anthropological Archaeology 24* (1), 25–61.

McIntosh, S. (2006) The Holocene prehistory of West Africa 10,000–1000 BP. In E. K. Akyeampong (Ed.), *Themes in West Africa's History*, pp. 11–32. Athens: Ohio University Press.

Neumann, K. (1989) Vegetationsgeschichte der Ostsahara im Holozän: Holzkohlen aus prähistorischen Fundstellen. In R. Kuper (Ed.), *Forschungen zur Umweltgeschichte der Ostsahara* (Africa Praehistorica 2), 13–182. Köln: Heinrich-Barth-Institut.

———. (1999) Charcoal from West African savanna sites: Questions of identification and interpretation. In M. van der Veen (Ed.), *The Exploitation of Plant Resources in Ancient Africa*, pp. 205–19. New York: Kluwer Academic/Plenum.

———. (2005)The romance of farming: Plant cultivation and domestication in Africa. In A. B. Stahl (Ed.), *African Archaeology: A Critical Introduction*, pp. 249–75. Malden: Blackwell.

Neumann, K., A. Fahmy, L. Lespez, A. Ballouche, and E. Huysecom (2009) The early Holocene palaeoenvironment of Ounjougou (Mali): Phytoliths in a multiproxy context. *Palaeogeography, Palaeoclimatology, Palaeoecology 276*, 87–106.

Neumann, K., K. Hahn-Hadjali, and U. Salzmann (2004) Die Savannen der Sudanzone in Westafrika: Natürlich oder menschengemacht? In K.-D. Albert, D. Löhr, and K. Neumann (Eds.), *Mensch und Natur in West-Afrika*, pp. 39–68. Weinheim: Wiley-VCH.

Neumann, K., S. Kahlheber, and D. Uebel (1998) Remains of woody plants from Saouga, a medieval west African village. *Vegetation History and Archaeobotany* 7, 57–77.

Neumann, K., and P. Müller-Haude (1999) Sols et végétation de forêts sèches au sud-ouest du Burkina-Faso. *Phytocoenologia 29* (1), 53–85.

Neumann, K., W. Schoch, P. Détienne, and F. H. Schweingruber (2001) *Hölzer der Sahara und des Sahel (Woods of the Sahara and the Sahel; Bois du Sahara et du Sahel)*. Bern: Paul Haupt.

Neumann, K., and R. Vogelsang (1996) Paléoenvironnement et préhistoire au Sahel du Burkina Faso. *Berichte des Sonderforschungsbereichs* 268, Band 7, 177–86.

Normand, D. (1950) Atlas des Bois de la Côte d'Ivoire, Tome 1. Nogent-sur-Marne: Centre Technique Forestier Tropical.

——— (1955) Atlas des Bois de la Côte d'Ivoire, Tome 1. Nogent-sur-Marne: Centre Technique Forestier Tropical.

——— (1960) Atlas des Bois de la Côte d'Ivoire, Tome 3.

Nouaceur, Z. (2001) Climat. In J. C. Arnaud (Ed.), *Atlas du Mali*, pp. 16–19. Paris: Jeune Afrique.

Ozainne, S., E. Huysecom, A. Ballouche, and M. Rasse (2004) Le site des Varves à Ounjougou (Mali): Nouvelles données sur le peuplement néolithique des zones subsahariennes en Afrique de l'Ouest. *Le Forum Suisse des Africanistes 4*, 265–81.

Ozainne, S., L. Lespez, Y. Le Drezen, B. Eichhorn, K. Neumann, and E. Huysecom (2009) Developing a chronology integrating archaeological and environmental data from different contexts: The Late Holocene Sequence of Ounjougou (Mali). *Radiocarbon 51* (2), 457–70.

Rasse, M., A. Ballouche, E. Huysecom, C. Tribolo, S. Ozainne, Y. Le Drezen, S. Stokes, and K. Neumann (2006) Évolution géomorphologique, enregistrements sédimentaires et dynamiques paléoenvironnementales Holocènes à Ounjougou (Plateau Dogon, Mali, Afrique de l'Ouest). *Quaternaire 17* (1), 61–74.

Rasse, M., S. Soriano, C. Tribolo, S. Stokes, and E. Huysecom (2004)La séquence Pléistocène supérieur d'Ounjougou (Pays dogon, Mali, Afrique de l'Ouest: Evolution géomorphologique, enregistrements sédimentaires et changement culturels. *Quaternaire 15* (4), 329–41.

Robert, A., S. Soriano, M. Rasse, S. Stokes, and E. Huysecom (2003) First chrono-cultural reference framework for the West African Palaeolithic: New data from Ounjougou (Dogon Country, Mali). *Journal of African Archaeology 1* (2), 151–69.

Russell, J., M. R. Talbot, and B. J. Haskell (2003) Mid-holocene climatic change in Lake Bosumtwi, Ghana. *Quaternary Research 60*, 133–41.

Salzmann, U. (2000). Are modern savannas degraded forests? A Holocene pollen record from the Sudanian vegetation zone of NE Nigeria. *Vegetation History and Archaeobotany* 9, 1–15.

Salzmann, U., and P. Hoelzmann (2005) The Holocene history of the Dahomey Gap: A climatic induced fragmentation of the West African rainforest. *The Holocene 15* (2), 190–99.

Salzmann, U., P. Hoelzmann, and I. Morczinek (2002) Late Quaternary climate and vegetation of the Sudanian zone of Northeast Nigeria. *Quaternary Research 58*, 73–83.

Salzmann, U., and M. Waller (1998) The Holocene vegetational history of the Nigerian Sahel based on multiple pollen profiles. *Review of Palaeobotany and Palynology 100*, 39–72.

Touré, A. S., M. M. Grandtner, and P. Y. Hiernaux (1994) Relief, sols et végétation d'une savane soudano-sahélienne du Mali Central. *Phytocoenologia 24*, 233–56.

Wheeler, E. A., P. Baas, and P. E. Gasson (Eds.) (1989) IAWA list of microscopic features for hardwood identification by an IAWA committee. *IAWA Bulletin New Series 10*, 219–332.

Wittig, R., K. Hahn-Hadjali, and A. Thiombiano (2000) Particularities of the Chaîne de Gobnangou in the Southeast of Burkina Faso. *Etudes sur la flore et végétation du Burkina Faso et des pays avoisinants 5*, 49–64.

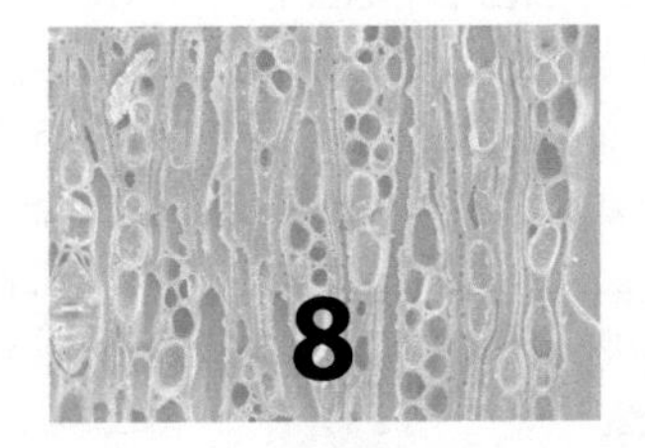

Early Agro-Pastoralism in the Middle Senegal Valley

The Botanical Remains from Walaldé

Shawn Sabrina Murray and Alioune Déme

Excavations at Walaldé began as part of the Middle Senegal Valley (MSV) Project, a joint effort between IFAN (Institut Fondamentale d'Afrique Noire) and Rice University (Houston). One of the primary goals of the Middle Senegal Valley Project was to investigate the settlement chronology of the Middle Senegal floodplain and to document change through time in material culture, subsistence, technology, and trade. Research at Walaldé was undertaken by Alioune Deme in 1999 to expand on the little-known Phase I occupations and to document whether initial colonisations were made by pastoralists or agro-pastoralists and if these populations derived from one or many source area(s) (Déme 2003; Déme and McIntosh 2006). As a floodplain, the Middle Senegal Valley is characterised as a reliable water source with a varying geography that enables a diversity of subsistence approaches that, as the findings at Walaldé suggest, might have lured early colonisers experiencing environmental stress elsewhere. Excavations revealed that Walaldé not only predates expected 1st century C.E. (Phase I) occupation but also straddles the transition from stone- to iron-based technology (ca. 800–500 B.C.E.), making it the earliest known site in the MSV floodplain. This site provides important data on early pastoralist incursions and an insight into the emergence of social complexity in a region thought to have given rise to the large-scale Takrur polity. This chapter reports on a small number of botanical samples collected from deposits at Walaldé, providing information on subsistence practices at this early transitional site.

Archaeology of African Plant Use by Chris J. Stevens, Sam Nixon, Mary Anne Murray, and Dorian Q Fuller, Eds., 97–101

Physical Environment and Traditional Land Use of the Middle Senegal Valley

The MSV is one of three great African floodplains located within the western part of the semi-arid Sahel (Figure 8.1). Situated within the 300 to 500 mm mean annual rainfall zone, the MSV has a climate characterized by high fluctuations in precipitation and temperature between years, decades, centuries, and millennia (Brooks 1986; Elouard 1962; Lezine 1987; 1989; Michel 1973; Nicholson 1994; Petit Maire 1979). The annual mean rainfall in the MSV between 1930 and 1960 was 327 mm (Goudiaby 1984; Michel 1973). Over 90% of the rainfall occurs between July and October, the amount that falls dependent on the fluctuating position of the moisture-bearing Inter Tropical Convergence Zone (ITCZ). Additionally, the 9-month-long dry season is characterised by the activities of the *harmattan*, a consequence of desiccating anticyclones that bring dry winds from the northeast and increases in temperature, the average December temperature at Podor 30.7°C rising to 41.4°C in May.

The negative effects of fluctuating rainfall and high temperature are compensated by a diverse hydrographic system dominated by the Senegal River and its distributaries. From its source at Bafoulabe to its outlet in the delta, the Senegal River, 1,784 km long, crosses three different climate zones—the subforest, Sudano-Sahelian, and Sahel-Saharan zones—and is fed by a catchment basin 335,000 km^2 in area (Elouard 1962, p. 22). Most of the water carried by the Senegal comes from the subforest

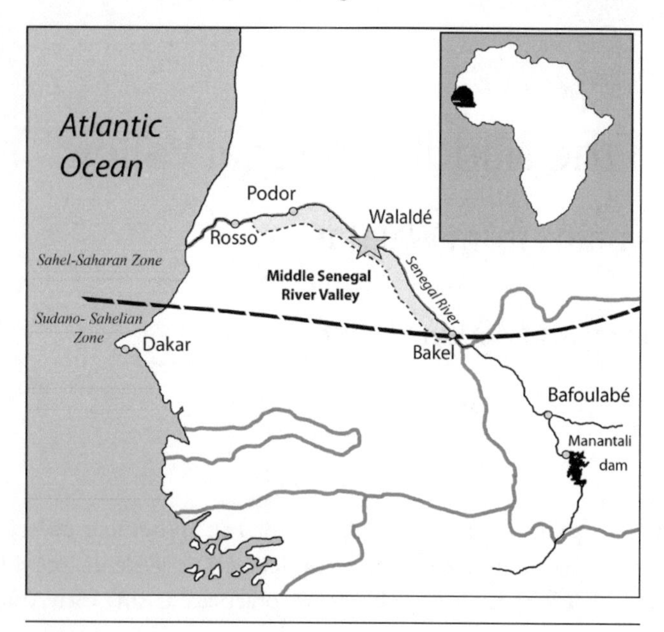

Figure 8.1 Map of the Middle Senegal River Valley showing the location of the site.

and the Sudano-Sahelian zones in the upper valley, an area geologically less suitable to agriculture (Elouard 1962; Michel 1973). The middle valley runs from Bakel downstream to Podor, taking an east-west orientation and is characterised by a low slope that increases the size of the floodplain from 2–5 km wide in the upper valley to 10–20 km in the middle valley. The delta begins downstream of Rosso, and it is here that the Senegal divides into several branches and takes a north-south orientation.

The flood is carried by the Senegal and its distributaries, beginning May to June in Guinea and progressing slowly downstream to reach the middle valley in July. The impact of the flood varies according to rainfall, topography, and soil type. Flooding and topography are two major elements of the land use system. A lateral cross section of the middle valley shows that before the construction of the Manantali dam, which altered the traditional land use system, the area was divided into three human exploitation zones: the floodplain (*walo*), the upper lands (*jeri*), and the transition zone between the two *jejeengol* (Figure 8.2). Each exploitation zone is characterised by its topography, its pedology, and its history of subsistence and settlement activities. The floodplain is the domain of high levees (*fonde*) and deep clay basins (*hollade*). Mostly occupied by

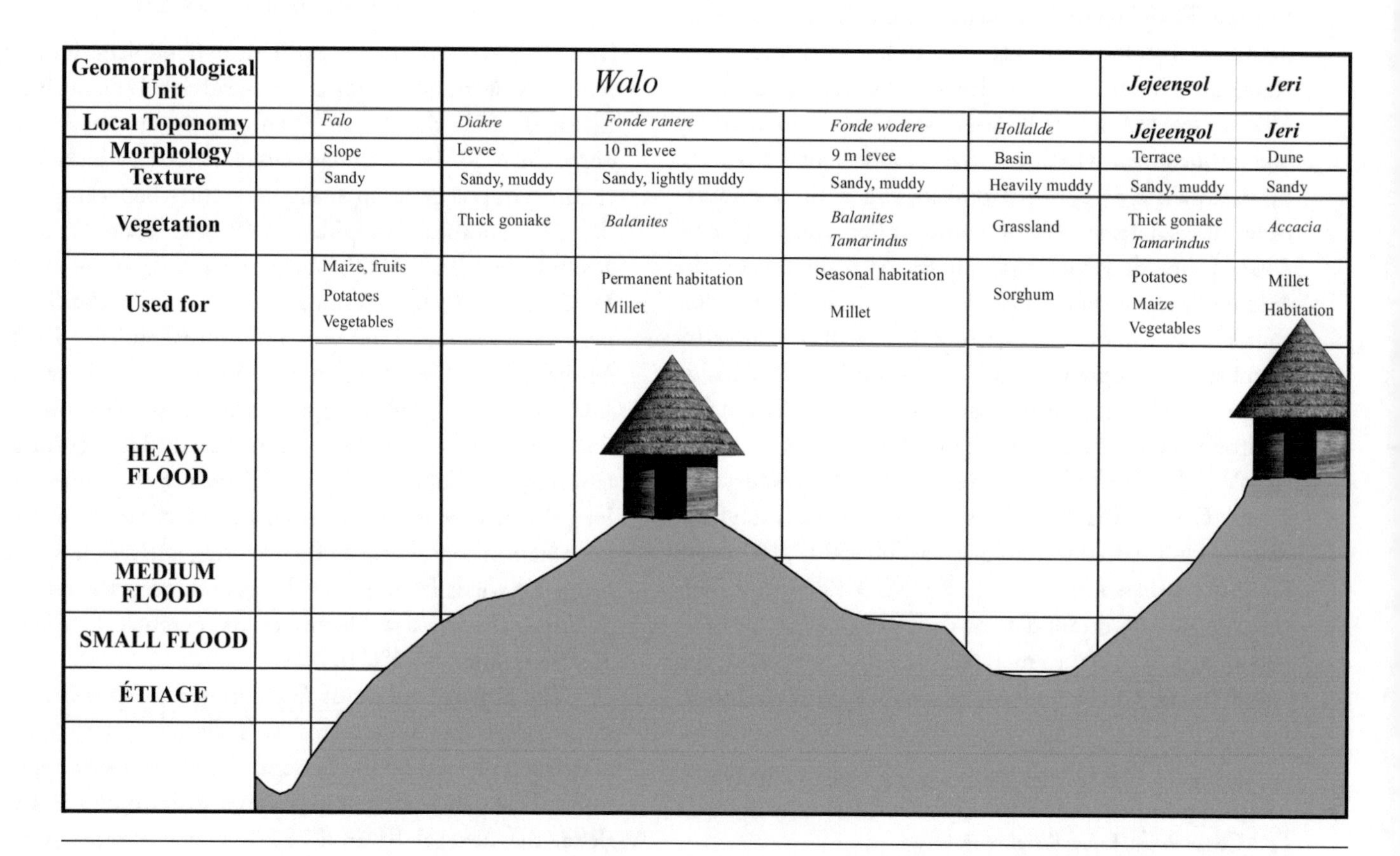

Figure 8.2 Topography and subsistence activities in the MSV (redrawn from Boutillier 1963, p. 13).

agriculturalists and fishermen, the *walo* is used for the cultivation of rain-fed maize and millet, the flood-recessional *décrue* cultivation of sorghum, the cultivation of fruits and vegetables, and for fishing activities. The *jeri* constitutes the upper lands surrounding the floodplain. Mostly occupied by pastoralists, the *jeri* is the domain of dunes and is most often used for pastoral activities, the cultivation of rain-fed millet, and as a refuge zone during high floods. The *jejeengol* is occupied by agriculturalists and agro-pastoralists. As a transition zone, the *jejeengol* is located in a strategic position that allows cattle herding and the cultivation of rain-fed millet in the *jeri* during the rainy season and the *décrue* cultivation of sorghum and rice in the *walo*.

The MSV offers two cultivation seasons a year: rain-fed and flood recession agriculture, while the surrounding areas are restricted to rainfed agriculture, which can be compromised by drought. Overall, the aridity of the Sahel makes areas of permanent water sources, such as the MSV, highly attractive to populations in search of better ecological conditions for subsistence farming, pastoral, and fishing activities.

The site of Walaldé is situated on a 9 m levée encircled by meander trains to the north and backswamp to the south. It lies in an area that likely acted as a refuge zone, especially during climatic extremes. The modern vegetation of the region is characterised primarily by *Acacia nilotica* (syn. *Acacia scorpioides*) and *Balanites aegyptiaca*, but *Acacia raddiana*, *A. seyal*, *A. senegal* and *Ziziphus mauritania* also occur (Déme 1991, p. 32). Other flora includes *Indigofera oblongifolia* and *Salvadora persica*, and grasses *Chloris prieurii*, *Vetiveria nigritana*, *and Cenchrus biflorus* (Elouard 1962, pp. 21–22).

THE ARCHAEOLOGICAL SITE OF WALALDÉ

Walaldé is a 5-hectare site with 1st millennium C.E. artefacts on its surface and deeply stratified deposits of about 4.5 m. It is among over 300 archaeological sites found in the MSV (Déme 1991; Déme and McIntosh [forthcoming]; Martin and Becker 1974, 1984). Excavations and radiocarbon dating indicate that the earliest occupation began ca. 800–550 cal B.C.E. and continued until ca. 200 cal B.C.E. The deposits straddle the Late Stone Age–Iron Age transition, and artefacts from the site corroborate this, with evidence for the use of iron and stone initially, and for iron production (smelting and forging) from 550–200 B.C.E. (Déme 1991; Déme and McIntosh 2006). The first occupants appear to have been primarily cattle herders, who also kept sheep/goat, as well as supplementing their diet through fishing and hunting. Their use of red-ochre in burials, their distinctive ceramics, and use of copper with a chemical signature of the Akjoujt mines in Mauritania distinguish them from other groups further east. The site appears to have reached its peak occupation during the early period (ca. 800–550 B.C.E.; Levels 20–25), as evidenced by high accumulation rates, and then activity fell off during Phase II (550–200 B.C.E.; Levels 1–17). The deposits in W1, a trench dug through the highest part of the mound, indicate multiple occupation and abandonment phases (Déme 2003).

ARCHAEOBOTANICAL RESEARCH AT WALALDÉ

Twenty-one litre flotation samples were collected from domestic features and inhumations in Unit W1, processed by flotation and analysed by S. Murray. Each sample was sorted under a binocular microscope at 10–40x magnification. Identification of the specimens was made by comparison with modern seed material, drawings, and photographs available at the Johann Wolfgang Goethe Universität Frankfurt am Main and with Kahlheber (2004).

Analysis of the samples revealed an assemblage typical of many Sahelian sites (Table 8.1, next page). This includes the charred parenchyma and seed remains of wild fleshy fruits, such as jujube (*Ziziphus* cf. *mauritania*), *Grewia*, *Sclerocarya birrea*, and *Celtis integrifolia*, as well as caryopses of wild grasses *Setaria* and *Paspalum* and weedy taxa such as *Trianthema portulacastrum*. Among these finds, however, were numerous grains of domesticated pearl millet (*Pennisetum glaucum*), identifiable by their diagnostic club-shape (Figure 8.3). This taxon appears to dominate the assemblage in general, but particularly in Levels 20–23, dated ca. 800–550 cal B.C.E.

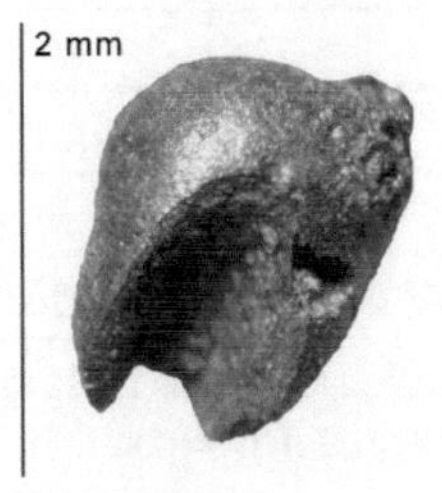

Figure 8.3 *Pennisetum glaucum*; specimen from Unit W1, Level 23 LRF 47, measuring 1.8 mm in height and 1.5 mm at greatest width (apex of caryopsis).

Table 8.1 Results of the Archaeobotanical Analysis from Waladé (*all specimens charred except *Celtis integrifolia*)

Walalde Botanical Remains		W1, Level 25, LRF 50	W1, Level 24, LRF 48	W1, Level 23, LRF 47	W1, Level 21, LRF 42	W1, Level 20 LRF 39	W1, Level 20, LRF 38	W1, Level 20, LRF 38, Pit 3	W1, Level 20, LRF 37, Hearth	W1, Level 20, LRF 34, Hearth	W1, Level 13, LRF 23, Hearth	W1, Level 7, LRF 13, Hearth	W1, Level 6, LRF 11	W1, Level 5, LRF 9, Hearth	W1, Level 4, LRF 7, Hearth 1	W1, Level 4, LRF 7, Hearth 2
Scientific Name	Part	Counts (charred only*)														
Celtis integrifolia	drupe	-	-	-	-	-	4	1	-	-	-	-	-	-	-	-
Cleome-type	seed	-	-	-	-	5	-	-	-	-	-	-	-	-	-	-
Convolvulaceae	seed	-	-	-	-	1	-	-	-	-	-	-	-	-	-	-
cf. Fabaceae	seed	-	-	-	-	1	5	-	-	-	-	-	-	-	-	-
cf. *Gisekia pharnaceoides*	seed	-	-	-	-	1	-	-	-	-	-	-	-	-	-	-
cf. *Grewia*	stone f.	-	-	-	-	-	2	-	-	-	-	-	-	-	-	-
Paniceae	cary	-	-	-	-	4	-	-	-	-	-	-	-	-	-	-
Paspalum-type	cary	-	-	-	-	1	-	-	-	-	-	-	-	-	-	-
cf. *Pennisetum*	cary	-	-	-	-	-	-	-	-	1	-	-	1	-	-	2
Pennisetum glaucum cf. *glaucum*	cary	-	-	30	-	1	1	-	4	-	-	-	-	-	-	-
Poaceae	seed m	-	-	9	-	-	-	-	-	-	-	-	-	-	-	-
cf. Poaceae	cary	-	-	-	-	-	-	-	2	-	-	-	-	-	-	-
cf. Poaceae	cary	1	-	-	-	-	-	-	-	-	-	2	-	1	-	-
cf. Sclerocarya birrea	oper	-	-	-	-	-	-	-	-	-	-	-	-	-	2	-
Setaria-type	cary	-	-	-	-	2	-	-	-	-	-	-	-	-	-	-
Solanum-type	seed	-	2	-	-	-	-	-	-	-	-	-	-	-	-	-
Solanaceae	seed	-	-	-	-	-	-	-	-	-	-	1	-	-	-	-
Trianthema portulacastrum	seed	-	-	-	-	-	-	-	-	-	-	-	-	-	1	1
Ziziphus cf. *mauritania*	stone	-	-	-	-	-	-	-	-	-	-	1	-	-	-	-
Ziziphus cf. *mauritania*	stone f.	-	-	-	-	14	-	-	-	-	-	-	-	-	1	-
Ziziphus cf. *mauritania*	seed	-	-	-	-	-	-	-	-	-	-	1	-	-	-	-
Unknown	emb	-	-	-	-	-	1	-	-	-	-	-	-	-	-	-
Unknown	fr f.	1	-	-	-	-	-	-	-	-	-	-	-	-	-	-
Unknown	nut sf.	-	-	4	4	-	-	-	-	-	-	-	-	-	-	-
Unknown	seed	-	-	-	-	-	-	-	-	2	1	-	1	-	-	-
Indeterminate	seed/sm/pt	-	-	-	-	5	-	-	-	4	-	-	-	1	-	-

Plant parts recovered key: stone f. = stone fragments, cary = caryopsis, seed m = seed mass, oper = operculi, emb = embryo, fr f. = fruit fragment, ns f. = nutshell fragment, pt = parenchymatous tissue.

Discussion

Analysis of the charred botanical remains points to use of both wild and domesticated plant resources, indicating that the earliest inhabitants of Walaldé were likely agro-pastoralists. Most of the 15-plus taxa recovered probably represent debris from food preparation activities, although it is possible that some of the charred seeds derive from the burning of dung as fuel. The majority of specimens were identified as domesticated pearl millet, indicating that the cultivation of this important grain had diffused into this region by ca. 800–550 cal B.C.E. and providing evidence that the early pastoralists at Walaldé were practicing agriculture. The location of the site suggests that people exploited the higher levees and/or surrounding dunes for cultivation of rain-fed pearl millet and the pasturing of their cattle but collected wild seeds and fruits from a variety of other environments. The absence of African rice and sorghum is notable and suggests that *décrue* cultivation in the floodplain was not yet practiced.

Acknowledgements

The Walalde research could not have been accomplished without funding from three sources: the Wenner-Gren Foundation, which provided a fellowship for graduate study at Rice; the National Science Foundation (Dissertation Improvement grant SBR 9820919); and the Bremen Stiftung für Geschichte, which provided a grant for five radiocarbon dates and other research expenses. Thanks are extended to all these foundations for their indispensable support.

References

Brooks, G. (1986) A provisional historical schema for Western Africa based on seven climatic periods (c. 9,000 BC to the 19th century). *Cahiers d'Etudes Africaines* 26, 43–62.

Déme, A. (1991) *Evolution Climatique et processus de mise en place du peuplement dans l'Ile à Morphil.* Mémoire de maîtrise. Faculté des Lettres et Sciences Humaines, Département d'Histoire, Université Cheikh Anta Diop de Dakar.

———. (2003) *Archaeological Investigations of Settlement and Emerging Complexity in the Middle Senegal Valley.* Ph.D. thesis, Department of Anthropology, Rice University, Houston.

Déme, A., and S. K. McIntosh (2006) Excavations at Walaldé: New light on the settlement of the Middle Senegal Valley by iron-using peoples. *Journal of African Archaeology* 4 (2), 317–47.

Déme, A., and R. J. McIntosh (forthcoming) Survey. In R. J. McIntosh, S. K. McIntosh, and H. Bocoum (Eds.), *Archaeological Excavations and Reconnaissance along the Middle Senegal Valley.* New Haven, CT: Yale University Publications in Anthropology.

Elouard, P. (1962) *Etude Géologique et hydrogéologique des formations sédimentaires du Guelba mauritanien et de la Vallée du Sénégal.* Mémoires du Bureau de Recherches Géologiques et Minières [France] 7. Paris: Thèse Sciences.

Goudiaby, A. (1984) *L'évolution de la pluviométrie en Sénégambie de l'origine des stations à 1983.* Mémoire de Maitrise, Université Cheikh Anta Diop.

Kahlheber, S. (2004) *Perlhirse und Baobab: Archaeobotanische Untersuchungen im Norden Burkina Fasos.* Doctoral Dissertation. Johann Wolfgang Goethe Universität Frankfurt am Main.

Lezine, A. M. (1987) *Paléo-environment végétaux d'afrique nord tropicale depuis 12000 B.P.: Analyze pollinique de séries sédimentaires continentales (Sénégal–Mauritanie).* Paris: Thèse d'Etat.

———. (1989) Late quaternary vegetation and climate of the Sahel. *Quaternary Research* 32, 317–34.

Martin, V., and C. Becker (1974) *Répertoire des sites protohistoriques du Sénégal et de la Gambie.* Kaolack: Ronéotypé.

———. (1984). *Inventaire des sites protohistoriques de la Sénégambie.* Kaolack. CNRS.

Michel, P. (1973) Les bassins des fleuves Sénégal et Gambie: Etude géomorphologique. Mémoire ORSTOM 63. Paris: ORSTOM.

Nicholson, S. E. (1994) Recent rainfall fluctuations in Africa and their relationship to past conditions over the continent. *The Holocene* 4 (2), 121–31.

Petit Maire, N. (Ed.) (1979) *Le Sahara atlantic à l'holocène: Peuplement et écologie.* Mémoires du Centre de Recherches Anthropologiques, Préhistoriques et Ethnographiques XXVIII. Alger: S. N. E. D.

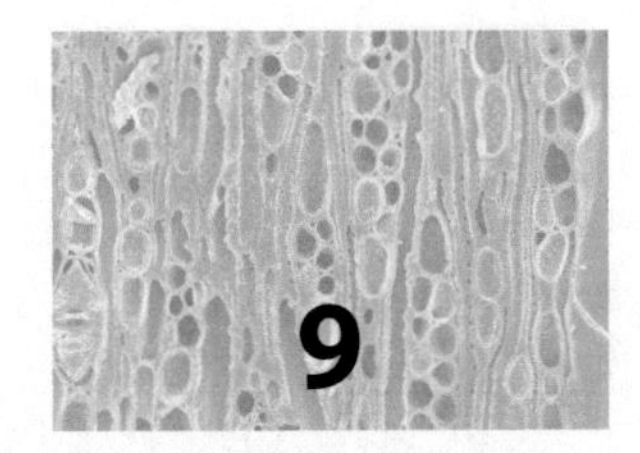

Humans and the Mangrove in Southern Nigeria

Emuobosa Akpo Orijemie and M. Adebisi Sowunmi

Palynological studies in coastal southwestern Nigeria were undertaken as part of an international project (The Dahomey Gap: Vegetation History and Archaeobotany of the Forest/Savannah Boundary in Bénin and Southwestern Nigeria; Sowunmi 2004). The Dahomey Gap is the unusual portion of the Guinean-Congolian forest zone, which, instead of forest vegetation, comprises a mosaic of the drier type lowland rain forest and savannah (Figure 9.1). Today it extends from the border between Nigeria and Bénin through Bénin and Togo to the easternmost part of Ghana. It constitutes a break in the Guinean-Congolian forest zone effectively partitioning it into western and eastern blocks.

The Gap would have formed part of the area from which the rain forest became drastically reduced, fragmented, and limited to 'refugia' during the arid phases of the Pleistocene (Sowunmi 2004). The history and cause of the Dahomey Gap are as yet not known with certainty, nor has the issue of the locations refugia been completely resolved. One of the main objectives of the project was to ascertain the late Quaternary history of the Gap using pollen analysis; work to date indicates that in southern Bénin and westernmost southwestern Nigeria it became closed in the middle Holocene, but open in the Late Holocene (Sowunmi 2004; Tossou 2002).

This chapter focuses on one unexpected outcome of the project from a palynological study of a core from Ahanve, in the Badagry area, in the westernmost part of southwestern Nigeria (Sowunmi 2004). The analysis of pollen from this core showed that in the Middle Holocene period, the coastal vegetation was dominated by *Rhizophora*; however, just before 1440–1310 cal B.C.E. (3109 ± 26 BP, KIA-17574) there was a sharp reduction of this mangrove, followed by its total disappearance. It is still absent from the Badagry area today but is present in the Lagos area, c. 60–90 km to the northeast. This abrupt and complete disappearance of the red mangrove and its continued absence from the Badagry area prompted a closer study of the distribution and history of the mangrove in southwestern Nigeria (Orijemie 2005).

The Importance of Mangroves

The Central African mangrove ecoregion subtends the Guinean-Congolian forest, extending from Nigeria through southwestern Cameroon south to the mouth of the Congo River, with outliers in Angola and small areas in Ghana (Tognetti 2001; Figure 9.1). For much of the West African coast, *Rhizophora* was highly abundant in the Early and Middle Holocene, becoming extensive during the maximum phase of the Nouakchottian transgression, ca. 3500 B.C.E. (5500 BP), declining about 2,000–2,500 years thereafter (Caratini and Giresse 1979; Sowunmi 1981; Tossou 2002).

Today, the Nigerian mangroves are the largest in this ecoregion, being most dense in the Niger Delta and covering an area of over 504,000 hectares, the third largest grove in the world. Significantly, in contrast to the luxuriant growth seen in the Niger Delta, the mangrove is scarce

Archaeology of African Plant Use by Chris J. Stevens, Sam Nixon, Mary Anne Murray, and Dorian Q Fuller, Eds., 103–112

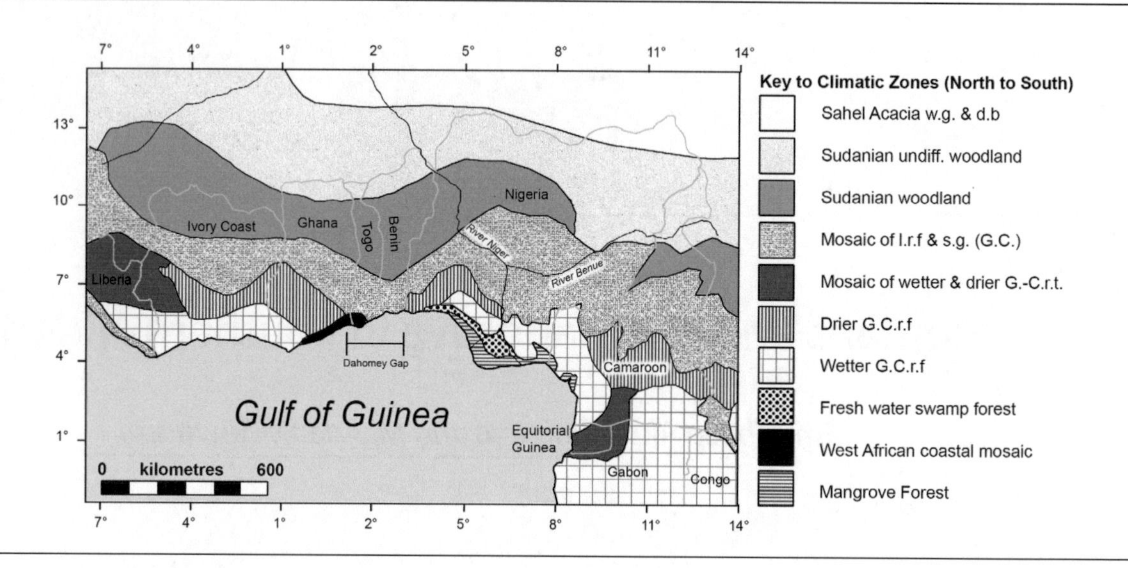

Figure 9.1 Vegetation map of West Africa (after Sowunmi 2004, Plate 2): G.-C.r.f. = Guineo-Congolian rain forest; l.r.f. & s.g. (G.-C.) = lowland rain forest and secondary grassland (Guineo-Congolian); w.g. & d.b. = wooded grassland and deciduous bushland.

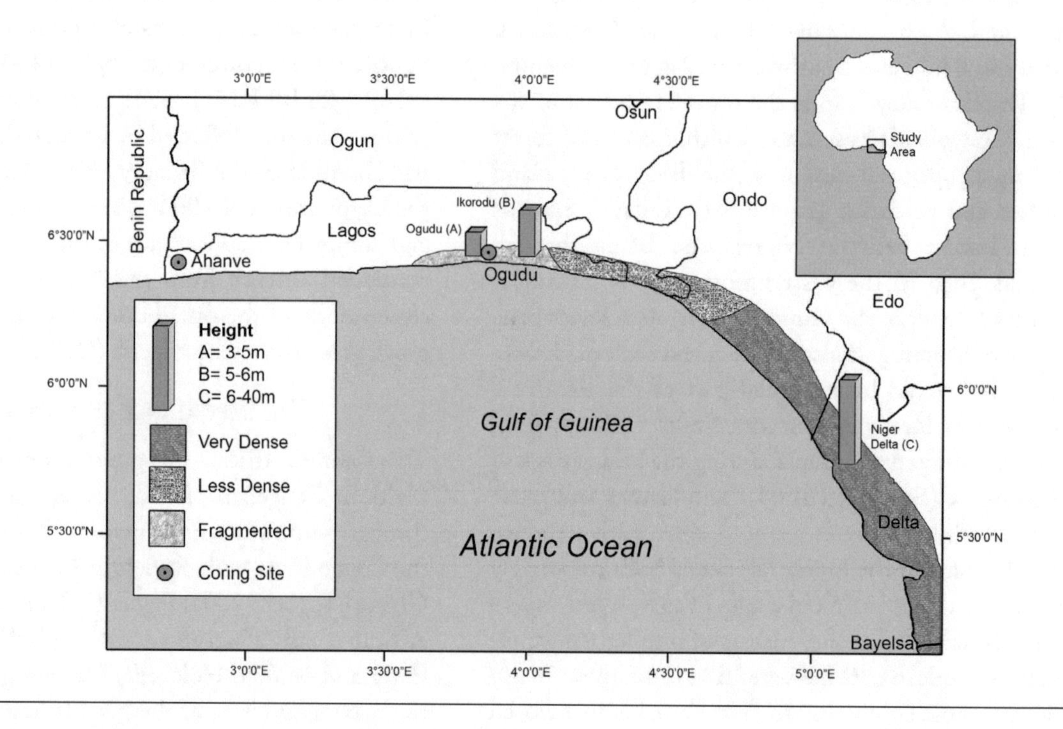

Figure 9.2 Map of study area, showing height range and density of *Rhizophora* in Southern Nigeria, vegetation survey sites, and coring sites.

in the Lagos area, the poorest in the country, and is represented only by patches of stunted trees of *Rhizophora* and the associated fern, *Acrostichum aureum* Figure 9.2).

Mangroves, the 'rain forest by the sea,' are of tremendous importance from socioeconomic, ecological, environmental, palaeoecological, and archaeobotanical perspectives. They sustain a rich variety of fauna, both aquatic and non-aquatic, and are a good spawning and breeding ground for fish, prawns, and crabs (Odum, McIvor, and Smith 1982). They have both dietary and economic significance, with the red mangrove (*Rhizophora* spp.) being a valuable resource for fuel wood, dyeing (the bark), herbal medicine, honey, salt production, and poles for house construction and boat building.

Through a process known as rhizofiltration, the mangrove forest neutralizes pollutants such as nitrogen, phosphorus, petroleum products, and halogenated compounds, which may poison and thus destroy corals (Khor 2005); root lenticels enhance microbial activities, which further break down such pollutants (Hayes-Conroy 2000). The tangled roots and density of mangrove vegetation also prevent erosion, trapping silt and halting its flow to the sea, and they provide protection against tropical storms, hurricanes, tidal waves, and tsunamis—for example, the tsunami of 2004 was notably more devastating in areas where mangroves and coral reefs had been destroyed in comparison to areas, such as Myanmar and Maldives, where they remained. Similarly, the 1960 tsunami that hit Bangladesh resulted in no loss of life, yet 31 years later, when most of the mangroves had been destroyed for shrimp farming and tourism, a tsunami of similar magnitude killed thousands (Devinder 2005).

Finally, mangroves are an important tool in palaeoecological and archaeobotanical studies, being excellent indicators of environmental changes—for example, sea level, climate, geomorphology, and vegetation. *Rhizophora* are particularly valuable indicators because of their predominance in mangrove vegetation, distinctive pollen morphology, prolific pollen production, and effective dispersal mechanism.

MATERIALS AND METHODS

Two cores were studied: the top 80 cm of an 11 m core from Ahanve (Sowunmi 2004) (6°25'58" N, 2°46'29" E) and a 200 cm core at Ogudu (Orijemie 2005), a small settlement on the outskirts of Lagos, about 60 km east of Ahanve (Figure 9.2). Surveys of the local vegetation were conducted.

Subsamples at 10 cm intervals were subjected to standard preparation treatment (60% HF, 36% HCl, 5% KOH, $ZnCl_2$/HCl [S.G. 2.0], and acetolysis). The number of pollen grains counted per subsample ranged from 102 to 6,476; charcoal counts were also made. Subsamples were analysed for colour and texture using the Munsell soil colour and grain-size analysis charts, respectively.

RESULTS: VEGETATION SURVEY

AHANVE

The cored site is a fresh-water swamp dominated by *Typha australis* and annually flooded by the Badagry Creek, which is fed by the River Yewa (Figure 9.3). The vegetation is burned annually in the dry season and at the time of coring (February, 2001) had been recently burned, the surface a thick cushion of charred vegetal material. Around its periphery are isolated clusters of the mangrove swamp fern, *Acrostichum aureum*, along with *Alchornea cordifolia* and *Mimosa pigra*, but the fern, *Cyclosorus striatus* is abundant. The site is also surrounded by *Cocos nucifera* and

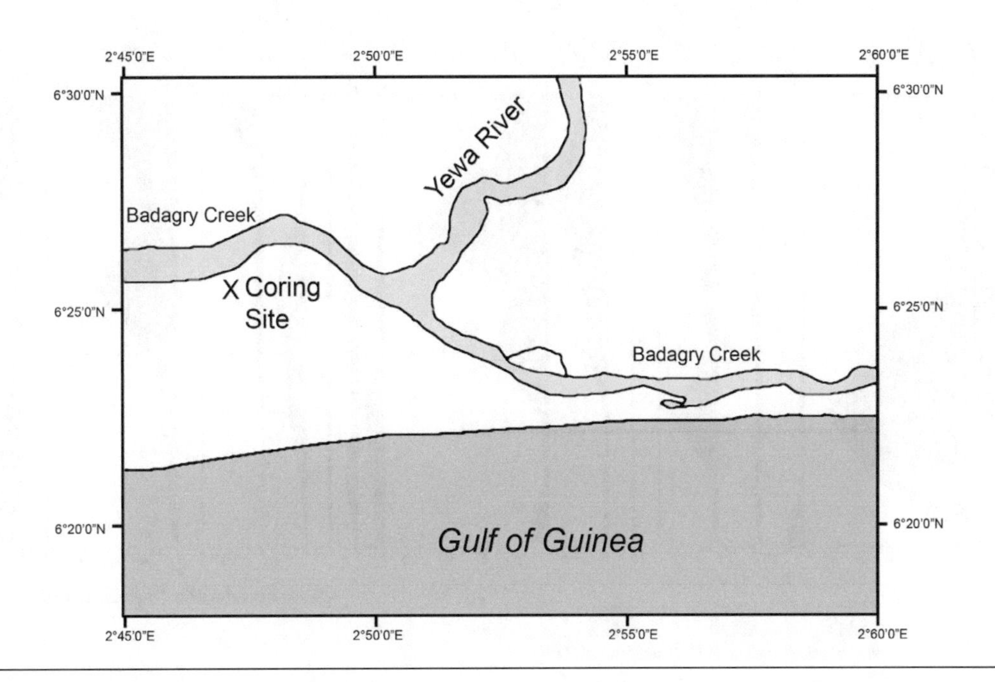

Figure 9.3 Map of Ahanve, Nigeria, coring site (after Sowunmi 2004; Figure 1); mosaic of vegetation types and scatt = mosaic of vegetation types and scattered cultivation.

Elaeis guineensis (coconut and oil palms, respectively), but *Rhizophora* is completely absent, being unknown by the local people today (neither was it there in the days of their ancestors).

Beyond this lies semideciduous and secondary forests, including *Newbouldia laevis, Pycnanthus angolensis, Alchornea cordifolia*, and *Elaeis guineensis*, the last being the most predominant; fresh-water swamp forest with *Phoenix reclinata, Raphia vinifera*, and abundant *Anthocleista, inter alia*; and herbaceous fresh-water swamp, dominated by *Thalia geniculata* (formerly *T. welwitschii* Ridl.) and the sedge *Cyperus articulatus*, while *Nymphaea lotus* occurs in pockets of open water. Farming is carried out at Ahanve using traditional West African bush-burning, cultivars including *Zea mays, Manihot utilissima, Artocarpus communis*, and *Cocos nucifera*.

Ogudu

In contrast to the Badagry area, the red mangrove is still present at Ogudu, although as stunted trees, along with patches of *Acrostichum aureum* and an abundance of grasses and sedges. Stands of cultivated bananas, *Musa* spp., impede the growth and expansion of the mangrove. The inhabitants of Ogudu had engaged in large-scale felling of the red mangrove trees, principally to use its wood as fuel, but no longer do so. However, the patchy occurrence and stunted nature of *Rhizophora* trees are indicative of the imminent disappearance of the mangrove vegetation.

Results: Pollen Analysis

Ahanve

Two pollen zones were recognised (Figure 9.4).

Zone I (80–50 cm) This zone is characterised by (1) the dominance of *Rhizophora* reaching a peak of 80% but declining toward the end of the zone; (2) the presence of other mangrove species, *Avicennia nitida* and *Acrostichum aureum*, at an initial value of 10.9% but similarly declining; (3) a high percentage of fern spores (45%) declining toward the end of the zone (5%); (4) forest species (9.5%), comprising mainly *Pycnanthus angolensis, Diospyros abyssinica*, and *Daniellia ogea*, declining toward the end of the zone (>3%); (5) initially low representation of secondary forest species, *Elaeis guineensis* (2.0%), but better representation of another major species, *Alchornea* (11.4%); (6) Poaceae initially comparatively low (6%), with fresh-water swamp forest species, such as *Spondianthus*

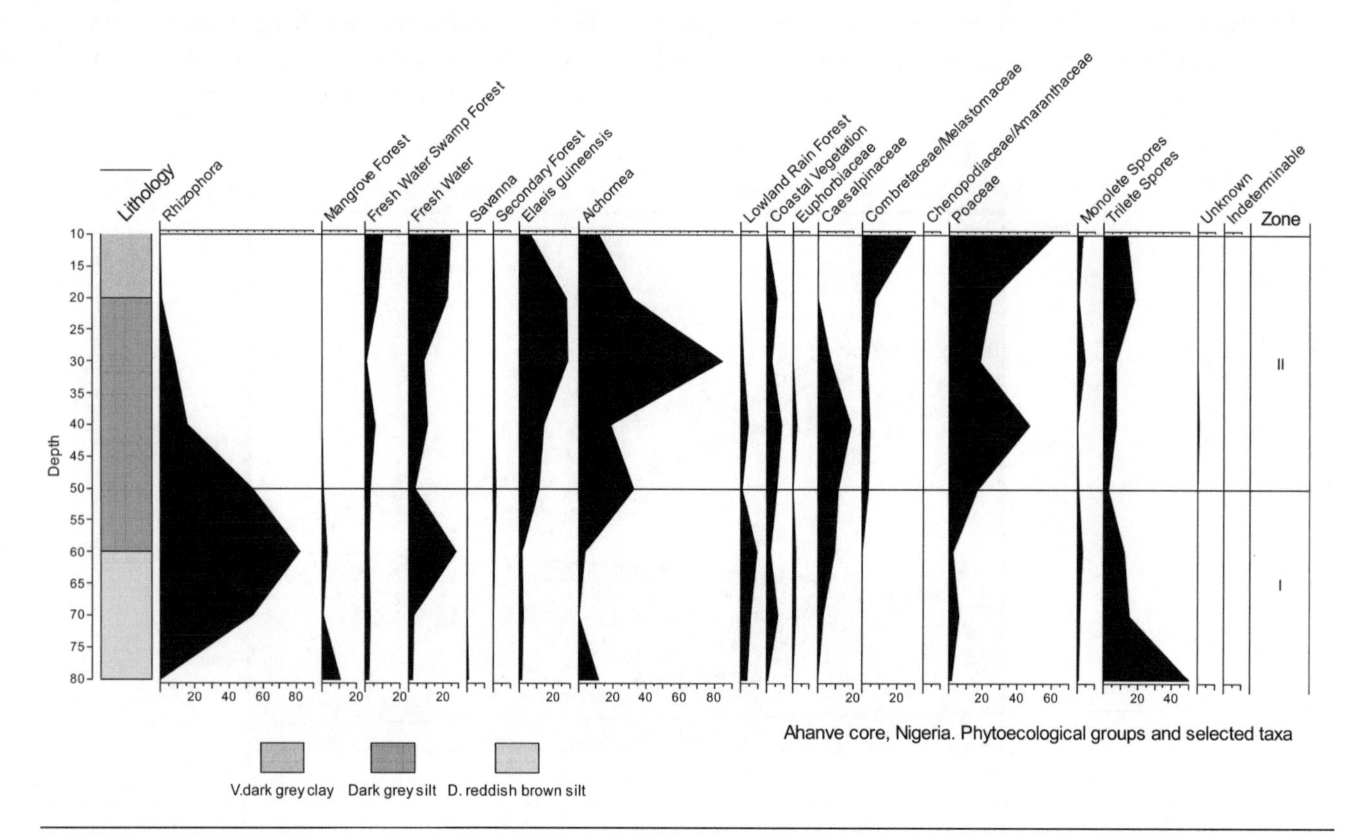

Figure 9.4 Pollen diagram from Ahanve, Nigeria.

preussii, *Nauclea* and *Uapaca* stable at ca. 6%. (7) Toward the end of the zone, with the reduction in rain forest and mangrove there were marked increases in secondary forest, notably *Elaeis* (11.6%), *Alchornea* (32.8%), and Poaceae (17%).

Zone II (50–10 cm) (1) A sharp decline followed by complete disappearance of mangrove forest, especially *Rhizophora* and *Avicennia nitida*; (2) sharp decrease of lowland rainforest species; (3) a phenomenal increase in secondary forest, especially *Elaeis* (29%) and *Alchornea* (84%); (4) similar rise in Poaceae (61.7% from an initial level of 6% in Zone I); (5), an increase in aquatic species such as *Eichornea crassipes*, *Lygodium*, and *Cyperaceae* toward the top of the zone.

OGUDU

Three pollen zones were designated (Figure 9.5).

Zone I (200–140 cm) (1) Mangroves dominate: *Rhizophora* (80%) and *Acrostichum aureum* (10.8%); (2) savannah species, such as *Phyllanthus* and *Bridelia* , are well represented, peaking at about 20%; (3) *Elaeis guineensis* (10.5%) and *Alchornea* (15.7%) are well represented; (4) lowland rain forest elements, such as *Canthium subcordatum*, *Pycnanthus angolensis*, and *Diospsyros* present (7.9%); (5) At ca. 180 cm the following developments were registered: (a) a drastic reduction in mangrove forest, *Rhizophora* declining to 19%, *Acrostichum* to 0.3%; (b) a severe reduction in *Alchornea* (0.9%) but less marked decline of *Elaeis guineensis* (6%); (c) rain forest species *Hymenostegia afzelii* type comparatively well represented (19.4%); (6) after this level *Rhizophora* and *Acrostichum aureum* recover dramatically (over 80% and 14%, respectively) to levels higher than before; (7) at the top of the zone, lowland rain forest further increases (28%).

Zone II (140–40 cm) (1) The abundance of mangrove is maintained, with slight fluctuations; (2) significant fluctuations in savannah, fresh-water swamp forest, and fresh-water swamp elements; (3) a high representation of *Elaeis* and *Alchornea*, until a decline toward the end of the zone; (4) a marked reduction in lowland rain forest species (<1%).

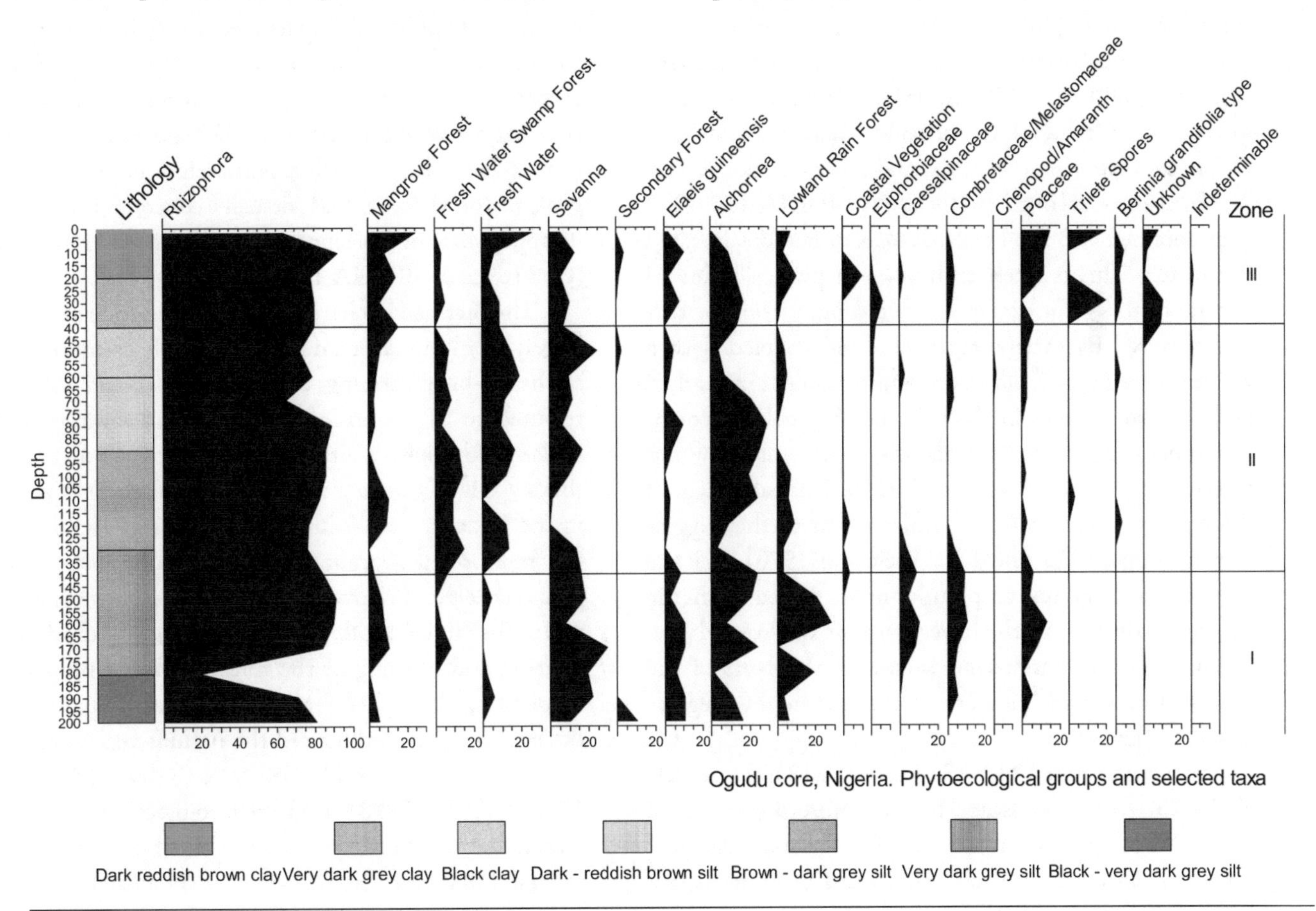

Figure 9.5 Pollen diagram from Ogudu, Nigeria.

Zone III (40–0 cm) (1) Mangrove remains dominant, reaching highest level, *Rhizophora* peaking at 92.5%, dropping to 64% in the uppermost level, with *Acrostichum aureum* at 24.6%; (2) savannah, *Elaeis* and *Alchornea* decreasing gradually, then drastically, with lowland rain forest species recovering then increasing; (3) Poaceae and other fresh-water swamp species, including *Lygodium*, increasing significantly (to 15.0% and 25.7%, respectively).

DISCUSSION

In the lowest part of the Ahanve core (80–50 cm), *Rhizophora* and two other mangrove species—*Acrostichum aureum* and *Avicennia nitida*—dominated the vegetation, along with lowland rain forest elements, including *Ouratea, Canthium, Celtis,* Moraceae complex, *Ceiba pentandra, Pycnanthus,* and *Paullinia pinnata.* The low percentages of Poaceae and secondary forest species suggest wet climatic conditions and closed forest. However, toward the end of the zone a reduction in mangrove, especially *Rhizophora,* began along with a decline in rain forest and a marked increase in secondary forest, especially *Elaeis guineensis, Alchornea,* and Poaceae, these three being components of coastal savannahs (Adejuwon 1970). There was a concomitant rise in fresh-water swamp species, especially the fern *Lygodium* (0.0% to 23.5%), along with fresh-water swamp forest (18.3% to 23.4%)—notably *Nauclea, Cynometra, Dalbergia, Piptostigma, Phoenix,* and other Arecaceae.

By 1440–1310 cal (3109 ± 26 BP, KIA-17574), a date obtained at 32–34 cm, *Rhizophora* had disappeared completely. This is likely to have been primarily due to natural factors—for example, reduced inundation by seawater caused by marine regression and coupled with a greater flooding by fresh water, which resulted in marked hydrological changes in the soil and the replacement of *Rhizophora* by coastal savannahs, fresh-water swamp forest, and swamp (Sowunmi 2004). The palynological evidence here constitutes a confirmation of the suggestion by Pugh 1953 (cited in Adejuwon 1970) that the greyish-black humus or pseudohumus mixed with the silvery, sandy soils in the lower parts of the Porto-Novo (Bénin)-Badagry beach-sand ridges are remnants of the mud swamp in which a more extensive mangrove vegetation once flourished.

Sometime after 1440–1310 cal B.C.E. (3109 ± 26 BP, KIA-17574), the landscape became more open, with a reduction in species diversity. While *Alchornea* declined there was an upsurge in oil palm, and Sowunmi (1985, 1999, 2002) has consistently maintained that its rapid spread in Ghana and Nigeria, from ca. 1800–1200 B.C.E. (3800–3200 BP) onward, is due to human action, in particular the opening up of forests for the cultivation of various crops, but not of *Elaeis guineensis* itself. This opening up of forest, which enhances the natural proliferation of the oil palm, entails burning of vegetation. In the same vein, Vincens and associates (1998) deduced that the earliest agricultural activities in the Congo region occurred around 1000 C.E. (1000 BP), based on the first appearance of *Elaeis guineensis* pollen in Lake Sinnda deposits, Southern Congo. Furthermore, Alabi (1999) highlighted both palynological and anthracological evidence from the West African Sahel and Ghana, which shows that human impact on the natural vegetation became noticeable from ca. 1000 B.C.E. (3000 BP).

Charcoal counts from Ahanve provide additional interesting results, rising phenomenally from an initial 75, through 775 and peaking at 2,525, the level just before that at which *Rhizophora* disappeared completely. At Ogudu there appears to be a delay between the disappearance of *Rhizophora* and when this disappearance is reflected in the pollen record. *Rhizophora* is known to have decreased drastically from Ogudu over the past 40 years or so, yet its pollen still registered a very high percentage (up to 92.5%) in the topmost section of the core. This discrepancy with historical accounts suggests that there is a time lag between the decline in *Rhizophora* and its reflection in the pollen record. It is therefore inferred that the peak in the charcoal was virtually concomitant with the disappearance of *Rhizophora* before 1440–1310 cal B.C.E. (ca. 3109 ± 26 BP, KIA 17574) at 32–34 cm.

The increase in charcoal is presumed to have resulted largely from human-made fires, probably connected with slash-and-burn farming. Such practices could have contributed to the extension of the coastal savannahs that replaced *Rhizophora,* although it is difficult to ascertain at this stage if *Rhizophora* was part of the fuel. At Apa, a site in the Badagry area 4 km east of Ahanve, Alabi (1999) has interpreted a ground stone axe along with charred palm kernels and charcoal found at a level above one dated to 1060–530 cal B.C.E. (2670 ± 90 BP) as indicating 'the clearing and burning of the forest probably preparatory to planting'. It is therefore possible that humans in the Badagry area also influenced the natural vegetation from sometime after 1440–1310 cal B.C.E. (3109 ± 26 BP, KIA-17574). Alabi (1998) from archaeological evidence estimated that the Badagry area must have been occupied for at least 3,000 years. Some farming is practised in the area today; however, it would be illuminating to ascertain more precisely the antiquity of both humans and farming on the basis of another independent line of evidence, and further work to this end is anticipated.

Results from Ogudu sharply contrast those of Ahanve. Unfortunately the two cores cannot confidently be correlated, since there are as yet no dates from Ogudu. There are, however, three notable differences between the patterns of occurrence of species in both cores that might shed more light on what occurred at Ahanve.

The first most outstanding difference is that *Rhizophora*, along with *Acrostichum*, remained consistently predominant throughout at Ogudu, except for a brief sharp decline at 180 cm (from 80% to 19% and 10.8% to 0.3%, respectively). Both mangrove species quickly recovered to previous levels, remaining abundant with only slight fluctuations. Indeed in one of the uppermost layers they reached their highest levels (92.5% and 14%, respectively). In contrast, after the sharp decline in *Rhizophora* at Ahanve, there was no recovery, and it disappeared completely.

Second, when the mangrove declined sharply at Ogudu both *Alchornea* and *Elaeis guineensis*, along with grasses, also decreased, but lowland rain forest representation was high. In contrast, when the mangrove declined at Ahanve, *Alchornea* and *Elaeis guineensis* registered phenomenal increases, and there was a significant increase in Poaceae along with sharp decreases of lowland rain forest species and pteridophytes. Finally, at Ogudu, the freshwater swamp species, including *Lygodium*, decreased at the time *Rhizophora* declined; but at Ahanve, in contrast, there were increases in the aquatics when *Rhizophora* declined.

These sharp contrasts in the patterns of occurrence of component species at both sites raise a number of questions: Why did *Rhizophora* recover so quickly at Ogudu but fail to recover at Ahanve? Why did coastal savannahs not develop at Ogudu when *Rhizophora* declined drastically? Why was there a reduction in aquatics at Ogudu but an increase in the same aquatics at Ahanve when *Rhizophora* declined? And could the presumably human-made fires at Ahanve have contributed to the establishment of coastal savannahs, especially the park savannah, dominated by *Elaeis guineensis* with thickets of *Alchornea cordifolia* above a ground layer of grasses?

Finally, the disappearance of *Rhizophora* at Ahanve can be attributed to a marine regression and the building up of beach-sand ridges during the late Holocene dry phase. Einsele and colleagues (1977) suggests there was a lowering of the sea level to −3.5 ± 0.5 m. along the Mauritanian coast around 2100 B.C.E. (ca. 4100 BP), and possibly one or two more oscillations between 2000 B.C.E. and 500 C.E. (4000 and 1500 BP)—that is, the Taffolian regression. Furthermore, a thick sandbar was built up on the shoreline along the estuary of the Senegal River area between 2000 B.C.E. and 200 C.E. (ca. 4000 and 1800 BP) (Faure et al. 1980). The marine regression and the building of beach-sand ridges are likely to have been regional phenomena. Hence one can surmise that the Porto-Novo (Bénin)-Badagry beach-sand ridges most probably developed during this period. This dry phase has been recorded for other parts of West Africa, (Sowunmi 2002; Tossou 2002). Do these discrepancies between Ogudu and Ahanve in their respective pollen sequences therefore mean that the effect of the late Holocene arid phase was somehow assuaged at Ogudu? And if so, how? Further comprehensive studies in these two areas are anticipated, since they might shed some light on these questions.

The results of the charcoal counts are both intriguing and illuminating, as can be seen from Table 9.1 (Ogudu charcoal count, Figure 9.6.). Except for 140 cm, which has an unusually high figure, all other counts from the lowest level studied (200 cm to 60 cm) are comparatively low

Table 9.1 Ogudu Charcoal Count

Depth (cm)	Charcoal Count
2	850
10	875
20	675
30	250
40	300
50	362
60	70
70	35
80	45
90	40
120	55
130	40
140	155
150	45
160	40
170	20
180	25
190	30
200	0

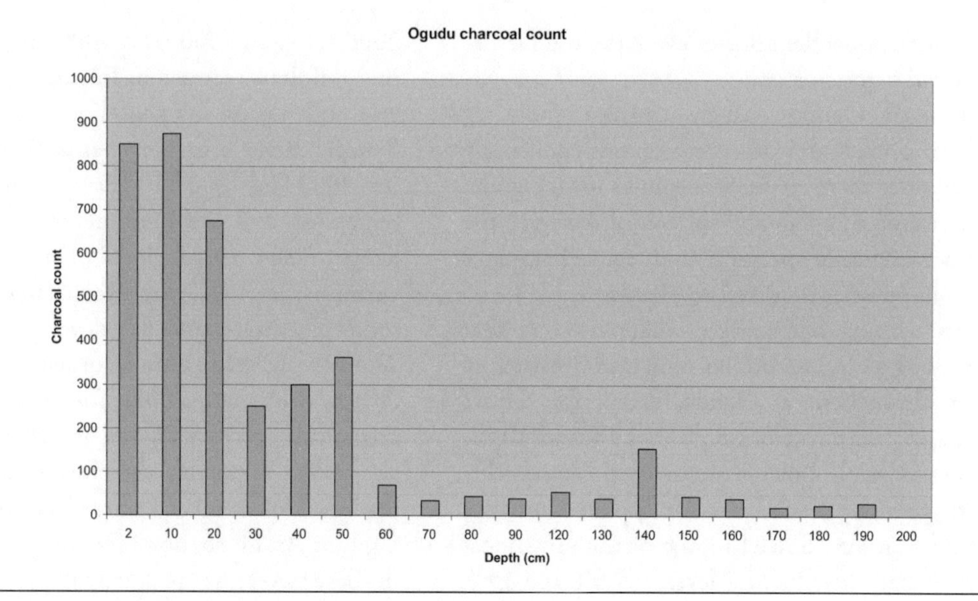

Figure 9.6 Ogudu charcoal count; age increases with depth from left to right.

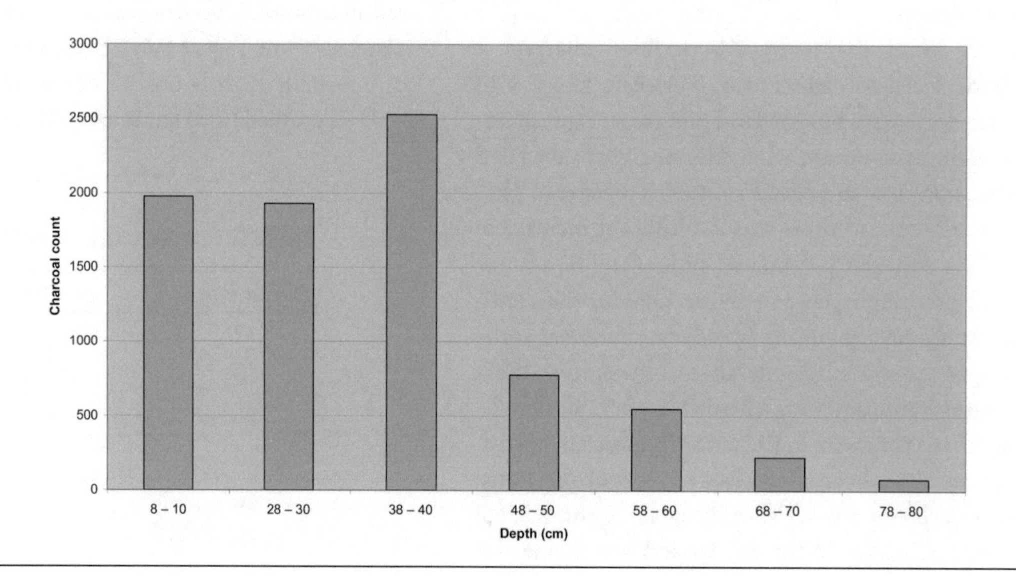

Figure 9.7 Ahanve charcoal count; age increases with depth from left to right.

Table 9.2 Ahanve Charcoal Count

Depth (cm)	Charcoal Count
8–10	1,975
28–30	1,925
38–40	2,525
48–50	775
58–60	550
68–70	225
78–80	75

(from 0 to 70). There was then a tremendous increase to 362 at 50 cm, and subsequent values were consistently high, reaching a peak of 875 at 10 cm. It can reasonably be inferred that human impact, through the making of fires, became significant from level 50 cm and continued until today. These charcoal counts seem to provide very strong support for the deduction made previously, that from about the time *Rhizophora* declined drastically at Ahanve (level 38 to 40 cm), the charcoal, which registered a marked abrupt upsurge, derived largely from human-made fires (Table 9.2; Figure 9.7).

CONCLUSION

This study has confirmed the evidence that *Rhizophora*, which had been the predominant component of the coastal vegetation of westernmost Southern Nigeria in the middle Holocene, disappeared completely by the beginning of the late Holocene and was replaced by coastal savannahs, fresh-water swamp forest, and swamp (Sowunmi 2004). In addition, the pollen study at Ogudu and the current imminent disappearance of *Rhizophora* have shed more light on the possible role of humans in the rapid expansion of these savannahs. The results of the charcoal counts for both sites support the conclusion that although 'climatic and geomorphological factors seem to be the prime causes of the late Holocene vegetation changes ... [in the Badagry area] ... palynological and archaeological evidence suggests that from sometime after 1440–1310 cal B.C.E. (ca. 3109 ± 26 BP—human action probably became an additional causative factor' (Sowunmi 2004, p. 214). Furthermore, Alabi's (1999) excavation of Apa in the Badagry area provides evidence for possible clearing and burning of the forest after 1060–530 cal B.C.E. (2670 ± 90 BP). It is possible that humans might even therefore have exacerbated the replacement of *Rhizophora* by coastal savannahs.

Much of the *Rhizophora* in the Lagos area (Ogudu, Ikorodu, and Lagos itself) has been destroyed for fuel wood and building construction, and ongoing sand dredging poses an additional serious threat to the remaining few clumps of mangrove trees. Similarly, current land reclamation for residential buildings entails the destruction of mangroves. But, although there is a much justified and commendably high level of concern and action about the destruction of the mangrove in the Niger Delta, with elaborate and comprehensive programmes aimed at its conservation by such organisations as the Nigerian Conservation Foundation (NCF), various Environmentalist Groups, and the Nigerian National Space Research and Development Agency (NASRDA), to the best of our knowledge nothing similar is being done for the mangrove in the Lagos area.

This study indicates that the mangrove in southwestern Nigeria is under severe threat of extinction. The lagoon at Ogudu should still be able to support a much denser mangrove forest. Consequently it is strongly advocated that a mangrove reforestation of the Lagos area, especially Ogudu and Ikorodu, should be undertaken immediately, to act as a buffer against possible tidal waves that could cause serious devastation to Lagos and its environs. Apart from its ecological and environmental importance, the mangrove is also a valuable socioeconomic resource. Thus urgent action is required to conserve it in the entire coastal part of southern Nigeria, its natural habitat.

ACKNOWLEDGEMENTS

The financial assistance from the British Academy, facilitated by Dr. Mary Anne Murray and Dr. Dorian Q Fuller, which enabled one of us (M. A. S.) to attend the Fifth International Workshop for African Archaeobotany, where this paper was presented, is gratefully acknowledged. Once again, thanks go to the Volkswagen-Stiftung of Germany for the generous financial grant toward the Dahomey Gap project and to Dr. Katharina Neumann of Johann Wolfgang Goethe Universität Frankfurt am Main, Germany, for her initiative and able coordination of the project. Thanks are also due Messrs. P. C. Opara and E. Nwagbara for technical assistance in both the field and the laboratory.

REFERENCES

Adejuwon, J. O. (1970) The ecological status of coastal savannas in Nigeria. *Journal of Tropical Geography 30*, 1–10.

Alabi, R. A. (1998) *An environmental archaeological study of the coastal region of southwestern Nigeria, with emphasis on the Badagry area.* Ph.D. thesis, University of Ibadan, Ibadan, Nigeria.

———. (1999) Human-environment relationship: A synthesis of ethnoarchaeological evidence of human impact on the environment of the Badagry coastal area of southwestern Nigeria. *Journal of Science Research* 5 (1), 25–31.

Caratini, C., and P. Giresse (1979) Contribution palynologique à la connaissance des environnements continentaua et marins du Congo à la fin du Quaternaire. *Comptes-Rendus Acadèmie des Sciences, Paris 288* série D, 379–82.

Devinder, S. (2005) Tsunami, mangroves and market economy. *India Together*. www.indiatogether.org/2005/jan/dsh-tsunami.html.

Einsele, G., D. Herm, and H. U. Schwartz (1977) Variation du niveau de la mer sur la plateforme continentale et la côte mauritanienne vers la fin de la glaciation du Würm et à l'Holocène. *Bulletin de liaison de l'Association sénégalaise pour l'Étude du Quaternaire africain 51*, 35–48.

Faure, H., J. C. Fontes, L. Hebrard, J. Monteillet, and P. A. Pirazzoli (1980) Geoidal change and shore-level tilt along Holocene estuaries: Sénégal river area, West Africa. *Science 210* (4468), 421–23.

Hayes-Conroy, J. (2000) Why the world needs its Mangroves: A look into the natural and human relationships with mangrove forests. *Biology 103. 2000. 3rd Web Report on Serendip,* http://serendip.brymawr.edu/biology/b103/foo/web3/havesconrovyj3.

Khor, M. (2005) Save the mangroves to fight Tsunami. *Third World Network*, www.twnside.org.sg/titlez/gtrends39.html.

Odum, W. E., C. C. McIvor, and T. J. Smith (1982) *The Ecology of the Mangroves of South Florida: A Community Profile*, FWS/OBS-81/24. Washington, D.C.: U.S. Fish and Wildlife Service/Office of Biological Services.

Orijemie, A. E. (2005) *Late Holocene vegetation history of the mangrove swamp forests in Lagos and Badagry areas, Nigeria*. M.Sc. thesis, Department of Archaeology and Anthropology, University of Ibadan, Ibadan. Nigeria.

Sowunmi, M. A. (1981) Late Quaternary environmental changes in Nigeria. *Pollen et Spores* 23 (1), 125–48.

———. (1985) The beginnings of agriculture in West Africa: Botanical evidence. *Current Anthropology* 26 (1), 127–29.

———. (1999) The significance of the oil palm (*Elaeis guineensis* Jacq.) in the Late Holocene environments of west and west central Africa: A reconsideration. *Vegetation History and Archaeobotany* 8 (3), 199–210.

Sowunmi, M. A. (2002) Environmental and human responses to climatic events in west and west central Africa during the late Holocene. In F. A. Hassan (Ed.), *Droughts, Food and Culture: Ecological Change and Food Security in Africa's Later Prehistory*, pp. 95–104. New York: Kluwer Academic/Plenum Publishers.

———. (2004) Aspects of Nigerian coastal vegetation in the Holocene: Some recent insights. In R. W. Batarbee, F. Gasse, and C. E. Stickley (Eds.), *Past Climate Variability through Europe and Africa*, pp. 199–218. Dordrecht: Springer.

Tognetti, S. (2001) *Terrestrial Ecoregions—Central African mangroves*. Washington, D.C.: World Wildlife Fund.

Tossou, M. (2002) *Recherche palynologique sur la vegetation Holocène du Sud-Bénin (Afrique de l'Ouest)*. Doctoral thesis, Université de Lome, Faculté des Sciences.

Vincens, A., D. Schwartz, J. Bertaux, H. Elenga, and C. de Namur (1998) Late Holocene climatic changes in Western Equatorial Africa inferred from pollen from Lake Sinnda, Southern Congo. *Quaternary Research* 50, 34–45.

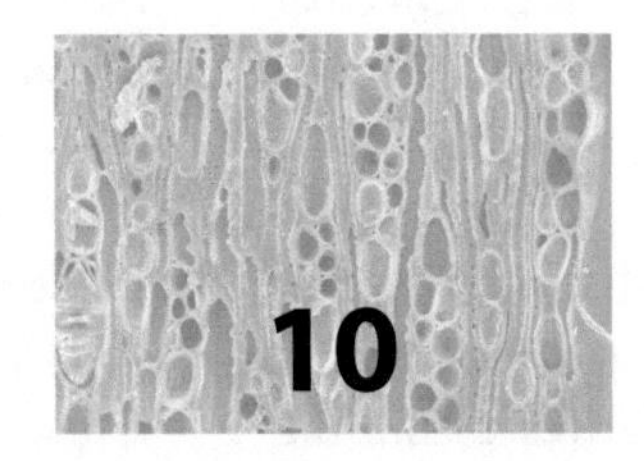

10

Plant and Land Use in Southern Cameroon 400 B.C.E.–400 C.E.

Stefanie Kahlheber, Alexa Höhn, and Katharina Neumann

During the 1st millennium B.C.E. a number of fundamental climatic, environmental, and cultural changes took place in sub-Saharan Africa. Pottery-producing societies started to settle the Central African rain forests, and iron production, a major innovation with a large impact on human culture and natural environments, became established. This appearance of pottery and iron is often associated with the immigration of Bantu-speaking populations (Wotzka 2001), which today occupy the greatest part of the equatorial rain forest zone and southern and eastern Africa (Blench 2006). Lexical reconstructions claim that the immigrants were farmers (for example, Bellwood 2002), and some historic linguists have constructed detailed pictures of farming systems and introduction routes of crops into the tropical African forests (for instance, Vansina 1990, 1994/1995). Supposedly yams configure the most likely candidates for staples in initial rain forest agriculture, since linguistic roots in northwestern Bantu languages indicate a great antiquity of yam exploitation (Philippson and Bahuchet 1994/1995; see also Bostoen this volume). By contrast, cereal cultivation was never considered as a traditional subsistence strategy of early Bantu speakers but as a late invention, adopted when they expanded to the savannahs East and South of the Central African rain forests. Lexical studies of Bostoen (2010), suggesting a greater time depth for grain crop agriculture, have changed this linguistic view only recently.

Archaeological sites providing evidence for human diet and plant exploitation in the Central African rain forest are few (Neumann 2005). The majority have not been subject to detailed archaeobotanical analyses, but delivered occasional finds of plant remains. Most common, and often associated, are endocarp remains of *Elaeis guineensis* (oil palm) and *Canarium schweinfurthii* (incense tree) (for example, Clist 2004/2005; D'Andrea, Logan, and Watson 2006; Eggert 1984; Lavachery 2001; Lavachery et al. 2005; Mbida et al. 2000; Mercader 2003; Stahl 1985). Remains of *Coula edulis* (African walnut) and *Antrocaryon klaineanum* are recorded for sites in the Ogooué Valley in northern Gabon (Oslisly 2001); those for *Coula edulis* and *Ricinodendron heudelotii* for Ale Mekudian in southwestern Cameroon (Mercader et al. 2006), and *A. micraster* and *Chytranthus macrobotrys* for Nkang in southern Cameroon (Mbida et al. 2000). All reported species are trees that occur wild. Although prehistoric arboriculture of some of these taxa has been discussed (Lavachery 2001; Mercader 2003), their domestic status remains unconfirmed. Thus far, there is only one archaeological site in Central Africa providing unequivocal evidence for cultivated plants: in Nkang, situated about 50 km to the north of the Cameroonian capital Yaoundé, banana phytoliths (*Musa* sp.) were recorded in pit sediments (pit F9) and in charred crusts adhering to ceramic sherds (pit F7NF)

Archaeology of African Plant Use by Chris J. Stevens, Sam Nixon, Mary Anne Murray, and Dorian Q Fuller, Eds., 113–128

(Mbida et al. 2000; Mbida Mindzie et al. 2001). Charcoal samples from pit F9 were radiocarbon dated to 840–370 cal B.C.E., and feature F7NF is thought to be contemporaneous (Mbida Mindzie et al. 2001, p. 5). Wild species of the genus *Musa* are not native to Central Africa (Lebrun and Storck 1995), and their characteristic phytoliths can be distinguished from those of close relatives within the family Musaceae (Mbida et al. 2005; Mbida Mindzie et al. 2001; Vrydaghs and De Langhe 2003). Therefore the phytoliths of Nkang must derive from introduced domesticated *Musa* species. Further evidence of *Musa* phytoliths from Munsa, Uganda, recovered from swamp sediments attributed to the 4th millennium B.C.E. (Lejju, Robertshaw, and Taylor 2006), remain chronologically unconfirmed and morphologically questionable (Neumann and Hildebrand 2009). Thus the isolated find from Nkang does not allow an assessment of the crop's importance in early rain forest subsistence systems.

Excavations carried out by an interdisciplinary German-Cameroonian research team in southern Cameroon unearthed archaeobotanical assemblages providing new information on agriculture and land use between 400–200 B.C.E. and 200–400 C.E. The finds modify the general assumption of a tuber crop and eventually banana-based rain forest agriculture. We present here an evaluation of fruits and seeds, charcoal and phytoliths from four archaeological sites—Bwambé-Sommet on the Atlantic coast and three more in the surroundings of Ambam further inland (Figure 10.1)—excavated between 2004 and 2006 under the direction of M. K. H. Eggert, University of Tübingen (Eggert et al. 2006).

RAIN FOREST VEGETATION AND AGRICULTURE

Today southern Cameroon is mostly covered by tropical evergreen lowland rain forests. These forests are supported by tropical rain forest climate (Köppen-Geiger climate type *Af*) at the coast and tropical monsoon climate (Köppen-Geiger climate type *Am*) in the interior (Peel, Finlayson, and McMahon 2007). Annual temperatures range between 24°C and 26°C with annual precipitation from 1,500–2,000 mm in the interior (for example, Ambam: 1,639.0 mm, 1951–1989) to 2,000–3,000 mm in the coastal region (for instance, Kribi: 2,905.1 mm, 1951–1990) (Hoare 1996–2007). Rainfall distribution is bimodal with two rainy seasons from March to June and from September to November. In the Bwambé area evergreen forests of the Atlantic Littoral and Atlantic Biafran district are growing (Figure 10.1), with coastal vegetation at the sandy shoreline (Letouzey and Fotius 1985). Near Ambam, the main vegetation type is a mosaic of mixed evergreen and semideciduous forests. In both areas the vegetation is strongly affected by humans, especially slash-and-burn practices of shifting cultivation systems, resulting in secondary formations.

Current rain forest agriculture in southern Cameroon is mainly based on plantain (*Musa* AAB group) and several plants yielding tubers, such as taro (*Colocasia esculenta*), tannia (*Xanthosoma sagittifolium*, *X. poeppigii*), sweet potato (*Ipomoea batatas*), yams (*Dioscorea* spp.), and, above all, cassava (*Manihot esculenta*). Occasionally the cereals maize (*Zea mays*) and Asian rice (*Oryza sativa*) are cultivated in small plots. Most of these crops are not indigenous to Africa, although *Musa* sp., Asian rice, and possibly also taro were distributed quite early from Asia to Africa (Matthews 2006; Mbida et al. 2000; Meertens 2006; Safo Kantanka 2004). Tannia, sweet potato, cassava, and maize are of tropical American origin, and their introduction dates back to colonial times only (Badu-Apraku and Fakorede 2006; Bohac, Dukes, and Austin 1995; Jansen and Premchand 1996; Jennings 1995). Only yams constitute true African tuber crops. Nearly 20 yam species occur naturally in the Cameroonian rain forest area, three of them being subject to a 'paracultivation' by hunter/gatherer groups (Dounias 2001). Four African yam species have been domesticated and are currently important food yams: *Dioscorea cayennensis* (yellow yam), *D. rotundata* (white yam), *D. dumetorum* (bitter yam), and *D. bulbifera* (aerial yam). The major yam cultivation area, however, is located in southern West Africa (Coursey 1976; Hahn 1995).

THE SITES

The four excavated sites are located in the south province of Cameroon. The site Bwambé-Sommet (Figures 10.1 and 10.2) is situated at the Atlantic coast near Kribi, whereas Abang Minko'o, Akonétyé, and Minyin are to be found about 160 km inland, in the vicinity of Ambam, a small administrative centre. All sites mainly consist of rather deep pit structures, having a maximum diameter of 3 m and a depth of up to 3.9 m, but they also include other structures, such as graves and ditches (Meister 2007). The radiocarbon dates currently available (Eggert et al. 2006, Table 4; Kahlheber, Boestoen, and Neumann 2009, Table 2; Meister and Eggert 2008, Tables 1 and 3) fall into the second half of the 1st millennium B.C.E. and the first half of the 1st millennium C.E. The material used for dating includes charcoal, charred fruit endocarps, and pearl millet grains. In detail, most dates from Akonétyé range between 1815 ± 21 BP (KIA-27030) and 1692 ± 29 BP (KIA-27026),

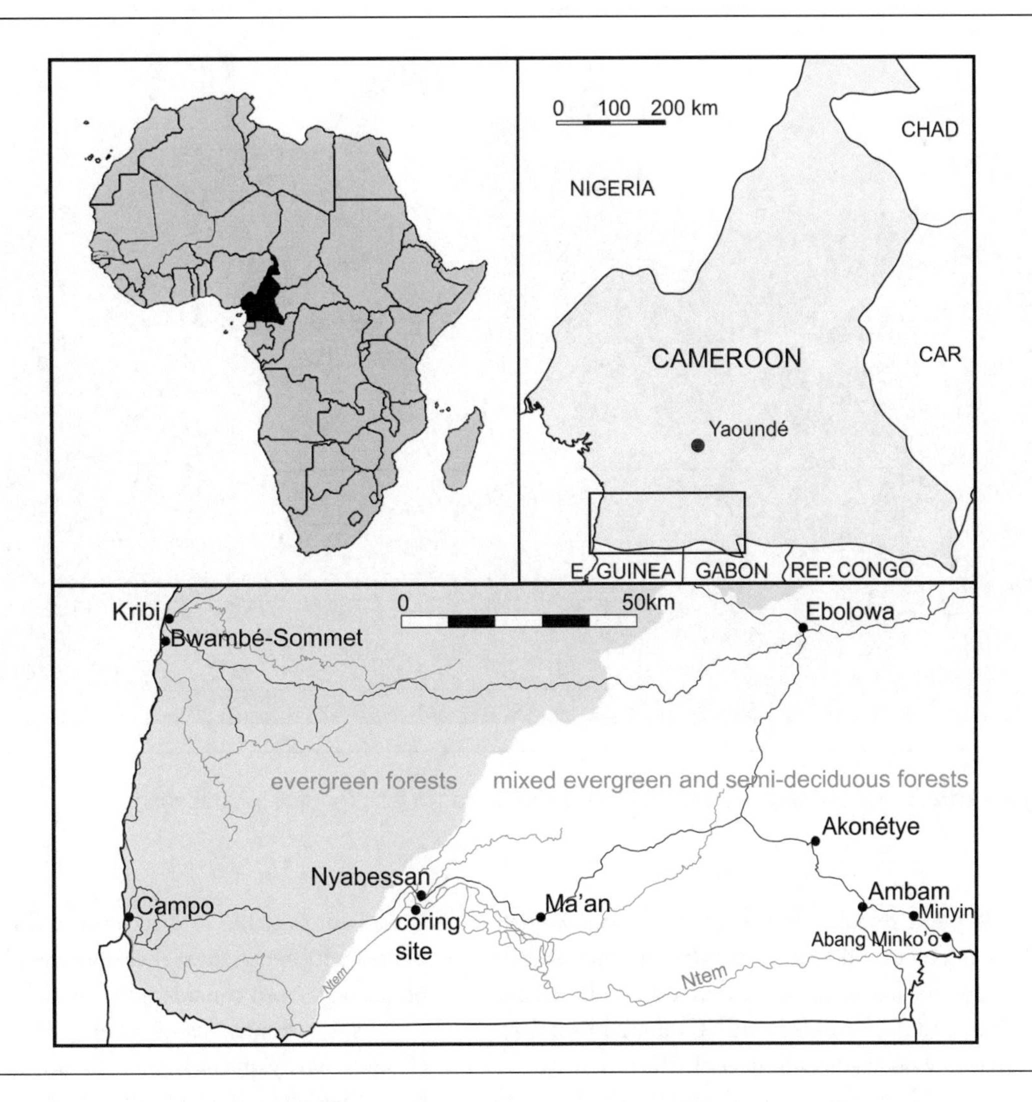

Figure 10.1 Map of southern Cameroon with localities and sites mentioned in the text.

two sigma calibrated to 130–420 C.E., with one date considered aberrant (Eggert et al. 2006, p. 285; Meister and Eggert 2008, p. 190). The cultural deposits from Minyin were dated between 50 cal B.C.E. and 390 C.E. (2001 ± 27 BP, KIA-29604 to 1739 ± 28 BP, KIA-29607). The sites Bwambé-Sommet and Abang Minko'o, with dates from 800–50 cal B.C.E. (2326 ± 86 BP, Erl-9176 to 2162 ± 60 BP, KIA-29697) and 110–410 cal B.C.E. (2286 ± 30 BP, KIA-29601 to 2181 ± 41 BP, KIA-29698), are regarded as largely contemporaneous, but calibration accuracy is partly hampered by the Hallstatt plateau. All sites are thus to be placed in the Central African Early Iron Age (Wotzka 2001), although they do not all include iron or slag. The ceramics inventories of Bwambé-Sommet and Abang Minko'o show a particular resemblance but differ from those of the younger sites, Akonétyé and Minyin (Eggert et al. 2006; Meister 2007). Akonétyé proved to be outstanding: two graves with rich ceramic and iron objects, including an axe, hoes, spearheads, a spoon, anklets, and bracelets, were excavated. They belong to the oldest graves with iron offerings in Central Africa (Meister and Eggert 2008).

METHODS

For the retrieval of botanical macroremains, bulk sediment samples of about 10 litres each were taken from all spits or artificial layers of the excavated structures. Following the excavation procedure (see Meister and Eggert 2008), cultural layers identified were sampled whenever possible. Processing included a combination of bucket flotation and wet sieving using 2.5, 1.0, and 0.5 mm meshes; large fruits, seeds, and charcoal fragments were handpicked during the excavation. Sorting took place in the Archaeobotanical Laboratory of the Goethe Universität, Frankfurt.

Figure 10.2 The site Bwambé-Sommet, surface and section of pit 1 (photos S. Kahlheber).

The charred plant remains were identified with the help of modern reference collections of African plant species as well as by the wood anatomical atlases of Normand (1950/1960) and the Frankfurt DELTA anatomical database of African woods (Neumann et al. 2001).

The phytolith study included eight samples: four samples from two pits in Bwambé-Sommet, three samples from two pits in Akonétyé, and one sample from the surrounding sediment of the pits in Akonétyé. Preparation using 5 g of sediment followed a standard procedure after Runge (1999) and Pearsall (2000). From each sample one slide was completely checked. Phytolith morphotypes were identified according to Runge (1999, 2000), Piperno (2006), Kealhofer and Piperno (1998), and Barboni and associates (1999).

Results

Fruit and Seed Remains

Up to now 85 flotation samples and 234 handpicked samples were examined from the four sites. Analyses focused on the identification of major components of human diet, leaving the identification of wild/weed taxa for further investigation. Altogether four tree taxa, two crop species, and one probable weed species (small-seeded *Vigna* sp.) are represented with charred fruits and/or seeds (Table 10.1).

It was not possible to discriminate distinct chronological phases within the sites; therefore the plant remain assemblages have been considered in total for interpretation.

As for domesticated plants, the cereal pearl millet (*Pennisetum glaucum*) was found at Abang Minko'o and Bwambé-Sommet; Bambara groundnut (*Vigna subterranea*), a pulse species, occurred exclusively in Akonétyé, whereas no crop plants were recovered from Minyin. Both crop species have been found in low quantities only: two seeds of *V. subterranea* in pit AKO 05/3 and about 40 caryopses of *P. glaucum* in several layers of the pits ABM 06/1 and ABM 06/2, BWS 04/1 and BWS 04/2. The caryopses are considerably smaller than modern *P. glaucum* (see Kahlheber, Boestoen, and Neumann 2009), but identification is quite confident: the obovate, club-shaped caryopses (Figure 10.3a) are characteristic for domesticated pearl millet, whereas all other species of the genus, including the wild variety of pearl millet (*P. violaceum*), have grains with elliptic, lanceolate, or ovate outlines. In fact, only 3 out of 13 species of the genus *Pennisetum* listed in the grass flora of Cameroon (van der Zon 1992) occur in the southern part of the country: *P. polystachion*, *P. hordeoides*, and *P. purpureum*. Neither the wild nor the cultivated variety of *P. glaucum* currently occurs in the area.

Table 10.1 Presence/Absence of Fruit and Seed Remains from South Cameroonian Sites (xx = frequent; x = present) (analysis S. Kahlheber)

Site/Feature	structure type	soil samples (studied/total)	handpicked samples (studied/total)	*Elaeis guineensis*	*Canarium schweinfurthii*	*Coula edulis*	*Raphia* sp.	*Pennisetum glaucum*	*Vigna subterranea*	*Vigna* sp., small	Poaceae, various species	herbs, various species
Abang Minko'o		13/16	33/33									
ABM 06/1	pit	6/9	15/15	xx	.	.	.	x	.	.	x	x
ABM 06/2	pit	7/7	17/17	xx	.	.	.	x	.	.	x	x
ABM 06/3	possible ditch	–	1/1	xx	.	.	.	.	.	.	.	.
Akonétyé		31/61	96/96									
AKO 05/1	pit	3/5	9/9	xx	x	.	.	.	.	.	x	x
AKO 05/2	grave	4/5	3/3	x	x	.	.	.	.	.	.	x
AKO 05/3	pit	5/10	24/24	xx	x	x	.	.	x	x	x	x
AKO 05/4	pit	4/9	18/18	xx	x	.	x	.	.	x	x	x
AKO 05/5	pit	8/17	15/15	xx	x	x	x	.	.	.	x	x
AKO 05/6	grave	4/9	16/16	xx	.	.	.	.	.	.	.	x
AKO 05/8	ditch	3/6	8/8	xx	x	.	.	.	.	.	.	x
AKO 05/9	pit	–	3/3	xx	.	.	.	.	.	.	.	.
Bwambé-Sommet		22/52	63/63									
BWS 04/1	pit	7/14	17/17	xx	.	.	.	x	.	.	x	x
BWS 04/2	pit	8/19	27/27	xx	x	.	.	x	.	.	x	x
BWS 04/3	pit	7/19	19/19	xx	x	.	.	.	.	.	x	x
Minyin		19/40	42/44									
MIY 06/1	pit	6/12	12/12	xx	x	x	x	.	.	x	x	x
MIY 06/2	pit	7/16	18/18	xx	x	.	.	.	.	.	x	x
MIY 06/3	pit	6/12	12/14	xx	x	x	x	.	.	x	x	x

Identification of *Vigna subterranea* (Figure 10.3b) was hampered by the poor preservation of the finds. The hilum, very characteristic for the species, is not preserved, but its position is recognizable. The nearly globular seed shape, however, allows excluding other crop species—such as, for instance, *Sphenostylis stenocarpa* and *Macrotyloma geocarpum*, which possess more laterally compressed seeds. The seed size of 6.5 mm (L), 4.8 mm (W), and 4.5 mm (H) is much smaller than the sizes given by Hepper (1963) both for wild (var. *spontanea*) and cultivated (var. *subterranea*) varieties. Reference material from the Frankfurt collection includes no wild varieties but cultivated West African landraces with comparatively small seeds, only about 7 mm long. Thus seed size does not allow attributing the find to the variety level. More meaningful as an argument for classification

Figure 10.3 Charred fruit and seed remains from South Cameroonian sites: (a) *Pennisetum glaucum*, caryopsis in dorsal and lateral view (BWS 04/2/I/9-10); (b) *Vigna subterranea*, seed in ventral view and insides of the two cotyledons (AKO 05/3/I/12-13); (c) *Canarium schweinfurthii*, endokarp in lateral view (AKO 05/5/I/10-11); (d) *Canarium schweinfurthii*, proximal end of endokarp (AKO 05/5/II/4); (e) *Canarium schweinfurthii*, distal end of endokarp (AKO 05/5/II/4); (f) *Elaeis guineensis*, endokarp in lateral views (AKO 05/5/6-7); (g) *Elaeis guineensis*, proximal end of endokarp (AKO 05/5/6-7); (h) *Coula edulis*, inner and outer surface of endokarp fragment (AKO 05/3/I/5-6); (i) *Raphia* sp., seed in ventral and dorsal view (AKO 05/4/II/5).

is the Sudanian distribution of the wild variety. The range of *V. subterranea* var. *spontanea* is restricted to a region south of Lake Chad (Hepper 1963; Jacques-Félix 1946; Pasquet and Fotso 1997), covering North Nigeria, North Cameroon, and the Central African Republic. In North Cameroon it seems to be widely distributed, although rare, between 7°N and 11°N (Pasquet and Fotso 1997, p. 121). Originating from beyond the distribution area of the wild variety, the Akonétyé finds have, accordingly, to be attributed to the cultivated form.

As for tree taxa, *Elaeis guineensis* (oil palm, Figure 10.3f–g) occurs in the assemblages of all four sites. *Canarium schweinfurthii*, the incense tree (Figure 10.3c–e), is much rarer but also present in three of them. *Coula edulis* (Figure 10.3h) and *Raphia* sp. (Figure 10.3i) were recorded only in Minyin and Akonétyé, which proved to be richest in taxa. The assemblages of the four sites are clearly dominated by the remains of these tree fruits. Especially in Akonétyé, endocarps and kernels of *Elaeis guineensis* make up the majority of the remains, whereas the other tree taxa are present in small quantities or as isolated finds. The large number of oil palm remains points to the use of its fruits and/or seeds for oil production. Oil palms are multipurpose trees, and all parts of the plants can be used, but palm oil extracted from the oil-bearing mesocarp and palm kernel oil gained from the seed inside the hard endocarp are regarded as the main products. The preservation of entire palm nuts in Akonétyé, however, indicates that the kernel was obviously not the main object of interest. As with *E. guineensis*, the fruits and seeds of *Canarium schweinfurthii*,

Coula edulis, and *Raphia* sp. are rich in lipids and constitute valuable resources for nutrition. *E. guineensis* and *C. schweinfurthii* grow in pioneer and secondary forests. The presence of their remains thus indicates disturbed habitats. *C. edulis* is a species of mature undisturbed forest, and *Raphia* palms grow in swamps and along rivers. It seems that particularly the people of Akonétyé used a wide range of habitats for the collection of oleaginous plant products.

Both crop species, *Vigna subterranea* and *Pennisetum glaucum*, originate from the northerly located savannah regions. They are adapted to drier environmental conditions and currently do not belong to the typical crop inventory of rain forest agriculture. Nevertheless, *V. subterranea* is capable of growing in wet areas, although humid climate brings out fungal diseases, which means that the plant needs careful handling. The lack of Bambara groundnut in modern southern Cameroonian crop inventories, as well as its decline in importance since the introduction of the South American peanut (*Arachis hypogaea*) about 400 years ago (National Research Council 2006), might be explained by cultural reasons (Pasquet and Fotso 1997). Although a pulse species, the grains of *V. subterranea* have a calorific value equal to a quality cereal (Westphal 1985, p. 303), which makes it suitable for use as a staple (National Research Council 2006).

As for *Pennisetum glaucum*, neither the wild nor the cultivated variety occurs currently in the Bwambé-Sommet and Abang Minko'o area, and pearl millet cultivation in the Central African rain forest is documented neither in oral traditions nor in early European written sources. Specimens of wild pearl millet (*P. violaceum*) have been collected in Northwest Cameroon, near the Mambila Plateau at the Nigerian border (van der Zon 1992), but the subspecies displays mainly a Sahelian pattern of distribution, whereas the cultivation of the domesticated form is confined to the drier savannah regions of Africa (Brunken 1977). The Sahelian origin of pearl millet is still visible in its physiological and ecological adaptations to a semi-arid environment, including C_4 photosynthesis, rapid development of a deep-reaching root system, a short growth cycle taking as few as 55–75 days, and precipitation needs of as little as 250 mm (Andrews and Kumar 2006). The crop copes with poor and acidic soils, but it does not tolerate waterlogging, and heavy rainfalls reduce seed set. Thus the humid climate of the rain forest region offers rather unfavourable conditions to pearl millet growth. Pearl millet is the only carbohydrate crop present in the archaeobotanical assemblages of Bwambé-Sommet and Abang Minko'o. It seems reasonable that it had been of some nutritional importance, although its role as a staple cannot be confirmed thus far. Yet the evidence of the cereal in the 1st millennium B.C.E. suggests dietary and farming conditions different from those of today.

CHARCOAL

In the charcoal samples from Akonétyé and Bwambé-Sommet, 13 arboreal taxa have been identified (Table 10.2). Owing to the lack of reference material,

Table 10.2 Presence/Absence of Charcoal Types from the Sites Bwambé-Sommet and Akonétyé (xx = frequent; x = present) (analysis A. Höhn/B. Eichhorn)

Charcoal Type	Plant Family	Bwambé-Sommet	Akonétyé
Dichostemma type	Euphorbiaceae	xx	
Malacantha/Manilkara type	Sapotaceae	xx	
Berlinia type	Caesalpiniaceae	x	
Maesobotrya/Protomegabaria type	Euphorbiaceae	x	
Pteleopsis type	Combretaceae	x	
Alchornea type	Euphorbiaceae		x
cf. *Caloncoba* type	Flacourtiaceae		x
Aucoumea/Canarium type	Anacardiaceae/Burseraceae		x
Cola sp.	Sterculiaceae		x
cf. *Coula* type	Olacaceae		x
Uapaca type	Euphorbiaceae	xx	xx
Parinari type	Chrysobalanaceae	x	xx
cf. *Lophira* type	Ochnaceae	x	x

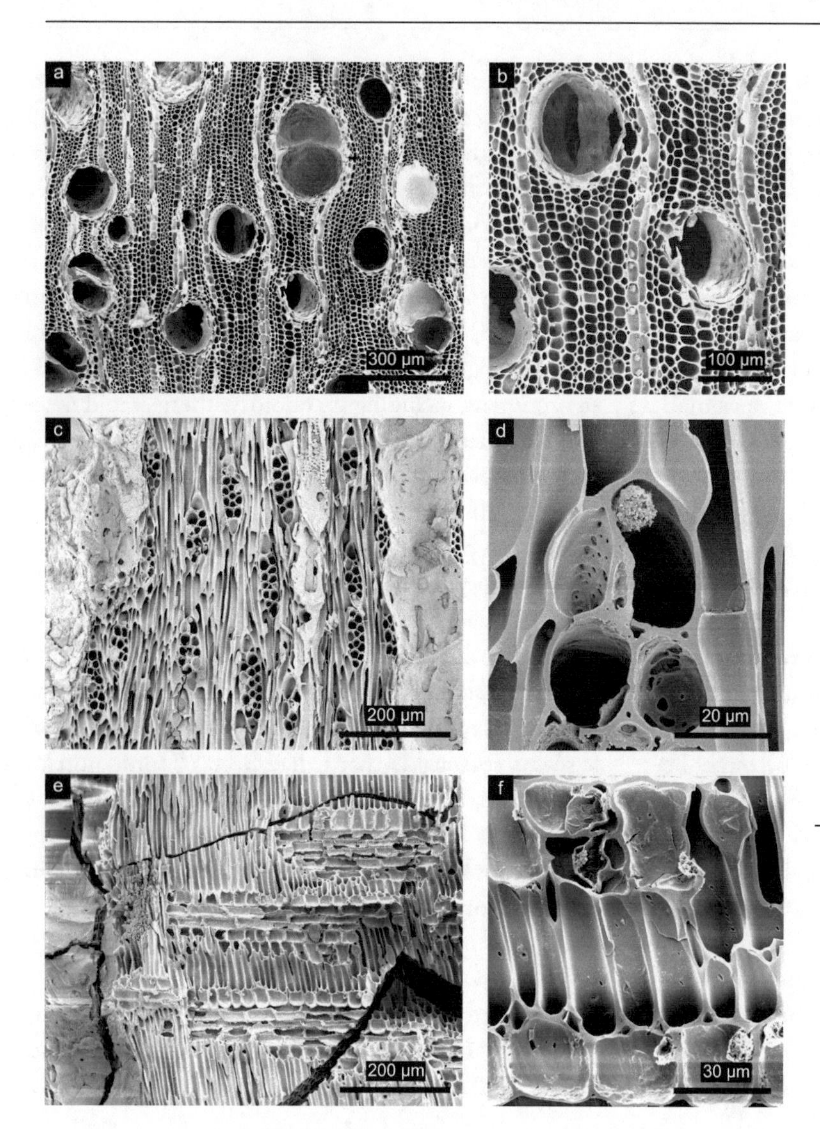

Figure 10.4 Photomicrographs of charcoal from *Aucoumea/Canarium* type (AKO 05/4/I/5-6): (a) transverse section; (b) transverse section, silica in ray cells; (c) tangential longitudinal section; (d) tangential longitudinal section, silica in upright ray cell; (e) radial longitudinal section, septate fibres; (f) radial longitudinal section, silica in marginal ray cells, crystals in chambered ray cell.

exact identification of the ecologically significant species level was rarely achieved. Yet there are clear differences between the two assemblages with taxa occurring in only one of the two sites to date; that is, charcoal of the types *Berlinia*, *Dichostemma*, *Maesobotrya/Protomegabaria*, *Malacantha/Manilkara*, and *Pteleopsis* are exclusively present in Bwambé-Sommet, whereas *Cola* sp. and the types *Alchornea*, cf. *Caloncoba*, cf. *Coula*, and *Aucoumea/Canarium* (Figure 10.4) were found only in Akonétyé. These differences are explained by the location of the sites in different forest types. In Bwambé-Sommet, taxa from the evergreen forest (*Berlinia* and *Pteleopsis* types) are more common, whereas at Akonétyé taxa typical for semi-evergreen forest (*Cola* sp., cf. *Caloncoba* and *Aucoumea/Canarium* types) dominate. Both sites include taxa of various ecological habitats ranging from pioneer communities to secondary and mature forests; for example, the types *Maesobotrya/Protomegabaria*, *Dichostemma*, and cf. *Coula* comprise shade-bearing taxa that grow in the understorey of mature rain forests, and the two nonpioneer light-demanding taxa, *Parinari* type and *Uapaca* type, point to the presence of disturbed mature forest. Further disturbed habitats are represented by pioneer taxa such as *Alchornea* type and *Malacantha/Manilkara* type. The number of fruit trees represented in the charcoal assemblages is remarkable; *Cola* sp., cf. *Coula* type, as well as *Malacantha/Manilkara* type and *Aucoumea/Canarium* type, comprise species with edible fruits and/or seeds.

The charcoal assemblages indicate that at both sites disturbed and undisturbed rain forest vegetation was present, reflecting the environment surrounding the former settlements. Consequently, fire wood was collected in more remote closed rain forest habitats as well as in the open vegetation in the immediate vicinity. No unambiguous savannah or dry forest species have been found, thus pointing to a rain forest environment. The high proportions of useful

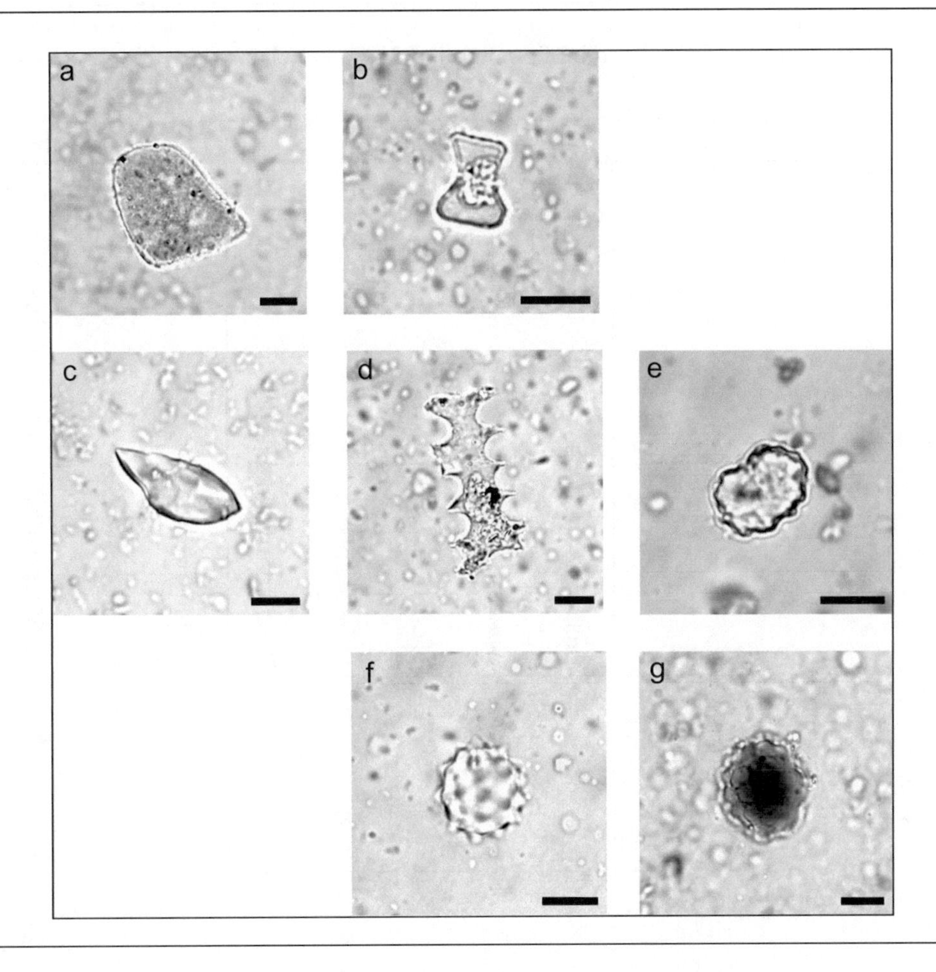

Figure 10.5 Phytoliths from Bwambé-Sommet and Akonétyé: Poaceae morphotypes: (a) bulliform; (b) bilobate; (c) point-shaped; (d) dendriform, globular phytoliths; (e) globular granulate; (f) globular echinate; (g) globular verrucate (scale bar 10 µm).

trees in the charcoal assemblages of both sites reveal that these taxa were well represented, if not overrepresented in the vegetation. This fact might suggest that the composition of mature forests was modified toward an enrichment of species valued for human subsistence.

Phytoliths

The most common morphotypes are globular echinate (B3) typical for palms (Piperno 2006; Runge 1999) and several morphotypes originating from grasses (bulliform, bilobates, point-shaped and long cells) (Figure 10.5; Table 10.3). Charred particles are equally abundant. The type globular granulate (B2), generally attributed to woody plants, is consistently present but less common. In addition, there are a number of unknown types that could not be identified owing to the lack of reference material. One characteristic morphotype was found in large quantities at Akonétyé but not in Bwambé-Sommet. It is a globular verrucate phytolith, produced in the leaves of the Marantaceae plant family (Piperno 2006, p. 72, plate 2.8d).

The composition of the seven samples taken from the pits clearly differs from that of the surrounding natural sediment. First, charred particles occur exclusively in pits, indicating that the pit fillings are likely to derive from anthropogenic activities. Second, grass phytoliths are common, demonstrating that the contents originate from open spaces, probably from the settlements themselves. The phytolith composition of the pit samples is also distinctly different from modern reference samples of rain forest soils, where morphotype B2 is usually present with percentages of 30–50%, and grasses are only weakly represented (Alexandre et al. 1997; Mercader et al. 2000). The presence of the Marantaceae phytoliths in the pits of Akonétyé indicates the extensive use of these leaves and merits further investigation. Marantaceae are very important in rain forest ecology and economy. Several species grow in thickets at the edge of the forest or in its

Table 10.3 Phytoliths from the Sites Bwambé-Sommet and Akonétyé (xxx = very frequent; xx = common; x = present) (analysis K. Neumann)

		Poaceae					sphericals					
site	sample	bulliform/fan-shaped	elongate/long cells	bilobate	point-shaped	dendriform/inflorencence	B1 spherical smooth	B2 spherical rugose	B3(B5) spherical spinulose/Arecaceae	spherical nodular/Marantaceae	B4/Burseraceae/Cucurbitaceae/Annonaceae	charred tissue
Bwambé-Sommet												
BWS 04/1, pit	PH 037	xxx	x					x	xxx			x
BWS 04/1, pit	PH 038	xx	x					xxx				
BWS 04/3, pit	PH 001	xxx	x		xxx		x	x	xxx			xxx
BWS 04/3, pit	PH 040	xxx	x		xx		x	x	xxx	cf		xxx
Akonétyé												
AKO 05/3, pit	PH 003	xxx	xxx		xx	x	xx	x	xxx	xxx		xxx
AKO 05/5, pit	PH 005	xxx		x	xx		xx		xxx	xxx		xx
AKO 05/5, pit	PH 041	xx	xx	x	x	x	xx	xx	xxx	xx		xx
AKO 05/5, sediment	PH 042	x	x					x			x	

understorey, where they play a significant role in the regeneration cycles of forest trees (Letouzey 1968; Maley 1990; White 2001). In modern times, Marantaceae leaves are used for many purposes, among them roof thatching and as wrapping material in food preparation (Figure 10.6; Burkill 1997; Dhetchuvi and Lejoly 1996).

DISCUSSION

HUMAN DIET AND PLANT USE

The archaeobotanical results of the four sites provide new information on plant use in southern Cameroon around 400–200 B.C.E. and 200–400 C.E., which seems, at least partly, to contradict the general assumption of a rain forest agriculture based on banana and tuber crops. As at other archaeological sites of the Central and West African rain forest, wild resources are well represented. The dominance of oil palm and the presence of *Canarium schweinfurthii*, *Coula edulis*, and *Raphia* sp. point to the significant role of

Figure 10.6 Snack with Marantaceae wrapper (photo S. Kahlheber).

vegetal oil and fat procurement, possibly to compensate for the scarcity of equivalent products from animals. No bones were preserved in the pits of the sites, but the archaeozoological results from Nkang near Yaoundé indicate that small livestock was present at the northern fringe of the Cameroonian rain forest by the middle of the 1st millennium B.C.E., albeit of minor importance (Mbida et al. 2000; van Neer 2000). Most probably, the oil palm fruits were harvested from wild stands, although it cannot be excluded that people protected and tended the trees and thus contributed to their dispersal. As demonstrated by the phytoliths of Marantaceae leaves found in Akonétyé, presumably a much wider range of wild plant species had been used.

The finds of *Pennisetum glaucum* from Bwambé-Sommet and Abang Minko'o and of *Vigna subterranea* from Akonétyé demonstrate that human nutrition comprised carbohydrates complementary to the oil and fat-containing fruits and seeds of wild woody plants. Other carbohydrate sources, such as the starchy tubers of yam or plantain, cannot be excluded, although the search for phytoliths from *Musa* sp. and for yam starch has not been successful so far. The presence of pearl millet and Bambara groundnut, however, indicates that farming was practiced. This cultivation of cereals and pulses is clearly not indigenous to the equatorial forest, evidently developing in the semi-arid West African savannahs. Archaeobotanical evidence as well as genetic studies suggest a west-Sahelian domestication of pearl millet (Kahlheber and Neumann 2007; Oumar et al. 2008; Tostain 1998), whereas *Vigna subterranea* originates from the region south of Lake Chad (Heller, Begemann, and Mushonga 1997; Hepper 1963). Both species are well adapted to arid conditions, and their cultivation in a rain forest, as indicated in the wood charcoal evidence, is surprising. The acquisition of these plant foods via trade is a possible explanation but seems at least for some sites unlikely; the archaeological inventories of Bwambé-Sommet and Abang Minko'o do not comprise any other item that might indicate trading contacts, and the Ambam region is rather remote and lacking geographically defined routes such as passable watercourses enabling long-distance trade.

Apart from our own archaeobotanical findings, pearl millet has been possibly identified at the site of Obobogo, a locality of greater Yaoundé. Obobogo yielded several pits containing *Elaeis guineensis* and *Canarium schweinfurthii*, and charred grains of *Pennisetum* sp. were found embedded in two pottery sherds (Claes 1985, cited in de Maret 1992, p. 245; de Maret 1994/1995, p. 321). It is not clear whether they represent caryopses of wild or of cultivated pearl millet. Currently, Obobogo is situated at the northern fringe of the forest belt in a mosaic of degraded rain forest and savannah (Eerens et al. undated), whereas the location of Bwambé-Sommet and Abang Minko'o in the rain forest offers rather difficult settings for pearl millet cultivation. Possibly, ancient and today extinct landraces of pearl millet were better adapted to the climatic conditions in humid tropical Africa. One of the four basic pearl millet races that have been distinguished by Brunken, de Wet, and Harlan (1977) principally by characteristics of the caryopsis, race *leonis*, seems to be more tolerant of high rainfall. *Leonis* millets are indigenous to Sierra Leone, and their elongated grain shape has been interpreted as a possible adaptation to high rainfall regimes. Another pearl millet type, said to be capable of coping with tropical climate, is grown in the humid areas of Ghana, Togo, and Benin (National Research Council 1996). Morphologically quite distinct from the northern pearl millet types, it was formerly treated as a separate species—*Pennisetum gambiense* Stapf and Hubb (Portères 1976). Today it is considered to belong to race globosum of *P. glaucum*. However, it is impossible to attribute the archaeobotanical grains of Bwambé-Sommet and Abang Minko'o to one of the pearl millet races. A more conclusive explanation for pearl millet cultivation is that the climate in the 1st millennium B.C.E. was much more seasonal. A longer dry season would have ensured a successful cultivation of the savannah species even under a humid climate. At the same time, dry seasons must have been short enough to sustain forest growth.

The pearl millet cultivators in Bwambé-Sommet and Abang Minko'o probably had to deal with severe ecological problems. Like all farmers they had to choose the appropriate location and schedule for cultivation and probably were forced to change their forest environment considerably to make pearl millet cultivation possible. Light sandy soils, as occurring close to the coast near Bwambé-Sommet, were especially suitable, because they prevent waterlogging. Large open spaces cleared of forest trees offered the necessary insolation. A fitting cultivation schedule would have allowed the crop to ripen in the short dry season. However, the scope of action for influencing the growing conditions by farming activities must have been restricted, and small changes in seasonality could have affected crop health and development. Changes in seasonality may explain the temporal pattern of ancient pearl millet cultivation in southern Cameroon, where pearl millet is present in the 1st millennium B.C.E. sites Bwambé-Sommet and Abang Minko'o; however, it seems to be missing in the later sites Akonétyé and Minyin dating to the beginning of the 1st millennium C.E.

Although this observation of 'negative evidence' still has to be confirmed, the number of examined sites being far from representative, it is conceivable that with less pronounced seasonality, dry seasons became too short and insufficient for grain ripening. Another explanation for the disappearance of pearl millet is that it turned out to be too inefficient, and farmers gave up cultivation in favour of better-suited and more promising crops forming the base for alternative, more productive forms of rain forest agriculture. However, there is no evidence for these crops as yet. Cultural changes might be seen as a third potential factor for the temporal pattern of pearl millet cultivation: the pearl millet farmers could have been replaced by a population with another, largely unknown crop inventory. This corresponds to the archaeological grouping of the sites based on ceramic assemblages and the presence of iron artefacts, with Bwambé-Sommet and Abang Minko'o differing considerably from the somewhat younger sites Akonétyé and Minyin (Eggert et al. 2006).

THE NATURAL ENVIRONMENT OF EARLY FARMING IN SOUTHERN CAMEROON

The results of charcoal analyses from Bwambé-Sommet and Akonétyé clearly indicate that the ancient settlements were situated in rain forests, in evergreen forests at Bwambé-Sommet, and in semi-evergreen forests at Akonétyé. These forests were, however, massively disturbed and partly substituted by pioneer plant formations, but not interspersed by savannah. Although macroremains and phytoliths from pit sediments of Bwambé-Sommet and Akonétyé point to the local existence of grasses, these presumably were present only as components of anthropogenic vegetation in the direct vicinity of the pits in which they became deposited.

The findings are supported by palaeoecological information provided by a pollen core taken near Nyabessan in the Interior Delta of the Ntem (Figure 10.1), about 110 km from Abang Minko'o and 80 km from Bwambé-Sommet, which covers the period between 1100 and 300 cal B.C.E. (Höhn et al. 2007; Ngomanda et al. 2009). In its upper part, the Nyabessan pollen diagram illustrates a distinct change in the local hydrology and a shift from wetter to drier conditions. Mature rain forests and swamp forests were massively disturbed and partly replaced by pioneer plant communities. The abrupt increase of the pioneer species *Trema orientalis* between 500 and 300 B.C.E. indicates in particular a change in annual rainfall distribution. However, grass pollen is absent, and there is no evidence that enclosed savannahs existed in the forest.

The lack of savannah taxa in the palaeoecological records contradicts the hypothesis of Maley (2004, p. 167), which favours a forest-savannah mosaic with dominating forest patches between the Atlantic coast and the Ebolowa/Ambam region during the 1st millennium B.C.E., while at the same time savannahs are supposed to have expanded in eastern Cameroon. The vegetation change in the Ntem delta has its parallels in numerous hydrological and palynological records from Atlantic Equatorial Africa (summarised by Maley 2004; Vincens et al. 1999), showing either a local opening of forest formations or their complete disappearance. Schwartz (1992) and others have postulated that the opening of the rain forest and its large-scale replacement by savannahs favoured the immigration of Bantu speaking agriculturalists into this hitherto closed environment. With the current archaeological, archaeobotanical, and palaeoenvironmental data from southern Cameroon we have to modify this view. Farming populations of the 1st millennium B.C.E. occupied a rain forest environment, which, however, was distinctly disturbed. The major factor for this rain forest disturbance is rather to be seen in a stronger seasonality than in a decrease of annual precipitation amounts.

CONCLUSION

The evidence from southern Cameroon demonstrates that some current ideas on Central African rain forest agriculture have to be revised. For more than a century linguists and historians as well as archaeologists have developed detailed pictures of farming (for example, Vansina 1994/1995, agriculture with yams, beans, and gourds) and the introduction of crops into the tropical African forests, almost without any hard material evidence for domesticated plants. Yet, the presence of pearl millet at Abang Minko'o and Bwambé-Sommet in the last centuries B.C.E. and of Bambara groundnut at Akonétyé in the first centuries C.E. strongly suggests some sort of contact between the West African savannah and the Central African rain forest, which seems to support the linguistic idea of a north-to-south movement of Bantu speakers. The archaeological record is, however, at present devoid of further material evidence to substantiate such a connection (Eggert et al. 2006). Likewise, the new archaeobotanical data do not fit into the reconstructions of chronological sequences and migration routes for grain crops proposed by some historical linguists (Vansina 1994/1995). We do not aim to ignore historical linguistics, which is an important tool for understanding African subsistence systems, and its integration with archaeological data is of great value (cf. Blench 2006; Bostoen this volume; Ehret this volume). At present, however, we see a special need to focus on archaeobotanical research in Central African sites in favour of filling the gaps of material evidence and providing more data for the modification of the existing picture of rain forest prehistory.

ACKNOWLEDGEMENTS

We would like to thank Barbara Eichhorn for her support in charcoal identification; Monika Heckner, Christoph Herbig, Jennifer Markwirth, Manfred Ruppel, and Barbara Voss for technical assistance; and Richard J. Byer for language editing. We appreciated geographical information and maps provided by Astrid Schweizer, Joachim Eisenberg, and Mark Sangen. Thanks are also due Barthélemy Tchiengué and Nolé Tsabang—during field work we benefited considerably from their vast botanical knowledge. Funding was provided by the *Deutsche Forschungsgemeinschaft* within the frame of FOR 510.

REFERENCES

Alexandre, A., J.-D. Meunier, A.-M. Lézine, A. Vincens, and D. Schwartz (1997) Phytoliths: Indicators of grassland dynamics during the late Holocene in intertropical Africa. *Palaeogeography, Palaeoclimatology, Palaeoecology 136*, 213–29.

Andrews, D. J., and K. A. Kumar (2006) Pennisetum glaucum (L.) R. Br. In M. Brink and G. Belay (Eds.), *Plant Resources of Tropical Africa, 1: Cereals and Pulses*, pp. 128–33. Wageningen: PROTA Foundation, Backhuys Publishers, CTA.

Badu-Apraku, B., and M. A. B. Fakorede (2006) Zea mays L. In M. Brink and G. Belay (Eds.), *Plant Resources of Tropical Africa, 1: Cereals and Pulses*, pp. 229–37. Wageningen: PROTA Foundation, Backhuys Publishers, CTA.

Barboni, D., R. Bonnefille, A. Alexandre, and J. D. Meunier (1999) Phytoliths as palaeoenvironmental indicators, West Side Middle Awash Valley, Ethiopia. *Palaeogeography, Palaeoclimatology, Palaeoecology 152*, 87–100.

Bellwood, P. (2002) Farmers, foragers, languages, genes: the genesis of agricultural societies. In P. Bellwood and C. Renfrew (Eds.), *Examining the Farming/Language Dispersal Hypothesis*. Oxford: Oxbow.

Blench, R. (2006) *Archaeology, Language, and the African Past*. Lanham, MD: AltaMira Press.

Bohac, J. R., P. D. Dukes, and D. F. Austin (1995) Sweet potato: Ipomoea batatas (Convolvulaceae). In J. Smartt and N. W. Simmonds, *Evolution of Crop Plants* (2nd ed.), pp. 57–62. Harlow: Longman.

Bostoen, K. (2010) Pearl millet in early Bantu speech communities in Central Africa: A reconsideration of the lexical evidence. *Afrika und Übersee 89* (2006/07), 183–213.

Brunken, J. (1977) A systematic study of Pennisetum sect: Pennisetum (Gramineae). *American Journal of Botany 64* (2), 161–76.

Brunken, J., J. M. J. de Wet, and J. R. Harlan (1977) The morphology and domestication of pearl millet. *Economic Botany 31*, 163–74.

Burkill, H. M. (1997) *The Useful Plants of West Tropical Africa* (2nd ed.), Vol. 4 (families M-R). Kew: Royal Botanic Gardens.

Claes, P. (1985) *Contribution à l'étude de céramiques anciennes des environs de Yaoundé*. M.A. thesis, Université Libre de Bruxelles.

Clist, B.-O. (2004/2005) *Des premières villages aux premiers Européens autour de l'Estuaire du Gabon: Quatre millénaires d'interactions entre l'homme et son milieu*, 2 volumes. Ph.D. thesis, Université Libre de Bruxelles.

Coursey, D. G. (1976) The origins and domestication of yams in Africa. In J. R. Harlan, J. M. de Wet, and A. B. L. Stemler (Eds.), *Origins of African Plant Domestication*, pp. 383–408. The Hague: Mouton.

D'Andrea, A. C., A. L. Logan, and D. J. Watson (2006) Oil palm and prehistoric subsistence in tropical West Africa. *Journal of African Archaeology 4* (2), 195–222.

De Maret, P. (1992) Sédentarisation, agriculture et métallurgie du Sud-Cameroun: Synthèse des recherches depuis 1978. In J.-M. Essomba (Ed.), *L'Archéologie au Cameroun: Actes du Premier Colloque International de Yaoundé (6–9 Janvier 1986)*, pp. 247–62. Paris: Karthala.

———. (1994/1995) Pits, pots and the far-west streams. In J. E. G. Sutton (Ed.), *The Growth of Farming Communities in Africa from the Equator Southwards*, Azania 29/30, pp. 318–23. Nairobi: British Institute in Eastern Africa.

Dhetchuvi, M. M., and J. Lejoly (1996) Les plantes alimentaires de la forêt dense du Zaïre, au nord-est du Parc National de la Salonga. In C. M. Hladik, A. Hladik, O. F. Linares, G. J. A. Koppert, and A. Froment (Eds.), *L'Alimentation en Forêt Tropicale: Interactions Bioculturelles et Perspectives de Développement*, pp. 301–14. Paris: Editions Unesco.

Dounias, E. (2001) The management of wild yam tubers by the Baka pygmies in southern Cameroon. *African Study Monographs 26*, 135–56.

Eerens, H., B. Deronde, J. van Rensbergen, and M. Badji (undated) *A New Vegetation Map of Central Africa: Update of the JRC-TREES Map of 1992 with SPOT_VEGETATION imagery of 1998*, www.geosuccess.net/geosuccess/documents.

Eggert, M. K. H. (1984) Imbonga und Lingonda: Zur frühesten Besiedlung des äquatorialen Regenwaldes. *Beiträge zur Allgemeinen und Vergleichenden Archäologie* 6, 247–88.

Eggert, M. K. H., A. Höhn, S. Kahlheber, C. Meister, K. Neumann, and A. Schweizer (2006) Pits, graves and grains: Archaeological and archaeobotanical research in southern Cameroon. *Journal of African Archaeology 4* (2), 273–98.

Hahn, S. K. (1995) Yams. Dioscorea spp. (Dioscoreaceae). In J. Smartt and N. W. Simmonds, *Evolution of Crop Plants* (2nd ed.), pp. 112–20. Harlow: Longman.

Heller, J., F. Begemann, and J. Mushonga (1997) *Bambara Groundnut: Vigna subterranea (L.) Verdc. Proceedings of the Workshop on Conservation and Improvement of Bambara Groundnut (Vigna subterranea [L.] Verdc.),* November 1995, Harare, Zimbabwe, Promoting the conservation and use of underutilized and neglected crops 9, Rome: IPGRI.

Hepper, F. N. (1963) Plants of the 1957–1958 expedition: II. The Bambara groundnut (*Voandzeia subterranea*) and Kersting's groundnut (*Kerstingiella geocarpa*) wild in West Africa. *Kew Bulletin 16*, 395–403.

Hoare, R. (1996–2007) *The Global Historical Climatology Network, Version 1*, www.worldclimate.com, accessed July 30, 2007.

Höhn, A., S. Kahlheber, K. Neumann, and A. Schweizer (2007) Settling the rain forest: The environment of farming communities in southern Cameroon during the first millennium BC. In J. Runge (Ed.), *Dynamics of Forest Ecosystems in Central Africa during the Holocene: Past—Present—Future, Palaeoecology of Africa 28*, pp. 29–41. London: Taylor and Francis.

Jacques-Félix, H. (1946) Remarques sur l'origine et la géocarpie du Voandzeia subterranea Thou. (Pap.). *Bulletin de la Société Botanique de France 93*, 360–62.

Jansen, P. C. M., and V. Premchand (1996) Xanthosoma Schott. In M. Flach and F. Rumawas (Eds.), *Plant Resources of South-East Asia, 9: Plants Yielding Non-Seed Carbohydrates*, pp. 159–64. Leiden: Backhuys Publishers.

Jennings, D. L. (1995) Cassava: Manihot esculenta (Euphorbiaceae). In J. Smartt and N. W. Simmonds, *Evolution of Crop Plants* (2nd ed.), pp. 128–32. Harlow: Longman.

Kahlheber, S., K. Bostoen, and K. Neumann (2009) Early plant cultivation in the Central African rain forest: First millennium BC pearl millet from South Cameroon. *Journal of African Archaeology 7* (2), 253–72.

Kahlheber, S., and K. Neumann (2007) The development of plant cultivation in semi-arid West Africa. In T. P. Denham, J. Iriarte, and L. Vrydaghs (Eds.), *Rethinking Agriculture: Archaeological and Ethnoarchaeological Perspectives*, pp. 320–46. Walnut Creek, CA: Left Coast Press.

Kealhofer, L., and D. Piperno (1998) *Opal Phytoliths in Southeast Asian Flora*, Smithsonian Contributions to *Botany* 88. Washington, D.C.: Smithsonian Institution Press.

Lavachery, P. (2001) The Holocene archaeological sequence of Shum Laka Rock Shelter (Grass fields, Western Cameroon). *African Archaeological Review 18*, 213–47.

Lavachery, P., S. MacEachern, T. Bouimon, B. G. Gouem, P. Kinyock, J. Mbairo, and O. Nkokonda (2005) Komé to Ebomé: Archaeological research for the Chad Export Project, 1999–2003. *Journal of African Archaeology 3* (2), 175–93.

Lebrun, J.-P., and A. L. Stork (1995) *Enumération des Plantes à Fleurs d'Afrique Tropicale III: Monocotylédones: Limnocharitaceae à Poaceae*. Genève: Editions des Conservatoire et Jardin botaniques.

Lejju, B. J., P. Robertshaw, and D. Taylor (2006) Africa's oldest bananas? *Journal of Archaeological Science 33*, 102–13.

Letouzey, R. (1968) *Etude Phytogeographique du Cameroun*. Paris: Editions Paul Lechevalier.

Letouzey, R., and G. Fotius (1985) *Carte Phytogéographique du Cameroun, 1: 500.000*. Toulouse: Institut de la Carte Internationale de la Végétation.

Maley, J. (1990) L'histoire récente de la forêt dense humide africaine: Essai sur le dynamisme de quelques formations forestières. In R. Lanfranchi and D. Schwartz (Eds.), *Paysages Quaternaires de l'Afrique Centrale Atlantique*, pp. 367–82. Paris: Editions de l'ORSTOM.

———. (2004) Les variations de la végétation et des paléoenvironnements du domaine forestier africain au cours du Quaternaire recent. In A.-M. Sémah and J. Renault-Miskovsky (Eds.), *L'Évolution de la Végétation Depuis Deux Millions d'Années*, pp. 143–78. Paris: Editions Artcom, Errance.

Matthews, P. J. (2006) How and when did taro become the most widespread starchy food crop in the world? In R. Torrence and H. Barton (Eds.), *Ancient Starch Research*, pp. 22–23. Walnut Creek, CA: Left Coast Press.

Mbida, C., H. Doutrelepont, L. Vrydaghs, R. Swennen, R. Swennen, H. Beeckman, E. De Langhe, and P. de Maret, (2005) The initial history of bananas in Africa: A reply to Jan Vansina, Azania 2003. *Azania 40*, 128–35.

Mbida, C. M., W. van Neer, H. Doutrelepont, and L. Vrydaghs (2000) Evidence for banana cultivation and animal husbandry during the first millennium BC in the forest of southern Cameroon. *Journal of Archaeological Science 27*, 151–62.

Mbida Mindzie, C., H. Doutrelepont, L. Vrydaghs, R. L. Swennen, H. Beeckman, E. De Langhe, and P. de Maret

(2001) First archaeological evidence of banana cultivation in central Africa during the third millennium before present. *Vegetation History and Archaeobotany 10*, 1–6.

Meertens, H. C. C. (2006) Oryza sativa L. In M. Brink and G. Belay (Eds.), *Plant Resources of Tropical Africa, 1: Cereals and Pulses*, pp. 112–20. Wageningen: PROTA Foundation, Backhuys Publishers, CTA.

Meister, C. (2007) Recent archaeological investigations in the tropical rain forest of South-West Cameroon. In J. Runge (Ed.), *Dynamics of Forest Ecosystems in Central Africa during the Holocene: Past—Present—Future, Palaeoecology of Africa 28*, pp. 43–58. London: Taylor and Francis.

Meister, C., and M. K. H. Eggert (2008) On the early Iron Age in southern Cameroon: The sites of Akonétyé. *Journal of African Archaeology* 6 (2), 183–202.

Mercader, J. (2003) Introduction: The paleolithic settlement of rain forests. In J. Mercader (Ed.), *Under the Canopy: The Archaeology of Tropical Rain Forests*, pp. 1–31. New Brunswick, NJ: Rutgers University Press.

Mercader, J., R. Marti, J. Wilkins, and K. D. Fowler (2006) The eastern periphery of the Yoruba cultural sphere. *Current Anthropology* 47 (1), 173–84.

Mercader, J., F. Runge, L. Vrydaghs, H. Doutrelepont, C. E. N. Ewango, and J. Juan-Tresseras (2000) Phytoliths from archaeological sites in the tropical forest of Ituri, Democratic Republic of Congo. *Quaternary Research* 54, 102–12.

National Research Council (1996) *Lost Crops of Africa. Vol. 1: Grains*. Washington, D.C.: National Academy Press.

———. (2006) *Lost Crops of Africa*. Vol. 2: *Vegetables*. Washington, D.C.: National Academy Press.

Neumann, K. (2005) The romance of farming: Plant cultivation and domestication in Africa. In A. B. Stahl (Ed.), *African Archaeology: A Critical Introduction*, pp. 249–75. Malden: Blackwell.

Neumann, K., and E. Hildebrand (2009) Early bananas in Africa: The state of the art. *Ethnobotanical Research and Applications* 7, 353–62.

Neumann, K., W. Schoch, P. Détienne, and F. H. Schweingruber (2001) *Woods of the Sahara and the Sahel: An Anatomical Atlas*. Bern: Paul Haupt Verlag.

Ngomanda, A., K. Neumann, A. Schweizer, and J. Maley (2009) Seasonality change and the third millennium BP rainforest crisis in Central Africa: A high resolution pollen profile from Nyabessan, southern Cameroon. *Quaternary Research 71*, 307–18.

Normand, D. (1950/1960) *Atlas des Bois de la Côte d'Ivoire*, Vols. 1–3. Nogent-sur-Marne: Centre Technique Forestier Tropical.

Oslisly, R. (2001) The history of the human settlement in the middle Ogooué Valley (Gabon): Implications for the environment. In W. Weber, L. J. T. White, A. Vedder, and L. Naughton-Treves (Eds.), *African Rain Forest Ecology and Conservation: An Interdisciplinary Perspective*, pp. 101–18. New Haven, CT: Yale University Press.

Oumar, I., C. Mariac, J.-L. Pham, and Y. Vigouroux (2008) Phylogeny and origin of pearl millet (*Pennisetum glaucum* [L.] R. Br) as revealed by microsatellite loci. *Theoretical and Applied Genetics 117*, 489–97.

Pasquet, R., and M. Fotso, (1997) The ORSTOM bambara groundnut collection. In J. Heller, F. Begemann, and J. Mushonga (Eds.), *Bambara Groundnut Vigna subterranea (L.) Verdc. Proceedings of the Workshop on Conservation and Improvement of Bambara Groundnut (Vigna subterranea [L.] Verdc.)*, November 1995, Harare, Zimbabwe, Promoting the conservation and use of underutilized and neglected crops 9, pp. 119–23. Rome: IPGRI.

Pearsall, D. (2000) *Palaeoethnobotany: A Handbook of Procedures* (2nd ed.). San Diego: Academic Press.

Peel, M. C., B. L. Finlayson, and T. A. McMahon (2007) Updated World Map of the Köppen-Geiger Climate Classification. *Hydrology and Earth Systems Sciences 11*, 1633–44.

Philippson, G., and S. Bahuchet (1994/1995) Cultivated crops and Bantu migrations in Central and Eastern Africa: A linguistic approach. In J. E. G. Sutton (Ed.), *The Growth of Farming Communities in Africa from the Equator Southwards*, Azania 29/30, pp. 103–20. Nairobi: British Institute in Eastern Africa.

Piperno, D. R. (2006) *Phytoliths: A Comprehensive Guide for Archaeologists and Palaeoecologists*. Lanham, MD: AltaMira Press.

Portères, R. (1976) African cereals: Eleusine, fonio, black fonio, teff, Brachiaria, Paspalum, Pennisetum and African rice. In J. R. Harlan, J. M. de Wet, and A. B. L. Stemler (Eds.), *Origins of African Plant Domestication*, pp. 408–52. The Hague: Mouton.

Runge, F. (1999) The opal phytolith inventory of soils in central Africa: Quantities, shapes, classification, and spectra. *Review of Palaeobotany and Palynology 107*, 23–53.

———. (2000) *Opal-Phytolithe in den Tropen Afrikas und ihre Verwendung bei der Rekonstruktion Paläoökologischer Umweltverhältnisse*. Paderborn: Books on Demand GmbH.

Safo Kantanka, O. (2004) Colocasia esculenta (L.) Schott. In G. J. H. Grubben and O. A. Denton (Eds.), *Plant Resources of Tropical Africa, 2: Vegetables*, pp. 206–11. Wageningen: PROTA Foundation, Backhuys Publishers, CTA.

Schwartz, D. (1992) Assèchement climatique vers 3000 B P et expansion Bantu en Afrique centrale atlantique: Quelques réflexions. *Bulletin de la Société Géologique de France 163*, 353–61.

Stahl, A. B. (1985) Reinvestigation of Kintampo 6 rock shelter, Ghana: Implications for the nature of cultural change. *African Archaeological Review* 3, 117–50.

Tostain, S. (1998) Le mil, une longue histoire: Hypothèses sur sa domestication et ses migrations. In M. Chastanet (Ed.), *Plantes et Paysages d'Afrique: Une Histoire à Explorer*, pp. 461–90. Paris: Karthala.

van der Zon, A. P. M. (1992) *Graminées du Cameroun. 2. Flore*, Wageningen Agricultural University Papers 92-1. Wageningen: PUDOC.

van Neer, W. (2000) Domestic animals from archaeological sites in West and Central Africa. In R. Blench and K. C. MacDonald, *The Origins and Development in African Livestock: Archaeology, Genetics, Linguistics, and Ethnography*, pp. 163–90. London: University College London Press.

Vansina, J. (1990) *Paths in the Rainforests: Toward a History of Political Tradition in Equatorial Africa*. London: James Currey.

———. (1994/1995) A slow revolution: Farming in subequatorial Africa. In J. E. G. Sutton (Ed.), *The Growth of Farming Communities in Africa from the Equator Southwards*, Azania 29/30, pp. 147–60. Nairobi: British Institute in Eastern Africa.

Vincens, A., D. Schwartz, I. Reynaud-Farrera, A. Alexandre, J. Bertaux, A. Mariotti, L. Martin, J.-D. Meunier, F. Nguetsop, M. Servant, S. Servant-Vildary, and D. Wirrmann (1999) Forest response to climate changes in Atlantic Equatorial Africa during the last 4000 years BP and inheritance on the modern landscapes. *Journal of Biogeography* 26, 879–85.

Vrydaghs, L., and E. De Langhe (2003) Phytoliths: An opportunity to rewrite history. *INIBAP Annual Report 2002*, 14–17.

Westphal, E. (1985) *Cultures Vivrières Tropicales avec Référence Spécial au Cameroun*. Wageningen: PUDOC.

White, L. J. T. (2001) Forest-savanna dynamics and the origins of Marantaceae forest in central Gabon. In W. Weber, L. J. T. White, A. Vedder, and L. Naughton-Treves (Eds.), *African Rain Forest Ecology and Conservation: An Interdisciplinary Perspective*, pp. 165–82. New Haven, CT: Yale University Press.

Wotzka, H.-P. (2001) Central African Iron Age. In P. N Peregrine and M. Ember, *Encyclopedia of Prehistory, 1: Africa*, pp. 59–76. New York: Kluwer Academic, Plenum Publishers.

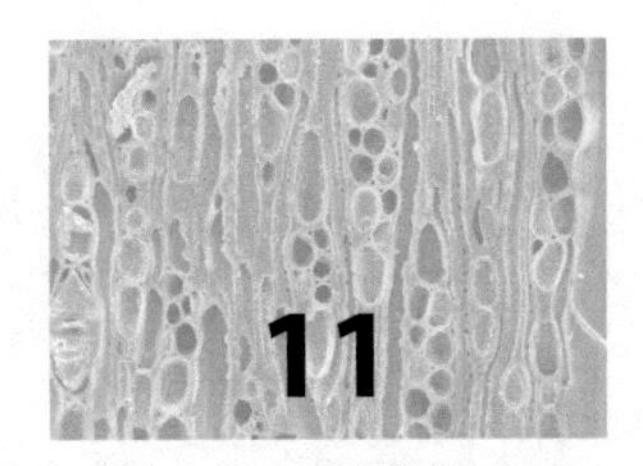

Wild Trees in the Subsistence Economy of Early Bantu Speech Communities

A Historical-Linguistic Approach

Koen Bostoen

The Bantu expansion and the spread of agriculture are often bracketed together as two distinct manifestations of the same historical macro-event (Renfrew 1992). Holden (2002, p. 793) expresses this view clearly: 'The Bantu language tree reflects the spread of farming across this part of sub-Saharan Africa between ca. 3000 B.C.E. and 500 C.E. Modern Bantu subgroups . . . , mirror the earliest farming traditions both geographically and temporally'. Several factors, however, argue in favour of a more careful approach to the problem. Linguistically, *the* Bantu language tree does not exist: genealogical trees differ according to the method used and the kind of language data considered (Bastin, Coupez, and Mann 1999). Neither a satisfactory complete internal classification of the Bantu languages nor entire agreement on the way they spread over their current distribution area has been achieved (Schadeberg 2003, p. 155). Besides, as regards the earliest farming traditions, it is increasingly acknowledged that the development of agriculture in sub-Saharan Africa was 'a slow revolution' (Vansina 1995) and that a 'dualistic concept of hunter-gathers and food producers as opposite and exclusive is not appropriate for Africa' (Neumann 2005, p. 249). If early Bantu speech communities did produce food, which plants did they cultivate, which part did these domesticates take in their nutrition, and what was their agricultural technology?

It is well known that direct archaeological evidence for early farming in West-Central Africa is scarce. Assumptions concerning the early reliance on plant food production are therefore generally founded on circumstantial evidence, including that provided by historical linguistics. However, lexical approaches to early food plant exploitation have been 'crop-centred' and have not sufficiently considered the possibility of mixed economies (Neumann 2005, p. 263). The biased search for early domesticates has led to a somewhat distorted picture of prehistoric economies. In this chapter, I assess what can be reasonably hypothesized regarding the agricultural knowledge of the earliest Bantu-speaking communities from reconstructed terms for crops and agricultural techniques. Then I consider indirect lexical evidence for the ancient exploitation, management, protection, and/or domestication of certain wild tree species.

Lexical Reconstruction and Early Agriculture: Potentialities and Shortcomings

Lexical reconstruction is capable of showing us pathways, where archaeology and archaeobotany have let us down so far. Even for food plants or agricultural tools that leave no archaeological traces, words for food plants can be reconstructed. Those lexical reconstructions provide

Archaeology of African Plant Use by Chris J. Stevens, Sam Nixon, Mary Anne Murray, and Dorian Q Fuller, Eds., 129–140

Figure 11.1 Map locating the Bantu homeland and the main Bantu subgroups according to Vansina (1995).

indirect evidence for the exploitation of these plants by the speakers of the most recent common Bantu ancestor language, provided that the meaning of these names did not change. In this respect, Proto-Bantu can be taken as an approximate point of reference in time, that is, more or less 4000 to 5000 years BP, and in space, that is, a homeland in the vicinity of the Grassfields region of Cameroon (Figure 11.1; Nurse and Philippson 2003).

The reconstruction of food crop lexicon in ancestral languages of different time depth yields information on successive changes in subsistence economy. The time depth of an ancestor language is not absolute, but relative. It depends on its position in the family tree (Figure 11.2): the higher the node, the older the ancestor language. To give one example, certain yam vocabulary reconstructible in Proto-Bantu goes even farther back in time, that is, at least to Proto-Benue-Congo, a language phase ancestral to Proto-Bantu (Figure 11.2; Maniacky 2005). The exploitation of yams can therefore be assumed to be older than the 4,000 to 5,000 years ago of Proto-Bantu. Hence, the linguistic ancestors of the present-day Bantu speech communities were in all likelihood familiar with yams before they reached the Bantu homeland. On the contrary, no names appear to be reconstructible in Proto-Bantu for any of the indigenous domesticated African cereals, such as pearl millet (*Pennisetum glaucum*), finger millet (*Eleusince coracana*), and sorghum (*Sorghum bicolor*). Terms for *pearl millet* are reconstructible, however, in Proto-East-Bantu as well as in the common ancestor language of the West-Coastal, Inner-Congo Basin and South-West Bantu languages (Figure 11.1), also known as Proto-Narrow Western Bantu (Vansina 1995). The relative time depth of these lexical reconstructions suggests that Bantu speakers acquired pearl millet only after Bantu languages started to expand over Central and Southern Africa and that this cereal was introduced independently in the eastern and western half of Bantu-speaking Africa (Bostoen 2006–2007; Kahlheber, Bostoen, and Neumann 2009; Neumann et al. 2012), in contrast to common belief that it was introduced to Bantu speakers from a single eastern centre of origin (Ehret 1974; Philippson and Bahuchet 1994–1995; Vansina 1990).

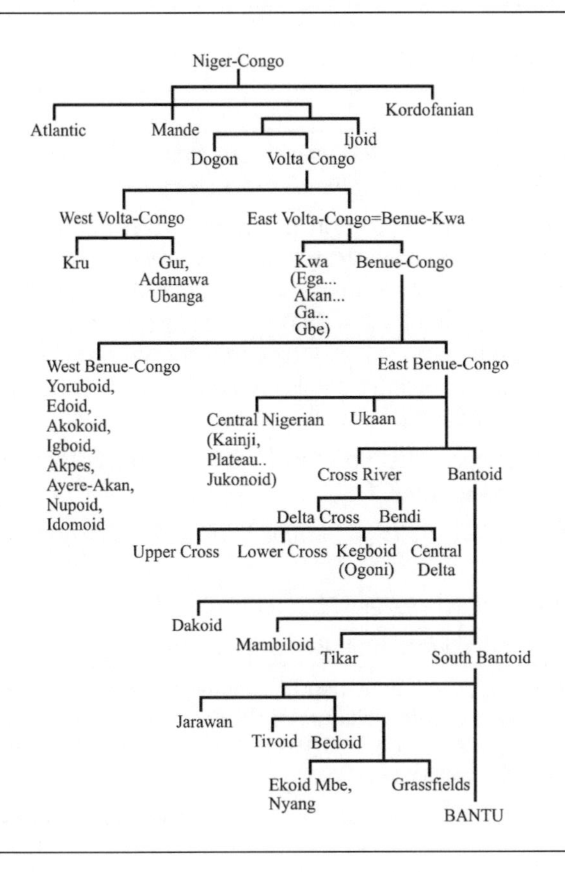

Figure 11.2 Position of different stages of evolution of the Benue-Congo language group within the greater Niger-Congo family tree, according to Williamson and Blench (2000) and adapted by Schadeberg (2003).

Nevertheless, the use of crop name reconstructions as evidence for early plant food production is not unproblematic. A major problem has to do with indigenous domesticates that were exploited in their wild form. Several lexical reconstructions have been proposed for yams (*Dioscorea* spp.), some of them reaching considerable time depths (Blench 1995; Connell 1998; Maniacky 2005; Williamson 1993). However, the earliest terms have been reconstructed in proto-languages assumedly situated in areas of western Africa, where several of the currently common domesticated yam species (*Dioscorea rotundata, Dioscorea cayennensis,* and *Dioscorea dumetorum*) might have originated. Linguistic evidence for the age of yam cultivation is thus doubtful, because the domestication of wild species did not necessarily lead to significant changes in the terminology (Blench 2006, p. 123; see Ehret, this volume). It is clear that such doubt is not justified for crop names reconstructed in proto-languages, which were supposedly spoken outside the natural range of the wild predecessor(s) of the crop concerned. Such is, for example, the case for the pearl millet vocabulary just discussed, given that the centre of origin of domesticated *Pennisetum* is situated in the south-central Sahara (Neumann 2003; see Manning and Fuller, this volume). A problem closely related to the distinction between names for wild and domesticated species is the impossibility to link a precise species to a reconstructed name. The same name may refer to a different species in different languages, and, vice versa, the same species may be designated by different names in different languages (Maniacky 2005). So even if one can assume that yams were cultivated in early Bantu speech communities, it is difficult to retrace with much certainty the precise species of yam domesticates that were cultivated.

As lexical evidence for early agriculture, the reconstruction of vocabulary related to farming tools and techniques seems less ambiguous, although not entirely unproblematic, as I discuss in the next section. There are lexical indications for utensils generated from food plants, such as containers made from *Lagenaria* fruits (Bulkens 1999a), and tools used to exploit food plants, such as grindstones, pestles, and mortars (Bulkens 1999b), but these do not unequivocally point toward plant cultivation. Proto-Bantu verbs meaning 'to cultivate' or 'to clear for cultivation' may suggest some form of agricultural development, but the vocabulary necessary to form a more precise picture of the farming traditions in question is generally lacking. The absence of such specialised vocabulary in Proto-Bantu suggests that the agricultural technology was not highly developed yet at that time, but the reason could also be that specialized technology does not necessarily imply specialized vocabulary (Bostoen 2004, 2005a). More comparative lexical research on agricultural technology is certainly needed (see Blench, this volume), but firm results are not guaranteed.

ASSESSING PROTO-BANTU LEXICAL EVIDENCE FOR EARLY FOOD CROP CULTIVATION

Lexical reconstructions that may indicate the existence of plant cultivation in the Proto-Bantu era include the names of domesticates (Table 11.1). Apart from the two *Vigna* species, Proto-Bantu reconstructions for domesticates are restricted to yam species. Although the great number of reconstructible yam names into Proto-Bantu does not provide proof for plant cultivation as such, it certainly suggests that the exploitation of different *Dioscorea* species was highly important. What is more, none of these reconstructions are Bantu innovations. All Proto-Bantu yam terms were inherited from an older language stage,

Table 11.1 Proto-Bantu Vocabulary for Currently Domesticated Food Plants

Rec.	Plant	Ref.	TD	Comments
*-kʊ̀á	*Dioscorea* spp.	M; P&B	PEBC	(V: 289): probably *Dioscorea cayennensis*, but difficult to confirm with certainty; td based on cognate in PLC **-gʷá, Dioscorea dumetorum* (C)
*-bàdá	Dioscorea spp.	M	PBC	Supposed td based on cognates in Igboid (WBC) (W); throughout Bantu, this stem is very regularly associated with *Dioscorea alata*. This is remarkable given the species' presumed Asiatic origin. Either the African exploitation of this tuber is older than generally assumed (Blench 1994), or the noun used to designate an indigenous species strongly resembling the *Dioscorea alata*, inducing a parallel semantic shift in the speech communities that adopted the newly arrived crop.
*-kódò	*Dioscorea* spp.	M	PEBC	td based on PEBC reconstruction **-kod*, 'yam' (DW) and PCD cognate **-kòlò 'Colocasia esculenta'* (C)
*-dɪ̀gà	*Dioscorea* spp.	M	PBC	td based on **-dìà* 'yam (general term)' in Edoid (WBC) (W) and **jíén* 'yam' in PDC (C)
*-kʊ̀ndè	*Vigna unguiculata*	P&B	PEBC	td based on cognates in several EBC languages (Bl)
*-jʊ̀gʊ́	*Vigna subterranea*	P&B	PB	td based on lack of correspondences with BC forms in (Bl)

Abbreviations: rec. = lexical reconstruction; ref. = reference; td = supposed time depth; Language stages: PB = Proto-Bantu; (P)BC = (Proto)Benue-Congo; PCD = Proto-Central-Delta; PDC = Proto-Delta-Cross; PEBC = Proto-East-Benue-Congo; PLC = Proto-Lower-Cross; WBC = West-Benue-Congo. Sources: Bl = Blench (2006); C = Connell (1998); DW = De Wolf (1969); M = Maniacky (2005); P&B = Philippson and Bahuchet (1994–1995); V = Vansina (1990); W = Williamson (1993)

The system used here to note vowels in Bantu reconstructions is as follows *i* ɪ *e a o* ʊ *u*. As regards tone, ` represents low tone, and ´ stands for high tone.

ranging between at least Proto-East-Benue-Congo and Proto-Benue-Congo. The time depth of this yam vocabulary strongly suggests that several yam species already belonged to the diet of Benue-Congo speakers before the latter reached the Bantu homeland (Blench 2006; Maniacky 2005).

As stated, the reconstruction of terms for farming tools and techniques provides firmer lexical evidence for early agriculture (see also, Blench this volume). For the time being, little such vocabulary has been reconstructed in Proto-Bantu. Such verbs as **-dɪ̀m-*, 'to cultivate (especially with hoe)', **-tém-*, 'to cut; cut down; clear for cultivation', and a noun such as **-gʊ̀ndà*, 'garden', unmistakably suggest the existence of certain forms of food production before the first Bantu language dispersals (Bastin et al. 2002; Dalby 1976; Vansina 1990). Nevertheless, the absence of unquestionable Proto-Bantu reconstructions for typical farming practices, such as planting, sowing, making mounds, weeding, and harvesting, or for typical farming utensils, such as the hoe, the digging stick, and the bush-knife, allows for debate on the state of agricultural advancement at that time (Jacquot 1991).

In sum, the state of lexical reconstructions indicates the existence of farming activities at the time of Proto-Bantu, including cultivation of yams and pulses. Nonetheless, this vocabulary is so far certainly not developed to an extent that the cultivation of food crops can be claimed to have been the main subsistence strategy

PROTO-BANTU LEXICAL EVIDENCE FOR THE EXPLOITATION OF WILD TREES

Because of the agricultural-Bantu-language-expansion hypothesis bias, little attention has been given to the role of wild food plants in the subsistence economies of early Bantu speech communities. Nonetheless, ethnobotanical studies have demonstrated that wild-plant exploitation constitutes a considerable part of all African subsistence economies (Neumann 2003, p. 83). Lexically, the existence of widespread Bantu words, such as **-ká*, 'to gather fruit', **-píp-*, 'to suck; extract juice', and **-kómb-*, 'scrape; dig; lick (food) with finger' (Bastin et al. 2002), suggests that more systematic comparative lexical research may yield knowledge on early food-gathering practices. For the time being, the most clear-cut lexical evidence for the exploitation of wild food plants comes from the reconstruction of names for the plants themselves. In this section, I concentrate on the comparative vocabulary related to five common wild or semidomesticated trees. Such a selection is necessary not only because 'the compilation and analysis of vernacular names for trees, with over ten thousand species in sub-Saharan Africa, remains a daunting task' (Blench 2006, p. 115); there is also a more trivial reason. Lexical sources are very often poorly documented as regards wild food plants, and they do not always offer good scientific identifications. Botanical sources, on the contrary, are rich in plant species but often poor in local names and language identifications.

ELAEIS GUINEENSIS JACQ.—OIL PALM (ARECAEAE)

The oil palm's early importance in the subsistence economies of west and west-central Africa is a well-known fact. Its carbonized endocarp is abundantly present in several Late Stone Age (LSA) sites from about 5000 BP onward. These vegetal remains are often found in association with other indicators of plant food-processing: pounding/grinding equipment, polished stone tools, such as axes and hoes, substantial quantities of pottery, and moreover, the hard pericarp of another oleaginous plant, that is, the *Canarium schweinfurthii* Engl. (D'Andrea, Logan, and Watson 2006; de Maret 1995; Lavachery 2001).

This high and early archaeological visibility is well-reflected linguistically. Two distinct names for the oil palm can be reconstructed into Proto-Bantu—that is, **-bídà* being the generic name—and **-téndé*, more precisely referring to a young oil palm tree (Bostoen 2005b). From a cultural historical point of view, the reconstruction of these nouns into Proto-Bantu suggests that the ancestral Bantu speech communities were familiar with the oil palm. What is more, both nouns go back to at least Proto-Benue-Congo. Proto-Bantu inherited them thus from this or even from an earlier parent language, pointing out that ancestral pre-Bantu speech communities must have known this tree long before the first Bantu language dispersals in the Central African rain forest.

Such is not true for a third oil-palm-related noun reconstructible into Proto-Bantu, that is, **-gàdí* 'palm oil'. This noun for a derived oil palm product has no cognates in other non-Bantu Benue-Congo languages. Thus intensification of oil palm exploitation probably started after the break-up of Proto-Benue-Congo and happened independently in different areas settled by different Benue-Congo speech communities. Supporting this conclusion is the fact that, in the Proto-Delta-Cross subgroup, none of the reconstructed oil-palm-related vocabulary, apart from the noun for the tree itself, is shared with other groups (Connell 1998).

CANARIUM SCHWEINFURTHII ENGL.—'BUSH-CANDLE', 'AFRICAN OLIVE', 'SAFOUTIER SAUVAGE' (BURSERACEAE)

The *Canarium schweinfurthii* tree is distributed throughout tropical Africa in rain forest, gallery forest, and transitional forest from Senegal to Cameroon and extending to Ethiopia, Tanzania, and Angola (Burkill 1985, p. 302). As mentioned, its pericarp is in archaeological sites often found together with carbonized oil palm husks, suggesting that both plants belonged to the same early food complexes. The outer pulp of the fruit is edible and oily. The resin burns easily and is therefore used as a bush candle. It also has medicinal, ritual, and craft-related functions (Carrière 1999; Koni Muluwa 2010; Koni Muluwa and Bostoen 2008; Ngila Bompeti 2000). Compared to the oil palm, however, the *C. schweinfurthii* currently has a lower economic importance and has been less documented.

Vocabulary related to the *C. schweinfurthii* is much less numerous and varied, as compared to the oil palm. The lexical information pertaining to the *C. schweinfurthii* is mostly limited to the tree name, sometimes extended with the fruit name (generally the same noun stem), and only very occasionally to the resin extracted from the bark or the oil extracted from the nut (Table 11.2).

Two recurrent nouns for *C. schweinfurthii* occur across the Bantu languages indicating that the economic importance of the tree is old. The root **-bídí* can be reconstructed as the proto-form of the first series (Bastin et al. 2002). Present-day forms of this noun stem occur in North-West-Bantu and two distinct groups of West-Bantu, that is, West-Coastal Bantu and Congo Basin Bantu. The noun stem **-pátù*, and not *°-pápú*, as proposed by Bastin and associates (2002), can be reconstructed as the proto-form of the second series. Present-day forms of this noun stem occur in North-West Bantu and in certain West-Coastal, South-West Bantu languages, and East-Bantu languages, spoken in areas adjacent to the tropical forest. What is more, as reported by Blench (2006, p. 117), **-pátù* probably has cognates in a number of Central Nigerian languages, which belong to the Plateau subgroup of the East-Benue-Congo group (Williamson and Blench 2000). The noun **-pátù* thus has a time depth considerably greater than Proto-Bantu. It can be reconstructed into a Proto-East-Benue-Congo at least from which Proto-Bantu must have inherited it. As regards **-bídí*, no cognates have so far been identified beyond Bantu, which denotes that it is a more recent term than **-pátù*. The fact that both terms have an almost entirely complementary distribution among the Bantu languages suggests that **-bídí* replaced **-pátù* at some point, probably once the Bantu Expansion had

Table 11.2 Two Comparative Series of Bantu Names for *Canarium schweinfurthii**

	Language	Country	Term	Source
(a)	Ewondo	Cameroon	*a-bel*	(Mallart Guimera 2003, p. 322)
	Duma	Gabon	*mu-bili*	(Raponda-Walker and Sillans 1961, p. 110)
	Aka	Congo	*bélé*	(Motte 1980, p. 83)
	Bolia	DR Congo	*bo-bélé*	(Ngila Bompeti 2000, p. 67)
	Turumbu	DR Congo	*o-bele*	(Gillardin 1959, p. 167)
	Kumu	DR Congo	*bili*	(Duchesne 1938, p. 192)
(b)	Ganda	Uganda	*òmù-wafù*	(Snoxall 1967, p. 229)
	Haya	Tanzania	*omu-bâfu*	(Kaji 2000, p. 99)
	Rwanda	Rwanda	*-hafú*	(Coupez et al. 2005, p. 711)
	Ciluba	DR Congo	*mu-pàfu*	(De Clercq and Willems 1960, p. 208)
	Cokwe	Angola	*mu-bafo*	(Gossweiler 1953, p. 175)
	Nomaande	Cameroon	*pu-hétú*	(Taylor and Scruggs 1983, p. 70)

*Tables 11.2–5 present only a representative sample of Bantu tree names belonging to each of the comparative series—not all comparative lexical data we could collect.

started and a number of Bantu languages having retained **-pátù* had already branched off. For a more extensive discussion of the possible climate-induced vegetation dynamics underlying this lexical innovation, see Bostoen, Grollemund, and Koni Muluwa (2013).

In sum, Bantu lexical evidence is in line with archaeological findings as regards the early economic importance of the *C. schweinfurthii*. The fact that one of these terms goes back to at least Proto-East-Benue-Congo suggests that this food plant was exploited long before the ancestors of Bantu speech communities reached the Bantu homeland.

DACRYODES EDULIS (G. DON) HJ LAM—'AFRICAN PLUM', 'AFRICAN PEAR', 'SAFOUTIER' (BURSERACEAE)

The natural range of the *Dacryodes edulis* is not well established. Although certain scholars consider it to be limited to the southern part of Nigeria (Keay 1989; Vivien and Faure 1985), it has been reported in the evergreen forests of several countries of West-Central Africa: Cameroon, Gabon, Equatorial Guinea, Congo, and DR Congo (Aubreville 1962; Burkill 1985; Tchoundjeu, Kengue, and Leakey 2002), and even from Uganda to Sierra Leone (Troupin 1950). If the tree is not indigenous to these regions, human translocation may have played a role in its diffusion. This tree is grown from either seeds or cuttings (Burkill 1985, p. 307). Farmers in southern Cameroon, for instance, have been cultivating the tree for generations through the protection of regenerated individuals or through planting. Its fruit (*safou*) is traditionally highly appreciated for its nutritional value, and all kinds of rituals and beliefs are related to its management and use (Boli Baboulé 2002, cited by Schreckenberg et al. 2002, pp. 16–17). As regards the history of its distribution, archaeology has unfortunately little to contribute. This situation is not astonishing, since the woody endocarp of the fruit is very thin and does not preserve easily (S. Kahlheber pers. comm.). Comparative linguistics is more telling.

In French-speaking Central Africa, the *Canarium* ('*safoutier sauvage*') is commonly considered as the wild variant of *Dacryodes edulis* ('*safoutier*'). In most Bantu languages, the tree generally has its proper name distinctive from that of the *C. schweinfurthii*. Lexical confusion between both trees is observed only in some rare cases, possibly because of bad botanical identification. What is more, as shown in Table 11.3, one common term is found in a large number of West-Bantu languages.

As the Lumbu, Kongo, and Kimbundu examples in Table 11.3 show, the fruit's popular French name has its origin in the languages spoken in the proximity of the Atlantic Coast in Gabon, Congo, and Angola. The comparative series to which these and the other forms of Table 11.3 belong can be regularly derived from a single ancestral form, that is, **-cákú* (Bastin et al. 2002; Vansina 1990). Given the distribution of the tree species itself, this term cannot be expected to occur all over the Bantu domain. Reflexes of **-cákú* are represented, however, in sufficient different Bantu subgroups in order to reconstruct this

Table 11.3 Comparative Series of Bantu Names for *Dacryodes edulis*

Language	Country	Term	Source
Duala	Cameroon	*bo-sáó*	(Helmlinger 1972, p. 52)
Londo	Cameroon	*di-sa*	(Kuperus 1985, p. 271)
Tsogo	Gabon	*o-sago*	(Raponda-Walker and Sillans 1961, p. 113)
Lumbu	Gabon	*mu-safu*	(Raponda-Walker and Sillans 1961, p. 113)
Bolia	DR Congo	*bo-háwú*	(Ngila Bompeti 2000, p. 71)
Kiluba	DR Congo	*mw-afu*	(Duchesne 1938, p. 192)
Kongo	Angola	*safu*	(Gossweiler 1953, p. 175)
Kimbundu	Angola	*safu*	(Gossweiler 1953, p. 175)

fruit and tree name into Proto-Bantu. The presence of this term in Proto-Bantu suggests that ancestral Bantu speech communities must have been familiar with the *D. edulis* before first Bantu dispersals into equatorial Central Africa. Beyond Bantu, no cognate terms have been identified yet. Until new non-Bantu data prove the opposite, this term can thus be considered as a Bantu innovation. Consequently, if *D. edulis* really has its natural origin in southern Nigeria (Tchoundjeu, Kengue, and Leakey 2002, p. 4), its southwestern expansion can be inferred to have been (at least partially) anthropogenic.

COLA SPP. 'COLA NUT TREE' (STERCULIACEAE)

In West and West-Central Africa, cola nuts are obtained from several species of cola trees growing wild, protected, and cultivated. The two most common species are *Cola nitida* (Vent.) Schott et Endl. and *Cola acuminata* (P. Beauv.) Schott et Endl. (Ibu et al. 1986). Their fruits do not really serve as food, but are chewed as a stimulant and have particular uses in religious customs and in social practices, for instance as a token of hospitality (Blench 2006; Koni Muluwa 2010; Koni Muluwa and Bostoen 2008). Given that cola nuts have been widely traded and are increasingly commercialized today as a non-timber forest product (Facheux et al. 2006), the current-day distribution can be partially explained as the result of diffusion along trade networks. Despite their high cultural and ethnographical prominence, the trees and their fruits have a low archaeological visibility. Linguistically, on the contrary, cola names turn out to be reconstructible to considerable time depths. Williamson (1993) shows that such terms could be reconstructed in the protolanguages of various West-African language groups. Blench (2006, p. 117) draws attention on one of these noun stems, common among the Benue-Congo languages of Nigeria and spreading into the Bantu area of Cameroon, and beyond (Table 11.4).

In as much as the sources mention the precise tree species, the names in Table 11.4 always designate the *Cola acuminata*, except for Pove, where it refers to the *Cola lateritia* K. Schum. The comparative series in Table 11.4 has a single ancestral form, that is, **-bèdú* (Bastin et al. 2002; Guthrie 1970). As the trees previously discussed, cola trees thrive in only a restricted part of the Bantu area. Logically, the present-day forms of **-bèdú* cannot be expected to occur in all major Bantu subgroups. Nevertheless, the distribution of this comparative series is very similar to that of **-cákú*, '*Dacryodes edulis*' and **-bídí*, '*Canarium schweinfurthii*'. This internal Bantu distribution and the fact that it is widely found among non-Bantu Benue-Congo languages allow for reconstructing it in Proto-Bantu—but at the same time for presuming that Proto-Bantu itself inherited the word from an older language phase. In cultural historical terms, the reconstruction of a cola tree name in Proto-Bantu suggests that early Bantu speakers exploited this tree. What is more, if the different cola tree species have their natural origin in the Guinean Gulf region, their human protection and/or cultivation may

Table 11.4 Comparative Series of Bantu names for *Cola* spp.

Language	Country	Term	Source
Gunu	Cameroon	*i-benú*	(Orwig 1989, p. 287)
Ewondo	Cameroon	*a-bël*	(Mallart Guimera 2003, p. 323)
Pove	Gabon	*-bèdù*	(Mickala Manfoumbi 2004, p. 624)
Bali	Congo	*bilu*	(Guthrie 1970, p. 43)
Bobangi	DR Congo	*li-bêlu*	(Whitehead 1899, p .146)
Ngombe	DR Congo	*bo-belú*	(Rood 1958, p. 19)
Mongo	DR Congo	*bo-elú*	(Hulstaert 1966, p. 176)

explain why their diffusion into West-Central-Africa may have coincided with the Bantu language dispersal.

PARINARI CURATELLIFOLIA PLANCH. EX BENTH.—'MOBOLA PLUM' (CHRYSOBALANACEAE)

Parinari curatellifolia is a savannah forest tree occurring from Senegal throughout the region to Cameroon and is widely dispersed across tropical Africa (Burkill 1985, p. 382). Contrary to the trees discussed so far, it does not prosper in the equatorial rain forest. Its fruit is commonly known as the mobola plum in southern Africa. Because of the value of this fruit, the trees are left when new agricultural land is being cleared. The fruit is tasty when ripe and is also used for making beer. The two nuts inside the endocarp are also edible, and the timber and bark have many uses in craftwork and/or medicine (van Wyk and Gericke 2000: 52). Ritual functions are reported, too (Koni Muluwa and Bostoen 2008, p. 37; 2010, p. 108). Despite its importance in certain African subsistence economies, its archaeological visibility is low (see, for instance, Greenfield, Fowler, and van Schalkwyk 2005 for attestations in Early Iron Age sites of the Thukela Basin in South Africa).

Linguistically, two distinct terms for this tree are widespread (Table 11.5). Contrary to the tree names discussed before, and logically given the tree's habitat, none of the comparative series is represented in the Bantu languages of the equatorial rain forest. Both are limited to one Bantu subgroup and can be reconstructed in, respectively, Proto-East-Bantu (**-bʊda*) (Bastin et al. 2002) and Proto-South-West-Bantu (**-cà*). In cultural historical terms, its reconstruction in Proto-East-Bantu means that the ancestors of the East-Bantu speech communities exploited this plum tree before they started spreading east and south from the Great Lakes region—widely believed to be the East-Bantu homeland—presumably during the last half of the 1st millennium B.C.E. (Nurse 1999, p. 2). The same holds for the ancestors of the South-West-Bantu speech communities, who are assumed to have begun their dispersal south and west from 'somewhere between the lower Kwilu River and the mouth of the Sakuru River, south of the lower Kasai River' 'well back in the 1st millennium B.C.E.' (Vansina 2004, p. 279–82). One can conclude that Bantu speech communities became familiar with this species only when they changed their habitat from a tropical rainforest environment to a savannah forest environment, both east and south of the equatorial rain forest. However, a possible cognate form of **-bʊda* is attested in Tiv, a South-Bantoïd Benue-Congo language spoken mainly in southeastern Nigeria—that is, *i-bua* (Keay 1989, p. 182), which could mean that the fruits of the *P. curatellifolia* already belonged to the diet of Proto-Bantu speech communities, which must have had access then to both rain forest and savannah ecotones. If so, then the ancestors of East-Bantu speech communities already knew the tree before they arrived in the Great Lakes, which would further imply that they did not cross the equatorial rain forest but must have moved through a rain forest-savannah transition belt north of the rain forest.

DISCUSSION AND CONCLUSIONS

Agriculture has been seen as one of the driving forces behind the Bantu Expansion (Diamond and Bellwood 2003; Holden 2002). Nevertheless, little is known of the precise nature of the food complexes of these early farming societies or on the role of wild foods. Linguistic research has so far been too crop-centred. The assessment of Proto-Bantu lexical evidence indicates that early food crop exploitation included different *Dioscorea* species,

Table 11.5 Two Comparative Series of Bantu Names for *Parinari curatellifolia*

	Language	Country	Term	Source
(a)	Swahili	Tanzania	*m-bura*	(Heine and Legère 1995, p. 352)
	Lozi	Zambia	*mu-bula*	(Storrs 1995, p.275)
	Yao	Mozambique	*mu-ula*	(Ngunga 2001)
	Shona	Zimbabwe	*mu-ura*	(Hannan 1974, p. 943)
	Venda	South Africa	*mu-vhúlà*	(van Warmelo 1989, p. 253)
(b)	Kimbundu	Angola	*no-xa*	(Gossweiler 1953, p. 257)
	Ngangela	Angola	*mu-ca*	(Maniacky 2003, p. 43)
	Lunda	Zambia	*mu-cha*	(Storrs 1995, p. 275)
	Mbukushu	Namibia	*ghu-tha*	(Legère and Munganda 2004, p. 144)

and the pulses *Vigna unguiculata* and *Vigna subterranea*. These data, however, need not imply that cultivation of food crops was the principal subsistence strategy of these communities. The lexical evidence presented in this chapter suggests that the exploitation of wild trees was maybe even more important than the production of food crops, if we go by the amount of reconstructible vocabulary. Work on further wild plant food names is needed.

Archaeological evidence for food plant production is even more limited. The Shum Laka rock-shelter, bearing witness of 30,000 years of human occupation from the Late Pleistocene to Late Holocene, is the principal archaeological site associated to the Bantu homeland (de Maret 1995; Lavachery et al. 1997). Its occupation layer, which coincides more or less in time with the presumed start of the Bantu language dispersal, indicates significant evolution in human activities—namely, the emergence of macrolithism and a growing significance of pottery (Lavachery 2001, pp. 226–33). The most significant food plant remains are from *Canarium schweinfurthii*, attested from 7000–6000 BP, but significantly more abundant from 5000–4000 BP, contrary to many other Ceramic Late Stone Age sites of West and West-Central Africa, where *Canarium schweinfurthii* remains are often found in association with those of *Elaeis guineensis*, oil palm nuts appeared only very recently (200–0 BP) at Shum Laka. Lavachery (1998, p. 403) links the intensified exploitation of the *Canarium schweinfurthii* to the development of macrolithic industries and sees it as a possible beginning of the arboriculture still typical for the Guinean Gulf today. The reconstruction of two distinct Proto-Bantu terms for both *Canarium schweinfurthii* and *Elaeis guineensis* is in line with archaeological findings, at Shum Laka and/or other Ceramic Late Stone Age sites. The time depth of lexical reconstructions for *Dacryodes edulis*, *Cola spp.*, and possibly *Parinari curatellifolia*, is unmatched, however in the archaeological record. Their archaeological invisibility is mainly due to the trees' low preservation qualities. Whereas the other species are rather typical of the evergreen forest, *Parinari curatellifolia* is a savannah forest species. Hence, Proto-Bantu speech communities probably lived in a transition zone between rain forest and savannah and were familiar with both ecotones. Such an ecological transition zone fits well with the assumed homeland in the Cameroonian Grassfields environment (Lavachery 2001; Maley and Brenac 1998). If we suppose that the Guinean Gulf was indeed the natural habitat of certain species discussed in this chapter, *Dacryodes edulis* and *Cola spp.* for instance, their diffusion in West-Central Africa can be considered anthropogenic.

At the same time, Proto-Bantu does not constitute a breaking point as regards the vocabulary for food plants thriving in a tropical forest environment. The terms reconstructible in Proto-Bantu for some trees as well as for *Vigna unguiculata* and different *Dioscorea* species have a time depth greater than Proto-Bantu. All these nouns are inherited from either Proto-East-Benue-Congo, or an even older language phase. The homeland of Proto-Benue-Congo is inferred to be around the Niger-Benue river confluence (Armstrong 1981; Blench 2006; Williamson 1989). It is interesting to note that part of the Proto-Bantu pottery vocabulary was likewise inherited from Proto-Benue-Congo (Bostoen 2005a). In the archaeological sense, macrolithic tools, polishing tools, and pottery became gradually mixed at Shum Laka with local preexisting microlithic LSA traditions from 7000–6000 BP onward. The appearance of these new technologies has been attributed to the small-scale immigration of more northerly communities fleeing the climatic deterioration that scourged the Sahara and Sahel around 7100–6900 BP (de Maret 1995, p. 320; Lavachery 2001, p. 240–41). In this respect, the Proto-Benue-Congo origin of part of the Proto-Bantu pottery vocabulary substantiates the hypothesis that these immigrants spoke Benue-Congo languages (Bostoen 2007). These newcomers not only introduced new technologies and new languages, but they also brought a forest plant food complex. However, certain vocabulary related both to certain food crops and trees (**-jʊ̀gʊ́*, '*Vigna subterranea*'; **-gàdí*, 'palm oil'; **-cákú*, '*Dacryodes edulis*'; and maybe **-bʊda*, '*Parinari curatellifolia*') and to pot-making (Bostoen 2005a, 2007) is reconstructible into Proto-Bantu but not into older language phases. This fact suggests that the food complexes and technologies underwent local adaptations and transformations before being diffused in West-Central Africa together with the Bantu languages. All this circumstantial linguistic evidence remains to be substantiated by archaeobotany.

ACKNOWLEDGEMENTS

Research for this chapter has been funded by the Belgian Federal Science Policy Office as part of the Words and Plants project of the Linguistics Service of the Royal Museum for Central Africa in Tervuren (finished in 2006). My thanks go to Roger Blench, Baudouin Janssens, Jacky Maniacky, and Anneleen van der Veken for commenting on a previous version of this article, which was first submitted in September 2006. In the course of the editorial process, an effort was done to keep it up to date by including references to scientific work that appeared since then.

REFERENCES

Armstrong, R. G. (1964) The use of linguistic and ethnographic data in the study of Idoma and Yoruba history. In J. Vansina, R. Mauny, and L. V. Thomas (Eds.), *The Historian in Tropical Africa*, pp. 127–38. London: Oxford University Press.

Aubreville, A. (1962) *Flore du Gabon: Irvingiaceae, Simaroubaceae, Burseraceae*. Paris: Museum National d'Histoire Naturelle, Laboratoire de Phanerogamie.

Bastin, Y., A. Coupez, and M. Mann (1999) *Continuity and Divergence in the Bantu Languages: Perspectives from a Lexicostatistic Study*. Annales, Série in 8, Sciences humaines 162. Tervuren: Royal Museum for Central Africa.

Bastin, Y., A. Coupez, E. Mumba, and T. C. Schadeberg (Eds.) (2002) Bantu lexical reconstructions 3/Reconstructions lexicales bantoues 3. Tervuren: Royal Museum for Central Africa, online database http://linguistics.africamuseum.be/BLR3.html (accessed September 4, 2012).

Blench, R. (1994) The ethnographic evidence for long-distance contacts between Oceania and East Africa. In J. Reade (Ed.), *The Indian Ocean in Antiquity*, pp. 417–38. London: Kegan Paul International/ British Museum Press.

Blench, R. (1995) Linguistic evidence for cultivated plants in the Bantu borderland. *Azania 29-30*, 83–102.

———. (2006) *Archaeology, Language and the African Past*. Lanham, MD: AltaMira Press.

Boli Baboulé, Z. (2002) Le safoutier dans le terroir Banen: savoirs locaux et traditions. In J. Kengue, C. Kapseu, and G. J. Kayem (Eds.), *Actes du troisième séminaire international sur la valorisation du safoutier et autres oléagineux non-conventionnels, Yaoundé -Cameroun, 3–5 octobre 2000*, pp. 562–67. Yaoundé: Presses universitaires d'Afrique.

Bostoen, K. (2004) The vocabulary of pottery fashioning techniques in Great Lakes Bantu: A comparative onomasiological study. In A. Akinlabi and O. Adesola (Eds.), *Proceedings of the 4th World Congress of African Linguistics*, pp. 391–408. Köln: Rüdiger Köppe Verlag.

———. (2005a) *Des mots et des pots en bantou : Une approche linguistique de l'histoire de la céramique en Afrique*. Frankfurt am Main: Peter Lang Verlag.

———. (2005b) A diachronic onomasiological approach to early Bantu oil palm vocabulary. *Studies in African Linguistics 34* (2), 143–88.

———. (2006–2007) Pearl millet in early Bantu speech communities in Central Africa: A reconsideration of the lexical evidence. *Afrika und Übersee 89*, 183–213.

———. (2007) Pots, words and the Bantu problem: On lexical reconstruction and early African history. *Journal of African History 48*, 173–99.

Bostoen, K., R. Grollemund, and J. Koni Muluwa (2013) Climate-induced vegetation dynamics and the Bantu expansion: Evidence from Bantu names for pioneer trees (*Elaeis guineensis, Canarium schweinfurthii* and *Musanga cecropioides*). *Comptes Rendus Geoscience*, accessed May 27, 2013, http://dx.doi.org/10.1016/j.crte.2013.03.005.

Bulkens, A. (1999a) Linguistic indicators for the use of calabashes in the Bantu world. *Afrikanistische Arbeitspapiere 57*, 79–104.

———. (1999b) La reconstruction de quelques mots pour 'mortier' en domaine bantou. *Studies in African Linguistics 28* (2), 113–53.

Burkill, H. M. (1985) *The Useful Plants of West Tropical Africa*. Vol. 1: *Families A-D*. Kew: Royal Botanical Gardens.

Carrière, S. (1999) *Les orphelins de la forêt: Influence de l'agriculture itinérante sur Brûlis des Ntumu et des pratiques agricoles associées sur la dynamique forestière du Sud Cameroun*. Thèse de Doctorat, Montpellier: Université des Sciences et Techniques du Languedoc.

Connell, B. (1998) Linguistic evidence for the development of yam and palm culture among the Delta Cross peoples of southeastern Nigeria. In R. Blench and M. Spriggs (Eds.), *Archaeology and Language II, Archeological Data and Linguistic Hypotheses*, pp. 324–65. London: Routledge.

Coupez, A., T. Kamanzi, S. Bizimana, G. Sematama, G. Rwabukumba, and C. Ntazinda (2005) *Dictionnaire Rwanda-Rwanda et Rwanda-Français: Inkoranya y'ikinyarwaanda mu kinyarwaanda nó mu gifaraansá*. Tervuren: Musée royal de l'Afrique centrale.

Dalby, D. (1976) The prehistorical implications of Guthrie's comparative Bantu, Part II: Interpretation of cultural vocabulary. *Journal of African History 17* (1), 1–27.

D'Andrea, A. C., A. L. Logan, and D. J. Watson (2006) Oil Palm and prehistoric subsistence in tropical Africa. *Journal of African Archaeology 4* (2), 195–222.

De Clercq, A. and E. Willems (1960) *Dictionnaire tshiluba-français*. Léopoldville: Imprimerie de la Société missionnaire de St Paul.

de Maret, P. (1995) Pits, pots and the far west streams. *Azania 29-30*, 318–23.

De Wolf, P. (1969) *Benue-Congo Noun Class System*. Leiden: West African Linguistic Society, Afrika Studiecentrum.

Diamond, J., and P. Bellwood (2003) Farmers and their languages: The first expansions. *Science 300*, 597–603.

Duchesne F. (1938) *Les essences forestières du Congo*. Bruxelles: Bothy.

Ehret, C. (1974) Agricultural history in Central and Southern Africa, ca. 1000 B.C. to A.D. 500. *Transafrican Journal of History 4* (1-2), 1–25.

Facheux, C., Z. Tchoundjeu, D. Foundjem, C. Mbosso, and T. T. Manga (2006) From research to farmer enterprise development in Cameroon: Case study of kola nuts. *ISHS Acta Horticulturae 699*, 181–88.

Gillardin, J. (1959) *Les essences forestières du Congo et du Rwanda-Urundi*. Bruxelles: Dafe.

Gossweiler, J. (1953) *Nomes indigenas de plantas de Angola*. Luanda: Imprensa Nacional.

Greenfield, H. J., K. D. Fowler, and L. O. van Schalkwyk (2005) Where are the gardens? Early Iron Age horticulture in the Thukela River Basin of South Africa. *World Archaeology 37* (2), 307–28.

Guthrie, M. (1970) *Comparative Bantu: An Introduction to the Comparative Linguistics and Prehistory of the Bantu Languages*, Vol. 3. London: Gregg International Publishers Ltd.

Hannan M. (1974) *Standard Shona Dictionary*. Salisbury: Mardon printers Ltd for Rhodesia Literature Bureau.

Heine, B., and K. Legère (1995) *Swahili Plants*. Köln: Rüdiger Köppe.

Helmlinger, P. (1972) *Dictionnaire duala-français, suivi d'un lexique français-duala*. Langues et Littératures de l'Afrique noire 9. Paris: Klincksieck.

Holden, C. (2002) Bantu language trees reflect the spread of farming across sub-Saharan Africa: A maximum-parsimony analysis. *Proceedings of the Royal Society of London Series B. 269*, 793–99.

Hulstaert, G. (1966) *Notes d'ethnobotanique Mongo*. Bruxelles: Académie royale des sciences d'Outre-Mer.

Ibu J.O., A. C. Iyama, C.T. Ijije, D. Ishmael, M. Ibeshi, and S. Nwokediuko (1986) The effect of *Cola acuminata* and *Cola nitida* on gastric acid secretion. *Scandinavian Journal of Gastroenterol Supplement 124*, 39–45.

Jacquot, A. (1991) Le nom de la houe dans les langues bantoues du nord-ouest. *Cahiers des sciences humaines 27*, 561–76.

Kahlheber, S., K. Bostoen, and K. Neumann (2009) Early plant cultivation in the Central African rain forest: First millennium BC pearl millet from South Cameroon. *Journal of African Archaeology 7*, 253–72.

Kaji, S. (2000) *A Haya Vocabulary*. Tokyo: University of Foreign Studies, ILCAA.

Keay, R. W. J. (1989) *Trees of Nigeria*. Oxford: Clarendon Press.

Koni Muluwa, J. (2010) *Plantes, animaux et champignons en langues Bantu: Etude comparée de phytonymes, zoonymes et myconymes en nsong, ngong, mpiin, mbuun et hungan (Bandundu, RD Congo)*. Bruxelles: Université libre de Bruxelles, thèse de doctorat.

Koni Muluwa, J., and K. Bostoen (2008) *Noms et usages de plantes utiles chez les Nsong (RD Congo, Bantu B85F)*. Göteborg Africana Informal Series 6. Göteborg: University of Göteborg, Department of Oriental and African Languages.

Koni Muluwa, J., and K. Bostoen (2010) Les plantes et l'invisible chez les Mbuun, Mpiin et Nsong (Bandundu, RD Congo): Une approche ethnolinguistique. *Sprache und Geschichte in Afrika 21*, 95–122.

Kuperus, J. (1985) *The Londo Word: Its Phonological and Morphological Structure*. Annales, Série in-8°, Sciences humaines 119. Tervuren: Royal Museum for Central Africa.

Legère, K., and R. Munganda (2004) *Thimbukushu-Thihingirisha/English-Thimbukushu: Manandorandathana ghoThikuhonga/Subject Glossaries (Language and Literature, Mathematics, Body and Health, Fauna and Flora)*. Windhoek: Gamsberg Macmillan.

Lavachery, P. (1998) *De la pierre au métal: Archéologie des dépôts holocènes de l'abri de Shum Laka (Cameroun)*. Ph.D. thesis, Université Libre de Bruxelles. Brussels.

———. (2001) The Holocene archaeological sequence of Shum Laka Rock Shelter (Grassfields, Cameroon). *African Archaeological Review 18*, 213–47.

Lavachery, P., E. Cornelissen, J. Moeyersons, and P. de Maret (1997) 30 000 ans d'occupation, 6 mois de fouille: Shum Laka, un site exceptionnel en Afrique centrale. *Anthropologie et Préhistoire 107*, 197–211.

Maley, J., and Brenac P. (1998) Vegetation dynamics, palaeoenvironments and climatic changes in the forests of western Cameroon during the last 28,000 years BP. *Review of Palaeobotany and Palynology 99*, 157–87.

Mallart Guimera, L. (2003) *La Forêt de nos ancêtres. Tome I: Le système médical des Evuzok du Cameroun, Tome II: Le savoir botanique des Evuzok*. Annales Sciences humaines vol. 167. Tevuren: Musée royal de l'Afrique centrale.

Maniacky, J. (2003) *Tonologie du ngangela—Variété du Menongue (Angola)*. Lincom Studies in African Linguistics 61. München: Lincom Europa.

———. (2005) Quelques thèmes pour 'igname' en Bantu. In K. Bostoen and J. Maniacky (Eds.), *Studies in African Comparative Linguistics with Special Focus on Bantu and Mande*, Human Sciences Collection 169, pp. 165–89. Tervuren: Royal Museum for Central Africa.

Mickala-Manfoumbi, R. (2004) *Lexique pové-français français-pové*. Libreville: Editions Raponda Walker.

Motte, E. (1980) *Les plantes chez les pygmées aka et les Monzombo de la Lobaye (Centrafrique)*. Paris: SELAF.

Neumann, K. (2003) The late emergence of agriculture in sub-Saharan Africa: Archaeological evidence and ecological considerations. In K. Neumann, A. Butler, and S. Kahlheber (Eds.), *Food, Fuel and Fields: Progress in African Archaeobotany*, Africa Praehistorica 15, pp. 71–92. Köln: Heinrich-Barth-Institut.

———. (2005) The romance of farming: Plant cultivation and domestication in Africa. In A. B. Stahl (Ed.), *African*

Archaeology: A Critical Introduction, Blackwell Studies in Global Archaeology, pp. 249–75. Oxford: Blackwell Publishers.

Neumann, K., K. Bostoen, A. Höhn, S. Kahlheber, A. Ngomanda, and B. Tchiengué (2012) First farmers in the Central African rainforest: A view from southern Cameroon. *Quaternary International 249*, 53–62.

Ngila Bompeti, P. (2000) *Expérience végétale bolia (République Démocratique du Congo): Catégorisation, utilisation et dénomination des plantes*, Grammatische Analysen Afrikanischer Sprache Band 14. Köln: Rüdiger Köppe.

Ngunga, A. (2001) *Ciyao Dictionary*. Online database www.cbold.ddl.ish-lyon.cnrs.fr.

Nurse, D. (1999) Towards a historical classification of East African Bantu languages. In J.-M. Hombert and L. M. Hyman (Eds.), *Bantu Historical Linguistics: Theoretical and Empirical Perspectives*, pp. 1–41. Stanford, CA: CSLI Publications.

Nurse D., and G. Philippson (2003) Towards a historical classification of the Bantu languages. In D. Nurse and G. Philippson (Eds.), *The Bantu Languages*, pp. 164–81. London: Routledge.

Orwig, C. (1989) Les extensions verbales en nugunu. In D. Barreteau and R. Hedinger (Eds.), *Description de langues camerounaises*, pp. 283–314. Paris: ACCT, ORSTOM.

Philippson G., and S. Bahuchet (1994–1995) Cultivated crops and Bantu migrations in central and eastern Africa: A linguistic approach. *Azania 29-30*, 103–20.

Raponda-Walker, A., and R. Sillans (1961) *Les plantes utiles du Gabon: Encyclopédie biologique*. Paris: Paul Lechevalier.

Renfrew, C. (1992) Archaeology, Genetics and Linguistic Diversity. *Man* (NS) 27 (3), 445–78.

Rood, N. (1958) *Ngombe-Nederlands-Frans Woordenboek. Dictionnaire ngombe-néerlandais-français*. Sciences de l'Homme, Linguistique vol. 21. Tervuren: Musée royal de l'Afrique centrale.

Schadeberg, T. C. (2003) Historical linguistics. In D. Nurse and G. Philippson (Eds.), *The Bantu Languages*, pp. 143–63. London: Routledge.

Schreckenberg, K., A. Degrande, B. Mbosso, Z. Boli Baboulé, C. Boyd, L. Enyong, J. Kanmegne, and C. Ngong (2002) The social and economic importance of *Dacryodes edulis* (G. Don) H. J. Lam in southern Cameroon. *Forests, Trees and Livelihood 12*, 15–40.

Snoxall, R. A. (1967) *Luganda-English Dictionary*. Oxford: Clarendon Press.

Storrs, A. E. G. (1995) *Know Your Trees: Some of the Common Trees Found in Zambia*. Lusaka: Regional Soil Conservation Unit (reprinted).

Taylor, C., and T. Scruggs (1983) *Lexique nomaandé-français, un parler de l'arrondissement de Bokito*. Yaoundé: Société Internationale de Linguistique.

Tchoundjeu Z., J. Kengue, and R. R. B. Leakey (2002) Domestication of *Dacryodes edulis*: State of the art. *Forests, Trees and Livelihood 12*, 3–13.

Troupin, G. (1950) Les Burseracées du Congo Belge et du Ruanda-Burundi. *Bulletin de la Société Botanique de Royale de Belgique 83*, 111–26.

van Wyk, B. E., and N. Gericke (2000) *People's Plants: A Guide to Useful Plants of Southern Africa*. Pretoria: Briza Publications.

Vansina, J. (1990) *Paths in the Rainforests*. Madison: University of Wisconsin Press.

———. (1995) New linguistic evidence and the Bantu expansion. *Journal of African History 36*, 173–95.

———. (2004) *How Societies Are Born: Governance in West Central Africa before 1600*. Charlottesville: University of Virginia Press.

van Warmelo, N. J. (1989) *Venda Dictionary: Tshivenda-English*. Pretoria: J. L. van Schalk.

Vivien, J., and Faure, J. J. (1985) *Arbres des forêts denses d'Afrique Centrale*. Paris: Ministère des Relations Extérieures, Agence de Développement et de Coopération Culturelle et Technique.

Whitehead, J. (1899) *Grammar and Dictionary of the Bobangi Language, as Spoken over a Part of the Upper Congo West Central Africa*. London: Kegan P., Trench, Trübner CO.

Williamson, K. (1993) Linguistic evidence for the use of tree and tuber food plants. In T. Shaw, P. Sinclair, B. Andah, and A. Okpoko (Eds.), *The Archeology of Africa: Food, Metals and Towns*, pp. 139–53. London: Routledge.

Williamson, K., and R. Blench (2000) Niger-Congo. In B. Heine and D. Nurse (Eds.), *African Languages: An Introduction*, pp. 11–42. Cambridge: Cambridge University Press.

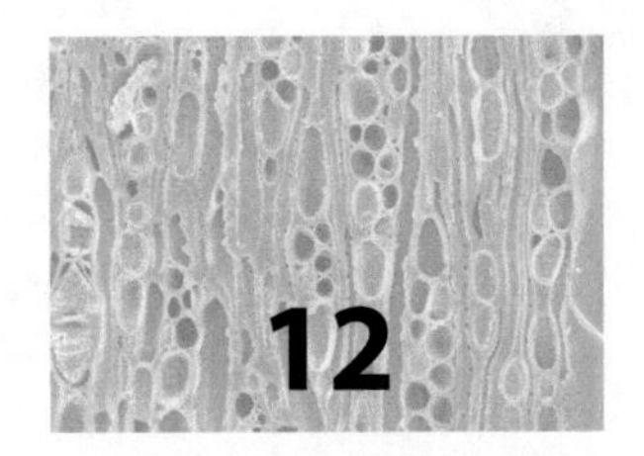

Archaeobotany of Two Middle Kingdom Cult Chambers at Northwest Saqqara, Egypt

Ahmed Gamal-El-Din Fahmy, Nozomu Kawai and Sakuji Yoshimura

In Egypt, plant macro-remains have been recovered as offerings from tombs or as the remains of food, fuel, animal fodder etc from ancient settlements. They have generated important contributions to the study of archaeobotany both in Egypt and beyond (de Vartavan and Asensi Amoros 1997; Murray 2000a, 2000b).

This study focuses on the recovery of plant macro-remains from inside two rock-cut chambers at Northwest Saqqara. The Saqqara necropolis was a significant site in Lower Egypt (Figure 12.1), housing the royal and elite cemeteries associated with the city of Memphis from the Early Dynastic period (3000 B.C.E.) into Graeco-Roman times. Excavations in 2001 on an eastern slope of a hillock uncovered a large layered stone structure constructed in the style of early Old Kingdom step or layer pyramids along with an associated, probably contemporary substructure in the form of a rock-cut chamber (AKT02) that had been subsequently enlarged and restored in the Middle Kingdom (Kawai 2011; Yoshimura, Kawai, and Kashiwagi 2005;). Approximately 20 m away, another rock-cut chamber AKT01 was revealed in the centre of the hills eastern face (Figure 12.2).

Archaeology of African Plant Use by Chris J. Stevens, Sam Nixon, Mary Anne Murray, and Dorian Q Fuller, Eds., 141–149

Rock-Cut Chamber AKT01

AKT01 included a forecourt from which a shallow rectangular shaft descended to allow access into a subterranean T-shaped chamber, comprising a transverse and inner hall (Figure 12.3). Excavation of the chamber indicated that it was never used for burial; instead a number of terracotta statue fragments of a lion goddess, probably Bastet or Sakhmet, and wooden statues of human and animal figures were recovered, as well as broken pottery associated with well-preserved, desiccated plant macro-remains (Yoshimura and Kawai 2003). The style of the statues indicated an Old Kingdom (2686–181 B.C.E.) date, yet the pottery sherds found with them date to the Middle Kingdom (ca. 1900–1785 B.C.E.) (Yoshimura, Kawai, and Kashiwagi 2005, pp. 389–96). It is assumed that Old Kingdom terracotta statues were reused and deposited after their usage in the Middle Kingdom.

Rock-Cut Chamber AKT02

AKT02 originally consisted of one subterranean chamber accessed by a shaft sealed with a portcullis stone (Figure 12.4). This eastern chamber contained ritual and votive objects as well as pottery sherds from two different periods: the Early Dynastic Period and Middle Kingdom, respectively. Apparently the chamber was reused in the Middle Kingdom. A clay statue of a lion goddess dating to the Middle Kingdom was uncovered in this chamber,

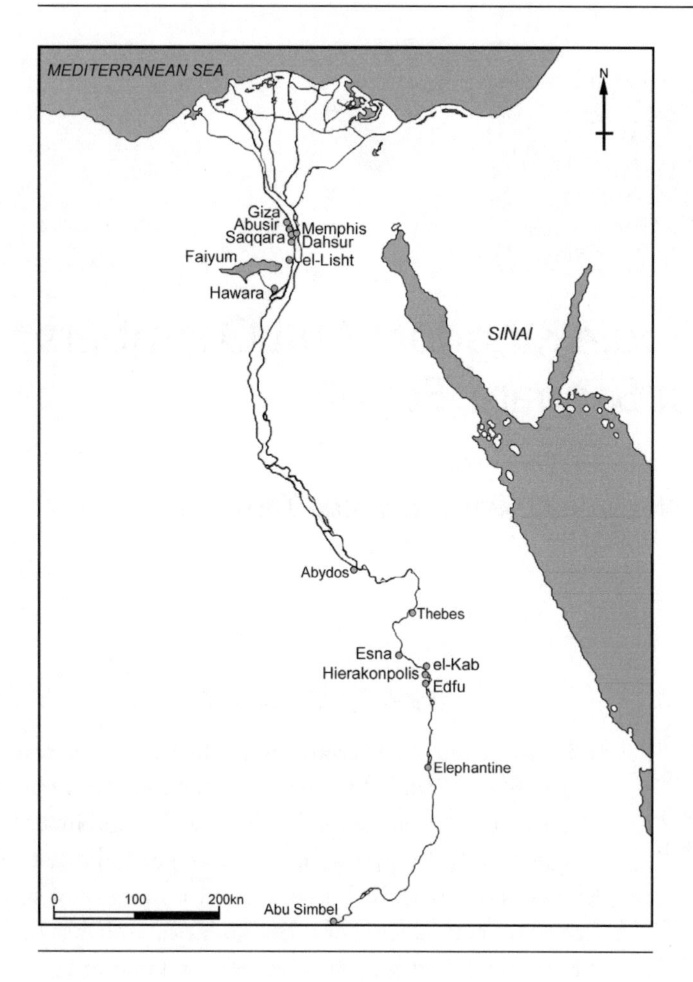

Figure 12.1 Map of Egypt showing location of archaeological sites in text.

which is stylistically similar to a clay statue of a lion goddess found in the rock-cut chamber AKT01. A west chamber was cut in the Middle Kingdom, and excavation revealed smaller votive objects along with food offerings that are almost *in situ*. Near the entrance was a reed mat, probably originally covered with linen, animal skins, and associated with Middle Kingdom (mid-12th–early 13th Dynasty) pottery. In the centre of the chamber was a concentration of animal bones and pottery containing organic material, mud, and natron. The bones included calf, duck, goose, and fish and were wrapped in linen; they had originally been placed inside the dishes or the conical shaped 'meat jars'.

The similarity of the statues and pottery assemblages in chambers AKT01 and AKT02 suggest that Middle Kingdom objects were deposited simultaneously in both rock-cut chambers for cultic purpose (Yoshimura, Kawai, and Kashiwagi 2005, pp. 398–400). Ultimately, all the chambers under study may have functioned as a cachet for cult statues.

The botanical material recovered is unique in context and provides an excellent opportunity to analyse, discuss, and evaluate methods of plant utilization in Middle Kingdom Saqqara. The current study aims at increasing our knowledge on the past relationship between man and plants with special reference to the economic aspects of this period. The study focuses also on the differences between the types of botanical offerings found in tombs of the Old to New Kingdoms and those placed in the statuary cache at Northwest Saqqara.

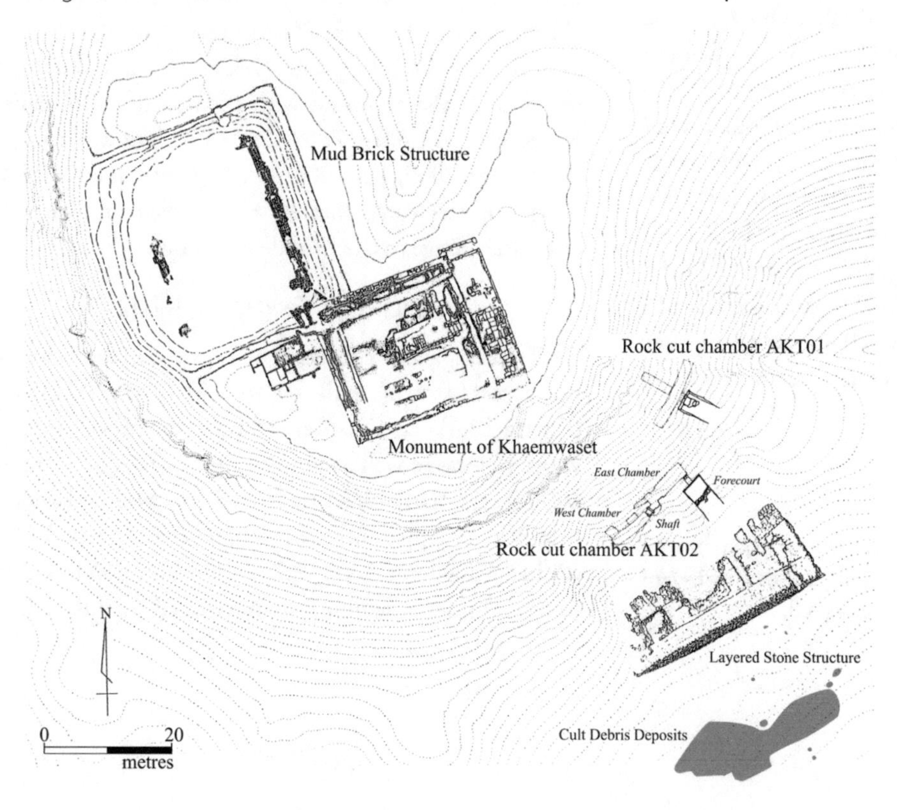

Figure 12.2 General map of site showing location of rock-cut chambers AKT01 and AKT02.

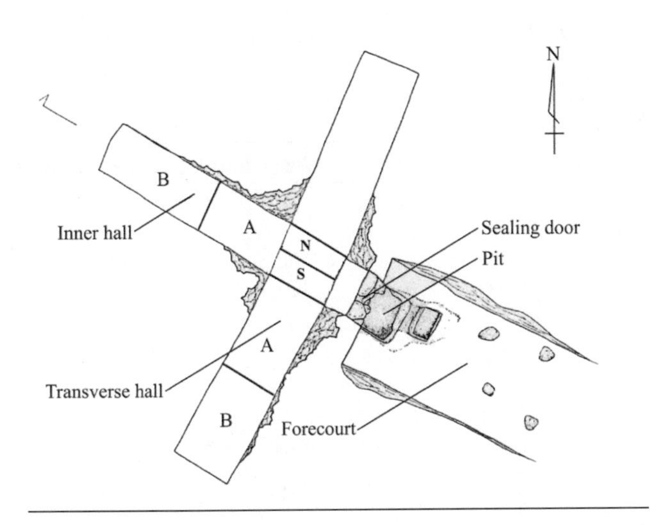

Figure 12.3 Plan of rock-cut chamber AKT01.

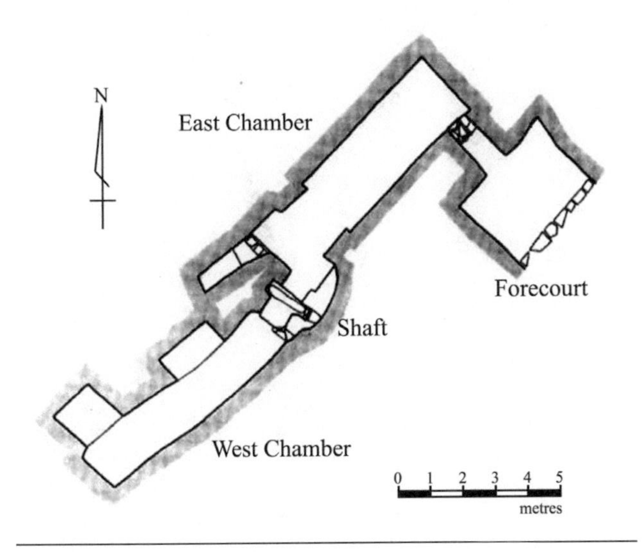

Figure 12.4 Plan of rock-cut chamber AKT02.

Material and Methods

Sixty-three soil samples of 1 to 3 litres were collected in the 2001 to 2002 excavations along with specimens handpicked by members of the excavation team. The samples were stored on site and examined in August 2005, being sieved through a 0.5 mm mesh and plant macro-remains sorted using a stereo-microscope, then photographed. Much of the material was identified to species level, and their confirmation was done at the Department of Botany and Microbiology, Faculty of Science, University of Helwan, Egypt.

Results

Morphological investigation of the plant macro-remains revealed the presence of desiccated cereal remains, seeds, fruits, and tubers as well as wood charcoal fragments. A total of 885 specimens belonging to 15 taxa have been identified from all sectors inside AKT01 and AKT02. The botanical assemblage was then classified into four major groups: cultivated cereals, edible fruits, arable weeds, and wild species (Table 12.1).

Distribution of Plant Macro-Remains

Table 12.1 shows the number and percentages of plant macro-remains in five different archaeological contexts inside the two chambers. In chamber AKT01, the 372 plant macro-remains were concentrated in the transverse hall, being dominated by hulled grains of barley, *Hordeum vulgare*. Stones of *Ziziphus spina-christi* were also relatively common, and seeds of *Lupinus digitatus* and tubers of *Cyperus esculentus* along with one seed of *Mimusops laurifolia* were also recorded. The presence of tubers at archaeological sites attests to conditions highly favourable for preservation during various taphonomic processes (Fahmy 2005; Hillman 1989), attributed to the extremely dry conditions prevailing inside the two rock-cut chambers.

A total of 416 plant macro-remains were examined from the west chamber (AKT02). Remains of cereals (emmer wheat and barley) dominated. Edible fruits included stones and pips of *Ziziphus spina-christi*, *Ficus* sp., and *Phoenix dactylifera*. Arable weeds are represented by four taxa: *Lathyrus hirsutus*, *Lupinus digitatus*, *Phalaris minor*, and *Vicia sp*. Only 96 seeds were recovered from the east chamber of AKT02, comprising 95 stones of *Z. spina-christi* and one fruit of *Neurada procumbens*.

Figure 12.5 shows that plant macro-remains were concentrated in definite areas, such as the transverse hall of chamber AKT01 (42%) and the west chamber of AKT02 (47%). Fewer plant macro-remains were recovered from the sample from the inner hall of AKT01 (1%) and the east chamber of AKT02 (10%). In comparison to AKT02, it is interesting to note the absence of *Triticum dicoccum*, *Ficus* sp., *Phoenix dactylifera*, *Lathyrus hirsutus*, *Phalaris minor*, and *Vicia* sp. from AKT01. These differences in the distribution and composition of the botanical material suggest different activities within or functions for these contemporary structures, perhaps related to different cult requirements.

Discussion

Remains of cereals from archaeological sites can provide us with evidence for reconstructing agricultural practices in the past. The samples under consideration are small in size and limited in diversity; however, we can make some observations on cereal production at the time the samples were deposited.

Table 12.1 Numbers (N) and Percentage of Plant Macro-Remains from Two Rock-Cut Chambers AKT01 and AKT02 in Saqqara, Egypt

	AKT01						AKT02			
Taxa	Transverse Hall		Inner Hall		Forecourt		West Chamber		East Chamber	
Cultivated Cereals	N	%	N	%	N	%	N	%	N	%
Hordeum vulgare L.										
Grains	312	84	–	–	–	–	101	24	–	–
Chaff	–	–	–	–	–	–	2	0.4	–	–
Triticum dicoccum Schrank										
Grains	–	–	–	–	–	–	4	1	–	–
Spikelets	–	–	–	–	–	–	110	26	–	–
Forks	–	–	–	–	–	–	30	8	–	–
Edible Fruits										
Ficus sp.										
Pips	–	–	–	–	–	–	12	3	–	–
Mimusops laurifolia (Forssk.) Friis										
Stone	1	0.2	–	–	–	–	–	–	–	–
Phoenix dactylifera L.										
Stones	–	–	–	–	–	–	4	1	–	–
Ziziphus spina-christi (L.) Desf.										
Stones	40	11	–	–	–	–	126	30	95	99
Fruits	5	1.3	–	–	–	–	5	1.2	–	–
Arable Weeds										
Lathyrus hirsutus L.										
Seeds	–	–	–	–	–	–	8	2	–	–
Lupinus digitatus Forssk.										
Seeds	6	1.5	1	100	–	–	2	0.5	–	–
Phalaris minor Retz										
Grains	–	–	–	–	–	–	2	0.5	–	–
Vicia sp.										
Seeds	–	–	–	–	–	–	5	1.2	–	–
Wild Species										
Cyperus esculentus L.										
Tuber	8	2	–	–	–	–	–	–	–	–
Cyperus sp.										
Tuber	–	–	–	–	–	–	5	1.2	–	–
Juncus culm	–	–	–	–	–	–	+	–	–	–
Neurada procumbens L.										
Fruit	–	–	–	–	–	–	–	–	1	1
Typha sp.										
Leaf fragments	+	–	–	–	+	–	+	–	+	–

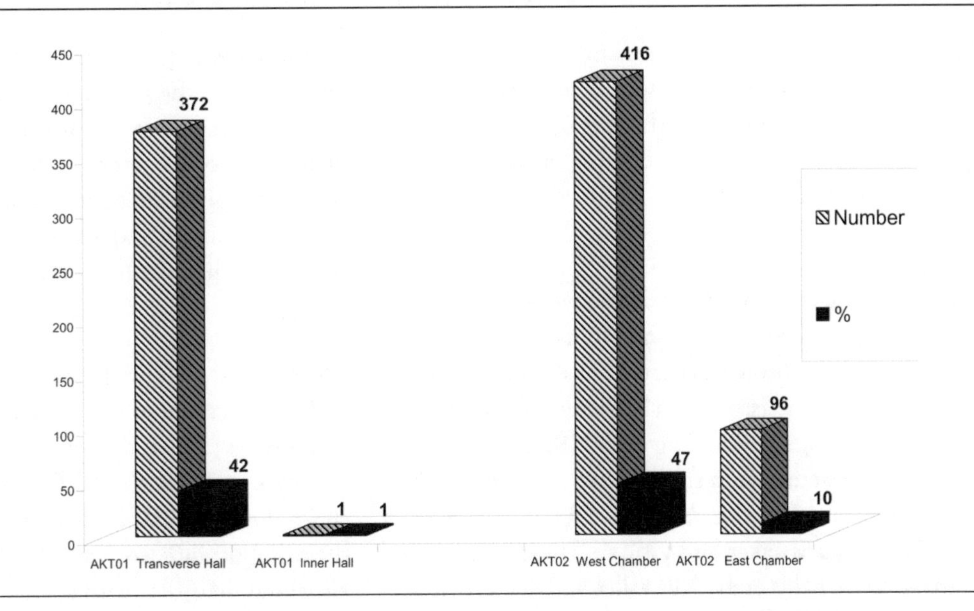

Figure 12.5 Distribution of plant remains in the archaeological sectors of rock-cut chambers AKT01 and AKT02.

CULTIVATED CEREALS

Grains of hulled barley were present in the transverse hall of AKT01, whereas a mixture of emmer wheat spikelets and several grains of hulled barley were retrieved from the west chamber of AKT02 (Figures 12.6 and 12.7). The absence of culm fragments and rachis internodes indicate that coarse/medium sieving occurred before the deposition of these spikelets and grains at the site. Murray (2000a, pp. 520–26) describes in detail the sequence of cereal processing in ancient Egypt conducted before the storage of grain, comprising four steps: harvesting, threshing, winnowing, and coarse to medium sieving.

The small samples available cannot provide us with the data necessary to determine whether emmer wheat was dominant over hulled barley or vice versa in the Middle Kingdom. At Ma'adi (3700–3500 B.C.E.), van Zeist and associates (2003) conclude that the arable economy of this Lower Egypt site was largely based on cultivation of emmer wheat, probably free threshing wheat as well as hulled barley. Extensive cultivation of emmer wheat over barley during the New Kingdom may be due to improvements in the irrigation system (*sensu* Hassan 1984).

From the Middle and New Kingdom, model granaries have been found within tombs. These granaries have been interpreted by Winlock (1955) as 'perennial grain stores' for the afterlife and may also represent daily life practices. Significantly, the models contained emmer wheat spikelets and hulled barley, as well as chaff and seeds of arable weeds. New Kingdom tomb scenes often show that harvested ears were collected in baskets before they were carried to the threshing enclosure (Murray 2000a, p. 522).

Figure 12.6 Spikelets of emmer wheat.

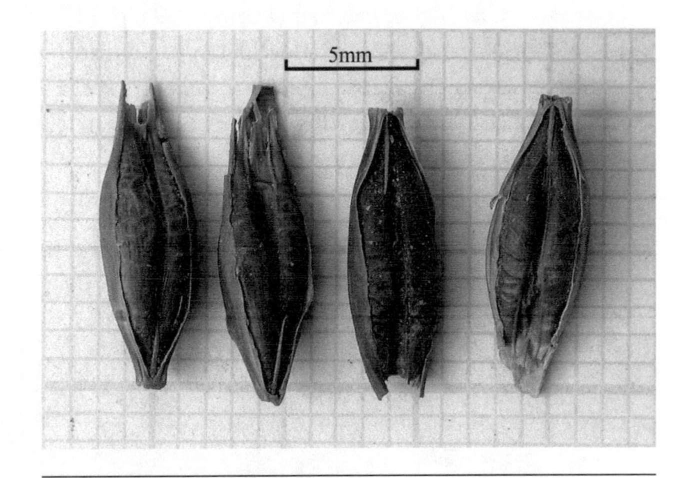

Figure 12.7 Hulled grains of barley.

That the archaeological evidence suggests the chambers functioned as a cache for cult statues implies that the cereal plant remains discovered within the rock-cut chambers provide a parallel to funeral offering rituals, rather than storage practices.

Arable Weeds

Analysis of the macro-remains of arable weeds retrieved from archaeological sites can provide an insight into past agricultural practices through aspects of past ecology, such as soil conditions or the association of groups of weed species with specific crops (Willerding 1986, 1991).

Seeds of four arable weed species were recovered with the cereal remains: *Lathyrus hirsutus, Lupinus digitatus* (Figure 12.8), *Phalaris minor,* and *Vicia* sp. These taxa commonly infest cereal fields in the Nile valley and delta (Amry 1981; El-Bakry 1982). There are archaeobotanical records of *Lathyrus hirsutus* and *Phalaris minor* from Predynastic to Graeco-Roman sites, whereas *Lupinus digitatus* is known only from the Pharaonic period and beyond (Cappers et al. 2007; Fahmy 1997, p. 243). Täckholm (1961) reported seeds of *Lupinus digitatus* from Middle Kingdom Abusir and Graeco-Roman Hawara. These seeds are relatively large and heavy, meaning they would not have been easily separated from the cereal crop by either winnowing or sieving operations that, from tomb scenes at Saqqara, took place at the threshing floor (Murray 2000a, p. 525).

The ratio of weed seeds to cereals is very low, (0.04%) at the site compared to the Predynastic site of Hierakonpolis (locality HK11C, 3800–3500 cal B.C.E.), where the ratio of arable weed remains to cereal waste was considerably higher (36.5%) (Fahmy 1995, p. 119). Such differences may result from differences in the method of harvesting the cereals, for example, cutting the culms at about 40 cm above the soil level and leaving the rest as stubble to fertilise the soil. Alternatively, it may be that farmers applied a more effective system of weeding or harvested the cereals by gathering the ears only, thus reducing the possibility of collecting the fruits and seeds of arable weeds together with the crop ears. An additional possibility, given that such remains represent offerings, is that the grain at Saqqara was deposited in a more processed and fully cleaned state.

Figure 12.8 Seeds of *Lupinus digitatus.*

Edible Fruits

The remains of four species of wild and cultivated edible fruits were retrieved from AKT01 and AKT02: fig (*Ficus* sp.), persea (*Mimusops laurifolia,* Figure 12.9), date palm (*Phoenix dactylifera*), and Christ's thorn (*Ziziphus spina-christi,* Figure 12.10). The presence of such fruits suggests their importance to the Middle Kingdom inhabitants, enriching their diet with several sources of carbohydrates and proteins.

Pips of fig were found in the west and east chambers of AKT02. Two species of fig are known from archaeological sites in Egypt—*Ficus sycomorus* L. and *Ficus carica* L.—and although morphological similarities can make identification to species difficult (Murray 2000b), some differences can be noted that may permit

Figure 12.9 Seed of *Mimusops laurifolia.*

Figure 12.10 Stones of *Ziziphus spina-christi.*

identification to species in the future (Cappers, Neef, and Bekker 2009).

The sycamore fig tree is native to the south Arabian peninsula and tropical east Africa and certainly was cultivated in the past (Zohary and Hopf 1993, p. 156), being recorded from the 1st Dynasty onward (Hepper 1990, p. 58).

One stone of persea (*Mimusops laurifolia*) was discovered in the transverse hall of chamber AKT01. Fruits, leaves, and wood of persea are frequently found on many archaeological sites in Egypt and are first known from the 3rd Dynasty in the tomb of Djoser in Saqqara (de Vartavan and Asensi Amoros 1997, pp. 173–76). Persea is native to the hills of Ethiopia and Yemen; hence it must have been deliberately cultivated in Egypt, where it became a popular tree in gardens (Hepper 1990, p. 15; Manniche 1999, p. 121). The persea also had religious significance during the Pharaonic period Murray (2000b, p. 625).

Date palm stones and fibres were recovered from the west chamber of AKT02. The palm tree is a multipurpose species and is known in Egypt as a cultivated palm (Hepper 1990, p. 62). Wild relatives of the date palm (*Phoenix dactylifera*) are distributed in the southern Near East and the northeast Saharan and north Arabian deserts (Zohary and Hopf 1993, p. 158). The leaves were used in ancient Egypt for thatching, basketry, sandals, nets, and similar items (Germer 1989, p. 233), and the fruits were eaten raw or fermented with wine. Remains of this plant are known from Egyptian archaeological sites from the Predynastic period onward (de Vartavan and Asensi Amoros 1997, pp. 193–99). Finds of date palm stones from Middle Kingdom sites are limited and small in number; the richest records are from the New Kingdom and post-Pharaonic periods (Murray 2000b, p. 619).

It may be that the Ancient Egyptians were not familiar with the artificial pollination of date palms before the Middle Kingdom (Täckholm and Drar 1950, p. 216). The disproportion of male to female trees within a population makes such fertilisation necessary in order to ensure that all female trees produce full fruits. Date palm fruits produced by wind pollination are very small and less fleshy.

Stones and fruits of Christ's thorn were recovered from both rock-cut chambers, usually mixed with tubers of tigernut grass (*Cyperus esculentus*), grains of hulled barley (*Hordeum vulgare*), and emmer wheat (*Triticum dicoccum*). A pure deposit of the fruits and stones of Christ's thorn was discovered in the east chamber of AKT02. Remains of Christ's thorn are common from the Predynastic period onward and have been found previously among offerings placed within tombs (de Vartavan and Asensi-Amoros 1997, pp. 281–83). The sacred role of this tree in ancient Egypt, a native to the Egyptian flora, is discussed briefly in Täckholm (1976). It grows in desert wadis, plains, and mountains and is now cultivated in villages and parks for its shade, timber, and fruits (Boulos 2000, p. 84).

Wild Species

This group includes the fruits, tubers, leaf, and culm fragments of four taxa—namely, *Cyperus* sp. (including *Cyperus esculentus*), *Juncus* sp., *Neurada procumbens*, and *Typha* sp. Eight tubers of tigernut (*Cyperus esculentus*, Figure 12.11) were retrieved from the transverse hall of rock-cut chamber AKT01. Five tubers from west chamber of AKT02 have also been attributed to the genus *Cyperus*. Tubers of *Cyperus* have been recorded by Hillman (1989) as food in the late Palaeolithic site at Wadi Kubbaniya in Upper Egypt. The archaeobotanical record of Egypt shows that *Cyperus* tubers are known from Predynastic, Pharaonic, and later sites (de Vartavan and Asensi Amoros 1997; Murray 2000b). The genus *Cyperus* includes 15 taxa considered to be water-loving species (Boulos 2005; Zahran and Willis 1992). Living populations of *Cyperus rotundus* have been monitored growing on terraces wetlands along the Nile Valley (Abdel Ghani and Fahmy 2001).

Culm fragments of *Juncus* were retrieved from the west chamber of AKT02 and probably originate from the mats reported to be there for the offerings. *Juncus* is common in the archaeobotanical record, with most finds refined to two species, *Juncus acutus* and *Juncus rigidus* (de Vartavan and Asensi Amoros 1997; Täckholm 1990), *J. rigidus* having more conspicuously furrowed culms (Greiss 1957). Both taxa were used widely by the ancient

Figure 12.11 Tubers of *Cyperus esculentus*.

Egyptians to make mats and baskets in the same way that they are used today (Täckholm and Drar 1950), as seen in particular at Predynastic Hierakonpolis (Fahmy 2003, 2005; Friedman et al. 2002).

CONCLUSION: OFFERINGS IN SAQQARA

In ancient Egypt, food was an essential commodity in death, as it was in life. Food offerings, models as well as the depictions of food offering, were placed in tombs to maintain the soul of the deceased. Emery (1962) reports on the discovery in tomb 3477, at Saqqara, of an intact cooked meal beside the burial of a noble woman of the 2nd Dynasty, comprising meat, bread, porridge of barley, stewed fruit (probably fig?), and Christ's thorn fruits. The tomb of King Tutankhamen (18th Dynasty, 1332–1322 B.C.E.) revealed food offerings including bread, beer, wine, honey, fruits, nuts, vegetables, and spices (Germer 1989; Hepper 1990; de Vartavan 1999), as well as a model granary with spikelets of emmer wheat and barley and seeds of *Lens culinaris*, *Trigonella foenum-graecum* L., *Cicer arietinum* L., and *Pisum* species (Germer 1989).

The deposition of botanical material inside the storage rooms for cult statues can be interpreted as part of special final ritual offering in honour of the statues, linked to religious/funerary practices during the Middle Kingdom in Saqqara.

The type and content of food offerings differ from those deposited in royal and elite tombs, with the botanical material from both chambers being relatively basic in form and limited in diversity. Wild fruits and tubers, as well as cultivated cereals, were a major constituent of the subsistence strategy prevailing during this period. Additionally, the tree fruits present—sycamore fig, persea, Christ's thorn, and date palm—all had religious significance and are all consistent with what was being cultivated during the Middle Kingdom (Murray 2000b). Distribution analysis of botanical material shows variation within different sectors of the chambers that may have functioned as special locations in which to set offerings.

REFERENCES

Abdel Ghani, M., and A. G. Fahmy (2001) Analysis of aquatic vegetation in islands of the Nile, Egypt. *International Journal of Ecology and Environmental Sciences 27*, 1–11.

Amry, M. (1981) *Plant Life in Minya Province, Egypt*. Ph.D. thesis, University of Cairo.

Boulos, L. (2000) *Flora of Egypt*, Vol. 2. Cairo: Alhadara Publishing.

———. (2005) *Flora of Egypt*, Vol. 4. Cairo: Alhadara Publishing.

Cappers, R. T. J., R. Neef, and R. M. Bekker (2009) *Digital Atlas of Economic Plants*. Groningen Archaeological Studies 9. Groningen: Barkhuis Publishing.

Cappers, R. T. J., L. Sikking, J. C. Darnell, and D. Darnell (2007) Food supply along the Theban desert roads (Egypt): The Gebel Romaa, Wadi el-Huôl, and Gebel Qarn el-Gir caravansary deposits. In R. T. J. Cappers (Ed.), *Fields of Change: Progress in African Archaeobotany*, Groningen Archaeological Studies 5, pp. 127–38. Groningen: Barkhuis Publishing.

de Vartavan, C. (1999) *Hidden Fields of Tutankhamun*. London: Triade Exploration.

de Vartavan, C., and V. Asensi Amoros (1997) *Codex of Ancient Egyptian Plant Remains*. London: Triade Exploration.

El-Bakry, A. (1982) *Studies on Plant Life on Cairo-Ismailiya Region*. M.Sc. thesis, University of Cairo.

Emery, W. B. (1962) *A Funerary Repast in an Egyptian Tomb of the Archaic Period*. Leiden: Nederlands Instituut Voor Het Nabije Oosten.

Fahmy, A. G. (1995) *Historical Flora of Egypt, Preliminary Survey*. Ph.D. thesis, University of Cairo.

———. (1997) Evaluation of the weed flora of Egypt from Predynastic to Graeco-Roman Times. *Journal of Vegetation History and Archaeobotany* 6 (4), 241–47.

———. (2003) Palaeoethnobotanical studies of Egyptian Predynastic cemeteries: New dimensions and contributions. In K. Neumann, A. Butler, and S. Kahlheber (Eds.), *Food, Fuel and Fields*, pp. 95–106. Köln: Heinrich-Barth-Institut

———. (2005) Missing plant macro remains as indicators of plant exploitation in Predynastic Egypt. *Journal of Vegetation History and Archaeobotany 14*, 287–94.

Friedman, R., E. Watrall, J. Jones, A. G. Fahmy, W. van Neer, and V. Linseel (2002) Excavations at Hierakonpolis. *Archeo-Nil 12*, 55–68.

Germer, R. (1989) *Die Pflanzenmaterialien aus dem Grab des Tutanchamun*. Hildesheim: Gerstenbergverlag.

Greiss, E. (1957) *Anatomical Identification of Some Ancient Egyptian Plant Material*, Mémoires de l'Institut d'Égypte, 55. Le Caire: Tsoumas.

Hassan, F. (1984) Toward a model of agricultural development in Predynastic Egypt. In L. Krzyzaniak and M. Kobusiewicz (Eds.), *Origin and Early Development of Food Processing Cultures in North Eastern Africa*, pp. 221–24. Poznan: Polish Academy of Sciences.

Hepper, F. N. (1990. *Pharaoh's Flowers: The Botanical Treasures of Tutankhamun*. London: HMSO.

Hillman, G. (1989) Late Palaeolithic plant foods from Wadi Kubbaniya in Upper Egypt: Dietary diversity, infant weaning and seasonality in a riverine environment. In D. R. Harris and G. Hillman (Eds.), *Foraging and Farming: The Evolution of Plant Exploitation*. London: Unwin Hyman.

Kawai, N. (2011) An early cult center at Abusir-Saqqara? Recent discoveries at a rocky outcrop in northwest Saqqara. In R. Friedman and P. N. Fiske (Eds.), *Egypt at Its Origins 3: Proceedings of the Third International Conference, Origin of the State. Predynastic and Early Dynastic Egypt*, London, July 27–August 1, 2008, Orientalia Lovaniensia Analecta, 205, pp. 801–30. Leuven: Peeters.

Manniche, L. (1999) *An Ancient Egyptian Herbal*. London: British Museum Press.

Murray, M. (2000a) Cereal production and processing. In P. T. Nicholson and I. Shaw (Eds.), *Ancient Egyptian Materials and Technology*, pp. 505–36. Cambridge: Cambridge University Press.

———. (2000b) Fruits, vegetables, pulses and condiments. In P. T. Nicholson and I. Shaw, *Ancient Egyptian Materials and Technology*, pp. 609–55. Cambridge: Cambridge University Press.

Täckholm, V. (1961) Botanical identification of the plants found at the monastery of Phoebammon. In C. Bachatly (Ed.), *Le monastère de Phoebammon dans la Thébaide, III*, pp. 1–38. Cairo: Societe d'Archaeologie Copte.

———. (1976) Ancient Egypt, landscape, flora, agriculture. In J. Rzoska (Ed.), *The Nile Biology of an Ancient River*, pp. 51–68. The Hague: Junk.

———. (1990) Botanical examination. In F. Debono and B. Mortensen (Eds.), *El Omari: A Neolithic Settlement and other Sites in the Vicinity of Wadi Hof, Helwan* 82, pp. 115–16. Mainz am Rhein: Philip Von Zabern.

Täckholm, V., and M. Drar (1950) *Flora of Egypt*, Vol. 2. Cairo: Fouad I University Press.

van Zeist, W., G. Roller, and A. G. Fahmy (2003) An archaeobotanical study of Ma'adi, a Predynastic site in lower Egypt. In W. van Zeist (Ed.), *Reports of Archaeobotanical Studies in the Old World*, pp. 167–207. Groningen: The Groningen Institute of Archaeology, University of Groningen.

Willerding, U. (1986) *Zur Geschichte der Unkräuter Mitteleuropas*. Neu Münster: Karl Wachholtz Verlag.

———. (1991) Präsenz, Erhaltung und Repräsentanz von Pflanzenresten in archäologischem Fundgut. In W. van Zeist, K. Wasylikowa, K.-E. Behre (Eds.), *Progress in Old World Palaeoethnobotany*, pp. 25–51. Rotterdam: Balkema.

Winlock, H. E. (1955) *Models of Daily Life in Ancient Egypt from the Tomb of Meketere at Thebes*. Cambridge, MA: Harvard University Press.

Yoshimura, S., and N. Kawai (2003) Finds of the Old and Middle Kingdoms at North Saqqara. *Egyptian Archaeology* 23, 38–40.

Yoshimura, S., N. Kawai, and H. Kashiwagi (2005) *A Sacred Hillside at Northwest Saqqara: A Preliminary Report on the Excavations 2001–2003*, Mitteilungen des Deutschen Archäologischen Instituts Abteilung Kairo 61. Mainz am Rhein: Verlag Philipp Von Zabern.

Zahran, M., and A. J. Willis (1992) *The Vegetation of Egypt*. London: Chapman and Hall.

Zohary, D., and M. Hopf (1993) *Domestication of Plants in the Old World* (2nd ed.). Oxford: Clarendon.

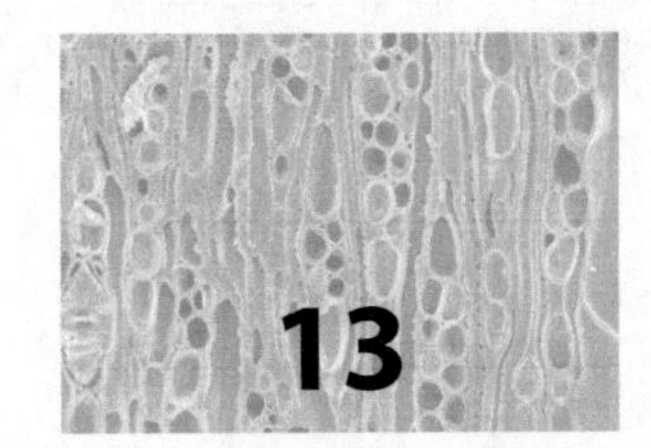

13

Botanical Insights into the Life of an Ancient Egyptian Village

Excavation Results from Amarna

Chris J. Stevens and Alan J. Clapham

Although preservation by desiccation occurs in more remote regions of the world, its relatively common occurrence on Egyptian sites provides a unique insight into many aspects of an ancient civilisation unseen elsewhere in the Old World.

Possibly the single greatest collection of desiccated botanical material from Egypt was that recovered from the Tomb of Tutankhamun, initially studied by Newberry (1927) and later by Germer (1989), de Vartavan (1990), and Hepper (1990). Although these studies provided an insight into the range of plants used in the New Kingdom, the study of desiccated plant remains recovered from excavations at Amarna, the probable birthplace of Tutankhamun, provided a unique opportunity to directly relate preserved botanical material to the everyday lives of the city's inhabitants.

The city of Akhetaten, Amarna, was founded by Akhenaten around 1341 B.C.E., between the two time-honoured capitals, Memphis and Thebes (Figure 13.1). Breaking custom with the established burial ground in the Valley of the Kings, Akhenaten situated the Royal Tombs in a wadi to the east of the city, and nearby those of his entourage—his family, advisors, and administrators. The work on these tombs led to the establishment of two villages, the Stone Village and the Workmen's Village, where the workers and craftsmen integral to the successful transition of the living into the afterlife were housed. In addition to these two desert villages were a series of palaces and temple complexes situated close to the city's suburbs and the Nile (Figure 13.1).

The occupation of Akhetaten lasted just 20–22 years, extending into the short-lived reign of Tutankhamun. After, the city was not only abandoned but much of it systematically destroyed (Kemp 1989, pp. 267–80). This short-lived settlement means that deposits from excavations can be directly related to the area or building from which they were recovered, and they have been little disturbed by later occupation. Such a scenario creates a high potential for relating archaeobotanical remains to the activities conducted within the areas from which they are recovered. Early excavations at the Workmen's Village had yielded many organic remains (Peet and Wooley 1923), but it was not until the excavations conducted by Professor Barry Kemp that sampling for botanical material was routinely conducted.

Between 1979 and 2006 over 300 samples from Pharaonic deposits were taken from excavations at the main city, a temple complex at Kom-el Nana, and the Workmen's Village; further samples from the Late Antique phases of the excavations were studied by Smith (2003). A preliminary analysis of material from the quarry site at the Workmen's Village was undertaken by Renfrew (1985), and subsequent studies have been conducted by Samuel (1994, 1995), concentrating on bread and beer making (1989, 1996, 1999, 2000).

Archaeology of African Plant Use by Chris J. Stevens, Sam Nixon, Mary Anne Murray, and Dorian Q Fuller, Eds., 151–164

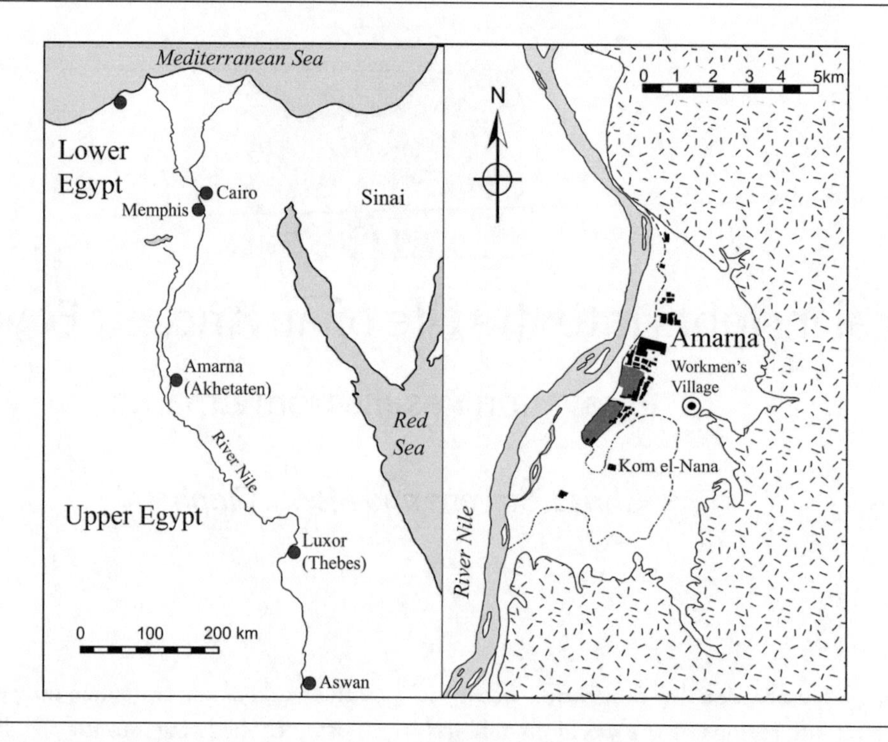

Figure 13.1 Map showing location of Tell el-Amarna (from Wilkinson and Stevens [2003]).

We began work in the 2000 season, and over subsequent field seasons the vast majority of the samples have been systemically studied (Clapham and Stevens 2009, 2012; Stevens and Clapham 2010; Wilkinson and Stevens 2003, pp. 275–94). This chapter provides a summary of our findings primarily from the Workmen's Village, but it brings into the discussion some of the desiccated remains from the temple complex at Kom-el Nana.

The Workmen's Village

The village can be divided up into six distinct areas or area types from which archaeobotanical samples were obtained. The following descriptions provide information on the role, possible function, and activities conducted within each area, as well as the potential for the preservation of plant remains.

Site X1 and the Zir Area

A trail of broken pottery delineated the passage that all goods took from the Central City (Renfrew 1987), including water, cereals, and animals, and this trail stopped at an area (Site X2, not shown) just short of a series of buildings to the southwest of the village: Site X1 (Figure 13.2). The function of these buildings was unclear, but given their position they may have had an administrative function, with scribes accounting for produce leaving and entering the village, and perhaps also housing the village guards (Kemp 1980, 1987b). Evidence for beadworking raised a further possibility that they were used for crafts (Kemp 1980), and dumps of horn-cores suggested the removal of horns (Luff 1994). Finally, the high presence of goat droppings indicated the housing of animals (Kemp 1987b). It was hoped the examination of botanical material might shed further light on the role of these buildings.

Past these buildings was the *Zir* area, named from the Arabic word for the large pottery vessels found there (Kemp 1984, pp. 60–80; Figure 13.2). Water arriving at the village was probably stored here, before being distributed among the villagers for their own consumption, for their animals, and for watering the chapel gardens (Kemp 1987b). Samples were taken from this area, although organic preservation was noted to be poor.

The Quarry

Sandwiched between the *Zir* area and animal pens, a quarry was dug in the early days of the village's existence to provide marl for the houses and chapels (Kemp 1984, pp. 81–88; 1987b, p. 28). During its subsequent life the quarry became the dumping ground for much of the village waste, and so it provides some of the best organic preservation seen on the site.

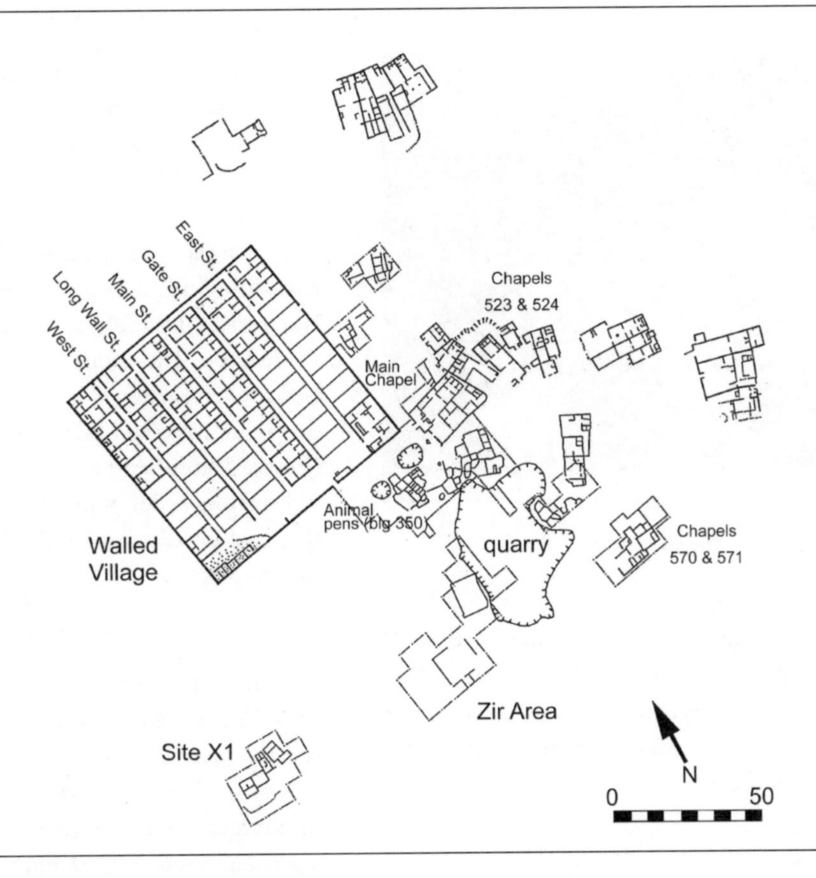

Figure 13.2 Map of the Workmen's Village (adapted from Kemp and Vogelsang-Eastwood [2001]).

Figure 13.3 Pig Pens at the Workmen's Village under excavation.

THE ANIMAL PENS

Skirting the edge of the quarry to the north and east were a number of small mudbrick buildings (about 7 groups) with troughs and low doorways interpreted as animal pens (Figure 13.3). Their form and layout, along with the identification of bristles and dung, indicated their use for raising pigs (Kemp 1984, 1986, 1987b, p. 40).

THE WALLED VILLAGE

To the north of the pens was an enclosed area measuring some 70 by 70 m, the Walled Village, comprising 5 streets and some 73 houses (Kemp 1984, pp. 1–13). On the ground floor each house comprised three segments, split into up to 6 rooms, with a distinct possibility that many also had a second storey (Figure 13.4). The front rooms appear to have been used for weaving, and the presence of mortar and quern emplacements implies their use also for cereal processing (Kemp and Vogelsang-Eastwood 2001), and troughs indicated the presence of animals (Kemp 1987a). Often one of the rear rooms appears to have served as a kitchen. Many of the houses had been looted or excavated by Peet and Wooley (1923), although several yielded relatively undisturbed deposits from which samples were obtained.

CHAPELS

On the east of the walled village a number of chapels were built by local families, perhaps in honour of traditional deities, some probably late in its history, possibly after Akhenaten's death (Bomann 1991). Their role seems to have extended beyond simply that of worship, and within

Figure 13.4 Workmen's Village House.

the courtyards animals were probably sacrificed and small crafts, including spinning, conducted (Hecker 1985; Kemp 1987b, pp. 33–36). Organic preservation varied but potentially provides the opportunity to examine the activities conducted within them.

Gardens

Associated with the chapels, or laid out next to the pens, were a number of garden plots. In some cases these were enclosed within walled courtyards (for example, Chapel 531, Kemp 1985, pp. 39–50; Figure 13.5), in others laid out in grid squares of 16–20 plots, each measuring some 0.5 to 1 m square (Kemp 1985, pp. 29–32; 1987a, pp. 47–55; Figures 4.2, 4.5, and 4.6, pp. 56–62). The fill of these plots was notably different from anything else encountered within the village, consisting of dark soils developed from the alluvial Nile silts. Such plots are commonly depicted in tomb scenes dating from the Old to the New Kingdom (Naville 1906, plate CXLII; Duell 1938, plate 21) and have been found on excavations in Nubia, at Mirgissa, (Vercoutter 1967–1968, plate LIV[b], 1970, plate Xa) and Amara West (Shinnie 1951), as well as at Amarna (Borchardt and Ricke 1980, p. 237; Griffith 1924, p. 303). They are usually thought to have been used for growing flowers and vegetables (Kemp 1987b), although no direct botanical evidence has been recovered to date. Organic preservation was generally poor, in part attributed to the fact that these plots are one of the few places in which soil formation and pedological processes (including the accompanying micro- and macro-fauna responsible for the breakdown of plant material) could take place. The presence of soil fauna coupled with the probability that much of the produce would have been removed meant that the potential for recovering the remains of the plants grown within them was poor.

Kom-el Nana

The complex at Kom-el Nana comprised a walled enclosure with a large temple, formal gardens, and several service buildings, including a bakery and probable workshops. The south and north pavilion of the temple had sunken garden plots within them, and additional garden plots were laid out in grids of up to 70 square plots (Figure 13.6). Samples were taken from all these plots, as well as from other buildings within the complex. However, this chapter discusses only those samples relating to the garden plots.

Sample Analysis

Samples averaged 2–7 litres, and processing varied according to their composition. Those rich in organic remains were dry-sieved through a stack of sieves to 300 microns, whereas less rich samples were floated in order to concentrate the organic remains. The resultant flots, again collected at 300 microns, were thoroughly dried before being bagged (see Smith 2003 for a more complete account).

Neither method was necessarily ideal, dry-sieving being labour intensive, especially in the sorting and extraction of finer fractions. Flotation, although quicker, had two distinct disadvantages. The first was that more fragile remains, for example, leaves, capsules, flowerheads, were likely to become disarticulated. Second, many samples contained large mudbrick fragments that disintegrated when floated, releasing plant material used as temper. Although the gradual disintegration of mudbrick in the millennia since the village had been abandoned no doubt released some organic material, the removal of mudbrick during dry-sieving serves to minimise such intrusive elements. The current preferred method is to dry-sieve the larger fractions and to float fractions less than 1–2 mm, according to the ratio of organic material to sand, stone, and mudbrick.

Botanical material was quantified and identified where possible using reference material of economic plants brought from England and Egyptian Markets, collected from the local farmer's fields and wadis, or in some cases brought from other parts of Egypt. Extensive use was also made of the various Egyptian floras, including Täckholm's *Student's Flora of Egypt* (1974) and the *Flora*

Figure 13.5 Garden plots in the courtyard of Chapel 531 showing excavation and reconstruction.

Figure 13.6 Kom el-Nana: Open garden plots after excavation with the dark Nile soil removed. Each plot is approximately 1 m square.

of Egypt by Loufty Boulos (1999, 2000, 2002, 2005). Table 13.1 lists all the species mentioned in the text, the authority, and the parts recovered.

Preservation and Taphonomic Considerations

The preservation of desiccated material in the samples was highly selective. It is not just the absence of water that allows such preservation but also the absence of the soil micro-organisms that are associated with the breakdown of plant material. However, many locally present larger organisms, such as birds, mice, rats, gerbils, and insects, would still have predated on fresh organic material. Consequently, the more readily digestible plant remains, such as grains, stems, and leaves, often did not survive, unless they were in deposits that had been rapidly buried. This situation provided an interesting contrast with charred material frequently containing cereal grains and occasionally those of lentil (*Lens culinaris*), which were generally absent from the desiccated remains.

A frequent observation on European sites has been that most charred assemblages consist of cereal grains, chaff, and the accompanying weed seeds (Knörzer 1971). The remains from Amarna provide a unique and direct opportunity to compare the two forms of preservation on a settlement site. Noticeably, with the exception of occasional isolated events, it was almost only cereal remains and wood that were recovered in charred state. When we are confronted with the vast array of species preserved by desiccation, this exclusive association of charred material with cereal remains, and by analogy with crop-processing, is clearly seen.

"Background Flora"

A good deal of variation was seen across the village in the botanical results, occurring not only in the state of preservation but also in the occurrence and association of certain species and categories of botanical remains with certain parts of the village. However, three factors could be seen to create what we might term a 'background flora'.

Plants Used in Building Construction

The first factor leading to the occurrence of more homogenous assemblages were species whose presence was largely a result of their incorporation in building materials, the principal source being mudbrick. To isolate this form of contamination and to study the use of plant material to temper mudbrick, a few bricks were collected from the village, broken down in water, and the organics floated off. The bricks yielded rachises of 6-row barley (*Hordeum vulgare* subsp. *vulgare*) but only occasional paleas and lemmas. Remains of emmer wheat (*Triticum turgidum* ssp. *dicoccum*) included whole spikelets, terminal spikelets, and basal rachis fragments but few individual glume bases or spikelet forks. Additionally, occasional stems of cereals, flax (*Linum usitatissimum*), and seeds of *Lolium* cf. *temulentum*, *Echium* sp., *Anagallis* sp., and *Ambrosia maritimus* were recovered. Given that cereals entered the village as spikelets or hulled grains to be dehusked within the villagers' houses (see sections 'Site X1 and the Zir Area' and 'The Walled Village'), the composition of the bricks indicate the use of threshing waste generated in processing immediately following harvest in the fields and agricultural villages, rather than waste rich in barley hulls or emmer glumes from the houses and bakeries of the city.

Roofing fragments from the Main Chapel (R20) contained reed and rope mesh impressions (Kemp 1985, p. 9), in the case of the former of *Arundo* sp. and/or *Phragmites* sp. and for the latter probably of halfa (*Desmostachya bipinnata*/*Imperata cylindrica*) or date palm fibre (*Phoenix dactylifera*). Fragments of date palm leaves and fibres, along with fragments of rope and twine, were common in many samples, as were the root stems of halfa grass. Although no attempt was made in this study to distinguish *Desmostachya* from *Imperata*, Hepper (1990) identified the former only within the Tomb of Tutankhamun. Fragments of plaster from the chapels were also noted to contain 6-row barley rachises.

Table 13.1 Species Recovered from the Botanical Samples at el-Amarna

Species	Parts	Family	Species	Parts	Family
Acacia cf. *nilotica* (L.) Delile	S, T, St, L	Leguminosae	*Fimbristylis bisumbellata*	S	Cyperaceae
Allium cepa L.	Tu	Liliaceae	*Francoeuria* sp. and/or *Pulicaria* sp.	Fl	Asteraceae
Allium sativum L.	Tu	Liliaceae	*Heliotropium bacciferum* Forssk.	S	Boraginaceae
Ambrosia maritima L.	S	Asteraceae	*Hordeum vulgare* subsp. *vulgare* L.	S, C	Poaceae
Amygdalus communis L.	S	Rosaceae	*Hyphaene thebaica* (L.) Mart.	S/F	Palmae
Asphodelus sp. L.	S	Asphodelaceae	*J. oxycedrus* L./*J. excelsa* M. Bieb. Possible *J. phoenicea* L.	S, cf. L	Cupressaceae
cf. *Anthemis pseudocotula* Boiss.	Fl	Asteraceae	*Lathyrus sativus* L.	S	Leguminosae
Apium graveolens L.	S	Apiaceae	*Lens culinaris* Medik.	S	Leguminosae
Artemisia judaica L.	L	Asteraceae	*Linum usitatissimum* L.	S, F, St	Lineaceae
Arundo sp. L./*Phragmites* sp. Adanson	Im	Poaceae	*Lolium* cf. *temulentum* L.	S, C	Poaceae
Balanites aegyptiaca (L.) Delile	S	Zygophyllaceae	*Lupinus albus* L.	cf. S	Leguminosae
Ocimum cf. *basilicum* L.	S	Lamiaceae	*Mimusops laurifolia* (Forssk.) Frilis *M. schimperi* Hochst.	S	Sapotaceae
Bassia muricata (L.) Asch	S	Chenopodiaceae	*Nigella sativa* L.	S	Ranunculaceae
Beta vulgaris L.	S	Chenopodiaceae	*Olea europaea L.*	S, L	Oleaceae
Carthamus tinctorius L.	S	Asteraceae	*Phalaris* cf. *minor L.*	S	Poaceae
Citrullus colocynthis (L.) Schrad.	S	Cucurbitaceae	*Phalaris paradoxa* L	S	Poaceae
Citrullus lanatus (Thunb.) Mats. & Nakai	S	Cucurbitaceae	*Phoenix dactylifera* L.	S, L, ?Fl	Palmae
Coriandrum sativum L.	S	Apiaceae	*Picris* sp. L.	S	Asteraceae
Cornulaca cf. monacantha Delile	S	Chenopodiaceae	*Ricinus communis* L.	S	Euphorbiaceae
Cucumis cf. *melo* L.	S	Cucurbitaceae	*Rumex dentatus* L.	S, F	Polygonaceae
Cyperus esculentus L.	Tu, cf. S	Cyperaceae	*Schouwia purpurea* (Fofssk.) Schweinf	S, F	Cruciferae
Crypsis sp.	S	Poaceae	cf. *Sinapis alba* L. *Raphanus* sp. L.	F	Cruciferae
Daucus cf. *carota* L.	S	Apiaceae	*Tamarix aphylla* (L.) H. Karst	L	Tamaricaceae
Desmostachya bipinnata (L.) Stapf/*Imperata cylindrica* (L.) Beauv	St	Poaceae	*Trigonella foenum-graecum* L.	S	Leguminosae
Echinops spinosisimus Turra	F/Fl	Asteraceae	*Triticum turgidum* L. ssp. *dicoccum* (Shrank) Thell.	S, C	Poaceae
Echium cf. *rubrum* Forssk. *E. angustifolium* (Mill.)	S	Boraginaceae	*Vitis vinifera* L.	S	Vitaceae
Echium rauwolfii Delile	S	Boraginaceae	*Withania somnifera* (L.) Dunal.	S, F	Solonaceae
Emex spinosa (L.) Campd.	S	Polygonaceae	*Zilla spinosa* (L.) Prantl.	S, cf. St	Cruciferae
Fagonia sp. L.	S, F	Zygophyllaceae	*Ziziphus spina-christi* (L.) Desf.	S, L, St, T	Rhamnaceae
Ficus sycomorus L.	S, F, cf. L	Moraceae	*Zygophyllum* spp. L.	S	Zygophyllaceae

Key: S = seed and/or stone; F = fruits; L = leaves; St = stems; T = thorns; Tu = tuber; C = chaff; Fl = flower-head; Im = impressions.

Material Blown by Wind

The second factor resulting in sample homogeneity was material distributed across the site by wind. Today in the city ruins, free-threshing wheat chaff readily gathers in the corners of buildings closest to the cultivation strips edging onto the Nile. Such modern material itself is not a potential source of contamination, in part because today cereals are grown that are different from those encountered in the past, but also because only secure deposits were sampled, and the Workmen's Village lies some 2 km into the desert. However, this modern observation demonstrates how cereal chaff could spread into locations away from areas where cereals were stored and processed. Given the absence of glumes within mudbrick, the redistribution of processing waste by the wind may explain the presence of emmer glumes within many contexts across the village.

Wadi Species and Locally Common Trees

The final factor is rogue plants growing in and around the village. Today, with the exception of the occasional plant of *Zilla spinosa*, the village is devoid of vegetation, although the wadi running a few hundred metres away is well vegetated. In the past, the bringing of large quantities of water into the village would have created ecological niches that many wadi species were able to colonise—growing as weeds in the chapel gardens, around the water troughs in the animal pens, and indeed in every area of the village where water was used and spilled.

In the samples were many species identifiable in the wadi today: *Echinops spinosisimus* (Figure 13.7), *Artemisia judaica, Zilla spinosa, Zygophyllum* spp., *Heliotropium bacciferum, Schouwia purpurea, Fagonia* sp., *Emex spinosa,* and *Echium rauwolfii*. Seeds of two other species, *Cornulaca* cf. *monocantha* and *Bassia muricata* var. *tencifolia,* were frequent in the samples but, despite being readily identifiable in the wild elsewhere, have not been recorded within the Amarna area itself.

Figure 13.7 Unopened flowerheads of *Echinops spinosisimus.*

Although the presence of wadi species signifies plants growing within the village, the frequent occurrence of thorns, leaves, and fruits of three locally common tree/shrub species, *Acacia* cf. *nilotica, Tamarix aphylla,* and *Ziziphus spina-christi*, is less easily deciphered. Such plants may have been brought for building or fuel but possibly come from trees deliberately planted within the village.

Activity Areas in and around the Workmen's Village: A Botanical Perspective

Site X1 and the Zir Area

The samples from Site X1 contained a few hundred spikelets of emmer and a large number of small rodent droppings. As might be expected, given the presence of these rodents along with insect pests (Panagiotakopulu 2001), few of the spikelets contained grain, but the remaining chaff, including the paleas and lemmas, was relatively intact, suggesting it had not yet been pounded (see Samuel 1994 for a description of rodent-gnawed spikelets). The presence of goat droppings might suggest it was destined to be used as animal fodder, yet examination of these droppings—although indicating the presence of *Lolium* sp., rushes, (Juncaceae/Cyperaceae) and leaves of date—provided no evidence for cereals (Hosey 1984). Rather, the presence of emmer spikelets and also flax capsules suggest the temporary storage of goods within the buildings, including crops and perhaps young goats, having been off-loaded and checked at Site X2. The administrative buildings also produced some 50 partially charred seeds of colocynth (*Citrullus colocynthis*). Seeds of this species are roasted to remove their bitterness, then eaten in manner similar to pumpkin seeds; they occur in earlier Egyptian sites (Zohary and Hopf 2000, p. 193).

The samples from the *Zir* area yielded very few botanical remains, other than occasional remains of cereals and flax and general background material, in part because there was little chance for the burial of such remains.

The Quarry

Samples from the quarry were among the first to be investigated (Renfrew 1985) and yielded the widest range of species found anywhere within the Amarna area. Finds from the current and previous study included seeds, stones, and nuts of grape (*Vitis vinifera*), dôm palm (*Hyphaene thebaica*), desert date (*Balanites aegyptiaca*), Persea tree (*Mimusops schimperi*), probable melon (*Cucumis* cf. *melo*),

Figure 13.8 Flowers of date palm (*Phoenix dactylifera*).

watermelon (*Citrullus lanatus*), lettuce (*Lactuca sativa*), fenugreek (*Trigonella foenum-graecum*), almond (*Amygdalus communis*), grass-pea (*Lathyrus sativus*), lentil, the epidermis, cloves, or bulbs of onion (*Allium cepa*) and garlic (*Allium sativum*), tubers of tiger-nuts (*Cyperus esculentus*), and stones and flowers of date palm (Figure 13.8).

The rapid burial of this material meant that desiccated edible seeds and cereal grains were present, whereas they were generally absent from most samples. Charred remains of cereal chaff, predominately emmer glume bases and spikelets, were abundant within the dump but found in quantity only in two other locations in the village.

The Animal Pens

The build-up of organic waste had led to good preservation of a large number of edible species, including frequent finds of fig seeds and whole or partial fragments of fruit (*Ficus sycomorus*). Other finds included seeds of probable melon and watermelon and occasionally grape pips, which were otherwise largely absent from the village. These finds suggest that food waste was fed to sows and piglets, although possibly nursing mothers were let out occasionally to roam the village for waste scraps. Large amounts of barley, emmer chaff, and spikelets suggest the pigs were also regularly fed cereals, found both in the pens and within fragments of pig dung.

The Walled Village

Five of the houses were sampled for botanical remains: Long Wall Street number 6, West Street numbers 2 and 3, and Gate Street numbers 8 and 9. The preservation of botanical material was generally poor, the most abundant remains being cereal chaff, along with fragments of flax capsules and seeds of fig, watermelon, and probable melon.

High numbers of desiccated emmer glumes and spikelet forks were found around the mortar emplacements in the houses, indicating emmer was brought into the houses as spikelets (Samuel 1989, 1994). Most houses contained a mortar, quern, and an oven, indicating that processing and baking activities were conducted separately within each household, although their absence in some houses may indicate that they were shared (Kemp 1994), perhaps by large extended families occupying multiple houses. The houses also produced more charred cereal remains than seen elsewhere within the village, other than the quarry and one of the chapel annexes. It appears that chaff from processing was either used as tinder for the ovens within the houses or burnt as waste, the rakings being disposed of within the quarry when the ovens were cleaned.

Chapels

These samples provided an interesting contrast with the rest of the village, the first observation being that leaves and flowerheads were relatively common. Many of the leaves were highly fragmented, making identification difficult. Leaves of Christ's Thorn (*Ziziphus spina-christi*) were identified by the distinctive three nerves, and the presence of leaves, along with probable thorns and stems within the annex courtyard of Chapel 571, might even suggest the possible presence of a tree of this species. However, given leaves of olive (*Olea europaea*) were also recovered, which is unlikely to have survived in the village; an alternative explanation is that the leaves relate to the presence and probable making of floral-garlands within the chapels. Hepper (1990) certainly records such use of leaves of olive, willow, date, pomegranate, and ox-tongue (*Picris radicata*).

Hepper (1990) also records the presence of flowers of mayweed (*Anthemis pseudocotula*) and fruits of Withania (*Withania somnifera*) within floral garlands. In this light the high incidence of Asteraceae seedheads in the chapels, probably *Francoeuria* sp. and/or *Pulicaria* sp., but also possibly mayweed, might indicate their use in such garlands. Furthermore, a large number of Solanaceae seeds were noted and some dried 'berries'. Both closely resemble *Solanum dulcamara*, a species generally not found in Egypt, however, Hepper notes that seeds and fruits of *Withania somnifera* are frequently misidentified as *Solanum dulcamara* within ancient Egyptian samples. It seems therefore probable that the presence of this species within the chapels is perhaps attributable to its use in floral garlands, although it also plays an extremely important role within traditional medicines (Puri 2003; Thakur, Puri, and Hussain 1987).

Another common element in the chapel samples, especially the Main Chapel and its annex (Annex 450), were immature capsules (Figure 13.9) and stems of flax

Figure 13.9 Immature capsule of flax (*Linum usitatissimum*).

Figure 13.10 Terminal nodes of flax and a stem showing partial separation of the fibres.

or linseed. In particular it was notable that in some samples terminal nodes (where the stem joins the capsule; Figure 13.10) far outnumbered finds of capsules, suggesting that the capsules had been removed before the stems entered the deposits. Finally, these samples occasionally contained stems in which the fibres were half separated or separated fibres that appeared to have been in the processes of being twisted into twine (see Stevens 2001 for a more detailed description).

It is unlikely that stems were retted within the village, and no evidence has been forthcoming. Although it is possible that the stems arrived at the village, with bundles of retted fibres, an intriguing alternative is that flax stems were processed dry within the chapels using a process known as decortification (see Kemp and Vogelsang-Eastwood 2001, p. 30).

It is possible the presence of fragments of mature capsules is associated with their removal in the preparation of fibres, although the fact that they are to be found over much of the village, sometimes in bulk, suggests that capsules of flax were brought into the village for the express purpose of extracting the seed.

The final observation concerning botanical evidence from the chapels was that whereas chaff was generally rare, in several samples the incidence of desiccated and charred glumes and spikelet forks of emmer was extremely high. These samples were often in close proximity to bread ovens, in particular Annex 450, where two types of oven were present: one for making the bread, the other for making the moulds in which the bread was baked, probably for use as offerings to Aten (Samuel 1989). The finds indicate that, along with baking, the processing of grain and flour would also seem to have taken place within the chapels.

Gardens

While the gardens initially looked unpromising, of all the areas of the village they have provided some of the most fascinating results. The samples were by no means rich in botanical remains; however, a large number of species were well represented in the plots compared with elsewhere in the village. Most predominant were large numbers of fruits (both whole schizocarps and mericarps) of coriander (*Coriandrum sativum*; Figure 13.11) and celery (*Apium graveolens*). Given that both species are cultivated predominately for their seed, it is perhaps unsurprising to find the seed within the garden plots, whereas evidence for species cultivated for fruits, such as melon and watermelon, are less likely to be present.

Associated with Chapel 531 was a small courtyard with several rectilinear garden plots hugging the edges of

Figure 13.11 Whole schizocarps of Coriander (*Coriandrum sativum*)

Figure 13.12 Capsules of possible white mustard (*Sinapis alba*) or radish (*Raphanus* sp.)

the courtyard wall and path. Samples taken from these plots yielded, in addition to coriander, seeds of black cumin (*Nigella sativa*), juniper (*Juniperus* sp.), beet (*Beta vulgaris*), possible basil (*Ocimum* cf. *basilicum*), and capsules of possible white mustard (*Sinapis alba*) or radish (*Raphanus* sp.) (Figure 13.12), with very few noncultigens present.

Given the low numbers of seeds of some these species it is uncertain whether all were grown within the courtyard. Berries of *Juniperus oxycedrus* or *J. excelsa* were found in the tomb of Tutankhamun (Manniche 1999) but are likely to have been imported from the Mediterranean. Although only some seeds and fruit fragments of *Ficus sycomorus* were found within the garden plot samples, it is possible that trees could have been grown in the courtyard.

Some of the garden plots, as with the chapel samples, contained flowerheads, leaves, and seeds of probable *Withania somnifera* and species of Asteraceae. Probably at least some of the species used within floral garlands were grown in the garden plots. That the garden plots also contained frequent stems and leaves of Christ's Thorn (*Ziziphus spina-christi*), thorns, pinnules, and occasional pods of Acacia (*Acacia* cf. *nilotica*) and remains of tamarisk (*Tamarix aphylla*) again raises the possibility that these tree species were deliberately grown in the gardens to provide shade. In particular, *Ziziphus* has the added benefit of providing edible fruits, and acacia pods can be used as a green fodder, for tanning leather, or to produce a blue dye.

With respect to the possibility that trees were grown within the garden plots we might consider several plots that were cleared sometime late in the village's history in order to build Chapels 570 and 571 (Kemp 1985, pp. 33–35). These plots yielded a large number of charred stems and seeds of wadi species, in particular *Artemisia judica* and *Bassia muricata*, implying that the plots had been abandoned and overgrown before they were cleared by burning. Of particular interest, however, is that one plot contained a large number of charred and uncharred pinnules of *Acacia nilotica* and a large number of thorns. Although it is possible that dead branches of Acacia may have been used to fuel the clearance, their presence raises the distinct possibility that a tree had once grown within this part of the garden. A further sample from an adjacent plot to the northeast of Chapel 571, had fruits, leaves, stems, and possible thorns of Christ's thorn (*Ziziphus spina-christi*), again suggesting the presence of a tree in the garden.

THE GARDEN PLOTS AT KOM-EL NANA

The results from Kom-el Nana provided clear distinctions from those from the Workmen's Village. The plots at both the western and eastern ends of the south pavilion produced abundant organic remains, including seeds of grape, safflower (*Carthamus tinctoria*), castor (*Ricinus communis*), *Coriandrum sativum*, *Beta vulgaris*, *Apium graveolens*, *Withania somnifera*, and *Ocimum* cf. *basilicum*. Notably, many of these were very well represented compared to the plots in the village. Also present but in lesser numbers were olive stones, fruits/mericarps of carrot (*Daucus* sp.), Persea, juniper, black cumin, melon, and watermelon. The assemblages from the western end, which were better preserved, also produced leaves of tamarisk and olive.

Unlike in the gardens at the Workmen's Villages, the origin of all these species is unclear, and it is improbable that all were grown within the garden plots. Although it is feasible that some plants were dumped in the plots after their abandonment, it is possible that some may originate from offerings.

The samples from the unenclosed grid plots yielded less material. However, one set of plots produced several more mericarps of *Daucus* sp. (Figure 13.13), which

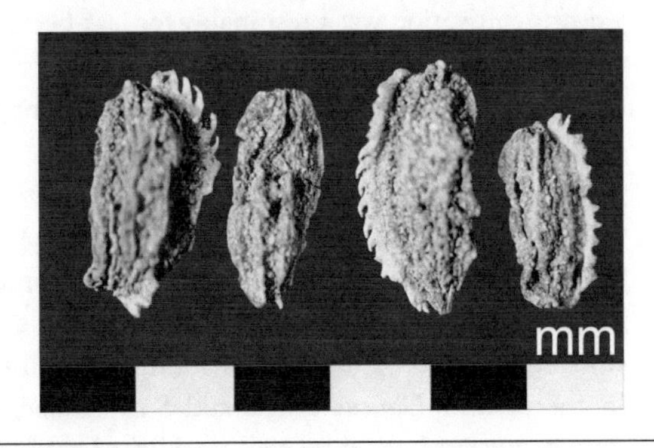

Figure 13.13 Possible fruits/mericarps of carrot (*Daucus* cf. *carota*). The specimen on the far right comes from the sunken garden plots, and the remainder come from a single open plot.

compare well to those of modern *Daucus carota*. Five wild species of *Daucus* sp. occur in Egypt today, and of these all but *Daucus syrticus* have prickles on the secondary ribs of the mericarp, a feature not seen on the archaeological specimens. However, in *Daucus syrticus* the prickles are equal in length to the width of the maricarp and appear to spread out at an acute angle to the ribs. In the archaeological specimen the prickles appear much shorter, and the angle of the prickles to the rib is acute, so that tips point to the apex rather than away from the seed (cf. Boulos 2000, pp. 182–84). Both of these features seen on the archaeological specimens are more characteristic of mericarps of carrot, although comparison with modern mericarps of all wild species of *Daucus* sp. has not yet been made.

CONCLUSION

Although the village at Amarna is far from typical, it still represents a rare glimpse into the lives of lower-status ancient Egyptians; this settlement, combined with the unique preservation of botanical material, provides many key insights into village life that have been hitherto unavailable. And in addition to these insights, the site has important implications for our understanding of the history of several cultivated species.

Many of the species recorded from Amarna had previously been recorded from the Tomb of Tutankhamun (de Vartavan 1990; Germer 1989; Hepper 1990; Newberry 1927). However, although earlier finds are known, the status of both beet (*Beta vulgaris*) and celery (*Apium graveolens*) as cultigens is relatively unclear (Murray 2000; Zohary and Hopf 2000). The finds of these species at Amarna provide important evidence that they had been brought into cultivation at this time. In the case of *Beta vulgaris* cultivation was presumably for leaf-beet, since the swollen root variety is unrecorded until the 16th century (Ford-Lloyds 1995). *Apium graveolens* was more probably cultivated for its seed than as a vegetable.

Three further species, carrot (*Daucus carota*), basil (*Ocimum* cf. *basilicum*), and potentially white mustard (*Sinapis alba*), are represented within the samples and in each case would provide the earliest records for their cultivation, although further clarification is still required. In the case of *Daucus carota* and *Sinapis alba*, classical sources suggest both were well established as cultigens by Hellenistic and Roman times (Zohary and Hopf 2000), but the finds from Amarna could push the potential domestication of these species back to before 1300 B.C.E. As for basil (*Ocimum* cf. *basilicum*), archaeological evidence is almost non-existent, although leaves of wild thyme (*Thymbra spicata*) were recovered from Tutankhamun's tomb (Hepper 1990).

ACKNOWLEDGEMENTS

We would like to thank Barry Kemp for his kind support and suggestions on many aspects of the findings within this report. Thanks also go to Gwil Owen for several of the site photographs and to Dorian Q Fuller for supplying some of the references. We would also like to thank Ruth Owen, Wendy Smith, and Delwen Samuel for the processing of the samples. Figure 13.1 was adapted from that of Keith Wilkinson. Funding for parts of the project has been provided by the McDonald Institute and the Egyptian Exploration Society.

REFERENCES

Bomann, A. (1991) *The Private Chapel in Ancient Egypt*. London: Kegan Paul International.

Borchardt, L., and H. Ricke (1980) *Die Wohnhäuser in Tell el-Amarna*, Wissenschaftliche Veröffentlichung der Deutschen Orient-Gesellschaft 91. Berlin: Man.

Boulos, L. (1999) *Flora of Egypt*, Vol. 1 (Azollaceae-Oxalidaceae). Cairo: Al Hadara Publishing.

———. (2000) *Flora of Egypt*, Vol. 2 (*Geraniaceae-Boraginaceae*). Cairo: Al Hadara Publishing.

———. (2002) *Flora of Egypt*, Vol. 3 (*Verbenaceae-Compositae*). Cairo: Al Hadara Publishing.

———. (2005) *Flora of Egypt*, Vol. 4: *Monocotyledons (Alismataceae-Orchidaceae)*: Cairo: Al Hadara Publishing.

Clapham, A. J., and C. J. Stevens (2009) Dates and confused: Does measuring date stones make sense? In S. Ikram and A. Dodson (Eds.), *Beyond the Horizon: Studies in Egyptian Art, Archaeology and History in Honour of Barry J. Kemp*, pp. 9–27. Cairo: Supreme Council of Antiquities.

———. (2012) *The plant remains from the Stone Village*. In A. Stevens (Ed.), *Akhenaten's Workers: The Amarna Stone Village Survey, 2005–2009*, Vol. 2: *The Faunal and Botanical Remains, and Objects*. Egypt Exploration Society Excavation Memoir 101, pp. 15–45. London: Egypt Exploration Society and Amarna Trust.

Duell, P. (1938) *The Mastaba of Mereruka, Part 1*. Chicago: Oriental Institute Publications XXXI.

Ford-Lloyd, B. V. (1995) Sugarbeet and other cultivated beets. In J. Smartt and N. W. Simmonds (Eds.), *Evolution of Crop Plants* (2nd ed.), pp. 35–40. New York: Longman.

Germer, R. (1989) *Die Pflanzenmaterialien aus dem Grab des Tutanchamun*. Hildesheim: Gerstenberg.

Griffith, Francis, Ll. (1924) Excavations at el-Amarnah, 1923–1924. *Journal of Egyptian Archaeology 10*, 299–305.

Hecker, H. (1985) Report on the excavation of floor [873] of the outer hall of Chapel 561/450. In B. J. Kemp (Ed.), *Amarna Reports I*. Occasional Publications 1, pp. 80–89. London: Egypt Exploration Society.

Hepper, F. N. (1990) *Pharaoh's Flowers: The Botanical Treasures of Tutankhamun*. London; HMSO/RBG Kew.

Hosey, C. (1984) Appendix: Analysis of coprolite sample from Site X1, in Chapter 4: The animal pens (Building 400). In B. J. Kemp (Ed.), *Amarna Reports I*. Occasional Publications 1, pp. 40–59. London: Egypt Exploration Society.

Kemp, B. J. (1980) Preliminary report on the el-'Amarna expedition, 1979. *Journal of Egyptian Archaeology 66*, 5–16.

———. (1984) *Amarna Reports I*. Occasional Publications 1. London: Egypt Exploration Society.

———. (1985) *Amarna Reports II*. Occasional Publications 2. London: Egypt Exploration Society.

———. (1986) *Amarna Reports III*. Occasional Publications 4. London: Egypt Exploration Society.

———. (1987a) *Amarna Reports VI*. Occasional Publications 5. London: Egypt Exploration Society.

———. (1987b) The Amarna Workmen's Village in retrospect. *Journal of Egyptian Archaeology 73*, 21–50.

———. (1989) *Ancient Egypt, Anatomy of a civilization*. London: Routledge.

———. (1994) Food for an Egyptian city: Tell el-Amarna. In R. Luff and P. Rowley-Conwy (Eds.), *Whither Environmental Archaeology?* Monograph 38, pp. 133–53. Oxford: Oxbow.

Kemp, B. J., and G. Vogelsang-Eastwood (2001) *The Ancient Textile Industry at Amarna*. London: Egypt Exploration Society.

Knörzer, K. H. (1971) Urgeschichte der Unkräuter im Rheinland, ein Beitrag zur Entstehung der Segetalgesellschaften. *Vegetatio 23*, 89–111.

Luff, R. (1994) Butchery at the Workmen's Village (WV), Tell-el-Amarna, Egypt, food for an Egyptian City: Tell el-Amarna. In R. Luff and P. Rowley-Conwy (Eds.), *Whither Environmental Archaeology?* Monograph 38, pp. 158–70. Oxford: Oxbow.

Manniche, L. (1999) *Sacred Luxuries: Fragrance, Aromatherapy and Cosmetics in Ancient Egypt*. Ithaca, NY: Cornell University Press.

Murray, M. A. (2000) Fruits, vegetables, pulses and condiments. In P. T. Nicholson and I. Shaw (Eds.), *Ancient Egyptian Materials and Technology*, pp. 609–55. Cambridge: Cambridge University Press.

Naville, E. (1906) *The Temple of Deir el Bahari, Part V*. Memoir of the Egypt Exploration Fund 27. London: Egyptian Exploration Fund.

Newberry, P. E. (1927) Appendix III Report on the floral wreaths found in the coffins of Tut-Ankh-Amen. In H. Carter and A. C. Mace (Eds.), *The Tomb of Tut-Ankh-Amen*, Vol. 2, pp. 189–96. London: Cassell.

Peet, T. E., and C. L. Wooley (1923) *The City of Akhenaten, Part I*. Boston: Egypt Exploration Society.

Panagiotakopulu, E. (2001) New records for ancient pests: Archaeoentomology in Egypt. *Journal of Archaeological Science 28*, 1235–46.

Puri, H. S. (2003) *Rasayana: Ayurvedic Herbs of Rejuvenation and Longevity*. London: Taylor and Francis.

Renfrew, A. C. (1987) Chapter 7, Report on the 1986 excavations and survey: The survey of Site X2. In B. J. Kemp (Ed.), *Amarna Reports IV*, Occasional Publications 5, pp. 87–102. London: Egypt Exploration Society.

Renfrew, J. M. (1985) Preliminary report on the botanical remains. In B. J. Kemp (Ed.), *Amarna Reports* II. Occasional Publications 2, pp. 175–90. London: Egypt Exploration Society.

Samuel, D. (1989) Their staff of life: Initial investigations on ancient Egyptian bread baking. In B. J. Kemp (Ed.), *Amarna Reports V*, Occasional Publications 6, pp. 253–90. London: Egypt Exploration Society.

———. (1994) Cereal food processing in Ancient Egypt: A case study of integration. In R. Luff and P. Rowley-Conwy (Eds.), *Whither Environmental Archaeology?* Monograph 38, pp. 153–58. Oxford: Oxbow.

———. (1995) Umbellifer fruits (*Trachyspermum copticum* [L.] Link) from the Workmen's Village. In B. Kemp (Ed.), *Amarna Reports VI*, Occasional Publications 10, pp. 372–83. London: Egypt Exploration Society.

———. (1996) Investigation of ancient Egyptian baking and brewing methods by correlative microscopy. *Science 273*, 488–90.

———. (1999) Bread making and social interactions at the Amarna Workmen's Village, Egypt. *World Archaeology 31*, 121–44.

———. (2000) Brewing and baking. In P. T. Nicholson and I. Shaw (Eds.), *Ancient Egyptian Materials and Technology*, pp. 537–76. Cambridge: Cambridge University Press.

Shinnie, P. L. (1951) Preliminary report on the excavations at 'Amarah West, 1989–1949 and 1949–1950. *Journal of Egyptian Archaeology 3*, 5–11.

Smith, W. (2003) *Archaeobotanical Investigations of Agriculture at Late Antique Kom el-Nana, (Tell el-Amarna)*. Egypt Exploration Society Excavation Memoir 70. London: Egypt Exploration Society.

Stevens, C. J. (2001) The finds of desiccated flax from the Workmen's Village, Amarna. In B. J. Kemp and G. Vogelsang-Eastwood (Eds.), *The Ancient Textile Industry at Amarna*, pp. 478–79. London: Egypt Exploration Society.

Stevens, C. J., and A. J. Clapham (2010) The botanical samples. In B. J. Kemp and A. Stevens (Eds.), *Busy Lives at Amarna: Excavation in the Main City (Grid 12 and the House of Ranefer, N49.18). Vol. 1: The Excavations, Architecture and Environmental Remains*, pp. 427–44, London: Egypt Exploration Society.

Tackholm V. (1974) *Students' Flora of Egypt* (2nd ed.). Cairo: Cairo University.

Thakur, R. S., H. S. Puri, and A. Hussain (1987) *Major Medicinal Plants of India*. Lucknow, India: CIMAP.

de Vartavan, C. (1990) Contaminated plant foods from the tomb of Tutankhamum: A new interpretative system. *Journal of Archaeological Science 17*, 473–94.

Vercoutter, J. (1967–1968) Excavations at Mirgissa III. *Kush 15*, 269–79.

Wilkinson, K., and C. J. Stevens (2003) *Environmental Archaeology*. Stroud: Tempus Publishing.

Zohary, D., and M. Hopf (2000) *Domestication of Plants in the Old World: The Origin and Spread of Cultivated Plants in West Asia, Europe, and the Nile Valley* (3rd ed.). Oxford: Clarendon Press.

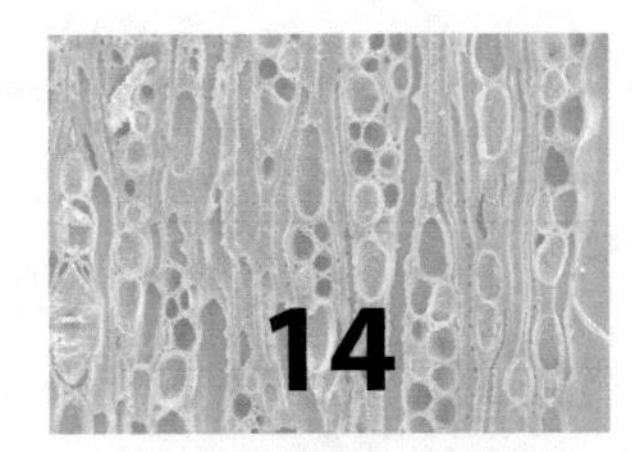

Agricultural Innovation and State Collapse in Meroitic Nubia

The Impact of the Savannah Package

Dorian Q Fuller

One of the great civilizations of Africa was the Meroitic state, or empire, which had its capital at Ancient Meroe, south of the Nile's Fifth Cataract, from the 4th century B.C.E. to the 4th Century C.E. It was a polity contemporary with the Hellenistic empire of Alexander, and his Ptolemaic successors in Egypt, as well as the Roman Empire and the Kingdom of Axum in Abyssinia. (For overviews of the Napatan and Meroitic era, see Edwards 1996, 2004; Shinnie 1967; Török 1997; Welsby 1996.) The Meroitic civilisation of the Sudan was an expansive and long-lived empire that left its distinctive archaeological mark on the landscape of the Middle Nile. The Meroitic period (ca. 350 B.C.E.–350 C.E.) shows general cultural and political continuity from the Napatan period, which began with Egyptian 25th-Dynasty rulers, who had come as conquerors from the south around 727 B.C.E.; then after roughly a millennium the Napatan-Meroitic tradition came to an end. Subsequent centuries provide evidence for three regional successor states—Noubadia, Makuria, and Alwa—at first pagan but Christianized in the 6th and 7th centuries (Figure 14.1). The transformation of the Terminal Meroitic period, often termed a collapse, remains a subject of scholarly speculation and debate. This chapter contributes to this discussion, not by discussing the chronology of the last Meroitic royal pyramids or temples, the role of military invasions from Axum (in modern Ethiopia), or the archaeology of the abandonment of Meroe city itself, but by considering the agricultural component of the Meroitic economy.

I argue that agricultural innovations in Lower Nubia, including some introduced from both north and south, contributed to the collapse of the Meroitic state, in as much as they established the economic and demographic basis for political and military power of a splinter polity (see also Fuller in press). The agricultural advances of the Meroitic period introduced a contagion for Meroitic state collapse or, more accurately, sowed the seeds for regional economic independence of the regional successor states. Of key importance is the development of improved irrigation through the adoption of the cattle-powered *saqia* (Persian waterwheel) in the 3rd Century C.E. The *saqia* provides a basis for expanding cultivation in the winter season, of such staples as wheat and barley, and in the summer season, of sub-Saharan crops (*Lablab purpureus*, *Vigna unguiculata*, *Sorghum bicolor*, and *Penniseutum glaucum*). Irrigation also provided a basis for the production of cash crops in Lower Nubia, including cotton and grapes. The changed productive potential of Lower Nubia provided the basis for the emergence of a local monarchy that increasingly monopolized wealth and power. This same period, the 3rd and 4th centuries C.E., saw increasing political and ecological instability in the central Sudan (at Meroe), but another important factor was the rise of

Archaeology of African Plant Use by Chris J. Stevens, Sam Nixon, Mary Anne Murray, and Dorian Q Fuller, Eds., 165–177

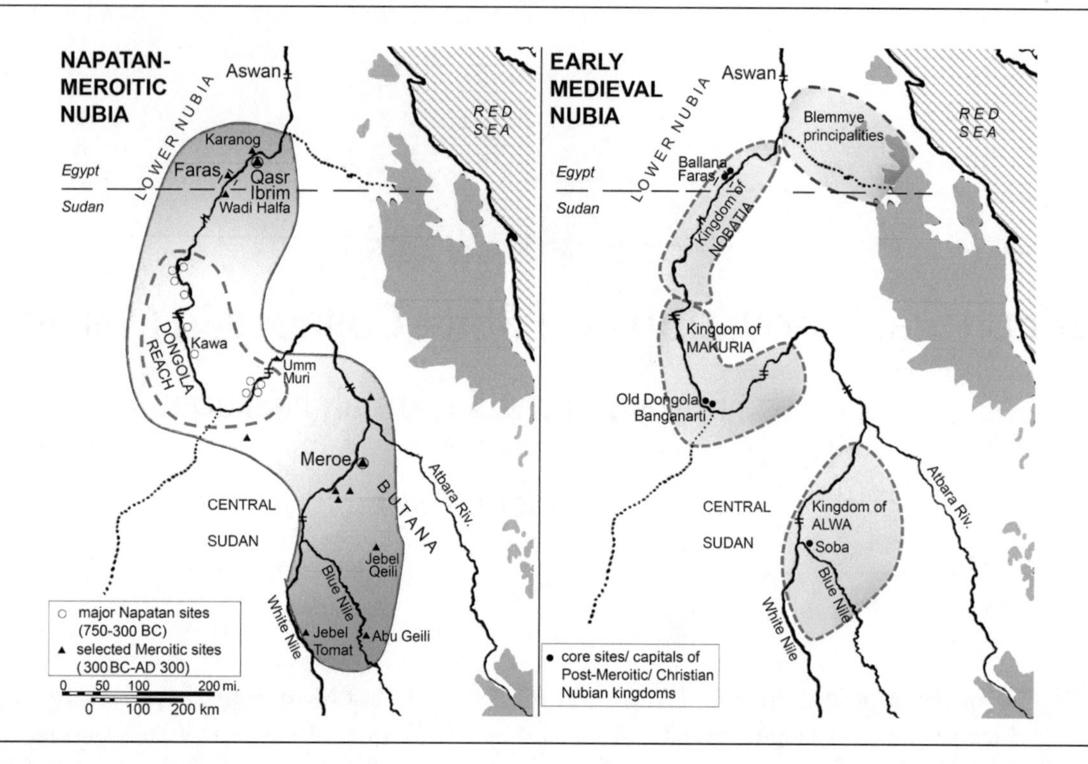

Figure 14.1 Maps contrasting Meroitic and Post-Meroitic political divisions. In the map on the left the Dongola Reach Core of the Napatan era (750–300 B.C.E.) Kingdom is outline by a dashed line, in contrast to the expanded Meroitic Kingdom (300 B.C.E.–300 C.E.). The map at the right shows the approximate extent of the three Post-Meroitic kingdoms (of ca. 500 C.E.), which became the kingdoms of Christian Nubia in the 7th century. Actual locations of boundaries are speculative.

splinter power in the north (Lower Nubia), which was made possible by agricultural transformations.

Geographic Expansion and Agricultural Variability

Although Meroe is generally regarded as a continuation of the Napatan Period (750–350 B.C.E.), there are important differences in settlement distribution, economy, and literacy that indicate that the Meroitic period was distinctive. The Meroitic period included a number of differences, or innovations, over the earlier Kushite period, in religion, with new deities such as the lion-headed Apedemak (Haaland and Haaland 2007; Žabkar 1975), new craft developments, such as 'eggshell'-ware ceramics made of kaolinitic clays, expanded iron working, and literacy in a locally developed script system of the indigenous Meroitic (from the 3rd century B.C.E.), which replaced an earlier use of Egyptian Hieroglyphic for royal monuments (Shinnie 1967; Török 1997). Together with these changes is a clear geographic expansion of the polity and enhanced economic productivity.

The Meroitic period included the expansion of settlement into new geographical and environmental zones (Figure 14.1), which is likely to imply new emphases in terms of primary economic production. By contrast to the earlier Napatan period, when settlement was focused on the Nubian Nile Valley, especially in the Dongola Reach, where several temple towns are known (see Welsby 1996), the Meroitic period includes a clear expansion southward and into the savannah zones away from the Nile (Edwards 1996, 2004). The Butana hinterlands, south and east of Meroe, became increasingly important; this area would have been the northern-most reach of the sub-Saharan savannahs that could have supported extensive herd grazing and probably some rain-fed monsoonal agriculture. The settlement evidence from the Butana indicates that many sites can be attributed to the Meroitic period, from the 3rd century B.C.E. or later. The only clear exception to this is Meroe city itself, on the Nile, which clearly has a long pre-Meroitic occupation back to early in the 1st millennium B.C.E. (Bradley 1984; Shinnie and Anderson 2004). Despite the data from the deep soundings, most of the evidence for extensive urban occupation is from the Meroitic era, after 300 B.C.E. (Grzymski 2004; Shinnie and Anderson 2004; Török 1997). This urban expansion must represent the culmination of a gradual process as nonsubsistence specialists, such as iron metallurgists, who gathered here from 500–600 B.C.E., joined an agricultural and trade community founded a few even centuries earlier.

In addition to settlements in the savannahs of the Butana, the same period witnessed the establishment of Faras and other communities in Lower Nubia and of Meroitic settlement in the Fourth Cataract. Northern (or Lower) Nubia has long been well known owing to three campaigns of survey and rescue excavations tied to the building of Aswan dams (see Trigger 1965, 1982). Recent salvage archaeology in the Fourth Cataract region has identified a Meroitic settlement on the island of Umm Muri (Edwards and Fuller 2005; Fuller 2004a; Payne 2005), which indicates that some sedentary agricultural population was established during this time. Previous eras, represented largely by burial complexes, suggest predominantly mobile, perhaps pastoral occupation of the region. So, too, it is likely that the Butana region had predominantly mobile forms of settlement only before the Meroitic period, whereas during the Meroitic some fixed centres, with towns and/or temple complexes, provided a settlement network for economic exploitation of the Butana by the state.

THE ARCHAEOBOTANICAL RECORD AND MEROITIC SORGHUM AGRICULTURE

Archaeobotanical evidence for agriculture in Northern Nubia is limited for any period (Figure 14.2). I have attempted to compile a comprehensive database of archaeobotanical evidence (up to and including records for 2010), often from casual finds rather than systematic sampling, for the Nubian region. This compilation identified some 57 sites or site phases with evidence for crops or likely wild foods dating from the Neolithic to the Late Christian period, ca. 1000 C.E. A few more recent finds, such as Beldados and Constantini's (2011) and my ongoing analyses of Meroitic material, are discussed shortly. Despite a wealth of settlement and cemetery evidence from Lower Nubia gained in the 1960s, this evidence contributed little information on agriculture, since this work was carried out before sieving was routine or field flotation had been developed. Much recent research has continued with a bias for tombs and temples, so archaeobotanical evidence for agriculture is rather patchy, with the notable exception of Qasr Ibrim (Clapham and Rowley-Conwy 2006, 2007; Rowley-Conwy 1989). Nevertheless, the overall picture appears to be one of continuity in the importance of barley and emmer wheat as cereal staples (Figure 14.3) in the Nubian Nile valley as far south as Meroe city itself, which is confirmed by wheat and barley finds there reported in Shinnie and Anderson (2004), as well as recent unpublished samples (from Meroe) studied by the author. These winter-grown cereals of Near Eastern origins were established by ca. 3000 B.C.E. (Chowdhury and Buth 2005; Hildebrand 2007) and, as in Egypt, would have been grown easily on the annual Nile flood silts.

It is the Meroitic period, when savannah crops with summer seasonality, especially sorghum, become evident (Clapham and Rowley-Conwy 2007; Fuller 2004a). Such crops would have been suited to summer rainfall in the

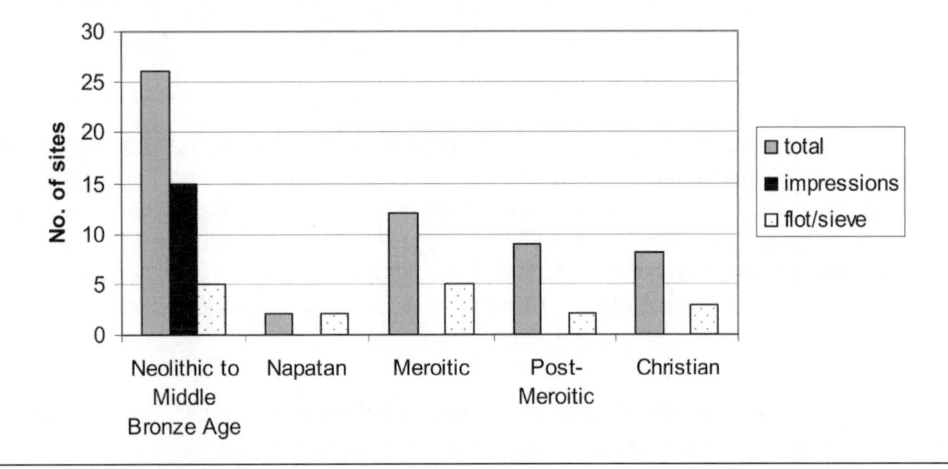

Figure 14.2 The Nubian Archaeobotanical Record: A summary of coverage, or lack thereof (based on a database of 57 published archaeobotanical datasets from Sudan/Egyptian Nubia).

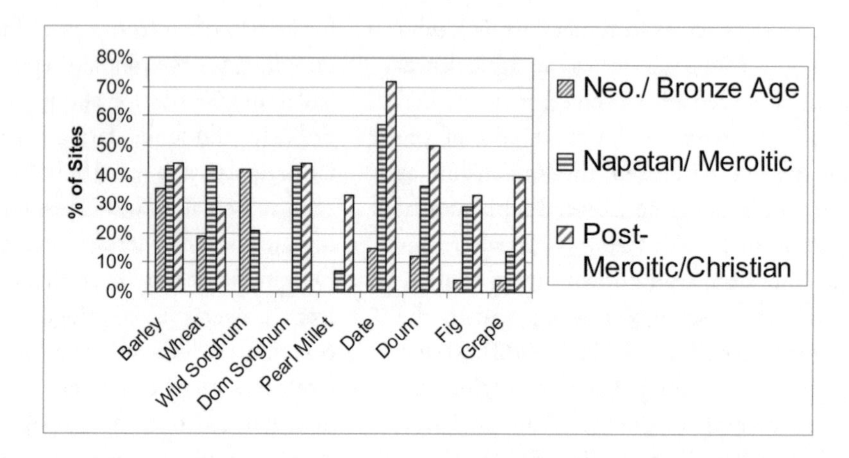

Figure 14.3 The regional presence of a select roster of crops over three broad eras: Neolithic-Bronze Age, Napatan-Meroitic, and Post-Meroitic/Christian. Note the rise in prominence of sorghum, pearl millet, and summer harvested fruits (from the same database as Figure 14.2).

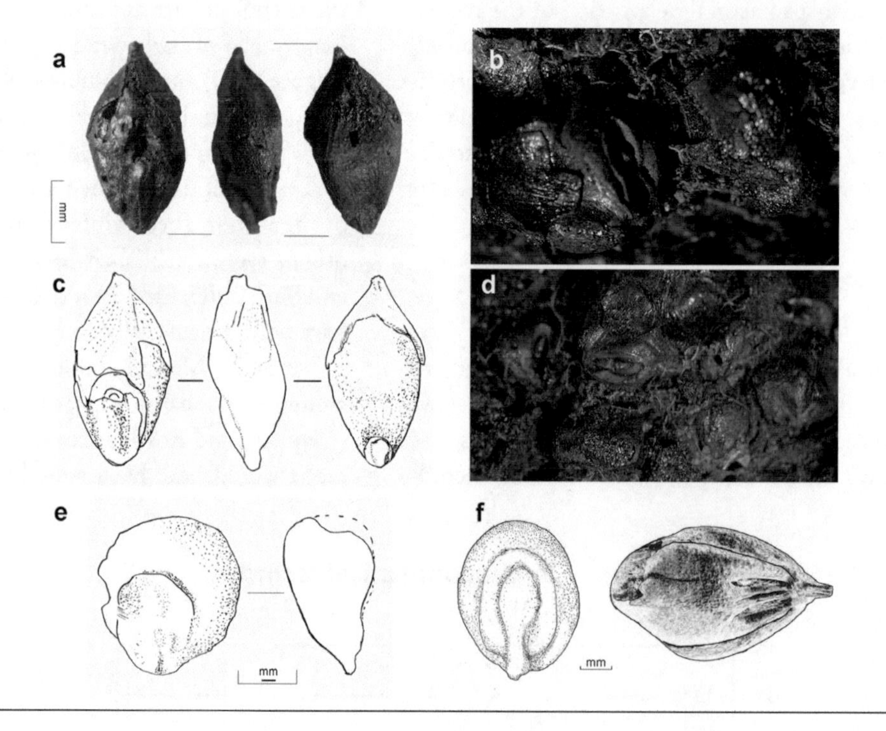

Figure 14.4 Examples of archaeological sorghum finds from the Meroitic and Napatan period: (a) carbonized grain of narrow, hulled race ***bicolor*** from Umm Muri, submitted for AMS dating. Note attached rachilla suggesting a domesticated form; (b) photo of charred mass of sorghum spikelets from Wellcome Expedition to Abu Geili in which grain is evident (scale approximately the same of 14.4A); (c) drawing of Umm Muri sorghum in A (after Fuller 2004a); (d) another photo of charred mass of sorghum from Abu Geili; (e) drawing of sorghum grain from Kawa (after Fuller 2004b) (note its plump, round form, suggestive of race *durra*); (f) drawing of sorghum grain and domesticated spikelet from Jebel Tomat (drawn from the photographs in Clark and Stemler 1975).

southern savannah zone or cultivation on irrigated lands above the Nile flood levels in northern Nubia. Small-scale flood season cultivation is suggested by my recently studied material from Umm Muri, a small Meroitic town on an island in the Fourth Cataract region (Figure 14.4a, c; Edwards and Fuller 2005; Fuller 2004a). Direct AMS-radiocarbon dating puts sorghum at Umm Muri back to the 1st–3rd century B.C.E. (BETA 194236, 2180 ± 40 BP, $\delta^{13}C$ −13.2‰). There is a some indication that even earlier Napatan period cultivation of this sort might have been carried out in the Napatan period, based on a few grains from Kawa (Figure 14.4e; Fuller 2004b), although the direct AMS date back to the Iron Age, 750–400 B.C.E. (BETA 194234, 2450 ± 40 BP) was accompanied by a $\delta^{13}C$ of −22.6‰, which is more in line with a C-3 plant than a C-4 plant such as sorghum (for this reason the identification should be accompanied by a question mark). Nevertheless, sorghum finds are especially prominent in the central Sudan, including at Meroe itself (Shinnie and Anderson 2004). In ongoing analyses I found sorghum from Hamadab, a site just south of Meroe (unpublished). North of the Nile-Atbara junction the site of Dangeil is known for a Meroitic temple to the god Amun, excavated over the past decade. Associated with the site are large quantities of sacrificial bread-mold ceramics, but analyses of preserved residue indicate use of a whole-grained sorghum product as a substitute of wheat-flour bread that would be expected of an Egyptian Amun temple (Anderson et al. 2007). The site of Abu Geili excavated by the Wellcome expedition in the 1920s also produced a lump of charred and adhering sorghum grains and spikelets, which is in the collections of the UCL Institute of Archaeology and has been examined by the author (Figure 14.4b, d). Although the excavation context within the site is unclear (cf. Addison 1949), it produced a direct AMS date of Meroitic age, ca. 2nd–4th centuries C.E. (BETA 194245, 1790 ± 40 BP, $\delta^{13}C$ −9.9‰). Farther south, a large sorghum assemblage recovered from 1970s excavations at Jebel Tomat is associated with five conventional charcoal radiocarbon dates between the 1st and 5th centuries C.E. (Clark and Stemler 1975). The distribution of these finds is plotted on the map in Figure 14.5, in which the increase in later Post-Meroitic reports from Lower Nubia in the north is evident.

Edwards (1996b) has suggested that sorghum was a socially important cereal during the Meroitic period and was a basis for making beers, which are inferred to be associated with distinctive necked jars. Although this inference is based on ethnographic parallel rather than hard scientific evidence for the identification of sorghum from jar residues, it remains plausible, and growing archaeobotanical evidence clearly indicates that sorghum was a central cereal of the Meroitic core region.

Most of these finds indicate that primitive race *bicolor* domesticates were widespread; that is, these are hulled sorghums, generally with smaller ovoid grains (Figure 14.4), although those from Jebel Tomat are somewhat rounder and plumper (Figure 14.4f). Finds from Meroe examined by sorghum expert Jack Harlan (in Shinnie and Anderson 2004) also suggest that some evolution of advanced, free-threshing *durra* race was underway, and indeed this race may already have been present at Napatan Kawa (Figure 14.4e; Fuller 2004b). The Qasr Ibrim evidence in Lower Nubia generally puts plump round grains and free-threshing forms of race *durra* into Post-Meroitic and Christian times (Clapham and Rowley-Conwy 2007).

The origins of Nubian sorghum remain obscure. Although the central Sudan is within the wild range of sorghum, early domesticated finds are missing. Based on the occurrence of morphologically wild sorghum impressions in Neolithic ceramics (6th–3rd millennium B.C.E.) across several sites, a hypothesis of early predomestication cultivation has been proposed (for example, Haaland 1995, 1999), but I find little merit in these arguments. Impressions of chaff and grain from the tempering of Neolithic ceramics from the Central Sudan include the evidence for morphologically wild sorghum and several other wild savannah millet grasses (*Panicum, Paspalum, Setaria, Pennisetum*) (Magid 1989; Stemler 1990). Haaland (1995, 1999) suggests cultivation of morphologically wild sorghum that did not lead to domestication traits primarily on the basis of large quantities of grindstones on Neolithic Nubian sites. Grindstones, however, are *post-harvest processing* tools that would have been equally necessary for the exploitation of collected wild grasses. The quantities suggest intensification of grass seed use but not cultivation. Her suggestion, following Magid (1989), that cross-pollination in *Sorghum bicolor* would have slowed down or stopped domestication by contrast to self-pollinating Near Eastern wheat and barley can now be rejected by improved archaeobotanical evidence for self-pollinating species (wheat and barley) and cross-pollinating species (for example, rice), since both kinds of cereals consistently show a protracted domestication process (Fuller 2007; Fuller, Allaby, and Stevens 2010; Purugganan and Fuller 2009). Furthermore, if cultivation of sorghum had begun at this time we would expect some evidence for shifts toward domesticated morphotypes.

The earliest evidence for cultivated and domesticated sorghum remains are those from India, where sorghum and other African grains were adopted in a piecemeal fashion by the early centuries of the 2nd millennium B.C.E. (Fuller 2003; Fuller and Boivin 2009). Although the appearance in India implies earlier cultivation, and probably morphological domestication in Africa in the 3rd millennium B.C.E. (or earlier), this must be posited to have taken place outside the Middle Nile valley. Ecologically the savannahs of northwestern Sudan, largely unstudied through archaeological

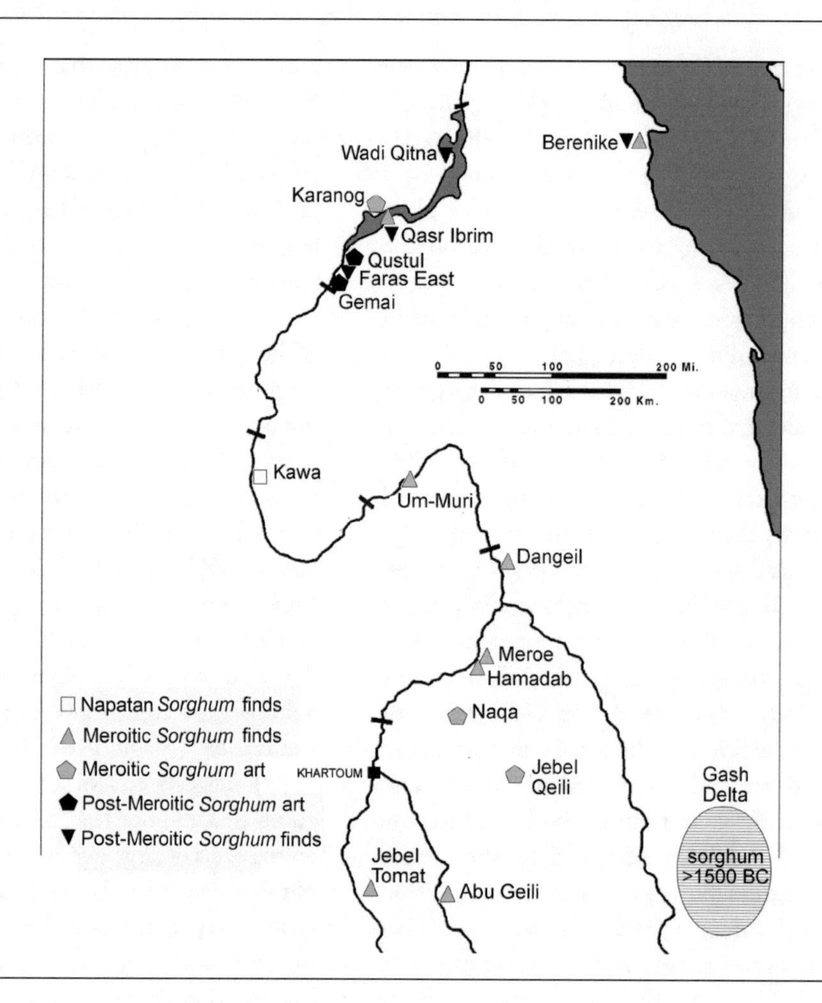

Figure 14.5 Distribution of archaeobotanical and art historical evidence for sorghum in Meroitic and Post-Meroitic Nubia.

excavation or archaeobotany, remain the likely candidate region. Another possibility is the eastern Sudan savannahs that extend across the borders to Eritrea and northwest Ethiopia. This latter region has gained new support from the study of ceramic impressions from the Gash Delta area near Kasala (Beldados and Costantini 2011). Here ceramics from the Gebel Mokram phase (1500–500 B.C.E.) at the site of Mahal Teglinos and some from nearby survey collections include sorghum impressions, in all likelihood including domesticated types. Because of the continuity from an earlier Gash Group ceramic phase (Sadr 1991), this should be a key region for sampling and documenting the early cultivation and perhaps domestication of sorghum. It was only in the 1st millennium B.C.E., during the Napatan period, that sorghum was introduced into the Middle Nile from savannah cultivation systems on its distant periphery, and so with the Meroitic expansion southward in the Butana grasslands, sorghum emerged as a key component of productive economy alongside cattle pastoralism and the trade in exotica.

The Meroitic Symbolism of Sorghum

Sorghum also had a symbolic importance in the Meroitic world, which can be seen in its occurrence in art known from several sites (Figure 14.6; mapped in Figure 14.5). Although we might expect this crop to have been important in subsistence in this Sahel/northern savannah zone, Meroitic art and material culture suggest a central symbolic role for this cereal within the Meroitic world. Notably, a monumental rock engraving of the Meroitic King Shorakaror (later 1st century C.E.; Hintze 1959; Török 1997, p. 205) shows the king offering bound captives to a deity emerging from the sun, who gives in exchange to the king a bouquet including a harvested sorghum ear (Figure 14.6b). This relief is located far to the southeast of Meroe, some 92 miles East of Khartoum, and is interpreted as a monumental commemoration of Meroitic conquest or pacification of these southern lands (Török 1997, pp. 466–67). The prominent depiction of sorghum as one of the divine rewards might be taken to suggest that this region could provide important tribute/tax in terms of sorghum. Indeed, we would

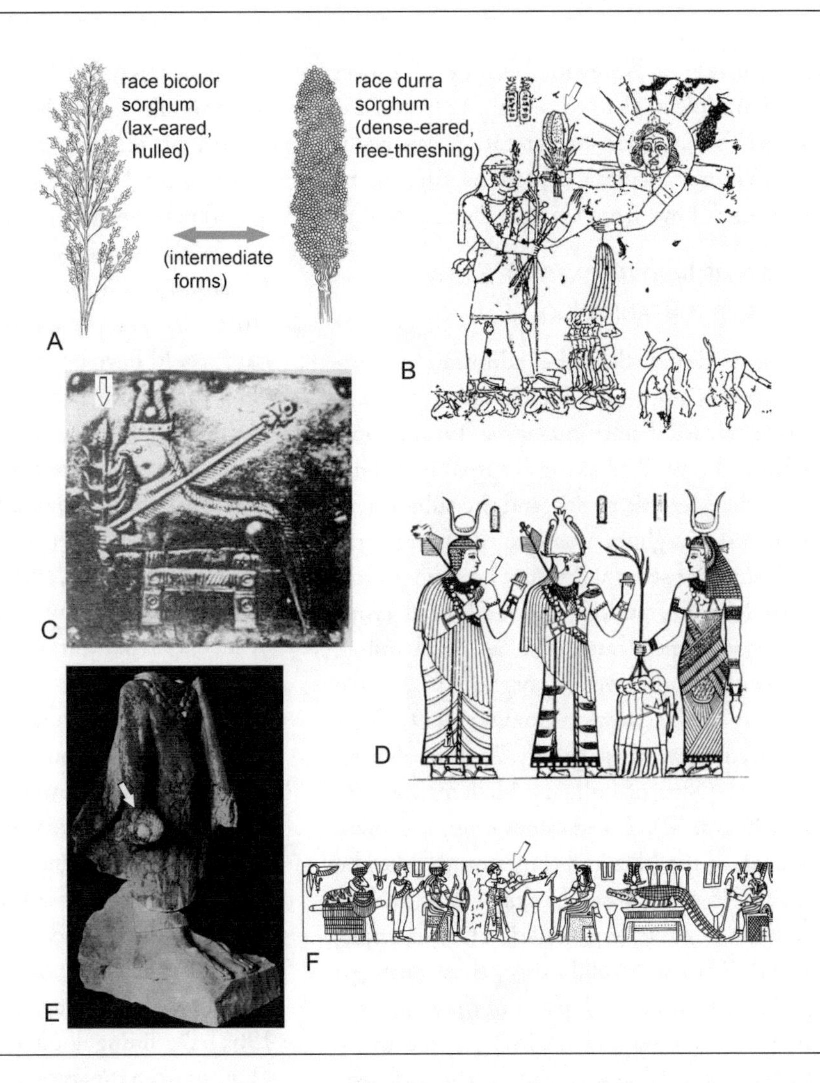

Figure 14.6 Art historical evidence with depictions identifiable as sorghum, with arrows indicating sorghum depictions: (A) sketches of typical race *bicolor* with lax-panicle and race *durra* dense-panicle sorghums for comparison; (B) the Jebel Qeili rock carving of King Shorakaror (after Hintze 1959); (C) silver plaque from Qustul 17, with composite deity holding dense-eared sorghum on culm, with leaves (after Emery and Kirwan 1938); (D) scene from the Naqa Lion Temple showing probable dense-eared sorghum in the hands of kings and queens (after Žabkar 1975); (E) ba statue from Karanog grave 203, attributed to viceroy Netewitar, mid-3rd century C.E. (after Woolley and Randall-MacIver 1910); (F) scene from bronze bowl from Gemai, early mid-4th century (after Bates and Dunham 1927).

expect this more southern location toward the Blue Nile to have been wetter savannah and thus able to support regular cultivation of rain-fed savannah crops. Other depictions at Meroitic temples, however, suggest that sorghum was also an icon of offering born by royalty to deities, as at the Lion Temple at Naqa (Figure 14.6d).

Other parallel depictions of sorghum come from Northern Nubia from the mid-3rd century–4th century C.E., suggesting a diffusion of the cultural symbolism associated with this crop. These included the depiction of a sorghum ear, with culm and leaves in the hand of a god (a composite sphinx deity with a horus-falcon head and a crocodile tail) from the Post-Meroitic period from a royal burial at Qustul, Tumulus 17, dated toward the end of the 4th century C.E. (Figure 14.6c; Török 1988; Williams 1991, p. 10). This ear appears to be a dense-panicle sorghum. This particular deity is known from other depictions, as at the Meroitic lion temple at Naqa in the central Sudan (Žabkar 1975, pp. 106–16) and on signet rings of Meroitic elites at Karanog (Woolley and Randall-MacIver 1910, plate 33, nos. 8080–8085). This find parallels objects held by Meroitic royalty at Naqa. A similar 'bouquet' is depicted in the hand of an elite ba statue from the Late Meroitic cemetery at Karanog in Lower Nubia

(Figure 14.6e) probably of Netewitar, viceroy of Lower Nubia toward the middle of the 3rd century C.E. (cf. Eide et al. 1998, p. 1018). A similar interpretation can probably be assigned to more schematic depictions, such as the example from a bronze bowl from Gemai (Figure 14.6f).

CROPS BEFORE INNOVATIONS: THE *SAQIA*, COTTON, AND WOOL

Current evidence indicates that the adoption of sorghum in the north was a protracted and gradual process. Nevertheless, an increased dietary and agricultural importance of sorghum in Lower Nubia can be attributed only to the very end of the Meroitic period and the subsequent Post-Meroitic period. Sorghum was one of several new crops, often with summer seasonality, that became prominent in the Post-Meroitic period. These adopted crops include sub-Saharan domesticates such as pearl millet, cowpea, and hyacinth bean, as well as sorghum (Clapham and Rowley-Conwy 2007; Rowley-Conwy 1989). This period also witnessed a new wheat variety, *Triticum durum*, which is part of a pattern of 'culinary Hellenization' in Egypt (Bagnall 1993, p. 32). The summer crops are sown during a low season of the Nile and thus would either be confined to just the lower exposed banks or require irrigation, especially for larger scale production (for example, Omer 1985, p. 31). Thus we would expect these crops to increase in significance with the adoption of the *saqia*.

Despite evidence from Egypt back to the 2nd century B.C.E. (Eyre 1994), a critical assessment of the archaeological evidence does not indicate the presence of *saqia* in Lower Nubia before the Terminal Meroitic period, when Meinarti was probably founded (Edwards 1996; Fuller 1999; Rose 1992; Williams 1991, p. 46). The earlier finds of sorghum indicate that the beginnings of summer crop cultivation came first, but subsequently increased evidence for their consumption comes from the Post-Meroitic period, once *saqia* irrigation was well-established.

That these summer crops became particularly important in the Post-Meroitic diet is indicated by the isotopic evidence from Nubian burials in the Wadi Halfa region, which show a marked shift toward the consumption of more C-4 plants (including sorghum and the millets) in the Post-Meroitic period (White and Schwarcz 1994). Isotopic evidence from mummified hair indicates the biseasonality of Post-Meroitic agriculture, and thus the importance of the summer-grown sorghum/millets (Schwarcz and White 2004; White 1993). This evidence suggests that the presence of sorghum in Lower Nubia during Meroitic times was insignificant in terms of consumption and probably played a role as a risk-buffering catch crop, whereas in Post-Meroitic times sorghum became an important element of consumption, along with other summer food crops. This expansion of summer-cropping would have been facilitated by the spread of the *saqia*.

With the introduction of the *saqia*, more crops could be produced, including staples and industrial or luxury crops (for example, cotton, dates, grapes), and more arable land would have been created out of old Pleistocene alluvium that was within the 8-m elevation range of the *saqia* (on the height range of the *saqia*, Tothil 1952, p. 627). The expanded arable land created by irrigation implies not only that new land may be valued and owned but also that there was new potential for conflict with pastoralists on what had previously been non-agricultural, grazing land. Such arable-pastoral conflicts may well be one of the contributing factors in the increased warfare evident from the mid-3rd century C.E. These new crops also imply a second season of cultivation, since the *saqia* allows irrigation of summer crops when the river is low. Thus there was both expansion of agriculture (more land under cultivation) and intensification (more labour invested in each bit of land)—for example, through two cropping seasons. This trend is all the more revolutionary, because the potential area of cultivation in the Nubian Nile had probably decreased at the end of the 2nd millennium B.C.E., after the end of the Egyptian New Kingdom (for example, Edwards 2004, p. 109; Trigger 1965). Declining flood levels from the late 2nd millennium B.C.E., or from the start of the 1st millennium B.C.E., through to the 5th or 4th century B.C.E. fit with evidence for East African climate change, including lower average temperatures and rainfall, but flood heights would have increased in the 3rd century B.C.E. (cf. Chalié and Gasse 2002; Gasse 2000; see also Edwards 2004, p. 209).

New irrigated lands and additional labour allowed another key revolution in northern Nubia, the emergence of a local cotton textile industry. Cotton is a thirsty crop but should not get waterlogged or get wet as the fruits form, and early varieties were likely to have been grown as shrubby perennials with life spans of several years (Brite and Marston 2013; Fuller 2008; cf. Nicholson 1960). While the cotton needed to be grown well above potential Nile flood levels, it also needed to be irrigated by lifting water to it. Potentially this might have been possible on a small scale with the *shaduf*, but the *saqia* made this much easier on a much larger scale. That cotton production was taking place locally in Nubia is implied by the evidence from Qasr Ibrim (25 B.C.E.–100 C.E.) for both seeds and capsules, with the latter less likely to be transported with raw cotton (Clapham and Rowley-Conwy 2009). This northern Nubian cotton presumably represents expansion from an earlier tradition of cotton production in the central Sudan, which may have been central to the Meroitic luxury goods economy. Although cotton production in the Meroitic world has long been implied by finds of textiles, spindle whorls, and loom weights (Griffith and Crowfoot 1934; Mayer-Thurman and Williams 1979), my current archaeobotanical investigation at Hamadab confirms quantities of cotton seeds, implying local deseeding and presumably cultivation. This finding supports the interpretation of the Ezana inscription (350–370 C.E.), as indicating an important role of cotton in the Meroitic economy at the time of the Axumite invasion of Ezana (Crowfoot 1911, p. 37). The expansion of cotton production, especially in northern Nubia, needs to be seen in the context of wider patterns of cotton production around the southern fringes of the Roman world (Figure 14.7), starting perhaps from the 1st century C.E., indicated by production in the Fazzan Libya, Dakleh, and Kargeh oases, Nubia and Saudi Arabia (Bouchaud, Tengberg, and Prà 2011; Pelling 2005; Wild, Wild, and Clapham 2007; see also Pelling, this volume). All the finds suggest a growth in cultivation in response to Roman demand but also probably made possible by agricultural innovations, especially in irrigation. In Nubia this probably involved the *saqia*, whereas in Libya this would have made use of the *qanāt* water system (see Drake et al. 2004; Pelling 2005). Brite and Marston (2013) argue that initial cotton adoption was on a small scale and connected to local innovations in agricultural practice but later grew in scale as states became interested in the wealth potential of this crop.

As with sorghum, the Meroitic adoption of cotton cultivation precedes *saqia* irrigation on current evidence. Thus rather than technological innovation allowing for new crops to be adopted, agricultural diversification apparently came first. Such diversification presumably was possible only on a small scale with the then existing techniques, with floodplain cultivation on the margins of the Nile focused on such winter crops as wheat and barley. Therefore, the new crops, the summer savannah package and cotton, provided an incentive for the adoption of new

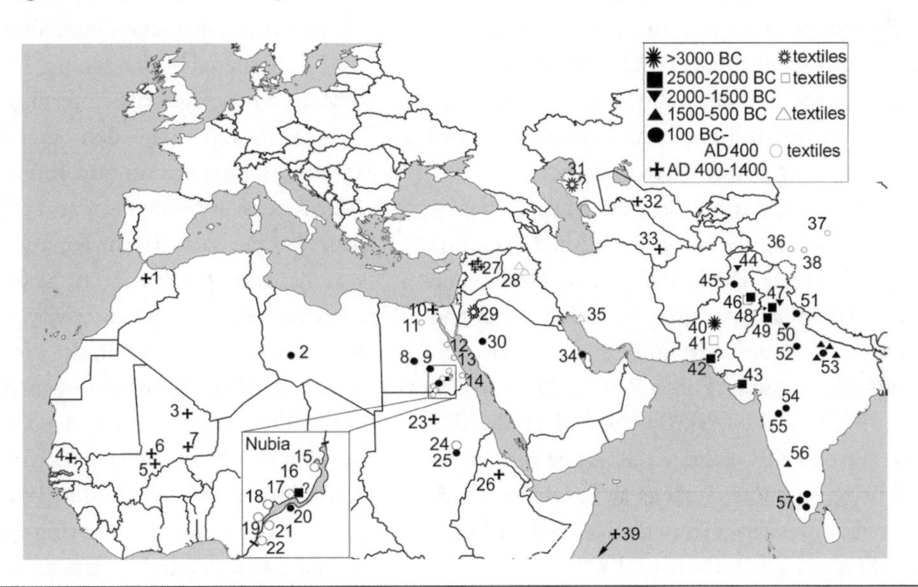

Figure 14.7 The archaeobotanical distribution of cotton finds in the Old World, highlighting the wide dispersal during the Meroitic/Roman horizon. For details of the finds, see Fuller 2008; Bouchaud, Tengberg, and Prà 2011; Brite and Marston 2013; Nixon, Murray, and Fuller 2011. Sites numbered: 1. Volubilis; 2. Jarma; 3. Essouk; 4. Ogo; 5. Tellem burial caves; 6. Dia; 7. Gao; 8. Dakhleh Oasis; 9. Khargeh Oasis; 10. Tinnis; 11. Karanis; 12. Abu Sha'ar; 13. Quesir al-Qadim; 14. Berenike; 15. Wadi Qitna; 16. Afyeh; 17. Karanog; 18. Arminna West; 19. Ballana; 20. Qasr Ibrim; 21. Qustul; 22. Second Cataract cemeteries; 23. Nauri; 24. Meroe West cemetery; 25. Hamabad; 26. Axum; 27. Sheheil, Guftan, Medād, Hrim; 28. Nimrud and Nineveh; 29. Dhuweila; 30. Madâ'in Sâlih; 31. Maykop kurgan, 'cotton-like' textiles, 3700–3200 B.C.E.; 32. Kara-Tepe; 33. Merv, seeds, 300–500 C.E.; 34. Qalat al Bahrain, 500–300 B.C.E.; 35. Arjan; 36. Sampul; 37. Ordek; 38. Niya; 39. arrow pointing in direction of Pemba island sites, Tumbe 600–1000 C.E. and Chwaka 1100–1600 C.E.; 40. Mehrgarh; 41. Mohenjodaro; 42. Balakot, pollen; 43. Kanmer; 44. Loebanhr; 45. Hund; 46. Harappa; 47. Sanghol; 48. Banawali; 49. Kunal; 50. Hulas; 51. Hund; 52. Hulaskhera; 53. middle Ganges: Sringaverapura, Kausambi, Charda, Imlidh-Khurd, Waina; 54. Paithan; 55. Nevasa; 56. Hallur; 57. Perur, Kodumanal, Mangudi.

irrigation technology, the *saqia*. Once the technology had been adopted, the productive potential for agriculture in general, and for these crops in particular, was greatly augmented, allowing for expansion and intensification of agriculture and demanding increased population (that could be fulfilled through increased birthrates as well as immigration).

A Basis for Post-Meroitic Independence

The outlined agricultural changes have important implications for the potential density of population, since more crops can support more people. They also increased the potential productivity of wealth in terms of cash crops, such as grapes or cotton, as well as in staple produce. But more land requires more labour, and two seasons of cultivation tie labourers to the land for a greater proportion of the year, thus likely decreasing the labourers' potential to be part-time specialists during the non-agricultural seasons (as potters, metallurgists, or part-time priests, all of which are plausible summer pursuits during the Meroitic period). The requirement for more labour and labourers and increase in produce to support them means that there was a 'population sink' in northern Nubia, which may have been a major 'pull factor' that encouraged immigration and settlement in the region and could be expected to have contributed to increasing ethnic heterogeneity.

It is during this same period (first half of the 3rd century C.E.) when communities were established (such as Arminna West, cf. Fuller 1997, 1999), and established lineages, such as the Wayekiye family of Gebel Adda and Kalabsha, asserted their authority in new ways (see various discussions—for example, Fuller 1997, 2003b; Millet 1981; Török 1997, pp. 471–75). By the end of the 3rd century or the start of 4th century C.E. other new communities were being founded, such as at Kalabsha/Wadi Qitna—presumably Blemmyes from the Red Sea Hills (as per Williams 1991, p. 157; cf. Barnard 2002).

With the founding of these new communities, and with the opening of new land to agriculture made possible with the *saqia*, we would expect there to have been a labour vacuum, creating a situation typical of 'internal frontiers,' in which new settlers were encouraged and welcomed. These frontier circumstances may have contributed to creating communities that included groups of people of more diverse backgrounds (Kopytoff 1987). In addition, because human reproductive strategies are often adjusted to perceived conditions (Shennan 2002, pp. 109–10; Voland 1998), perceived resource surplus and colonisation opportunities should have encouraged increased reproductive rates (increased child mortality from disease also encourages increased reproduction). Immigration might have also been influenced by 'push' factors; the settling of populations with ancestry in the Red Sea Hills (Blemmyes) in part of Lower Nubia, especially around Kalabsha must also be considered in relation to a probable decline in average rainfall in the semidesert region.

There is a clear case to be made for a decline in rainfall in northeastern Africa, which should have affected the Central Sudan and Red Sea Hills in the early to mid-1st millennium C.E. Various palaeoclimatic datasets suggest a decline in rainfall in Ethiopia over the course of the first few centuries C.E., perhaps focused on the 3rd and 4th centuries C.E., in contrast to the wetter conditions of later 1st millennium B.C.E. (Chalié and Gasse 2002; Gasse 2000, Figure 9: Lake Abhe; Machado, Pérez-González, and Benito 1998). This decline in Ethiopia is likely to be connected to less extensive rains in the Sudan, penetrating less far northward into the Sahel and desert. Thus in thinking through the social changes of the Late to Post-Meroitic periods we need to seriously consider demographic factors, which are likely to have contributed to increased population, punctuated by disease and by war related kill-offs, and increased heterogeneity in terms of cultural backgrounds.

Taken together—demographic expansion with the agricultural expansion and intensification—the basis for a local peasantry emerges, and this growth of a peasantry provided the basis for the emergence of a local state. There was increased population to be ordered hierarchically, and there was increased productivity providing the surplus. The increases in population in the North, which may well have included more ethnic heterogeneity and therefore less ethnic loyalty to Meroe, created a decreasing practical need to be reliant on Meroe, even if its ritual/symbolic importance, suzerainty persisted (Fuller 1997, 2003b). There seem to have been elites competing for power, influence, and surplus, as suggested by the multiple Terminal Meroitic/early Post-Meroitic centres of wealth and grandiose burials in the 4th century in northern Nubia (Kalabsha, Qasr Ibrim, Qustul, Gemai), but it was ultimately the elites of Qustul and especially Ballana who succeeded (see Fuller 2003b, In press; Török 1988, pp. 221–26).

Whereas savannah crops may have been central to the economic power of Meroe, the expansion of savannah crops into routine double-cropping the Nubian Nile region ultimately worked against Meroitic power. Perhaps a role was also played by increasing aridity over East Africa and the central Sudan in the 3rd–4th centuries C.E., but such climatic oscillations were felt in the subsistence economy that was already fracturing between the northern Nubian Nile and Meroitic heartland. Climatic factors therefore would have exacerbated a trend toward regionalism that was already underway.

ACKNOWLEDGEMENTS

Thanks to Pawel Wolf for archaeobotanical samples from Meroe and Hamabad, which are currently under analysis but mentioned in this chapter. My research at Umm Muri was supported by the Sudan Archaeological Research Society (SARS, U.K.), and samples from Kawa were made available by Derek Welsby of the British Museum. I am grateful to Michael Brass and Frank Winchell for drawing to my attention new evidence from the Kasala region. This chapter has been improved by editorial comments by Chris Stevens and Sam Nixon.

REFERENCES

Anderson, J. R., A. C. D'Andrea, A. Logan, and Salah Mohamed Ahmed (2007) Bread moulds from the Amun Temple at Dangeil, Nile State—An addendum. *Sudan & Nubia 11*, 89–93.

Bagnall, R. S. (1993) *Egypt in Late Antiquity*. Princeton, NJ: Princeton University Press

Barnard, H. (2002) Eastern desert ware, a first introduction. *Sudan and Nubia 6*, 53–57.

Bates O., and D. Dunham (1927) *Excavations at Gammai*, Harvard African Studies 8. Cambridge, MA: Harvard University Press.

Beldados, A., and L. Costantini (2011) Sorghum exploitation at Kasala and its environs, North Eastern Sudan in the second and first millennia BC. *Nyame Akuma 75*, 33–39.

Bouchaud, C., M. Tengberg, and P. D. Prà (2011) Cotton cultivation and textile production in the Arabian Peninsula during antiquity: The evidence from Madâ'in Sâlih (Saudi Arabia) and Qal'at al-Bahrain (Bahrain). *Vegetation History and Archaeobotany 20*, 405–17.

Bradley, R. (1984) Meroitic chronology. In F. Hintze (Ed.), *Meroitistische Forschungen 1980*, Meroitica 7, pp. 195–211. Berlin: Akademie-Verlag.

Brite, E. B., and J. M. Marston (2013) Environmental change, agricultural innovation, and the spread of cotton agriculture in the Old World. *Journal of Anthropological Archaeology 32*, 39–53.

Chalié, F., and F. Gasse (2002) Late Glacial-Holocene diatom record of water chemistry and lake level change from the tropical East African Rift Lake Abiyata (Ethiopia). *Palaeogeography, Palaeoclimatology and Palaeoecology 187*, 259–83.

Chowdhury, K. A., and G. M. Buth (2005) Plant remains from excavation of terraces of the Nile at Afyeh, Nubia and Egypt. *Purattatva 35*, 154–59

Clapham, A., and P. Rowley-Conwy (2006) Rewriting the history of African agriculture. *Plant Earth* (Summer 2006), 24–26.

———. (2007) New discoveries at Qasr Ibrim, Lower Nubia. In R. Cappers (Ed.), *Fields of Change*. Proceedings of the 4th International Workshop for African Archaeobotany, pp. 157–64. Groningen: Barkhuis and Groningen University Library.

———. (2009) The archaeobotany of cotton (*Gossypium* sp. L) in Egypt and Nubia with special reference to Qasr Ibrim, Egyptian Nubia. In A. Fairbairn and E. Weiss (Eds.), *From Foragers to Farmers: Papers in Honour of Gordon C. Hillman*, pp. 244–53. Oxford: Oxbow Books.

Clark, J. D., and A. Stemler (1975) Early domesticated sorghum from Central Sudan. *Nature 254*, 588–91.

Crowfoot, J. W. (1911) *The Island of Meroe. Part I. Sôba to Dangêl*. Archaeological Survey of Egypt 19th Memoir. London: Egyptian Exploration Society.

Drake, N., A. Wilson, R. Pelling, K. White, D. Mattingly, and S. Black (2004) Water table decline, springline desiccation and the early development of irrigated agriculture in the Wadi al-Ajal, Libyan Fazzan. *Libyan Studies 35*, 95–112.

Edwards, D. N. (1996a) *The Archaeology of the Meroitic State: New perspectives on Its Social and Political Organisation*, Cambridge Monographs in African Archaeology 38. BAR International Series 640. Oxford: Tempus Repartum.

———. (1996b) Sorghum, beer and Kushite society. *Norwegian Archaeological Review 29*, 65–77.

———. (2004) *The Nubian Past: An Archaeology of Sudan*. London: Routledge.

Edwards, D. N., and D. Q Fuller (2005) Excavations and survey in the Central Amri-Kirbekan Area, Fourth Cataract, 2003–2004. In H. Paner and S. Jakobielski (Eds.), *Gdansk Archaeological Museum African Reports*, Vol. 4, pp. 21–29. Gdansk: Gdansk Archaeological Museum.

Eide, T., T. Hägg, R. H. Pierce, and L. Török (1998) *Fontes Historiae Nubiorum: Textual Sources for the History of the Middle Nile Region between the Eighth Century BC and the Sixth Century AD*, Vol. 3, *From the First to the Sixth Century AD*. Bergen: Universitetet i Bergen.

Emery, W. B., and L. P. Kirwan (1938) *The Royal Tombs of Ballana and Qustul*. Cairo: Government Press.

Eyre, C. J. (1994) The water regime for orchards and plantations in Pharaonic Egypt. *Journal of Egyptian Archaeology 80*, 57–80.

Fuller, D. Q (1997) The confluence of history and archaeology in Lower Nubia: Scales of Continuity and Change,

Archaeological Review from Cambridge (for 1995) 14 (1), 105–28.

———. (1999) A parochial perspective on the end of Meroe: Changes in cemetery and settlement at Arminna West. In D. A. Welsby (Ed.), *Recent Research on the Kingdom of Kush,* Occasional Papers of the British Museum, pp. 203–17. London: British Museum Press.

———. (2003a) African crops in prehistoric South Asia: A critical review. In K. Neumann, A. Butler, and S. Kahlheber (Eds.), *Food, Fuel and Fields. Progress in Africa Archaeobotany,* Africa Praehistorica 15, pp. 239–71. Köln: Heinrich-Barth-Institut.

———. (2003b) Pharaonic or Sudanic? Models for Meroitic society and change. In David O'Connor and Andrew Reid (Eds.), *Ancient Egypt and Africa (Encounters with Ancient Egypt series,* Peter Ucko, Ed.), pp. 169–84. London: UCL Press.

———. (2004a) The Central Amri to Kirbekan survey: A preliminary report on excavations and survey 2003–2004. *Sudan and Nubia 8,* 4–16.

———. (2004b) Early Kushite agriculture: Archaeobotanical evidence from Kawa. *Sudan and Nubia 8,* 70–74.

———. (2007) Contrasting patterns in crop domestication and domestication rates: Recent archaeobotanical insights from the Old World. *Annals of Botany 100* (5), 903–24.

———. (2008) The spread of textile production and textile crops in India beyond the Harappan zone: An aspect of the emergence of craft specialization and systematic trade. In T. Osada and A. Uesugi (Eds.), *Linguistics, Archaeology and the Human Past Occasional Paper 3,* pp. 1–26. Kyoto: Indus Project, Research Institute for Humanity and Nature.

———. (In press) The economic basis of the Qustul splinter state: Cash crops, subsistence shifts, and labour demands in the Post-Meroitic transition. In M. Zach (Ed.), *The Kushite World. Proceedings of the 11th International Conference for Meroitic Studies.* Beiträge zur Sudanforschung Beiheft series. Wien: Universität Wien.

Fuller, D. Q, R. G. Allaby, and C. J. Stevens (2010) Domestication as innovation: The entanglement of techniques, technology and chance in the domestication of cereal crops. *World Archaeology 42* (1), 13–28.

Fuller, D. Q, and N. Boivin (2009) Crops, Cattle and Commensals across the Indian Ocean: Current and potential archaeobiological evidence. In G. Lefevre (Ed.), *Plantes et Societes* [special issue], *Etudes Ocean Indien* 42–43, 13–46.

Gasse, F. (2000) Hydrological changes in the African tropics since the Last Glacial Maximum. *Quaternary Science Reviews 19,* 189–211.

Grzymski, K. (2004) Meroe, the capital of Kush: Old problems and new discoveries. *Sudan and Nubia 9,* 47–58.

Griffith, F. Ll., and G. M. Crowfoot (1934) On the early use of cotton in the Nile Valley. *Journal of Egyptian Archaeology 20,* 5–12.

Haaland, R. (1995) Sedentism, cultivation, and plant domestication in the Holocene Middle Nile Region. *Journal of Field Archaeology 22,* 157–74.

———. (1999) The puzzle of the late emergence of domesticated sorghum in the Nile Valley. In C. Gosden and J. Hather (Eds.), *The Prehistory of Food: Appetites for Change,* pp. 397–418. London: Routledge.

Haaland, G., and R. Haaland (2007) God of war, worldly ruler, and craft specialists in the Meroitic Kingdom of Sudan: Inferring social identity from material remains. *Journal of Social Archaeology 7,* 372–92.

Hildebrand, E. A. (2007) The significance of Sai Island for early plant food production in Sudan. *CRIPEL 26,* 173–81.

Hintze, F. (1959) Preliminary report on the Butana expedition. *Kush 7,* 171–96.

Kopytoff, I. (1987) The internal African frontier: The making of African political culture. In I. Kopytoff (Ed.), *The African Frontier: The Reproduction of Traditional African Societies,* pp. 3–84. Bloomington: Indiana University Press.

Machado, M., J. A. Pérez-González, and G. Benito (1998) Paleoenvironmental changes during the last 4,000 years in the Tigray, Northern Ethiopia. *Quaternary Research* 49, 312–21.

Magid, A. A. (1989) *Plant Domestication in the Middle Nile Basin an Archaeobotanical Case Sudy.* BAR International Series 523. Oxford: British Archaeological Reports.

Mayer-Thurman, C. C., and B. B. Williams (1979) *Ancient Textiles from Nubia, Meroitic, X-Group and Christian Fabrics from Ballana and Qustul.* Chicago: Oriental Institute

Millet, N. B. (1981) Social and political organization in Meroe. *Zeitschrift für Aegyptische Sprache und Altertumskunde 108,* 124–41.

Nicholson, G. E. (1960) The production, history, uses and relationships of cotton (*Gossypium* spp.) in Ethiopia. *Economic Botany 14* (1), 3–36.

Nixon, S., M. A. Murray, and D. Q Fuller (2011) Plant use at an early Islamic merchant town in the West African Sahel: The archaeobotany of Essouk-Tadmakka (Mali). *Vegetation History and Archaeobotany 2,* 223–39.

Omer, El Haj Bilal (1985) *The Danagla Traders of Northern Sudan: Rural Capitalism and Agricultural Development.* London: Ithaca Press.

Payne, J. (2005) Excavations of the Late Kushite and Medieval Settlement on Umm Muri. *Sudan and Nubia* 9, 9–13.

Pelling, R. (2005) Garamantean agriculture and its significance in a wider North Africa context: The evidence of plant remains from the Fazzan Project. *The Journal of North African Studies 10*, 397–411.

Purugganan, M. D., and D. Q Fuller (2009) The nature of selection during plant domestication. *Nature 457*, 843–48.

Rose, P. J. (1992) *The Aftermath of the Roman Frontier in Lower Nubia*, Ph.D. thesis, University of Cambridge.

Rowley-Conwy, P. (1989) Nubia AD 0–550 and the 'Islamic' agricultural revolution: Preliminary botanical evidence from Qasr Ibrim, Egyptian Nubia. *Archeologie du Nil Moyen 3*, 131–38.

Sadr, K. (1991) *The Development of Nomadism in Ancient Northeast Africa*. Philadelphia: University of Pennsylvania Press

Schwarcz, H. P., and C. D. White (2004) The grasshopper or the ant? Cultigen-use strategies in ancient Nubia from C-13 analysis of human hair. *Journal of Archaeological Science 31*, 753–62.

Shennan, S. (2002) *Genes, Memes and Human History: Darwinian Archaeology and Cultural Evolution*. London: Thames and Hudson.

Shinnie, P. L. (1967) *Meroe: Civilization of the Sudan*. London: Thames and Hudson.

Shinnie, P. L., and J. Anderson (2004) *The Capital of Kush 2: Meroe Excavations, 1973–1984*, Meroitica 20. Wiesbaden: Harrassowitz.

Stemler, A. B. (1990) A scanning electron microscopic analysis of plant impressions in pottery from Sites of Kadero, El Zakiab, Um Direiwa and El Kadada, *Archeologie du Nil Moyen 4*, 87–106.

Török, L. (1987) The historical background: Meroe, North and South. In F. Hintze (Ed.), In *Meroitistische Forschungen 1980*, Meroitica 7, pp. 139–229. Berlin: Akademie-Verlag.

———. (1988) *Late Antique Nubia: History and Archaeology of the Southern Neighbor of Egypt in the 4th–6th Centuries AD*. Antaeus 16. Budapest: Archaeological Institute of the Hungarian Academy of Sciences.

———. (1997) *The Kingdom of Kush: Handbook of the Napatan-Meroitic Civilization*. Leiden: Brill.

Tothill, J. D. (1952) *Agriculture in the Sudan: Being a Handbook of Agriculture as Practised in the Anglo-Egyptian Sudan*. London: Oxford University Press.

Trigger, B. G. (1965) *History and Settlement in Lower Nubia*. New Haven, CT: Yale University Publications in Anthropology.

———. (1982) Reisner to Adams: Paradigms in Nubian cultural history. In J. M. Plumley (Ed.), *Nubian Studies*, pp. 223–26. Warminster: Aris and Phillips.

Voland, E. (1998) Evolutionary ecology of human reproduction. *Annual Review of Anthropology 27*, 347–74.

Welsby, D. (1996) *The Kingdom of Kush: The Napatan and Meroitic Empires*. London: British Museum Press.

White, C. D. (1993) Isotopic determination of seasonality in diet and death from Nubian mummy hair. *Journal of Archaeological Science 20*, 657–66.

White, C. D., and H. P. Schwarcz (1994). Temporal trends in stable isotopes for Nubian mummy tissues. *American Journal of Physical Anthropology 93*, 165–87.

Wild, J. P., F. C. Wild, and A. Clapham (2007) Irrigation and the spread of cotton growing in Roman times. *Archaeological Textiles Newsletter 44*, 16–18.

Williams, B. B. (1991) *Noubadian X-Group Remains from Royal Complexes in Cemeteries Q and 219 and from Private Cemeteries q, r, v, w, b, j, and m at Qustul and Ballana*, Oriental Institute Nubian Expedition IX. Chicago: Oriental Institute, University of Chicago.

Woolley, L., and D. Randall-MacIver (1910) *Karanòg: The Romano-Nubian Cemetery*. Philadelphia: University Museum.

Žabkar, L. V. (1975) *Apedemak Lion God of Meroe: A Study in Egyptian-Meroitic Syncretism*. Warminster: Aris and Phillips.

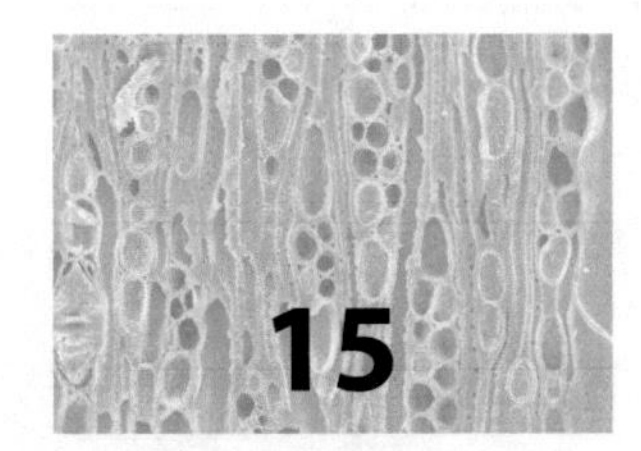

Islands of Agriculture on Victoria Nyanza

Andrew Reid and Ceri Ashley

There is obviously huge potential for archaeobotanical research throughout much of Africa. In this chapter we draw attention to some largely ignored and fairly enigmatic archaeology that may well need a concerted archaeobotanical programme in order to understand the remains we encounter. These are sporadic archaeological remains encountered on a number of islands on the Victoria Nyanza, the world's second largest freshwater lake (by area).

In examining the archaeology of the lake, researchers must recognise that modern political geography has framed perceptions of past activity. In colonial times the lake was viewed as problematic, harbouring dissidents and disease and requiring intervention to make its fishing commercially profitable (for example, Hoppe 1997). Since Independence, despite the shared boundaries and mutual ownership of the lake waters by Kenya, Uganda, and Tanzania, the focus has been on individual nation states, with the lake perceived to be marginal to the core of social activity in the respective capitals and land-focussed polities. Indeed, the lake has even come to symbolise a politically liminal periphery where it has played a prominent role in illicit economies. The islands on the northeastern shores of the lake, for example, have played key roles in smuggling of basic commodities such as coffee, sugar, and staple foodstuffs, and it is not uncommon for suspects/criminals to find sanctuary from the police on the many unregulated, unknown, and/or uninhabited islands, and piracy occurs not infrequently. Considering these problems along with the unpredictable and often dangerous weather on the lake, one does not wonder that archaeologists have largely dismissed the archaeology of the lake as a cultural backwater that has played little role in the historical developments of the region.

Recent archaeological research, however, has demonstrated that this viewpoint is misleading; the lake has instead played an active role in fostering local activity, and agriculture may well have been one of the most significant features of the societies that thrived on the lake.

A discussion of the geography of the lake is necessary before going on to discuss the island archaeology and the possible contribution of agriculture.

The Geography of the Victoria Nyanza

The Victoria Nyanza is undoubtedly a huge body of water, yet it masks a number of idiosyncrasies. It covers 68,800 km^2 yet is remarkably shallow and at its deepest is only 84 m deep (Figure 15.1). Tectonic uplift created a basin with a slow overspill, currently at Jinja, into the Nile. It is understandable that such a shallow basin straddling the equator sees significant fluctuations in lake levels over time owing to changing rates of water supply. These fluctuations, and the resultant division into separate lakes, have given rise to the evolution of a huge diversity of cichlid species, second only to Lake Malawi for the range of endemic species (Johnson et al. 1996, p. 1091). Between 2007 and 2010 the lake level fell over a metre owing to a mixture of low rainfall in its catchment and increased

Archaeology of African Plant Use by Chris J. Stevens, Sam Nixon, Mary Anne Murray, and Dorian Q Fuller, Eds., 179–188

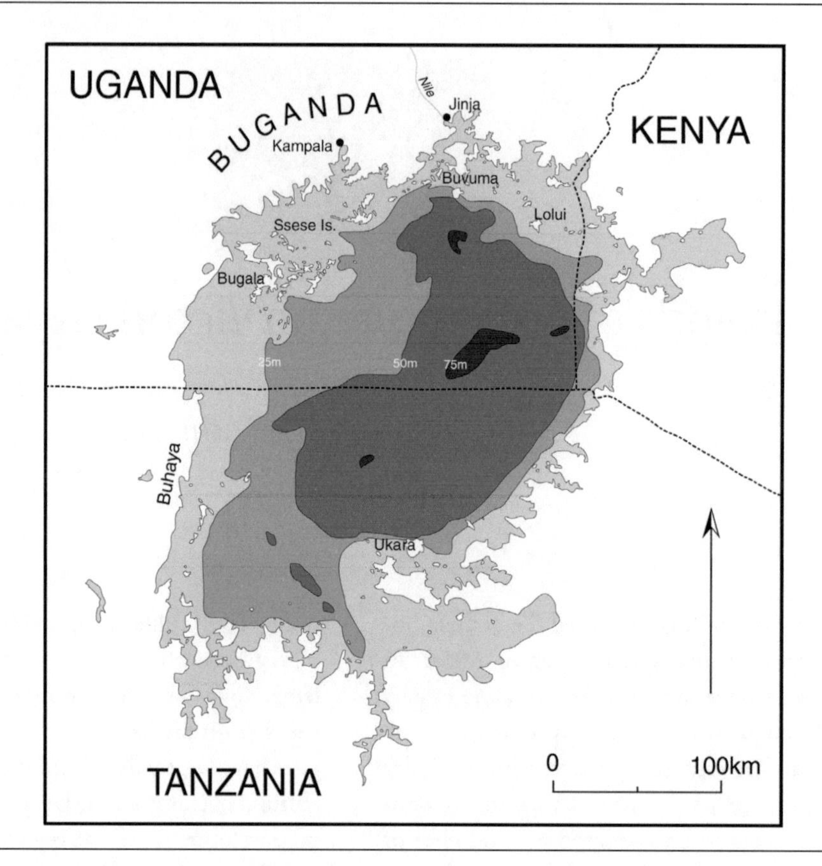

Figure 15.1 Map of the Victoria Nyanza, showing depth in metres.

release for hydroelectric power. In locations where recent observations have been made, such as the beach front at Lutoboka on Bugala Island, this situation has exposed significant wooden remains, such as plank sewn boats, wooden jetty foundations, and fish traps, which suggest that a huge archaeobotanical resource is being exposed and destroyed without being recorded. Farther back in time there have been pronounced fluctuations with one, possibly two, complete desiccations in the last 18,000 years, a low water period as late as 1200–600 years BP, and at least three phases of higher-than-present water levels, as evidenced by raised palaeoshores at 3 m, 12 m, and 18 m above the current shoreline (Johnson et al. 1996; Stager, Cumming, and Meeker 2003; Stager, Mayewski, and Meeker 2002). Although there are important patterns in lake level established for the last 1,500 years through correlation with the Rodah Nilometer in Egypt (Nicholson 1998), it is still very difficult to associate these fluctuations with archaeological settlement on the islands.

The number of islands on the lake depends on the criteria used to distinguish *island* from *outcrop* but certainly runs into the hundreds if not thousands. These islands are dispersed around the periphery of the lake, in conjunction with a large number of pronounced peninsulas and inlets. This geography is extremely important in determining population movements. When Speke walked around the western side of the lake in 1862, a journey of more than 250 km, he saw the lake only on four occasions, because the land routes simply could not follow the meandering lakeshore and/or penetrate the heavy shoreline vegetation (Speke 1863). Stanley subsequently used this information to justify his expedition to circumnavigate the lake, to prove that it was indeed a single body of water (Stanley 1879). In the 19th century, therefore, it was much easier to move on the lake than on its shores, and this is likely to have been an important factor in archaeological habitation of the lake.

Given the number of islands, their great difference in size, and their geographical spread, researchers are not surprised that they are very different in character and potential. Among the Ssese Islands, for instance, although most islands were considered fertile, a few were recognised as not being productive. Across the entire lake the most important climatic variable was no doubt the distribution of rainfall (Figure 15.2). With prevailing winds from the southeast, the heaviest rainfall is focussed on the

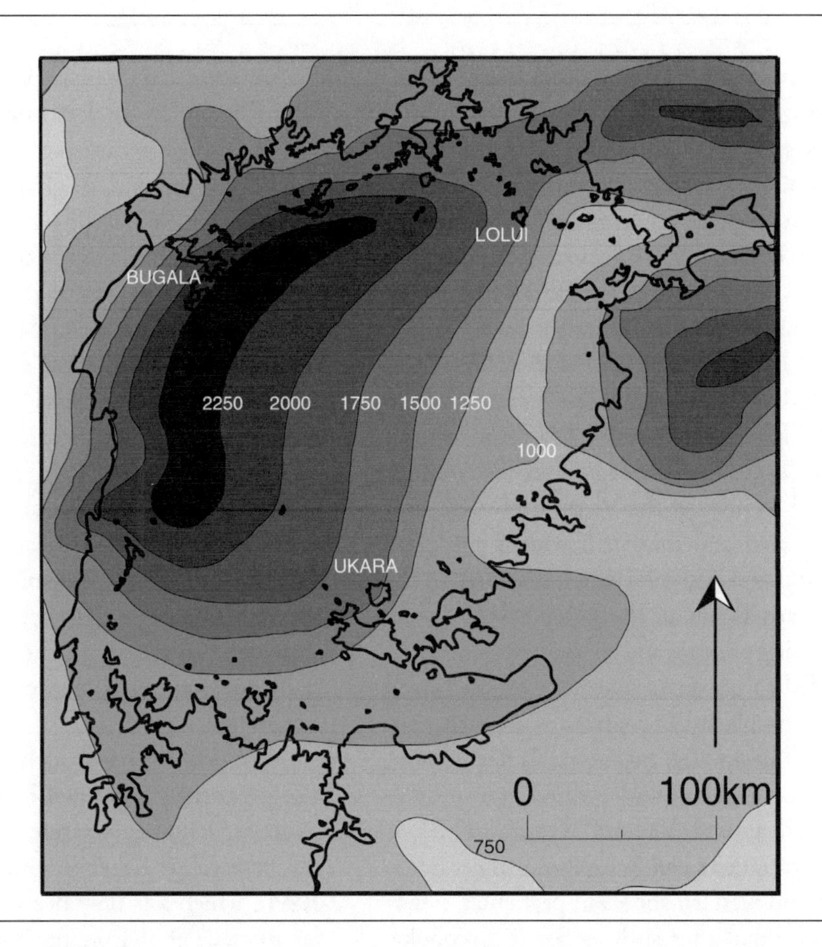

Figure 15.2 Rainfall distribution of the Victoria Nyanza (figures show annual rainfall in mm).

northwestern shores of the lake, which experience over 2 m of rain a year. This densely forested zone contrasts markedly with the grasslands of the eastern margins of the lake, with less than 1 m of rain a year.

THE ARCHAEOLOGY OF THE LAKE

Although isolated work has been undertaken on the margins of the lake, most notably in Buhaya, Tanzania (Schmidt 1978, 1997; Schmidt and Childs 1985), and in Kenya (for example, Robertshaw et al. 1983), in the last case actually spreading onto some of the major inshore islands, the archaeology of the islands of the lake itself has been largely ignored. Some survey and limited excavation data from the late colonial/early Independence eras were produced but generally were not then used as a platform for further research (Fagan and Lofgren 1966; Nenquin 1971; Soper and Golden 1969).

In recent years there has been a slow return of archaeological investigation to the lakeshore area after a significant hiatus, with research projects in Uganda (Kiyaga-Mulindwa 2004; Namono 2010, 2013) and Tanzania (Kwekason 2005; Kwekason and Chami 2003), incorporating the lakeshore region within their wider research remit. Further archaeological investigation involving the authors and specifically targeting the lake region has also been undertaken by University College London and the British Institute in Eastern Africa, in Uganda and Kenya, respectively (for example, Ashley 2010; Ashley and Reid 2008; Lane, Ashley, and Oteyo 2006; Lane et al. 2007; Reid 2001, 2002; Reid and Ashley 2008). This research consisted of a number of surveys followed up by excavation at select sites. As a result, it is now possible to recognise that there has been considerable archaeological evidence of past human activity on the lake.

Research, for example, has reiterated the known presence of Urewe ceramic-using communities around the lake. This distinctive ceramic is found across the wider Great Lakes region dating from ca. 500 B.C.E.–800 C.E. and is typically associated with a suite of socioeconomic characteristics, including the first long-term settlements, farming, and iron-working. For this reason is usually attributed to the Early Iron Age, as distinct from the preexisting

Late Stone Age hunter-gather communities (Reid 1994–1995). Significantly, though, this research has demonstrated the widespread presence of this ceramic on islands as well as the mainland lakeshore, with the dated site of Entebezamikusa on Bugala Island producing the earliest secure ^{14}C date for Urewe use in Uganda of 85–238 C.E. (Pta-9030) and also indicating an early affinity and understanding of maritime movement and settlement.

More unusual however, was the evidence from this research that clearly demonstrated heightened activity in and around the lake in the late 1st and early-to-mid-2nd millennia C.E. Typical archaeological reconstructions of this period tend to portray the Victoria Nyanza hinterland as depopulated and marginal, with a perceived shift in sociocultural and economic emphasis away to the grasslands of western Uganda. The latter habitats were suitable for the rearing of cattle and specialist pastoralism, which gradually led to unequal ownership of assets and emergent political complexity. In contrast to the picture of a lacustrine cultural backwater, this research has demonstrated that there was a vibrant and dynamic transitional-Urewe using period around Victoria Nyanza, which is best evidenced by variations and innovations in ceramics. Across the north and western shores in particular, newly discovered and excavated sites such as Sanzi, Lutoboka, and Malanga-Lweru, as well as reexamined collections from Lolui, Luzira, and Hippo Bay Cave (Figure 15.3), have shown a fluorescence in ceramic styles, with a clear diversification from the preexisting homogeneity of the ubiquitous Urewe pottery. Ceramic styles become localised and simpler in comparison to the highly decorated and well crafted Urewe. This decorative shift is believed to be part of a wider sociopolitical transformation, as emphasis shifts from a domestic, familial scale toward individual

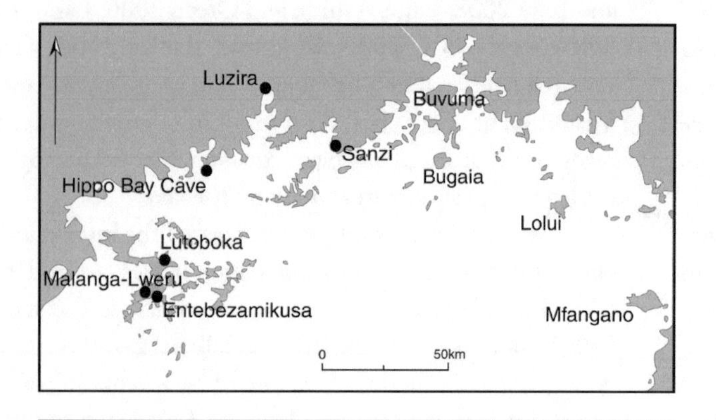

Figure 15.3 Archaeological sites of the northern Victoria Nyanza.

political and economic power (Ashley 2010). The case of Transitional Urewe illustrates this argument: first identified from the northeastern island of Lolui by Posnansky in the 1960s, and originally termed Devolved Urewe (Posnansky 1967, 1973), early recognition by Posnansky of internal variation in the ceramics was generally ignored. Reappraisal as part of the current research exercise (see Posnansky, Reid and Ashley 2005; Ashley 2005, 2010) validated this distinction and also identified analogous ceramics within the Luzira Head assemblage, allowing the redating of these enigmatic human figurines first found in 1929 (Ashley and Reid 2008; Reid and Ashley 2008; Wayland, Burkitt and Braunholtz 1933). Poorer in execution quality than the typically exquisite Urewe, and associated with a unique assemblage of clay human figurines, this innovation in clay technology has been interpreted as a demonstration of the growing importance of the individual and vested authority, perhaps embodied by the Luzira figurine itself.

The impetus for such changing social practice is also potentially explained by the new lacustrine-oriented societies. Another ceramic recovered from obscurity by this research is Entebbe pottery (Brachi 1960; Marshall 1954), which has been identified from 43 locations (an increase of 38 from previous knowledge) and has been tentatively dated, through association with the transitional-Urewe ceramics and the ^{14}C dating of Hippo Bave Cave, to the first half of the 2nd millennium C.E. This large, heavy, roulette and comb-scored ceramic is significantly found only in the immediate vicinity of the lake, and in noticeable density on the islands investigated (for example, Lolui, Malanga-Lweru on Bugala), with locales being a maximum of around 7 km from the lakeshore. In addition, a number of sampled ceramics show the presence of sponge spicules in the clay matrix, which were presumably provenanced from the swampy clays that fringe the lake/islands. This spatial discretion is in fact replicated in the transitional-Urewe ceramics as well, which demonstrate a preference for lake-shore areas and cannot be a result of sampling bias, since substantial survey was focussed on the interior, where no such ceramics were found. Thus we can argue that it is not only the maritime islands of the lake that are 'insular' but that the mainland lake fringe needs also to be incorporated in this newly defined phase of past human activity and be seen as mainland 'islands'. This distinct distribution pattern clearly indicates an active preference for the lake environment, in contrast to preceding projections, which tend to portray the lake as marginal, leaving populations at the precarious edge of human sustainability.

Instead, there is clear evidence from the exploitation of the lacustrine and mainland islands that the lake was a positive agent in local dynamics, facilitating activity and contact and acting as a conduit to social development. One possible facet of this lacustrine economy may have been trade, and the ease of movement and transport. Evidence for participation in longer distance trade is provided by a small number of rare glass beads found at Malanga-Lweru, which must have ultimately come from the east African Swahili coast trade and may have been intended for the pastoral communities of the grasslands interior. A small number of similar beads were excavated at Ntusi (Reid 1996) and Munsa (Robertshaw 1997). To support such trade superstructures, it can be argued that a strong and sustainable subsistence economy underpinned the communities of this period, fully exploiting the lake resources. The following discussion explores this posited agricultural foundation.

CULTIVATION ON THE LAKE

During the course of these excavations various different attempts were made to recover archaeobotanical remains from these sites, involving a variety of wet-sieving and flotation strategies. The most intensive of these was undertaken by Ruth Young at the site of Sanzi on the Bukunja peninsula. This sampling produced large amounts of wood charcoal but seemingly very little in the way of agricultural remains. The one exception to this situation was the presence on all the Ugandan sites of large quantities of canarium husks. Canarium, known in luganda as *empafu*, is a large tree species that produces an edible fruit consumed at the time of harvest, within which is a seed with an extractable kernel that can be stored for future use. The regular recovery of carbonised canarium seeds presumably indicates the widespread exploitation of these kernels and indeed the desirability of encouraging the growth of canarium trees in the immediate vicinity of habitation, as is the case today. Canarium seeds are also recovered on early forest sites in West Africa (Andah 1993; see Bostoen Chapter 11, this volume), and it is uncertain whether the proliferation of canarium in the Victoria Nyanza area is the result of transfer or whether canarium occurred wild in the forests surrounding the lake.

The survival of the canarium husks in the archaeological record is clearly related to the manner in which the husks were processed, possibly being fire-cracked and/or the husks subsequently disposed of in fires, leading to their carbonisation and preservation. Such conducive taphonomic conditions seem to have been absent from the processing of other plant remains in eastern Africa, such as finger millet and sorghum, which are prepared away from potential sources of carbonisation, and it has been suggested that this factor as well as the subsequent humid conditions in the ground actively obstruct the survival of agricultural plant remains (Young and Thompson 1999).

There are, however, other indirect indications of agricultural activity on the shores of the northern part of the lake. Although direct archaeological evidence for these crops is almost non-existent, historical reconstructions based on linguistics indicate that a whole range of crops including cereals and pulses has had a long and fairly complex history of use, going back around 2,500 years (for example, Schoenbrun 1998). Indirect archaeological evidence also includes Posnansky's examination of Lolui Island, where he recognised hundreds of worn hollows on flat rock surfaces, where they are of sufficient frequency, length, and wear to suggest that they were used for the preparation of grain crops, such as sorghum and finger millet (Figure 15.4). These grinding grooves are often found in clusters, and there was clearly a desire to make the most of suitable flat rock surfaces, with the high numbers of hollows possibly also relating to the quality of the stone or to social practices revolving around grinding activities. For example, there seems to have been preference for surfaces outside settlements where they can be readily accessed without having to prepare and transport suitable grinding stone to the settlement itself. Additionally, excavation on Lolui showed that huge numbers of quernstones were used to construct one of the piled stone cairns found in the interior of the island (30% of a total of 800 excavated stones), with over 4,000 quernstones recovered from the

Figure 15.4 Grain-grinding hollows on Lolui Island.

wider cairn area as a whole (Posnansky, Reid, and Ashley 2005). It is impossible to determine whether these differing processing strategies are contemporaneous or unrelated, because of a lack of absolute dating.

Additional grinding surfaces have been recorded near the site of Sanzi. These surfaces are situated on a pronounced grass-covered slope above the shoreline forest fringe in which the site is situated. There are also other rock-engraved features, consisting of rows of small 5–10 cm circular depressions, almost certainly forming mancala gaming boards (known locally as emyeso). These emyeso are scattered across the hillside, with no obvious concentration in their distribution, and probably represent the presence of herders spending countless hours grazing their stock from Sanzi and any other habitation sites since then. By contrast, the grinding hollows are located only in the vicinity of the Sanzi site itself and, given the need to limit the distance walked to and from the grinding platform, were probably used only by the inhabitants of the Sanzi site.

Grinding hollows were noticeably absent from the surveys conducted on Bugala Island, the largest of the Ssese group situated in the wettest part of the lake. Despite the presence of early farmers on the island, grinding hollows were not noticed. However, a different kind of ground feature, a groove, 10–20 cm wide and up to 60 cm in length was noted. Although these were occasionally encountered on the Bukunja peninsula, one rock surface on Bugala featured around 30 grooves. It is not immediately obvious what action may have produced these grooves, although some form of sharpening of metal implements may be a possibility.

Island Field Systems

A final discussion is required of possible field systems that have been mentioned from a number of islands. These notes are usually extremely brief and far from conclusive, but taken together they suggest considerable activity. Included in these notes are terraces on Bugaia (McFarlane 1967; Nenquin 1971) and Mfangano (Kenny 1979), which warrant further examination. By far the best evidence, however, comes once again from Lolui. Besides the grinding hollows and quernstone-filled cairns already noted there are also separate areas of banks and stone lines. The banks or bunds are found on the northwestern side of the island (Figure 15.5). These banks, which are never more than a metre in height, run roughly parallel to one another along contours for a length of several hundred metres. They were presumably designed to prevent erosion, although the slope is not pronounced in this area. Similar features were recognised by McFarlane (1967) on Buvuma and Bugaia Islands and are apparently recognisable on oblique air photographs.

The stone lines on Lolui are also very striking for their regularity. They consist of individual large stones (many of them quernstones), laid upright so as to stand proud of the ground by 5–12 cm with more than 20 cm buried below the surface. Each stone is set between 2 and 15 m apart and forms into a grid of regular straight lines (Figure 15.6). Once again these stone lines for the most part run roughly parallel, but this time perpendicular to the contours, measured examples running between 55 and 390 m in length but only between 5 and 15 m apart. It seems likely that

Figure 15.5 Banks on Lolui Island.

Figure 15.6 Stone lines on Lolui Island.

together the banks and stone lines would combine to form a close-knit field system. Stone lines were also found in the centre of the island around the cairns. Unfortunately, although Posnansky in the 1960s was able to explore an uninhabited island, Lolui has now become a significant population centre, benefiting from lucrative fishing and trading opportunities. This new settlement is situated in the northwestern part of the island, where the banks and stone lines were concentrated and almost certainly will now have been disturbed, as buildings are constructed and use is made of available stone. Lolui suggests the scale and complexity that some field systems would have acquired, and the human encroachment underlines the importance of visiting other possible field systems on such places as Bugaia and Mfangano, before they, too, are destroyed.

These pronounced field systems were not noted in the recent surveys of Bugala Island. Although Fagan and Lofgren (1966) noted embankments in their brief text, these could not be relocated, and most probably represent residue of recent/historic compound activity. One area of possible field systems was, however, encountered (Figure 15.7). This is close to the site of Malanga-Lweru, which features Lutoboka Urewe (part of the Transitional Urewe group) and Entebbe pottery as well as four glass beads, presumably derived from trade on the Indian

Figure 15.7 Possible field systems at Malanga-Lweru.

Ocean. A large area near the site, which includes the rock surface with the grooves, mentioned previously, is covered in low cairns and occasional quernstones. Occasionally the cairns are linear rather than circular, suggesting that they may have been made to line a boundary or may have accumulated along its edge. It is possible that these cairns represent an episode of field clearance.

UKARA

The productive potential of these possible field systems on the northern shores of the lake is suggested by a unique record of agricultural production on the island of Ukara, in the southeastern part of the lake (Thornton and Rounce 1936). In 1936 a complex network of field systems and crop rotations was supporting a population density for the whole island of around 250 people per square kilometre, although less than 70% of the island was suitable for cultivation. The agricultural system consisted of almost continuous cropping and removal of crop refuse, and therefore a number of strategies were practised to regularly replenish nutrient levels in the fields.

The main food crop throughout the island was *Pennisetum*, followed by either Bambara groundnuts (*Voandzeia subterranea*) or less frequently sorghum or cassava. Very often these would be interplanted, so that in one season three crops were produced by the same piece of land. Small numbers of stock were kept on the island, and these were stall fed, with only very occasional grazing. This system both prevented the animals from straying into crops and enabled the creation of regular quantities of manure, which was applied to the fields. In addition, two indigenous legumes were grown as green manure: *Crotalaria striata* and *Tephrosia* sp. These were normally interplanted with the *Pennisetum* once the latter was established and allowed to grow on after settlers had harvested the grain and removed the stalks. Once the legumes had matured at around eight to nine months, they were then dug in. Some weed species were also allowed to grow with the *Pennisetum* and were subsequently dug in. Rice was also being grown at the time, but in selective valleys and lake shore locations, and its purpose was as a cash crop with which to pay colonial taxes.

Awareness of the ownership of both cropped land and grazing was acute, with no trespass allowed. Stock were even muzzled to prevent them from damaging crops. Besides the considerable amounts of crop waste fed to stock, other sources of fodder were grasses cultivated on the lake shore, tree loppings, and cuttings of *Crotalaria*. Cultivated grasses included *Vossia cuspidata*, *Echinocloa* sp., and *Pennisetum purpureum* (Elephant grass). Pits over a metre in depth were dug to enable irrigation from the lake or for the roots to access the water level. Fodder was then cut and fed to the stock at regular intervals through the year. A further grass species, *Trichopteryx kagerensis*, was found in the reserved grassland in the centre of the island. Used for thatching purposes by the whole island, the grass was sold by the individuals who owned the grassland.

In addition, trees were carefully conserved and found only around human habitation, where they were deliberately planted and maintained. Every tree had a recognised owner who would use it for building or for fodder. Trees were never felled, but branches selectively removed. The most prevalent tree on the island was *Markhamia platycalyx*, recognised for its building poles, hoe handles, and quick growth and free pollarding. However, in total 56 different species of tree were noted on the island, of which 9 were used for stock feed.

Soil erosion was noted as a significant problem in the sandy soils of the island. Measures used against erosion included terracing without stones, stone wall terraces, and the creation of ridge paths between holdings. Considerable efforts had also been expended on controlling streams and preventing pronounced gullying. These measures were observed to be extremely successful in countering erosion.

Although Thornton and Rounce were clearly impressed by the agriculture on Ukara they do note that the practise of dividing land among male heirs was creating smaller and smaller units of land and threatened the integrity of the system as a whole. Thornton and Rounce concluded that the amazingly intricate system of cultivation and rejuvenation of the soils was the result of overpopulation and necessity rather than a successful innovation grown too popular. Nevertheless, the complexity of the Ukara agricultural system demonstrates the potential complexity of the island systems on other parts of the lake, particularly considering the relatively low rainfall on Ukara compared with islands in the northwest.

Discussion

Although hard evidence is markedly lacking from this examination, it has clearly been demonstrated that there are a range of features on the islands of the Victoria Nyanza that suggest intensive agricultural systems and that warrant concerted archaeobotanical investigation. These archaeological phenomena are as yet undated, but the Lolui ceramics suggest that the field systems date back between 500 and 1,000 years at the very least and up to a possible 1,000 years beyond that.

There are also issues concerning the crops that would have been grown in these field systems. At Lolui the numerous grinding hollows suggest grain crops such as sorghum and finger millet. The stone lines are particularly interesting, because they suggest small field boundaries of a densely packed crop. Finger millet in particular tends to be grown in very small stands in the region today. In areas of tightly packed cultivation, field boundaries may have been very important in defining possession of a particular location (for example, see inheritance issues on Ukara). The Ukara evidence from the early 20th century also throws in additional possibilities, with a pronounced focus on producing fodder for livestock. It is debateable whether such prominence for cattle would have been present 1,000 years ago on the northern shores of the lake. At this time specialist cattle-keeping communities were only just emerging in the western grasslands of Uganda, and the earlier Pastoral Neolithic phenomenon of western Kenya only rarely penetrated coastal areas. However, a number of sites discussed here do appear to have been situated to take advantage of both forest and open grass-covered hilltops, suiting both the needs of cultivation and animal husbandry.

Lolui and Ukara experience fairly similar rainfall regimes, which perhaps explains their similar patterns of agricultural terracing. Further questions can be raised about the agricultural systems that may have existed on the Ssese islands in the wettest part of the lake. Would conditions have been appropriate for grain crops, or could other crops have been featured here? According to Schoenbrun (1998) there is an explosion of banana terms around 800 C.E. on the northwestern shores of the lake, which he believes represents the transformation of bananas from a peripheral occasional crop to a plantation cultivated staple.

This staple and its means of propagation rapidly spread throughout this area of the lake, and the Buganda kingdom eventually emerged in around the 16th century by being able to link up isolated pockets of banana cultivation into a centralised political system (Wrigley 1989). Almost certainly this rapid dispersal of the new crop would have occurred primarily by means of the lake, taking advantage not only of the better mobility and access the lake provides but also of the enhanced capacity that boats offered in transportation, potentially moving entire plantations of suckers/saplings in one journey. It is intriguing to note that some Baganda traditions intimate that bananas originated from the Ssese Islands (Wrigley 1989). This occurrence may indicate the appearance of the enhanced plantation systems at around this time, coming from the lake, and certainly suggests that targeted archaeobotanical work on the islands needs to be undertaken.

It is also important to consider the effects that such a dramatic shift in agricultural practice had on both human populations and soils. The rapid replacement of a protein-rich staple (for example, finger millet) by a carbohydrate-based crop (banana) must have had a significant impact on human health. Malnutrition is likely to have become a significant problem faced by populations, but at the same time dental health is likely to have markedly improved. These factors will be important to explore as and when human skeletal material becomes available for physical anthropological studies. The impact of such transitions is still occurring in Uganda today as the banana crop fails in the core cultivation areas and cultivation of bananas has spread to areas (particularly in the west) that did not historically produce bananas, fuelled by demand from the capital, Kampala. This archaeology, therefore, has the potential to contribute to contemporary issues that concern the region today, and they further illustrate the importance of turning archaeobotanical attention to the islands of the lake. Far from being isolated backwaters on the margins of society that contributed nothing to historical developments, they can be seen as being extremely important to the changes that have taken place.

REFERENCES

Andah, B. W. (1993) Identifying early farming traditions of West Africa. In T. Shaw, P. Sinclair, B. Andah, and A. Okpoko (Eds.), *The Archaeology of Africa: Food, Metals and Towns*, pp. 240–54. London: Routledge.

Ashley, C. Z. (2005) *Ceramic Variability and Change: A Perspective from Great Lakes Africa*. Ph.D. thesis, University of London.

———. (2010) Towards a Socialised Archaeology of Great Lakes Ceramics. *African Archaeological Review 27*, 135–63.

Ashley, C. Z., and A. Reid (2008) A reconsideration of the figures from Luzira. *Azania XLIII*, 95–123.

Brachi, R. M. (1960) Excavation of a rock shelter at Hippo Bay, Entebbe. *Uganda Journal 26*, 62–70.

Fagan, B., and L. Lofgren (1966) Archaeological reconnaissance on the Ssese Islands. *Uganda Journal 30* (1), 81–86.

Hoppe, K. A. (1997) Lords of the fly: Colonial visions and revisions of African sleeping-sickness environments on Ugandan Lake Victoria, 1906–1961. *Africa: Journal of the International African Institute 67* (1), 86–105.

Johnson, T. C., C. A. Scholz, M. R. Talbot, K. Kelts, R. D. Ricketts, G. Ngobi, K. Beuning, I. Ssemmanda, and J. W. McGill (1996) Late Pleistocene desiccation of Lake Victoria and rapid evolution of Cichlid fishes. *Science 273* (5278), 1091–93.

Kenny, M. G. (1979) Pre-Colonial trade in Eastern Lake Victoria. *Azania XIV*, 97–107.

Kiyaga-Mulindwa, D. (2004) The archaeology of the riverine environments of the Upper Nile Valley in Uganda. In F. Chami, G. Pwiti, and C. Radimilahy (Eds.), *The African Archaeology Network: Reports and Review*, Studies in the African Past 4, pp. 38–56. Dar es Salaam: Dar es Salaam University Press.

Kwekason, A. (2005) *Rock Paintings and New Dates from Southwest of Lake Victoria*. Unpublished paper presented at 12th Congress of the PanAfrican Archaeological Association for Prehistory and Related Studies, July 3–10, 2005, University of Botswana, Gaborone.

Kwekason, A., and F. Chami (2003) Archaeology of Muleba, southwest of Lake Nyanza. In F. Chami, G. Pwiti, and C. Radimilahy (Eds.), *Climate Change, Trade and Modes of Production in Sub-Saharan Africa*, Studies in African Past 3, pp. 59–85. Dar es Salaam: Dar es Salaam University Press.

Lane, P. J., C. Z. Ashley, and G. Oteyo (2006) New dates for Kansyore and Urewe Wares from Northern Nyanza, Kenya. *Azania 41*, 123–38.

Lane, P. J., C. Z. Ashley, O. Seitsonen, P. Harvey, S. Mire, and F. Odede (2007) The transition to farming in Eastern Africa: New faunal and dating evidence from Wadh Lang'o and Usenge, Kenya. *Antiquity 81*, 62–81.

Marshall, K. (1954) The Prehistory of the Entebbe Peninsula. *Uganda Journal 18*, 44–57.

McFarlane, M. (1967) Some observations on the Prehistory of the Buvuma Island Group of Lake Victoria. *East African Freshwater Fisheries Resource Organisation Annual Report 1967*, 49–54.

Namono, C. (2010) *Surrogate Surfaces: A Contextual Interpretive Approach to the Rock Art of Uganda*. Ph.D. thesis, University of the Witwatersrand.

Namono, C. (1913) Dumbbells and circles: Symbolism of Pygmy rock art of Uganda. *Journal of Social Archaeology 12* (3), 404–25.

Nenquin, J. (1971) Archaeological prospection in the Islands of Buvuma and Bugaia, Lake Victoria Nyanza (Uganda). *Proceedings of the Prehistoric Society 37*, 381–418.

Nicholson, S. E. (1998) Historical fluctuations of Lake Victoria and other lakes in the Northern Rift Valley of East Africa. In J. T. Lehman (Ed.), *Environmental Change and Response in East African Lakes*, pp. 7–36. London: Kluwer Academic.

Posnansky, M. (1967) The Iron Age in East Africa. In W. Bishop and J. D. Clark (Eds.), *Background to Evolution in Africa*, pp. 629–49. Chicago: University of Chicago Press.

———. (1973) Terminology in the Early Iron Age Eastern Africa with particular reference to the dimple-based wares of Lodui Island, Uganda. *PanAfrican Congress for Prehistory as Associated Studies V*, 577–79.

Posnansky, M., A. Reid, and C. Ashley (2005) Archaeology on Lolui Island, Uganda 1964–1965. *Azania XL*, 73–100.

Reid, A. (1994–1995) Early settlement and social organisation in the interlacustrine region. *Azania 29-30*, 303–13.

———. (1996) Ntusi and the development of social complexity in southern Uganda. In G. Pwiti and R. Soper (Eds), *Aspects of African Archaeology*, pp. 621–28. Harare: University of Zimbabwe Press.

———. (2001) Bananas and the archaeology of Buganda. *Antiquity 75*, 811–12.

———. (2002) Recent archaeological discoveries in Buganda and their implications for Archaeological Heritage Management. *Uganda Journal 48*, 87–103.

Reid, A., and C. Z. Ashley (2008) A context for the Luzira Head. *Antiquity 82*, 99–112.

Robertshaw, P. T. (1997) Munsa earthworks: A preliminary report on recent excavations. *Azania 32*, 1–20.

Robertshaw, P. T., D. Collett, D. Gifford, and N. B. Mbae (1983) Shell midden on the shores of Lake Victoria. *Azania 18*, 1–44.

Schoenbrun, D. L. (1998). *A Green Place, A Good Place: Agrarian Change, Gender, and Social Identity in the Great Lakes Region to the 15th Century*. Oxford: James Currey.

Schmidt, P. (1978) *Historical Archaeology: A Structural Approach in an African Culture*. Westport, CT: Greenwood Press.

———. (1997) Archaeological views on a history of landscape change in East Africa. *Journal of African History 38*, 393–421.

Schmidt, P., and S. T. Childs (1985) Innovation and industry during the Early Iron Age in East Africa: KM2 and KM3 sites in Northwest Tanzania. *African Archaeological Review 3*, 53–94.

Soper, R., and B. Golden (1969) An archaeological survey of Mwanza region, Tanzania. *Azania 4*, 15–79.

Speke, J. H. (1863) *Journal of the Discovery of the Source of the Nile*. Edinburgh: Blackwood and Sons.

Stager, J. C., B. F. Cumming, and L. D. Meeker (2003) A 10,000-year high-resolution diatom record from Pilkington Bay, Lake Victoria, East Africa. *Quaternary Research 59*, 172–81.

Stager, J. C., P. A. Mayewski, and L. D. Meeker (2002) Cooling cycles, Heinrich event 1, and the desiccation of Lake Victoria. *Palaeogeography, Palaeoclimatology, Palaeoecology 183*, 169; 78.

Stanley, H. M. (1879) *Through the Dark Continent*. London: Sampson Low, Marston and Company.

Thornton, D., and N. V. Rounce (1936) Ukara Island and the agricultural practices of the Wakara. *Tanganyika Notes and Records 1*, 25–32.

Wayland, E. J., M. Burkitt, and H. J. Braunholtz (1933) Archaeological discoveries at Luzira. *Man 33*, 25–29.

Wrigley, C. (1989) Bananas in Buganda. *Azania 24*, 64–70.

Young, R., and G. Thompson (1999) Missing Plant Foods? Where is the archaeobotanical evidence for sorghum and finger millet in East Africa? In M. van der Veen (Ed.), *The Exploitation of Plant Resources in Ancient Africa*, pp. 63–72. New York: Kluwer Academic/Plenum Publishers.

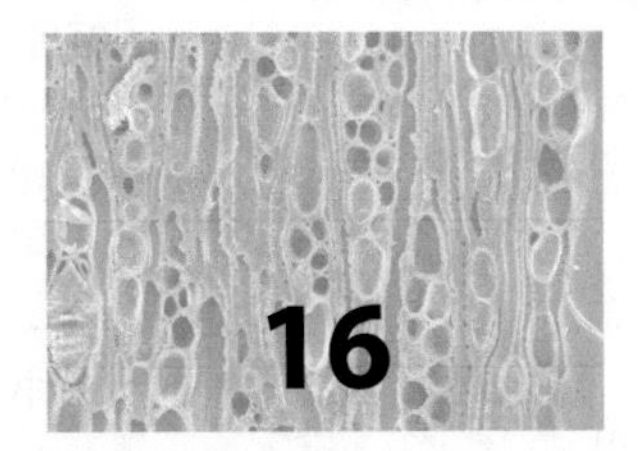

16

Archaeobotanical Investigations of the Iron Age Lundu State, Malawi

Ingrid Heijen

The research presented here was part of the Lundu Archaeology Project (Leiden University), developed to gain a better understanding of the 'identity' of the Iron Age Lundu state, which existed in what is now Malawi from at least 1500 to ca. 1860 C.E. (Welling 2002a, 2002b, 2003). Between 2001 and 2005 archaeological research was carried out in the Lower Shire Valley, the location of the Lundu state, at Mbewe ya Mitengo. The interdisciplinary project sought to investigate the Lundu state's internal dynamics of authority in relation to such external factors as trade, war, slavery, migration, drought, and tribute systems. Archaeological and palaeoecological data were recovered to complement to existing documentary, ethnographic, and oral sources. This chapter focuses on the first phase of the analysis of the archaeobotanical samples: those recovered from 2002 and 2003 (Heijen 2005) and limited further analysis in 2006. Whereas Welling (2002a, 2002b, 2003) surveyed and excavated various parts of the Lower Shire Valley, my archaeobotanical research was concentrated in the village of Mbewe ya Mitengo (Figure 16.1).

In Malawi, but also elsewhere in southeastern Africa, there is a considerable lack of archaeobotanical investigation. The Lundu Archaeology Project's aim was to raise awareness of the importance and potential of archaeology in Malawi and included archaeobotany from its inception. More broadly, though, this archaeobotanical research seeks to indicate the necessity of archaeobotany in other areas of southeastern Africa as well. Archaeobotanical research in Malawi is challenging, however, because researchers face a lack of comparative literature and adequate comparative reference collections.

Historical and Archaeological Background: The Lundu and Mbewe ya Mitengo

The *Lundu* is the king of the *Mang'anja* tribe, a group who live today in the Lower Shire Valley, southern Malawi. According to oral tradition, the first Lundu, Mbona, established the first powerful centralized state in the Lower Shire Valley in the 16th century C.E. Mbona played an important role in the tribe's rituals. During the second half of the 16th century large groups of people in the Lower Shire Valley and in neighboring countries were faithful to Mbona and his worship. Mbona came to be seen as the rain god, and the legend of his life and death is still passed on and celebrated annually. The central person in the Mbona cult was and continues to be the Lundu, and he plays a principal role when a new wife for Mbona has to be chosen (Schoffeleers 1992).

The village of Mbewe ya Mitengo is the home of the current Lundu. Mbewe ya Mitengo is situated on the southeast edge of Lengwe National Park and is so to speak the entrance to this park. The park and Mbewe ya Mitengo are presently separated from the river Shire by an extensive

Archaeology of African Plant Use by Chris J. Stevens, Sam Nixon, Mary Anne Murray, and Dorian Q Fuller, Eds., 189–194

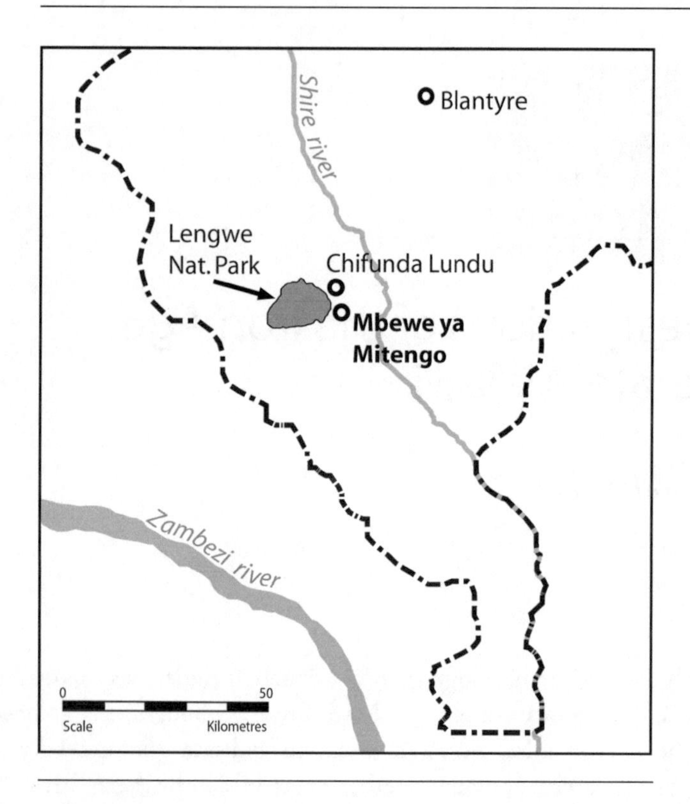

Figure 16.1 Malawi: The Lower Shire Valley and the site Mbewe ya Mitengo.

sugar-cane estate (Illovo). Mbewe ya Mitengo is not a conventional village, being mostly a sacred forest where sacrifices and initiations took place. According to oral sources from the 20th century there must have been a Lundu in Mbewe ya Mitengo as far back as the 17th century, although where exactly is not known. According to the archaeological data, however, any historic Lundu at Mbewe ya Mitengo may not have lived in the present forest of Mbewe ya Mitengo but rather some distance outside the forest.

In the area of Mbewe ya Mitengo five units (A–E) were excavated (Figure 16.2). Also, 23 test pits were located in the middle of the forest in a row. The excavations evidenced just one layer dating to the 19th century C.E. The archaeological material was varied, including glass beads (for example, Venetian whitehearts), vessel glass, a large amount of pottery, stone, charcoal, burned clay, and bone (Welling 2002a, 2003).

MATERIAL AND METHODS

In total 25 archaeobotanical samples were collected in Mbewe ya Mitengo, all obviously coming from the only archaeological level evidenced: 11 in unit A (only 10 of these samples were analysed); 6 in unit B; 2 in unit C;

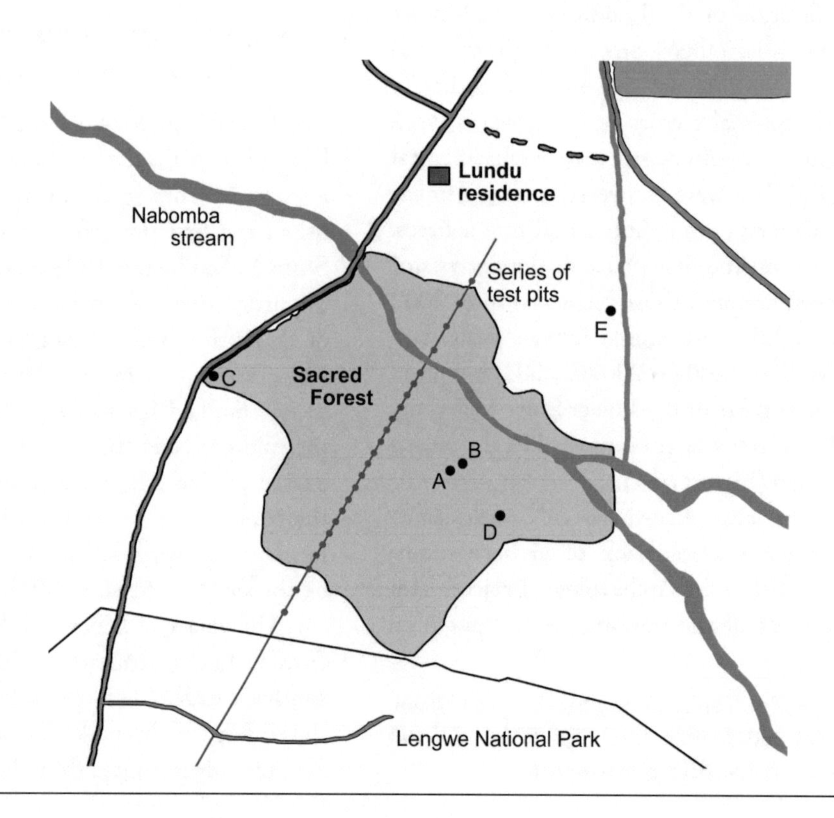

Figure 16.2 The site Mbewe ya Mitengo with the testpits and the five excavated units (A–E).

2 in unit D; and 4 in unit E. Although archaeobotanical samples were also taken in the units C, D, and E, the poor results led to the decision to focus on units A (2 × 2 m) and B (4 × 2 m). Where possible, sample size was 15 litres. The samples were taken from areas with the richest archaeological concentrations or places where a concentration of seeds was clearly visible. The samples were first dry-sieved over 2- and 1-mm sieves. The organic fraction was then separated by flotation. After drying in the sun, the samples were transported to Leiden for sorting and identification at the laboratory of the Faculty of Archaeology.

Given that in Leiden there is very limited relevant reference material, the current study was limited in the diversity of taxa that could be identified. As a preparation to tackle this problem, I collected a large number of seeds and fruits in Malawi, in Lengwe National Park, around the excavation area in Mbewe ya Mitengo, and at the local market. Ultimately 60 botanical parts of plants were taken. Although the material proved to be of limited value in seed identification, it is part of a growing collection that can assist future studies. After initial study, the samples were sent to Frankfurt University for comparison with the larger African collection there. In addition to limited reference materials, many remains were highly fragmented and unidentifiable.

RESULTS

All seed types, apart from *Sorghum bicolor*, that were not immediately recognized were given a type number, which appears in the data table (Table 16.1). Although we have not been able to provide a precise account of the plant taxa recovered, we do have a good broad definition of the plant remains present from an area about which we have known very little so far.

The plant remains recovered are almost entirely from units A and B, the other three units providing poor recovery. Units A and B are both 19th century and are situated next to each other in the sacred forest of Mbewe ya Mitengo. It is highly likely that not only are the plant remains from the same period but that they relate to the same household. Regarding variation between units A and B, any variation we see is spread between the different samples, and no clear pattern could be discerned; indeed, all species that could be identified were evident in both units A and B. The number of remains per sample was often limited to 10 or fewer, so relative proportions are not very significant. However, what is interesting is the relatively great variation of species encountered in these samples.

CEREALS: SORGHUM, FINGER MILLET, AND MAIZE

Sorghum (*Sorghum bicolor* subsp. *bicolor*) is by far the most dominant taxon recovered. The grain and chaff are present in large quantities and in most samples. The quantities of chaff suggests pounding (deshusking) on a regular basis, of what is assumed to have been a typical race bicolour, husked sorghum variety. The large amount of sorghum remains in both units A and B suggest that sorghum was a very important part of daily food. Whether the macroremains were carbonized through disposal of processing waste in fires or during a conflagration that included the structure(s) in units A and B is unclear. It is worth noting that some small sorghum grains were initially referred to as '*Panicum miliaceum*', although it subsequently became clear that these were a very small sorghum type. Sorghum is known to have been a crop in Southern Africa since the 1st millennium C.E., with Iron Age finds reported from the Shashe-Limpopo Basin and the Thukela Basin (South Africa), the Tsodilo Hills and Eastern Botswana, and Zimbabwe (Mitchell 2002, p. 274).

Some grains of *Eleusine coracana* (finger millet) were found among the sorghum, in sample AB1. Finger millet has also been established as a crop in Eastern and Southeast Africa since the 1st millennium C.E. (Mitchell 2002).

The samples contained only a small amount of *Zea mays* (maize). Although nowadays maize is considered to be a common food in most countries of Africa (Juwayeyi 1981; McCann 2001; Smale 1995), this represents a relatively recent development in Malawi. Previous staples were mainly sorghum and finger millet (Juwayeyi 1981; Williamson 1956). Written sources of travellers in the 18th and 19th centuries demonstrated that sorghum and finger millet were the earliest grain crops in Malawi. Lacerda does mention maize in 1798, as does Livingstone 60 years later, but both indicate that sorghum and finger millet were the principal crops (Burton 1873; Livingstone 1857). To be precise, sorghum and the millets were considered to be the staple diet until approximately the second and third decades of the 20th century, when maize took over (Juwayeyi 1981; McCann 2001; Smale 1995).

OTHER TAXA

Although found in small quantities, savannah tree fruits were recovered, which may come from managed or planted trees. These fruits comprised seeds of baobab (*Adansonia digitata*), tamarind (*Tamarindus indica*), and *Syzygium* sp., and a few seeds of Cucurbitaceae, including two types of *Cucumis* sp. (the genus that includes melon, cucumber,

Table 16.1 Archaeobotanical Results of Mbewe ya Mitengo (CK18) (key: X = 10s; XX = 100s; A = present)

site/context		CK18-A9	CK18-A9	CK18-A11	CK18-A12	CK18-A13	CK18-A15	CK18-A17	CK18-A18	CK18-A20	CK18-A26	CK18-B5	CK18-B8	CK18-B9	CK18-B10	CK18-B13	CK18-B16	CK18-C5	CK18-C6	CK18-D4	CK18-D5	CK18-E6	CK18-E7	CK18-E12	CK18-E13	
bag# & AB-#.		5087 AB5	5094 AB7	5108 AB8	5120 AB9	5139 AB12	5149 AB13	5171 AB14	5176 AB16	5180 AB15	5188 AB17	6038 AB4	6055 AB1	6078 AB2	6098 AB3	6124 AB6	6136 AB11	7034 AB20	7035 AB21	8114 AB18	8115 AB19	9114 AB26	9115 AB27	9135 AB28	9165 AB29	
volume (l)		8	6	15	15	15	15	15	15	15	15	0.2	50	17	15	15	15	15	15	15	15	15	15	15	15	
level below surface (cm)		10–50	40–60	26–31	31–35	40–50	44–50	57–62	56–61	54–59	73–86	22–??	30–40	25–35	40–45	50–60	60–70	22–26	26–32	30–35	35–40	35–38	39–46	72–75	73–76	**total**
taxon	**type**																									
Sorghum bicolor			4	47	81	6	9		1	3			207	157	29	11	1									556
S. bicolour with chaff													1	1												2
S. bicolor chaff			5	26	182	48	2		3	21			106	43	30	12	2									480
Eleusine coracana													A													A
Zea mays					A	A	A																			A
Gossypium sp.	1&7			1								1	5													7
Tamarindus sp.	2												1													1
Syzygium spec.	3&4												3	1	1											5
Fabaceae	5		1										1													2
Adansonia digitata	6												3													3
Cucumis sp.	12												3													3
Solanaceae	13												3													3
Aizoaceae/ Portulacaceae	14		1	14	21	13	1	1	4				17	17	4	4	1									98
Cucumis sp.	15												1													1

site/context		CK18-A9	CK18-A9	CK18-A11	CK18-A12	CK18-A13	CK18-A15	CK18-A17	CK18-A18	CK18-A20	CK18-A26	CK18-B5	CK18-B8	CK18-B9	CK18-B10	CK18-B13	CK18-B16	CK18-C5	CK18-C6	CK18-D4	CK18-D5	CK18-E6	CK18-E7	CK18-E12	CK18-E13	
Cucurbitaceae	23				1																					1
stem fragment	24		8		1				2				2													13
Fabaceae	31			4	1	4	3																			12
??	33															1										1
Fabaceae	35		1																							1
Lamiaceae	37													1												1
Aizoaceae	40										1		1					1						1		4
Lamiaceae	46		1																							1
??	48		1																							1
Aizoaceae	53					1																	1		1	3
??	54													1												1
??	56				1																					1
??	59														1											1
cf. Cannabis	60												2													2
indet carbonized & noncarbonized		10	X	X	X	X	X	21	55	X	35	9	X	X	X	X	44	1	3	35	10	5	2			XX

and several sub-Saharan wild species). All these are likely to include food plants.

Two samples contained seeds identified as *Gossypium*. Although wild *Gossypium* is known from the region (Saunders 1961), these likely come from cultivated cotton and indicate the cultivation and processing of cotton fibres.

In terms of quantity another significant taxon is Aizoaceae/Portulacaceae, which is most likely a weed of sorghum or millet.

CONCLUSIONS

As well as demonstrating the dominance of Sorghum, other cereals, tree fruits, possible cucurbitaceous vegetables, and cotton were recovered. Maize appears far less prominent than today. It is also worth noting the absence of rice, which is a fairly common crop in parts of southeastern African today. The results provide a good comparative dataset to be compared against documentary historical, oral, linguistic, and ethnographic evidence, allowing us to build up an improved picture of plant usage within the Lundu state.

ACKNOWLEDGEMENTS

I would like to thank C. C. Bakels, M. Welling, W. Kuijper, S. Kahlheber, Universiteit Leiden, and the Johann Wolfgang Goethe Universität Frankfurt am Main.

REFERENCES

Burton, R. F. (1873) *The Lands of Cazembe: Lacerda's Journey to Cazembe in 1798*. London: John Murray.

Heijen, I. C. E (2005) *Sorghum-eters in Malawi: Archeobotanische Vondsten uit de 15e–19e Eeuw te Mbewe ya Mitengo*. Master's thesis, University of Leiden.

Juwayeyi, Y. M. (1981) *The Later Prehistory of Southern Malawi: A Contribution to the Study of Technology and Economy during the Later Stone Age and Iron Age Periods*. Ph.D. thesis, University of California.

Livingstone, D. (1857) *Missionary Travels and Researches in South Africa*. London: John Murray.

McCann, J. (2001) Maize and grace: History, corn and Africa's new landscapes, 1500–1999. *Comparative Study in Society and History* 43 (2), 246–72.

Mitchell, P. (2002) *The Archaeology of Southern Africa*. Cambridge: Cambridge University Press.

Saunders, J. H. (1961) The *Wild Species of Gossypium and Their Evolutionary History*. Oxford: Oxford University Press.

Schoffeleers, J. M. (1992) *River of Blood. The Genesis of a Martyr Cult in Southern Malawi, c. 1600 AD*. Madison: The University of Wisconsin Press.

Smale, M. (1995) 'Maize is life': Malawi's delayed Green Revolution. *World Development* 23 (5), 819–31.

Welling, M. (2002a) *Landscapes of Trade, Production, and Sacred Power: An Interdisciplinary Inquiry into the Identity of the Lundu State in the Lower Shire Valley, c. 1500–1863 AD*, unpublished Annual Progress Report. The Hague: WOTRO.

———. (2002b) *Landscapes of Trade, Production, and Sacred Power: An Interdisciplinary Inquiry into the Identity of the Lundu State in the Lower Shire Valley, c. 1500–1863 AD*, unpublished Ph.D. research proposal. Leiden: CNWS.

———. (2003) *Landscapes of Trade, Production, and Sacred Power: An Interdisciplinary Inquiry into the Identity of the Lundu State in the Lower Shire Valley, c. 1500–1863 AD*, unpublished Annual Progress Report. The Hague: WOTRO.

Williamson, J. (1956) *Useful Plants of Malawi*. Zomba: The Government Printer.

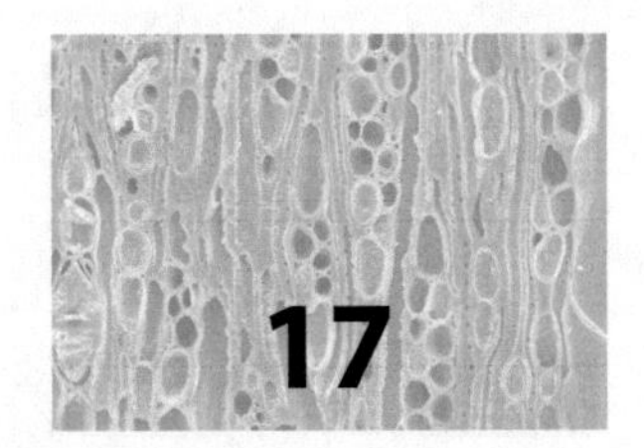

Prehistoric Plant Use on La Palma Island (Canary Islands, Spain)

An Example of the Disappearance of Agriculture in an Isolated Environment

Jacob Morales, Amelia Rodríguez, and Águedo Marrero

La Palma is a mountainous and remote island placed in the Canarian Archipelago (Spain) (Figure 17.1). The earliest colonisers arrived to the island from northern Africa approximately in the 1st millennium B.C.E., but it is assumed that they stayed practically isolated from the rest of the archipelago and the mainland for almost two thousand years. When modern Europeans first made contact with the indigenous people from La Palma, in the 15th century C.E., they reported that the indigenes, 'Auaritas', did not practise navigation or cultivation. At this time, they obtained food mainly from the livestock and gathered plant resources (Abreu [1602] 1977; da Zurara [1448] 1998). By contrast, farming was reported as practised in the rest of the archipelago during the pre-Hispanic times (Morales 2010).

Previous evidence for the use of plants during the pre-Hispanic settlement of La Palma comes from ethnohistorical texts written by Portuguese and Spanish explorers during the 15th and 16th centuries C.E. However, to date few archaeobotanical studies have been done in the Canaries, let alone La Palma, with most focused on the analysis of wood charcoal (Machado 1998; Morales 2003).

The main purpose of this work is to analyse and interpret collected archaeobotanical data from two archaeological sites of La Palma in order to improve our understanding about (a) the role of food plants during the pre-Hispanic period, (b) the use of other plant resources, and (c) the possible absence of cereal crops.

Archaeology of African Plant Use by Chris J. Stevens, Sam Nixon, Mary Anne Murray, and Dorian Q Fuller, Eds., 195–204

The Pre-Hispanic People from La Palma: General Features

The natural environment of La Palma is highly distinctive, being a volcanic island in the northwestern of the Canaries, characterised by high mountains and deep ravines. The climate is dominated by the Azores anticyclone that causes the trade winds during most of the year, with rainfall varying from 300 to 1,000 mm annually. Additionally, the ecological conditions and geographical isolation of the island have led to a tremendous botanical and animal biodiversity (Martín et al. 2001; Santos 1983).

The earliest known human settlement on La Palma, La Palmera, has been radiocarbon dated to the 3rd century B.C.E. (Navarro and Martín 1987). However, most sites are dated to the 1st or 2nd millennium C.E. The origin of the earliest migrants is unknown. Most scholars agree that the language and material culture of these people had much in common with the Tamazigh cultural sphere (Berber), although another colonisation process by people from the Sahara (Martín 1992) has been suggested on ceramic grounds. Recent mitochondrial DNA analysis of 38 aboriginal remains of La Palma show that the majority of lineages (93%) were from West Eurasian origin

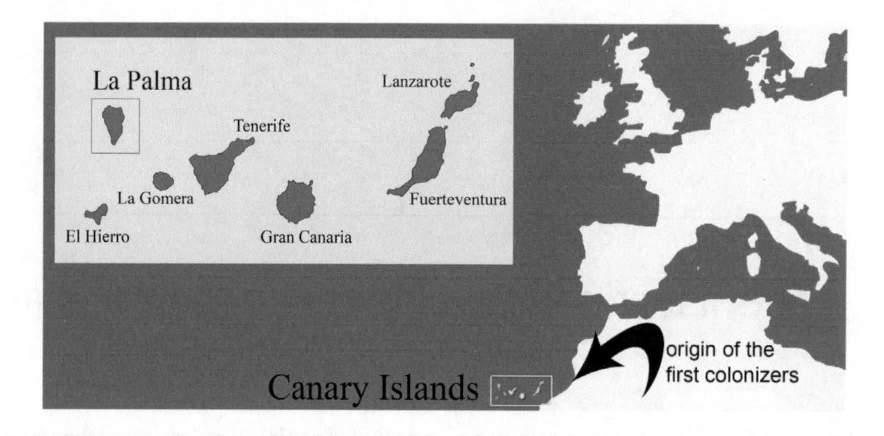

Figure 17.1 Location of the Canary Islands and possible place of origin of the first colonisers of the archipelago.

(Fregel et al. 2009). The bulk of the aboriginal haplotypes had exact matches in North Africa (70%). However, the indigenous Canarian subtype U6b1, also detected in La Palma, has not been found in North Africa yet, the cradle of the U6 expansion. This lack of evidence for the indigenous Canarian haplotype means that the exact region from which the ancestors of the Auaritas came has not been sampled yet or that they have been replaced by later human migrations. The high gene diversity found in La Palma is of the same level of that previously found in the central island of Tenerife and goes against the idea of an island-by-island independent maritime colonisation without secondary contacts. The data better fit to an island colonisation model with frequent migrations between islands (Fregel et al. 2009).

During the 15th century C.E., after a long process of conquest, the Kingdom of Castilla occupied the island, and La Palma became part of Spain, the native Auritas being killed or taken into slavery. Before the arrival of the Spanish the Auaritas were shepherds—with goats, sheep, pigs, dogs, and cats—who also carried out fishing, hunting, and gathering wild plants and marine shellfish. The bones of sheep and goat have been recovered in large numbers in almost every domestic site from La Palma (Pais 1996), and the importance of these animals is confirmed by dietary studies (González and Arnay 1992; Pérez 2000).

According to ethnohistorical texts, the Auaritas collected the fruits of *mocán* (*Visnea mocanera* L. f.), an endemic tree, harvested seeds from the *amagante* (*Cistus* section *Macrostylia* Willk.), and collected roots from ferns (*Pteridium aquilinum* [L.] Kuhn) to make *gofio* (Abreu [1602] 1977). (*Gofio* is a traditional staple of the Canaries; seeds are roasted and then milled to produce a type of flour that can be eaten with water, milk, soup, honey, fruit, and so on.)

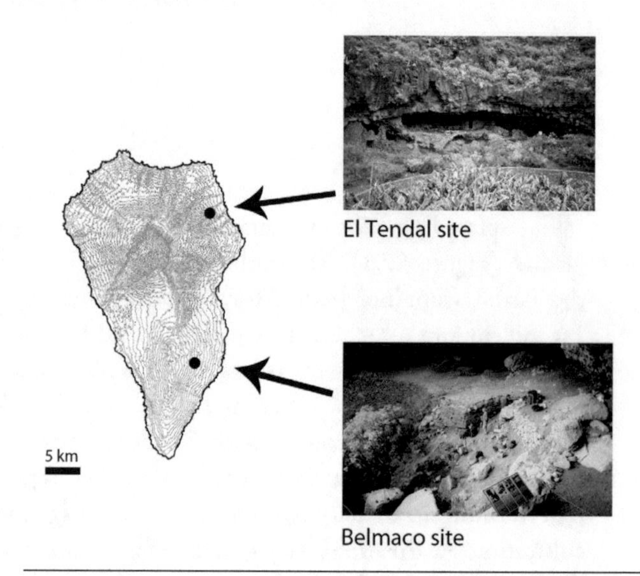

Figure 17.2 Location and photo of El Tendal and Belmaco sites.

THE ARCHAEOBOTANICAL EVIDENCE: SITES AND SEEDS

The archaeobotanical evidence presented here comes from two archaeological sites on La Palma, El Tendal and Belmaco.

EL TENDAL

El Tendal is a cave site, near the dry bed of a ravine, and is located in the northeast of La Palma at 150 m above the sea level (Figure 17.2). The climate is characterised by mild temperatures and a mean annual rainfall of approximately 600 mm. The cave has been excavated since 1981, but the archaeobotanical assemblage comes only from the last excavations in 1987. Archaeological remains revealed evidence for occupation site and activity (Navarro, Martín, and Rodríguez 1990). Radiocarbon dates on charcoals indicate that the

Table 17.1 List of Macrobotanical Remains from El Tendal Site (* = floated samples)

	ZONE B								ZONE C				ZONE C*		
Sector	E/1	E/0	E/1	E/1	E/1	E/2	E/2	D/00	V/9	V/10	V/11	V/10	V/11	V/10	
Level	III d	III f	IV c	IV d	IV e	IV d	IV e	V b	XX II	XX VI	XX VII	XX VIII	XX VI	XX VII	
Cereals															**TOTAL**
Hordeum vulgare subsp. *vulgare*, grain	16	–	18	15	28	2	3	3	–	12	19	3	5	43	167
Hordeum vulgare subsp. *vulgare*, rachis	–	–	–	–	–	–	–	–	–	–	–	–	6	39	45
Triticum sp., grain	11	1	–	2	–	1	–	–	–	28	9	–	5	102	159
Triticum turgidum subsp. *durum*, rachis	–	–	–	–	–	–	–	–	–	–	–	–	6	35	41
Pulses															
Vicia faba	1	–	–	–	–	–	–	–	–	–	1	–	–	1	3
Lens culinaris	–	–	–	–	–	–	–	–	–	–	1	–	–	3	4
Wild Plants															
Cedronella canariensis	–	–	–	–	–	–	–	–	–	–	–	–	–	1	1
Erica arborea, leaf	–	–	–	–	–	–	–	–	–	–	–	–	10	30	40
Erica arborea, inflorescense	–	–	–	–	–	–	–	–	–	–	–	–	2	16	18
cf. *Erica arborea*, inflorescense	–	–	–	–	–	–	–	–	–	–	–	–	–	1	1
Hypericum grandifolium, capsule	–	–	–	–	1	–	–	–	–	–	–	–	–	–	1
cf. *Ocotea foetens*	–	–	–	–	–	–	–	–	–	–	1	–	–	–	1
Pinus canariensis, cone fragment	–	–	–	–	–	–	–	–	–	–	–	–	2	3	5
Retama rhodorhizoides	–	–	–	–	–	–	–	–	–	–	–	–	–	5	5
cf. *Retama rhodorhizoides*	–	–	–	–	–	–	–	–	–	1	–	–	–	2	3
Ilex sp.	–	–	–	–	–	–	–	–	4	–	–	–	–	–	4
Leguminous indet.	–	–	–	–	–	–	–	–	–	–	1	–	–	–	1
Weeds															
Amaranthus sp.	–	–	–	–	–	–	–	–	–	–	–	–	18	5	23
Anagallis arvensis	–	–	–	–	–	–	–	–	–	–	–	–	–	5	5
Apiaceae indet.	–	–	–	–	–	–	–	–	–	–	–	–	–	1	1
Carex cf. *divulsa*	–	–	–	–	–	–	–	–	–	–	–	–	1	1	2
Chenopodium murale	–	–	–	–	–	–	–	–	–	–	–	–	4	1	5
Compositae indet.	–	–	–	–	–	–	–	–	–	–	–	–	–	2	2
Emex spinosa	–	–	–	–	1	–	–	–	–	–	–	–	–	–	1
Galium aparine	1	–	–	–	–	–	–	–	–	–	1	–	2	7	11
Gramineae indet.	–	–	–	–	–	–	–	–	–	–	–	–	1	4	5
Malva sp.	–	–	–	–	–	–	–	–	–	–	–	–	–	1	1
Small leguminous	–	–	–	–	–	–	–	–	–	–	–	–	5	21	26
Phalaris sp.	–	–	–	–	–	–	–	–	–	–	–	–	4	13	17

(Continued)

Table 17.1 Continued

	ZONE B								ZONE C				ZONE C*		
Sector	E/1	E/0	E/1	E/1	E/1	E/2	E/2	D/00	V/9	V/10	V/11	V/10	V/11	V/10	
Level	III d	III f	IV c	IV d	IV e	IV d	IV e	V b	XX II	XX VI	XX VII	XX VIII	XX VI	XX VII	TOTAL
Plantago cf. *lagopus*	–	–	–	–	–	–	–	–	–	–	–	–	1	4	5
cf. *Ranunculus sardous*	–	–	–	–	–	–	–	–	–	–	–	–	1	–	1
Rumex sp.	–	–	–	–	–	–	–	–	–	–	–	–	4	7	11
Silene gallica	–	–	–	–	–	–	–	–	–	–	–	–	7	9	16
Silene sp.	–	–	–	–	–	–	–	–	–	–	–	–	–	4	4
Sherardia arvensis	–	–	–	–	–	–	–	–	–	–	–	–	–	5	5
Solanum nigrum	–	–	–	–	–	–	–	–	–	–	–	–	29	22	51
Solanaceae indet.	–	–	–	–	–	–	–	–	–	–	–	–	–	1	1
Indeterminated	–	–	–	1	–	–	–	–	–	–	–	–	15	29	45
Total no. of seeds	29	1	18	18	30	3	3	3	4	41	33	3	128	423	737

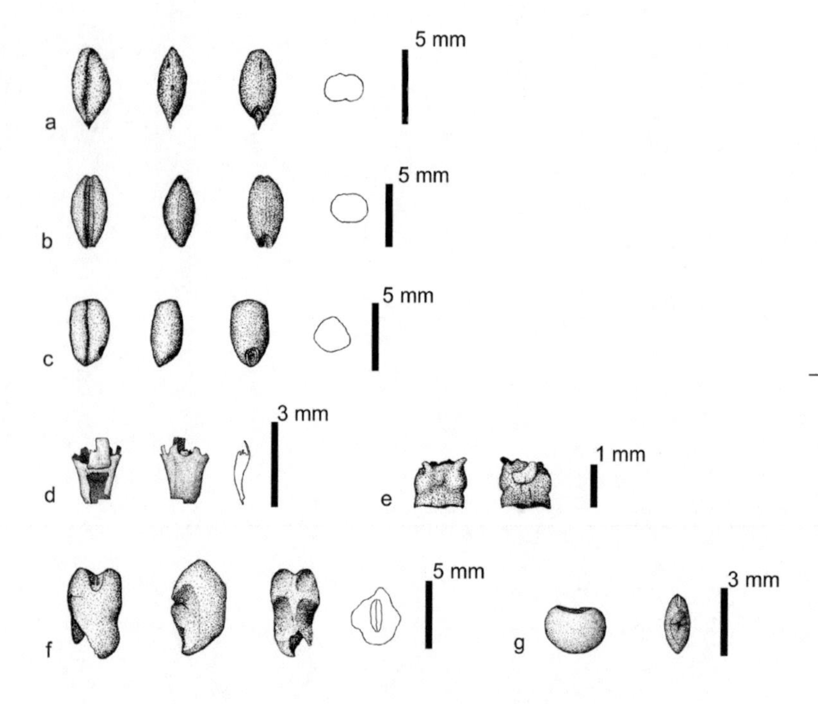

Figure 17.3 Cultivated plant remains from El Tendal site: (a) *Hordeum vulgare* subsp. *vulgare*, twisted grain; (b) *Hordeum vulgare* subsp. *vulgare*, straight grain; (c) *Triticum* sp., grain; (d) *Hordeum vulgare* subsp. *vulgare*, rachis segment; (e) *Triticum turgidum* subsp. *durum*, rachis segment; (f) *Vicia faba*, seed; (g) *Lens culinaris*, seed.

El Tendal site was used at least since the 3rd century C.E. until the 11th century C.E. (Soler et al. 2002).

Systematic sampling was not used in the 1987 archaeological excavations, and thus the archaeobotanical assemblage studied was recovered mainly through dry-sieving with 2- and 1-mm meshes. However, two samples of 7 litres from a hearth were collected and floated. In spite of these problems the assemblage is relatively large, and we have counted 737 charred, well-preserved macro-botanical remains (Table 17.1). The nomenclature of the scientific plant names follows Zohary and Hopf (1993) for crops and Acebes-Ginovés and associates (2010) for the wild plants in the Flora of the Canary Islands.

Conversely to previous data, which indicated absence of agriculture in La Palma, we have identified domesticated cereals and pulses in El Tendal. Hulled 6-row barley (*Hordeum vulgare* L. subsp. *vulgare*) and hard wheat (*Triticum turgidum* subsp. *durum* [Desf.]) were the most abundant remains recovered, with both grains and abundant rachis fragments present, as well as two legumes, lentils (*Lens culinaris* Medik.) and faba bean (*Vicia faba* L.) (Figure 17.3).

We have identified a significant number of macro-botanical remains from wild native plants from La Palma that could be gathered for different purposes, mainly as fuel. The most abundant are the inflorescences and leaves from heather (*Erica arborea* L.), together with cone fragments from pine (*Pinus canariensis* Sweet ex Spreng.). In addition, we have recovered other remains from endemic species from the Canaries as *Ilex* sp., *Hypericum grandifolium* Choisy, *Retama rhodorhizoides* Webb and Berthel., *Cedronella canariensis* (L.) Webb and Berthel., and cf. *Ocotea foetens* (Aiton) Baill.

Remains from weeds were retrieved, mainly from the floated samples. Some 20 species common to the archipelago were determinate; among them *Solanum nigrum* L. is the most abundant, but also numerous were *Amaranthus* sp., small legumes, *Silene gallica* L., and *Phalaris* sp. The occurrence of weeds seeds is very interesting; they were probably associated with crops, supporting the contention that agriculture was practiced by the inhabitants of El Tendal.

BELMACO

A shallow cave was excavated at Belmaco, renowned for its pre-Hispanic engravings. It is located in the southeast part of the island, at 400 m above the sea level, the climate conditions being similar to the El Tendal site (Figure 17.2). The cave was originally excavated during the 1960s and the 1970s (Hernández 1999). Conventional radiocarbon dates from charcoals dated the site between the 7th and the 11th centuries (Hernández 1999), whereas ceramics and coins imported from Europe indicate that the site could be have been used from the mid–late 1st millennium until the 15th century C.E. In 2000 a reexamination of the deposits in cross-section revealed levels where completely charred vegetal remains could be seen at a glance. Samples of 90 litres were collected from these and floated.

The quantity of macro-botanical remains is smaller than at El Tendal. In total, 301 charred items were recovered (Table 17.2). Only a few remains from crops were identified, consisting of three grains and three rachis segments from 6-row hulled barley (*Hordeum vulgare* subsp.

Table 17.2 List of Macrobotanical Remains from Belmaco Site

Cross–Section	1		2					3				4					
Level	III	II	V	IV	III	II	I	IV	III	II	I	VI, VII	IV, V	III	II	I	Total
Volume (in litres)	1	1	0.5	0.5	1.5	1.5	2	4	7	9	4	22	5	9	12	9	89
No of samples	1	1	1	1	1	1	1	9	8	7	6	5	4	3	2	1	52
Cultivated plants																	
Hordeum vulgare subsp. *vulgare*, seed	–	–	–	–	–	–	–	–	–	–	–	3	–	–	–	–	3
Hordeum vulgare subsp. *vulgare*, rachis	–	–	–	–	–	–	–	–	–	–	–	3	–	–	–	–	3
Gathered plants																	
Juniperus turbinata subsp. *canariensis*, seed	2	–	–	–	3	7	–	–	10	18	3	–	8	3	9	12	75
Juniperus turbinata subsp. *canariensis*, leaf	–	–	–	–	2	4	–	2	1	3	1	–	2	–	3	3	21
Juniperus turbinata subsp. *canariensis*, fruit	–	–	1	–	–	–	–	–	–	–	–	–	–	–	1	3	5
Olea cerasiformis	–	3	–	–	–	1	–	1	2	5	–	–	–	–	1	3	16
Retama rhodorhizoides	–	–	1	1	1	1	–	–	–	4	–	–	1	–	2	3	14
Teucrium heterophyllum	–	–	–	–	–	–	–	–	–	–	–	–	–	–	1	–	1
Weeds																	
Amaranthus sp.	–	–	–	–	–	–	–	–	–	1	1	1	1	–	–	–	4
Anagallis arvensis	–	1	–	–	–	–	–	–	–	–	–	1	–	–	–	–	2
cf. *Bromus* sp.	–	–	–	–	–	–	–	1	–	–	–	–	–	–	–	–	1

(Continued)

Table 17.2 Continued

Cross–Section	1		2					3				4					
Level	III	II	V	IV	III	II	I	IV	III	II	I	VI, VII	IV, V	III	II	I	Total
Volume (in litres)	1	1	0.5	0.5	1.5	1.5	2	4	7	9	4	22	5	9	12	9	89
No of samples	1	1	1	1	1	1	1	9	8	7	6	5	4	3	2	1	52
Cultivated plants																	
Chenopodium murale	–	–	2	1	1	–	–	8	5	3	5	4	–	1	14	1	45
Chrysanthemum/ Argyranthemum sp.	–	–	–	–	–	–	1	–	2	6	–	1	–	–	–	1	11
Compositae indet.	–	–	–	–	–	–	–	1	–	–	1	–	–	1	1	–	4
Cyperaceae	–	–	–	–	–	–	–	–	–	–	–	–	–	–	3	–	3
Fumaria sp.	–	–	–	–	–	–	–	–	–	1	–	1	–	–	–	–	2
Galium aparine	–	–	–	–	–	–	–	1	1	2	1	–	–	–	–	–	5
Gramineae	–	–	1	–	–	–	–	1	–	–	–	–	–	–	–	–	2
Malva sp.	–	–	–	–	–	–	–	1	–	–	1	–	–	–	1	–	3
Patellifolia procumbens	–	–	–	–	–	–	1	1	2	–	–	–	–	–	2	1	7
Small leguminous	–	–	–	–	–	–	–	–	2	–	–	–	–	–	1	–	3
Plantago cf. *lagopus*	–	–	–	–	–	–	–	–	–	–	–	1	–	–	–	–	1
Rumex sp.	–	–	1	–	–	–	–	6	3	1	–	1	–	–	–	1	13
Sherardia arvensis	–	–	–	1	–	–	–	–	–	–	–	–	–	–	–	–	1
Silene gallica	–	1	–	–	–	–	–	–	1	1	–	3	–	1	1	–	8
Solanum nigrum	–	–	–	–	1	–	–	1	2	8	1	1	–	–	4	3	21
Indeterminated	–	–	2	3	–	–	–	8	4	5	1	1	2	–	1	–	27
Total no. of seeds	2	5	8	6	8	13	2	32	35	58	15	21	14	6	45	31	301
Density per litre	2	5	16	12	5.3	8.6	1	8	5	6.4	3.75	1	2.8	0.6	3.75	3.4	3.4

Figure 17.4 Charred fruit and seeds of juniper (*Juniperus turbinata* subsp. *canariensis*) from Belmaco site.

Figure 17.5 Charred seeds of wild Canarian olive (*Olea cerasiformis*) from Belmaco site.

vulgare). It is interesting to note that these were collected in the deeper level, and in recent strata crop remains are completely absent.

Macro-botanical remains from wild gathered plants were exceptionally important. Juniper (*Juniperus turbinata* Guss. subsp. *canariensis* [A. P. Guyot] Rivas-Mart., Wildpret and P. Pérez) is the most abundant plant, with charred leaves, seeds, and whole fruits present (Figure 17.4), whereas seeds of the Canarian wild olive (*Olea cerasiformis* Rivas-Mart. and del Arco) were identified in many samples (Figure 17.5). In addition, two endemic, locally growing inedible shrubs, *Retama rhodorhizoides* and *Teucrium heterophyllum* L'Hér., were also recognised. Weed seeds were uncommon in the samples, being mainly concentrated in the older levels, with *Chenopodium murale* L., *Solanum nigrum*, *Silene gallica*, and *Rumex* sp. being the most frequent taxa.

CULTIVATED PLANTS AND AGRICULTURE DURING THE PRE-HISPANIC PERIOD OF LA PALMA

To establish the antiquity of the cereal remains, several grains were subjected to AMS radiocarbon dating (Table 17.3). At El Tendal three barley grains were dated. The earliest dated to 260–290 and 320–450 cal C.E. (1660 ± 40 BP, Beta-206154); the second to 410–580 cal C.E. (1570 ± 40 BP, Beta-206156), the third to 600–680 cal C.E. (1400 ± 40 BP, Beta-206155). These results corroborate the conventional radiocarbon dates obtained from charcoals (Soler et al. 2002) and indicate that farming was practised at least during the 1st millennium C.E.

At Belmaco site, the earliest determination was on a barley grain recovered in the deeper strata, level VII-VI, dating to 680–880 cal C.E. (1250 ± 40 BP, Beta-206151). A juniper seed from upper strata, level V-IV, was dated to 1040–1260 cal C.E. (870 ± 40 BP, Beta-206150).

The archaeobotanical remains recovered from these sites provide the first conclusive proof that pre-Hispanic people from La Palma cultivated cereals in the 1st millennium C.E. and would seem to offer conflicting evidence to the ethnohistorical accounts that state that the native population did not practise agriculture. However, that all the cereal dates fall within the 1st millennium C.E. raise the possibility that the pre-Hispanic population on La Palma abandoned agriculture at some point before the arrival of Portuguese and Spanish explorers in the 15th century.

Hulled barley was the main crop recovered from the two analysed sites, and it might be noted that, for the rest of the Canaries, barley is the most frequent cereal identified from archaeological sites (Arco et al. 1990; Morales 2010). Adaptation to adverse and dry conditions, high productivity, resistance to pests, and cultural traditions are the probable reasons for the preference of barley (Morales 2010). Hard wheat was also identified, with grains and rachis segments recovered at El Tendal, but was absent at Belmaco (Figure 17.3 and Figure 17.6). In other archaeological sites from the archipelago, wheat is also very scarce, and it has been recovered in very few samples (Arco et al. 1990; Morales 2010). In addition to taphonomic factors, the low presence of wheat can be linked to cultural choices and its more demanding ecological requirements.

Figure 17.6 Charred seeds of wheat (*Triticum* sp.) from El Tendal site.

Besides cereals, two different cultivated legumes are recorded at El Tendal, lentils and faba bean, which also have been documented at Gran Canaria and Tenerife (Arco et al. 1990; Morales 2010). However, finds of leguminous crops are very rare and recovered only from a few samples, perhaps linked to their lower probability of becoming charred and because they are less productive than cereals.

The absences of hulled wheat (*Triticum monococcum* L.; *T. turgidum* L. subsp. *dicoccum* [Shrank] Thell, or *T. aestivum* L. subsp. *spelta* [L.)] and African cereals (*Sorghum bicolor* [L.] Moench. and *Pennisetum glaucum* [L.] R. Br.) indicate that the pre-Hispanic assemblage of crops was more likely introduced to the Canaries at the end of the 1st millennium B.C.E. and beginnings of the 1st millennium C.E. Hulled wheat is displaced by naked wheat from the Mediterranean at the end of the

Table 17.3 Radiocarbon Determinations on Charred Seeds from El Tendal and Belmaco Sites

Site	Plant Species	Lab No.	Uncalibrated BP	Calibrated C.E.
El Tendal	*Hordeum vulgare*	Beta-206154	1660 ± 40	260–290 and 320–450
El Tendal	*Hordeum vulgare*	Beta-206156	1570 ± 40	410–580
El Tendal	*Hordeum vulgare*	Beta-206155	1400 ± 40	600–680
Belmaco	*Hordeum vulgare*	Beta-206151	1250 ± 40	680–880
Belmaco	*Juniperus turbinata*	Beta-206150	870 ± 40	1040–1260

1st millennium B.C.E., with African cereals spread over the north of Africa during Islamic times (Pelling 2003; van der Veen 1995, 2011). These facts support the theory that the crops recovered were introduced by the Berbers from North Africa when they first colonised the islands in the late 1st millennium B.C.E.

THE USE OF WILD PLANTS

The occurrence of wild fruits in the analysed samples is of some interest, especially since they are recovered in abundance in the most recent levels, where cultivated plants are absent, and because such remains are more seldom preserved by charring. At El Tendal most of the gathered plants seem to be used as fuel; the only berries identified were *Ilex* sp., and the fact that charcoals from *Ilex canariensis* Poir. were common in the samples (Machado 1998) suggests they arrived with fuel wood. The inflorescences and leaves from heather (*Erica arborea*), together with cone fragments from pine (*Pinus canariensis*), were the most abundant macro-remains and also the most frequent charcoal (Machado 1998).

In contrast, at Belmaco most of the wild plants recorded have edible fruits that could be used as food. Notably, however, they are absent from the older levels, from which barley remains were recovered. The most important gathered plant at this site was the Canarian juniper, along with Canarian wild olive, and charred remains of both have also been recovered from El Guincho and El Roque de los Guerra, also in La Palma (Morales Unpublished).

Canarian juniper is similar to the Mediterranean *Juniperus phoenicea* L., and shepherds in the Canaries have traditionally eaten juniper fruits (Morales and Gil in press). They are sweet and astringent but may be processed to improve the taste and reduce the toxicity. The Canarian wild olive is an endemic tree from the Canary Islands and produces small and bitter fruits that traditionally are soaked in salt water for days or weeks to render them edible (Morales and Gil In press). Both juniper and wild olive trees were frequent in the vicinities of the sites and could contribute to the pre-Hispanic diet in replacement of cultivated plants.

Also significant were the remains of *Retama rhodorhizoides* at both sites. Seeds from this non-edible shrub are also abundant on sites from Gran Canaria, La Gomera, and El Hierro (Morales 2010). No ethnohistorical or ethnographic information exists concerning the consumption of the seeds, and nowadays *Retama rhodorhizoides* is utilised for its beautiful aromatic flowers (Kunkel 1991). It is possible that the archaeological seeds were processed by heating to reduce toxicity, which provoked the carbonisation of the recovered remains, but otherwise the significance of this plant at the sites remains unknown. No seeds of *mocán* (*Visnea mocanera*) or *amagante* (*Cistus* section *Macrostylia* Willk.) were recovered despite being recorded as gathered by the Auaritas in the first European texts (Abreu [1602] 1977).

CONCLUSIONS

The data analysed in this study come only from two, partially sampled archaeological sites and cannot be regarded as representative of the whole island. However, the results substantially modify our knowledge about the pre-Hispanic occupation of La Palma. The records indicate the cultivation of four crops on the island: hulled barley (*Hordeum vulgare* L. subsp. *vulgare*), hard wheat (*Triticum turgidum* subsp. *durum*), lentil (*Lens culinaris*), and faba bean (*Vicia faba*) in the 1st millennium C.E. It also raises the interesting possibility that agriculture was abandoned by the Auaritas shepherds during the 2nd millennium C.E. in favour of the collection of wild plant foods and in keeping with the first accounts of European explorers in the 15th century C.E. (Martín 1992; Pais 1996).

The abandonment of agriculture is a very unusual and unknown historical event. We know only of some isolated cases as recorded in Iceland during medieval times (Guðmundsson 1996) and in the Chatham islands, New Zealand (Diamond 1997). La Palma was also an isolated island, but the natural conditions are more favourable to agriculture than Iceland or the Chatham islands, which are at the colder extreme of cereal tolerance, and it seems likely that other factors should be taken into account to explain the disappearance of crops in La Palma. Possible causes include, for example, extreme droughts or locust plagues, as historically recorded in the Canaries (Viera [1773] 1982). In addition, volcanic eruptions are very common in the history of La Palma (Pais 1996). At the site of El Roque de los Guerra an abandonment episode linked to a volcanic eruption was recorded in the archaeological levels, with reoccupation dated by palaeomagnetism in the 1st century B.C.E. (Navarro and Martín 1987).

Archaeological studies in general suggest that during the late 1st millennium C.E. there was a transformation in the economy and in the material culture of pre-Hispanic people from La Palma. During this event, many dwellings were abandoned, and new sites appeared. It has been argued that this change resulted from new colonisation by shepherds from the Sahara, although it may be a consequence of a dry climatic period (Martín 1992). The case for the abandonment of agriculture presented here should then be considered as part of a larger transformation that happened in La Palma at the end of the 1st millennium C.E.

ACKNOWLEDGEMENTS

The research leading to these results has benefited from the project HAR2010-19328 funded by the Ministry of Economy and Innovation of Spain. Fieldwork in La Palma Island was funded by Arqueocanaria s.l. We are very grateful to the editors of this volume and the two anonymous reviewers for their valuable comments.

REFERENCES

Abreu, J. de [1602] (1977) *Historia de la Conquista de las Siete Islas de Canaria*, A. Cioranescu (Ed.). Santa Cruz de Tenerife: Goya ediciones.

Acebes-Ginovés, J. R., M. C. León-Arencibia, M. L. Rodríguez-Navarro, M. del Arco-Aguilar, A. García-Gallo, P. L. Pérez de Paz, O. Rodríguez-Delgado, V. E. Martín-Osorio, and W. Wildpret (2010) Pteridophyta, Spermatophyta. In M. Arechavaleta, S. Rodríguez, N. Zurita, and A. García (Eds.), *Lista de Especies Silvestres de Canarias. Hongos, Plantas y Animales Terrestres* 2009, pp. 119–72. Tenerife: Gobierno de Canarias.

Arco, M. C., M. M. Arco, E. Atienzar, and M. Hopf (1990) Estudio de los restos vegetales de la cueva de Don Gaspar y algunas anotaciones sobre la agricultura prehistórica de Tenerife. *Investigaciones Arqueológicas en Canarias* 2, 13–25.

da Zurara, G. E. [1448] (1998) *Crónica del Descubrimiento y Conquista de Guinea*: Estudio crítico a cargo de Manuel Hernández González, J. A. Delgado Luis (Ed.), A través del Tiempo 16. La Laguna: Ayuntamiento Villa de Orotava, Excmo/Ayuntamiento Puerto de la Cruz.

Diamond, J. (1997) *Guns, Germs, and Steel: The Fates of Human Societies*. New York: W. W. Norton and Company.

Fregel, R., J. Pestano, M. Arnay, V. M. Cabrera, J. M. Larruga, and A. M. González (2009) The maternal aborigine colonization of La Palma (Canary Islands). *European Journal of Human Genetics* 17, 1314–24.

González, E., and M. Arnay (1992) Ancient skeletal remains of the Canary Islands: Bone histology and chemical analysis. *Anthropologischer Anzeiger* 50, 201–15.

Guômundsson, G. (1996) Gathering and processing of lyne-grass (*Elymus arenarius* L.) in Iceland: An ethnohistorical account. *Vegetation History and Archaeobotany* 5, 13–23.

Hernández, M. (1999) *La Cueva de Belmaco. Mazo, Isla de La Palma*, Estudios Prehispánicos 7. Madrid: Gobierno de Canarias Dirección General de Patrimonio Histórico.

Kunkel, G. (1991) *Flora y vegetación del Archipiélago Canario. Tratado Florístico (2ª parte): Dicotiledóneas*. Las Palmas de Gran Canaria: EDIRCA.

Machado, M. d. C. (1998) Approche paléoécologique et ethnobotanique du site archéologique 'El Tendal' (N-E de l'île de La Palma, Archipel des Canaries). In G. Camps (Ed.), *L'Homme préhistorique et la mer. 120 congrés*, pp. 179–86. Aix-en-Provence: CTHS.

Martín, E. (1992) *La Palma y los Auaritas*. Santa Cruz de Tenerife: Centro de la Cultura Popular Canaria.

Martín, J. L., I. Izquierdo, M. Arechavaleta, M. A. Delgado, A. García, M. C. Marrero, E. Martín, L. Rodríguez, S. Rodríguez, and N. Zurita (2001) Las cifras de la biodiversidad taxonómica terrestre de Canarias. In I. Izquierdo, J. L. Martín, N. Zurita, and M. Arechavaleta (Eds.), *Lista de Especies Silvestres de Canarias (Hongos, Plantas y Animales Terrestres)*, pp. 15–26. Tenerife: Consejería de Política Territorial y Medio Ambiente, Gobierno de Canarias.

Morales, J. (2003) Islands, plants and ancient human societies: A review of archaeobotanical works on the Canary Isles Prehistory. In K. Neumann, A. Butler, and S. Kahlheber (Eds.), *Food, fuels and fields: Progress in African Archaeobotany*, pp. 139–48. Köln: Heinrich-Barth-Institut.

———. (2010) *El uso de las Plantas en la Prehistoria de Gran Canaria: Alimentación, Agricultura y Ecología*. Gran Canaria: Cabildo Insular de Gran Canaria.

———. (Unpublished) *Análisis preliminar de los restos carpológicos procedentes de los yacimientos de Roque de los Guerra y Cueva de los Guinchos, La Palma*. Unpublished report.

Morales, J., and J. Gil (In press) Gathering in a new environment: The use of wild food plants during the first colonization of the Canary Islands, Spain (3–2nd BC to 15th AD). In A. Chevalier, E. Marinova, and L. Peña-Chocarro (Eds.), *Plants and People: Choices and Diversity through Time. EARTH Series* Vol. 1. Oxbow: Oxford Books.

Navarro, J. F., and E. Martín (1987) La prehistoria de la isla de La Palma (Canarias): Propuesta para su interpretación. *Tabona* 6, 147–84.

Navarro, J. F., E. Martín, and A. C. Rodríguez (1990) La primera fase del programa de excavaciones arqueológicas en Cuevas de San Juan y su aportación a la diacronía de la prehistoria de La Palma. *Investigaciones Arqueológicas en Canarias* 2, 187–201.

Pais, F. J. (1996) La *economía de producción en la prehistoria de la isla de la Palma: La Ganadería*. Santa Cruz de Tenerife: Viceconsejería de Culturay Deportes, Gobierno de Canarias.

Pelling, R. (2003) Medieval and early modern agriculture and crop dispersal in the Wadi el-Agial, Fezzan, Lybia. In K. Neumann, A. Butler, and S. Kahlheber (Eds.), *Food, Fuels and Fields: Progress in African Archaeobotany*, pp. 129–38. Köln: Heinrich-Barth-Institut.

Pérez, E. (2000) *La Dieta de los Habitantes Prehispánicos de la Isla de La Palma: El Análisis de los Elementos Traza.* Master's Dissertation, Universidad de La Laguna.

Santos, A. (1983) *Vegetación y Flora de La Palma.* Santa Cruz de Tenerife: Interinsular Canaria.

Soler, V., J. F. Navarro, E. Martín, and J. A. Castro (2002) Aplicación contrastada de técnicas de datación absoluta al yacimiento 'Cueva de Tendal', isla de La Palma, (Islas Canarias). *Tabona 11*, 73–86.

van der Veen, M. (1995) Ancient agriculture in Libya: A review of the evidence. *Acta Paleobotanica* 35 (1), 85–98.

———. (2011) *Consumption, Trade and Innovation: Exploring the Botanical Remains from the Roman and Islamic Ports at Quseir al-Qadim, Egypt.* Frankfurt: Africa Magna Verlag.

Viera, J. [1773] (1982) *Noticias de la Historia General de las Islas Canarias.* Santa Cruz de Tenerife: EDIRCA.

Zohary, D., and M. Hopf (1993) *Domestication of Plants in the Old World.* Oxford: Oxford University Press.

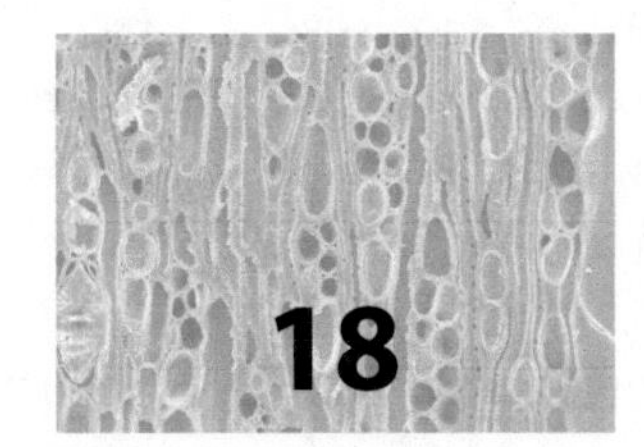

Patterns in the Archaeobotany of Africa

Developing a Database for North Africa, the Sahara, and the Sahel

Ruth Pelling

The five International Workshops of African Archaeobotany held since 1994 have presented a significant body of new archaeobotanical data from the continent and have provided a forum for discussion of issues associated with the history of plant use in Africa (*Acta Palaeobotanica* 35 [1] 1995; Cappers 2007, this volume; Neumann, Butler, and Kahlheber 2003; van der Veen 1999). Although the number of systematically sampled sites has increased dramatically since the 1980s and several overviews exist that refer to or include the data from a range of sites, and plot occurrences of certain crops (for example, Neumann 2003; Wetterstrom 1998), there has been little attempt to explore the empirical data on a pan-regional basis. The syntheses that have been undertaken have tended to focus on the origins of domestication and introduction of key crop species. Consequently, less attention has been directed toward wild taxa and the formation of anthropogenically influenced arable weed flora, with the exception of work within the Nile Valley (for example, El-Hadidi 1992; Fahmy 1997).

This chapter uses a database of archaeobotanical records from sites in North Africa, the Sahara, and the Sahel and offers a new direction for archaeobotany in Africa and a means of exploring pan-regional patterns. The project had its origins in my work on archaeobotanical remains from Garamantian settlements in Fazzān, Southern Libya, and specifically the observation that the Fazzān crop repertoire includes species of Mediterranean/Near Eastern origin and sub-Saharan African taxa (Pelling 2005, 2007, 2013).

Fazzān: The Archaeobotanical Setting

Modern-day Fazzān comprises the vast southwest desert province of Libya. The oases of the region are generally small but concentrated in dense linear bands. Human settlement has traditionally fallen into three broad bands of oases between 24° and 28° latitude running approximately east-west: the Wādī ash-Shati in the north; the Wādī al-Ajāl (also known as the Wādī al-Hayāt) and the Wādī Barjūj/ Wādī 'Utba in the centre; and the Murzuq/ al-Hufra and ash-Sharqīyāt depressions in the south (Mattingly 2003; Figure 3).

Archaeobotanical data from sites in the southwest of Fazzān, centring on the Akakus Mountains and adjacent broad depression of the Wādī Tanezzuft, have provided a detailed picture of plant exploitation during the Neolithic and Pastoral period based largely on savannah grasses (Castelletti et al. 1999; Mercuri 2001; Wasylikowa 1992). The grasses exploited were dominated by genus such as *Echinochloa*, *Brachiaria*, *Setaria*, *Cenchrus*, *Dactyloctenium*, and *Panicum*.

Soon after 1000 B.C.E., local populations in northern Fazzān, particularly in the Wādī al-Ajāl, adopted a sedentary lifestyle. Social and settlement development was radical and wide reaching, affecting settlement distribution and

Archaeology of African Plant Use by Chris J. Stevens, Sam Nixon, Mary Anne Murray, and Dorian Q Fuller, Eds., 205–223

form, the introduction of arable agriculture and successive waves of new crop and domestic animal introductions associated with new irrigation technology, and the formation of a Garamantian state, lasting until around 700 C.E. (Di Lernia, Manzi, and Merighi 2002, pp. 292–302; Mattingly 2003, 2010; Mattingly, Reynolds, and Dore 2003, pp. 339–62).

The extent of pre-Islamic trans-Saharan trade, and its significance to Garamantian development, has long been a matter of debate (for example, Daniels 1970; Wheeler 1954). Mattingly (2003, 2010) and Liverani (2000a, 2000b) argue that the evolution of the Garamantian kingdom is closely associated with the evolution of trans-Saharan trade, linking sub-Saharan Africa with the Classical Mediterranean world. Liverani (2000a, 2000b) suggests that both trade and the evolution of the Garamantian culture/polity should be seen as adaptations to the aridification of the Sahara, in which pastoralism was no longer possible. Although evidence for trade of Mediterranean goods is abundant at Garamantian settlements (Mattingly, Reynolds, and Dore 2003), artefactual evidence for trade with sub-Saharan regions remains largely elusive, making any assumptions concerning trade with southern populations speculative.

Archaeobotanical evidence from the promontory settlement of Zinkekra suggests that arable agriculture had been established early in the 1st millennium B.C.E., based on Near Eastern cereals (*Triticum dicoccum* and *Hordeum vulgare* with some *Triticum aestivum* type) supported by perennial fruits (van der Veen 1992; van der Veen and Westley 2010). This evidence for agriculture is now complimented by a large dataset from the site of Jarma (dating from the late 1st millennium B.C.E. to the early 20th century C.E.), the ancient capital of the Garamantes, and a grain rich deposit from the late 1st millennium B.C.E. promontory settlement of Tinda B (Pelling 2005, 2007, 2013). During the late 1st millennium B.C.E. settlement focus moved to the centre of Wādī al-Ajāl, where new irrigation technology (*foggara*), which utilised subterranean fossil water, appears to be associated with the introduction of three significant summer crops: pearl millet (*Pennisetum glaucum*), sorghum (*Sorghum bicolor*), and cotton (*Gossypium* sp.) (Pelling 2005, 2007, 2013).

A radiocarbon determination on *Pennisetum glaucum* from Jarma returned a date of 350–170 cal B.C.E. (2160 ± 40 BP, Beta-194242), and a specimen of sorghum, most closely resembling grain of race caudatum, returned a date of 360–190 cal B.C.E. (2180 ± 40 BP, Beta–194236). *Pennisetum* type caryopses have also been recorded in Classic Garamantian (Roman era) levels at Aghram Nadharif in the southern Fazzān (Mercuri et al. 2005). Elsewhere in North Africa pearl millet is recorded from Roman period deposits at Qasr Ibrim (Clapham and Rowley-Conwy 2007) but is absent on earlier Nubian sites and those of the Egyptian Oases, implying that it may have arrived in Fazzān via a more southerly route.

The early history of sorghum is more problematic given the paucity of finds in Africa and the continued uncertainty about centres and timing of origin (Haaland 1995, 1999; Harlan 1992a, b; Rowley-Conwy 1991; Rowley-Conwy, Deakin, and Shaw 1999; Wigboldus 1990, 1996). The Fazzān data would support an early rather than late date of domestication (Harlan 1992a, 1992b) or, more likely, multiple centres of origin.

A radiocarbon date from a seed of cotton (*Gossypium* sp.) showed it to be present by 230–325 C.E. (1770 ± 40 BP, Beta-194240). Cotton appears to have been introduced from Nubia and the Egyptian oases. It was present at Qasr Ibrim during the 1st century C.E. (Alan Clapham, pers. comm.) and at Kellis, Dakhla oasis (Thanheiser 1999) and North Kharga Oasis (Clapham, pers. comm.) during the Roman period. The cultivation of cotton was also recorded at Dakhla in the Kellis Agricultural Accounts Book (Bagnall 1997). It may have reached Fazzān fairly rapidly from the oases. In the early Islamic period cotton is well attested at sites at the northern and southern ends of the western trans-Saharan trade routes, with large deposits of seeds at Volubilis in Morocco in the 9th century C.E. (Fuller and Pelling Forthcoming), at Dia in the Middle Niger Delta of Mali in Horizon 4 deposits (1000–1600 C.E.) (Murray 2005) and elsewhere in Mali by the 11th century C.E. (Bedaux 1991; Chavane 1985). That cotton bolls seem to have been traded is supported by finds of cotton seed from Islamic deposits at Essouk-Tadmakka, in the Malian Sahara, a major trans-Saharan trade settlement in a region where it is extremely unlikely to have been cultivated (Nixon, Murray, and Fuller 2011).

The crop data from Jarma and Tinda B clearly demonstrate that the Fazzānese were receptive to new crops throughout the period of sedentary arable occupation, which also includes such Mediterranean fruits as olive, pomegranate, almond, and, later, peach. Hard or durum wheat (*Triticum durum*) was cultivated by 100–400 C.E.. New World crops were seen in the uppermost postmedieval deposits (chilli/sweet pepper and maize) at the site.

It has been speculated that the Garamantes represent a mixture of indigenous pastoralists and 'palaeoberbers' (non-negroid Mediterranean peoples) who migrated into the Sahara during the 2nd and 1st millennia B.C.E. (Brett and Fentress 1996; Camps 1980; Muzzolini 1984). Mattingly uses the term *made in Fazzān* (2010, p. 523) to stress the fact that, culturally and genetically, the Garamantes were the product of both indigenous and non-native peoples, traditions and technologies that formed a unique central Saharan culture. Although the Mediterranean influences on the region are evident in terms of artefactual remains, the sub-Saharan influences are harder to establish. The introduction of key sub-Saharan crops and their associated weed seeds, some pre-dating their arrival in the Nile Valley, is potentially a key tool in tracing relationships with sub-Saharan Africa. We should therefore include arable crops and food traditions within this same cultural fusion.

THE EVOLUTION OF WEED FLORA THROUGH THE INTRODUCTION OF NEW CROPS

Arable weed communities are, by definition, artificial in that they are the product of humans' manipulation of nature. Although most arable weeds may have originated from already existing vegetation that have adapted to the new plots created by agriculture, it is possible that many 'alien' species will be introduced with new seed crops. El-Hadidi (1992, p. 145) considered that 170 Egyptian weed species arose during the transition to agriculture during the Neolithic/Predynastic period (4500–3100 B.C.E.), increasing to 225–255 species for the Pharonic period (3100–332 B.C.E.). This increase in species diversity is related to an increase in the number and types of crops cultivated. Furthermore, he suggests that the weed flora reflects both the native flora (comprising the Saharo-Sindian, Sudano-Zambezian, and Palaeotropical communities), which became incorporated into the arable weed repertoire over time, and species of Mediterranean, Irano-Turanian, and cosmopolitan origin, which were introduced with the cultivated plants from the southern Levant (El-Hadidi 1992, Table 1). A similar mix of indigenous and imported species is potentially more marked for the central Sahara, given the highly artificial nature of irrigated agriculture in an arid zone lacking in fresh surface water. Although many of the species may be locally derived, their close association with cultivated crops may be a more recent evolution.

It was envisaged that a broad regional archaeobotanical database could then aid in tracing the evolution of the crop and weed flora in the Fazzān through time and space and in properly situating it in a pan-regional context. Specifically, it was hoped that such an approach could detect broader changes in the general Fazzān flora associated with the introduction of new crops from different regions.

DEVELOPING A DATABASE

To meet the immediate needs of the Fazzān investigation and as a tool for such future regional investigations, a database was assembled containing published and unpublished records of plant taxa, and C^{14} dates where available, from 61 sites (Pelling 2007). The region covered spans the North African coast down to and including the Sahel zone to the south of the Sahara, and includes sites in Morocco, Algeria, Tunisia, Libya, Egypt, Sudan, northern Nigeria, Mali, Burkina Faso, Mauritania, and Ghana. In total, the sites encompass 103 separate archaeological phases taking in the period from 10,000 BP to the postmedieval period (Table 18.1). Although the Nile Valley data are important, it was beyond the scope and resources of current study to include all Nile Valley sites. Therefore, only a selection of Nile Valley sites from the major periods has been included.

Data are entered by site and phase rather than sample (see Figure 18.1 for data model, Table 18.2 for phases at Jarma). Where counts were available, the total counts of each taxon and plant part (seed, rachis, and so on) are entered by each major cultural phase; where counts weren't available, species presence is recorded as 1. Species ubiquity was also recorded as the percentage of samples per phase in which a taxon is present. A total of 4,187 entries have been made, consisting largely of charred or desiccated remains, with some waterlogged and occasionally mineralised items or ceramic seed impressions. Sites for which there were multiple excavations and detailed phases have been condensed into major phases.

The level of identification was extremely variable, reflecting either the poor preservation at many sites, the difficulties of identification of many taxa (grasses for example), or the lack of familiarity with the flora generally in such a new subject. Owing to the differing levels of identification used in the database (Pelling 2007) some entries were combined using codes (for example, *Chenopodium* cf. *murale* and *Chenopodium murale* are combined as Chemur; *Hordeum vulgare* and *Hordeum* cf. *vulgare* are combined as *Hordeum vulgare* [Horvul] or *Triticum durum*; *Triticum aestivum* and *Triticum* free-threshing are combined as *Triticum aestivum/durum* Tria/d). Where archaeobotanists have used their own codes for recognisable but as yet unidentified taxa, these have been entered as discrete records. Items identified to species, genus, tribe, or family are recorded separately, resulting in 673 items in the

Table 18.1 Sites in the Database with Major Phases and Date Range of Archaeobotanical Samples Where Available. (Dates in brackets are for the archaeological period; see Figure 18.2 for location of sites.)

Site Name	Region	Period	Date Range	Publication
Abu Ballas, Eastpans	Egyptian Oasis	Neolithic	6200–6000 BP	Barakat and Fahmy 1991
Aghram Nadharif	Fazzān	Classic Garamantian	1st–2nd century C.E.	Mercuri and Bosi 2005
Al-Basra	Morocco	Late Roman/Islamic	800–1050 C.E.	Mahoney 1994; Pelling 2007; Pollock 1983–1984;
'Ayn-Manawir	Egyptian Oasis	Roman	5th century B.C.E. 2nd century C.E.	Newton et al. 2006
Badis	Morocco	Islamic		Pollock 1983–1984
Berenike	Nubia/Sudan	Roman	275 B.C.E.–early 6th century C.E.	Cappers 1999, 2006
Birimi	Ghana	Late Stone Age/ Neolithic	2nd millennium B.C.E.	D'Andrea, Klee, and Casey 2001
Carthage	Tunisia	Punic–Medieval	7th–4th century B.C.E.–11–13th century C.E.	Ford and Miller 1978; Hoffman 1981; Stewart 1984; van Zeist 1994; van Zeist et al. 2001
Daima	Sahel	Islamic/Early Medieval	550 B.C.E.–1150 C.E.	Connah 1981
Dhar Nema	Sahel	Late–Early Tichitt	1750–1600 B.C.E.	MacDonald et al. 2003
Dhar Tichitt	Sahel	Late Stone Age	1000–900 B.C.E.	Munson 1971, 1976, 1986
Dia Mara and Dia Shoma	Sahel	Iron Age–Post Medieval	800 B.C.E.–1900 C.E.	Murray 2005
El Omari	Egyptian Nile ValleyAmar	Neolithic/Pre-Dynastic	4600–4400 B.C.E.	Barakat 1990
el-'Amarna	Egyptian Nile Valley	New Kingdom	(1350–1370 B.C.E.)	Renfrew 1985
el-Hibeh	Egyptian Nile Valley	Late Dynastic–Ptolemaic	1st millennium B.C.E.	Wetterstrom 1984
el-Nana	Egyptian Nile Valley	Byzantine	5th–early 7th century C.E.	Smith 2003
Epiphianus, Thebes	Egyptian Nile Valley	Coptic	7th century C.E.	Winlock and Cram 1926
Eusperides	Northern Libya	Greco-Libyan	6th–3rd century B.C.E.	Pelling and al-Hassey 1998
Farafra, Hidden Valley	Egyptian Oasis	Neolithic	8000–7000 BP	Barakat and Fahmy 1991
Gajiganna	Nigeria	Stone Age	1800 B.C.E.–800 B.C.E.	Klee and Zach 1999
Gao	Sahel	Islamic	6/7th–16th century C.E.	Fuller 2000
Jarma*	Fazzān	Early Garamantian–ca. 1900 C.E.	4th–2nd century B.C.E.–19th/20th century C.E.	Pelling 2007
Jenné-Jeno	Sahel	Late Iron Age–Late Islamic	400 B.C.E.–1510 C.E.	McIntosh 1995
Karanis	Egyptian Nile Valley	Roman	late 3rd–early 5th century C.E.	Bartlett 1933; Leighty 1933
Karkarichinkat	Sahel	Late Stone Age	2000–1300 BP	Smith 1992
Kawa	Nubia/Sudan	Napatan	1st millenium B.C.E.	Fuller 2004
Kellis, Dakhla oasis	Egyptian Oasis	Roman	1st–4th century C.E.	Thanheiser 1999

Table 18.1 Continued

Site Name	Region	Period	Date Range	Publication
Kom el-Hisn	Egyptian Nile Valley	Old Kingdom	(2700 B.C.E.–2250 B.C.E.)	Moens and Wetterstrom 1988
Kursakata	Nigeria	Late Stone Age–Iron Age	1000 B.C.E.–100 C.E.	Klee, Zach, and Neumann 2000
Lahun	Egyptian Nile Valley	Middle Kingdom	(2040–1640 B.C.E.)	Germer 1998
Lambeisis	Algeria	Roman		Pelling 2007
Lepcis Magna	Northern Libya	Roman	?4/5th century C.E.	Pelling 2007
Leptiminus	Tunisia	Roman–Late Antique	1st/3rd–7th century C.E.	Jezik 2000; Smith 2001
Mege	Sahel	Late Stone Age–present	1000 B.C.E.–20th century C.E.	Klee and Zach 1999
Minshat Abu Omar	Egyptian Nile Valley	Early Dynastic	(4th–3rd millennium B.C.E.)	Thanheiser 1992b
Mons Claudianus	Egyptian Nile Valley	Roman	late 1st–2nd century C.E.	van der Veen 2001
Nabta Playa	Nubia/Sudan	Early Neolithic	8095–7950 BP	Wasylikowa 1997; Wasylikowa and Dahlberg 1999; Wasylikowa et al. 1993, 1995, 1997
Nakur	Morocco	Islamic		Pollock 1983–1984
Nauri	Nubia/Sudan	Christian	13th century C.E.	Fuller and Edwards 2001
North Kharga Oasis	Egyptian Oasis	Roman	3rd–4th century C.E.	Clapham and Dorri Unpublished
Oued Chebbi (Dhar Oualata)	Sahel	Neolithic	2740 ± 160 BP	Amblard Pernès 1989
Oursi	Burkina Faso	Late Stone Age–Early Iron Age	ca. 1000 B.C.E.–ca. 200 C.E.	Kahlheber, Albert, and Höhn 2001
Qasr es-Seghir	Morocco	Islamic to Portuguese		Pollock 1983–1984
Qasr Ibrim	Nubia/Sudan	Napatan–Medieval	7th century B.C.E. – 850 C.E.	Clapham and Rowley-Conwy 2007
Saouga	Burkina Faso	Medieval	10–11th century C.E.	Kahlheber 1999
Semna	Nubia/Sudan	New Kingdom	1500–1480 B.C.E.	van Zeist 1983
Setif	Algeria	Late Roman–Islamic	5/6th–12th century C.E.	Palmer 1991
Sijilmasa	Morocco	Islamic		Mahoney Unpublished
Soba East	Nubia/Sudan	Post-Meoritic–Medieval	13th century C.E.	Cartwright 1998; van der Veen 1991
Tell Ibrahim Awad	Egyptian Nile Valley	Late pre-Dynastic/Early Dynastic	(3050 B.C.E.)	Thanheiser 1992a
Tinda B	Fazzān	Early Garamantian	350–50 cal B.C.E.	Pelling 2007
Ti-N-Torha/Two Caves	Fazzān	Early and Late Acacus (Early Holocene)	ca. 6600–6400 uncal B.C.E.	Wasylikowa 1992
Uan Afuda	Fazzān	Early and Late Acacus	ca. 10,000 –8000 uncal B.C.E.	Castelletti et al. 1999
Uan Muhuggiag	Fazzān	Early–Late Pastoral (Mid Holocene)	ca. 4000–2900 B.C.E. and ca. 1800–200 uncal B.C.E.	Wasylikowa 1992

(*Continued*)

Table 18.1 Continued

Site Name	Region	Period	Date Range	Publication
Uan Tabu	Fazzān	Early–Late Acacus (Early Holocene)	ca. 9900–8500 uncal B.C.E.	Mercuri 2001
ULVS (UNESCO Libyan Valleys Survey sites)	Northern Libya	Roman–Byzantine	1st–7th century C.E. and +15/16th century C.E.	van der Veen et al. 1996
Volubilis	Morocco	Late Roman–Islamic	4th–14th century C.E.	Fuller & Pelling Forthcoming
Winde Koroji	Sahel	Late Stone Age	2100–1100 B.C.E.	MacDonald 1996
Zinkekra	Fazzān	Early Garamantian	900–400 B.C.E.	van der Veen 1992
Daima	Sahel	Islamic/Early Medieval	550 B.C.E.–1150 C.E.	Connah 1981
Oursi	Sahel	Late Stone Age–Early Iron Age	ca. 1000 B.C.E.–ca 200 C.E.	Kahlheber, Albert, and Höhn 2001
Saouga	Sahel	Medieval	10th–11th century C.E.	Kahlheber 1999

*Detailed Jarma phases shown in Table 18.2.

Table 18.2 Phases Used for the Jarma/Tinda Archaeobotanical Data

Phase	Date Range	Historical Phase
1–2	16th–early 20th century C.E.	Late Medieval/Early Modern
2		Late Medieval/Early Modern
3		Islamic
4		Post Garamantian/Early Arab
5	3rd–9th century C.E.	Late Garamantian
6	2nd–6th century C.E.	Classic Garamantian
7	1st century B.C.E.–3rd century C.E.	Early Garamantian
8	4th–1st centuries C.E.	Early Garamantian
9–10	4th–2nd century B.C.E.	Early Garamantian

taxa table. A further 57 unidentified taxa were given separate identification codes by the archaeobotanists and are included in the database but not any subsequent analysis.

During the course of multivariate analysis of the data it was decided to give a second code to enable easy condensing of the data to allow for variation in levels of identification between different sites. Combining taxa was particularly significant for some of the small seeded grasses, which were identified to type. For example, *Setaria* sp., *Setaria/Brachiaria* sp., or *Setaria/Brachiaria/Echinochloa* sp. are combined by Code 2 to 'Set/Bra' (*Setaria/Brachiaria/Echinochloa* types). Where items were identified more fully the species identification is retained, so *Brachiaria ramosa/deflexa* remains Braram/def and is not amalgamated under Set/Bra, although *Brachiaria ramosa* and *B. deflexa* are combined. Although the oversimplification of taxa and levels of identification is not ideal it was felt to be a reasonable balance given the inherent difficulties and potential over identification of some items and the fact that many taxa would otherwise be so rarely represented they would have to be removed from the analyses.

Ecological data have been included only for some of the more prominent taxa and those identified from Fazzān sites. There are obvious problems when compiling ecological data for weeds over such a vast geographical area: many arable weeds can behave differently away from their naturalised habitats and the botanical floras available for each country/region provide varying levels of detail. Given these difficulties in assigning ecological characteristics, it was decided to limit any ecological analysis to a very basic level, and very broad habitat groups are used (arable weed, plant of waste ground, steppe/desert species, species of wet ground or irrigation channel, and unknown/catholic). In addition, seasonality is also provided (summer or winter crop or summer or winter weed). This simplified ecological grouping also reduced the impact of combining taxa as just described.

LIMITATIONS AND UNEVENNESS OF THE DATA

The unevenness of the data is the most significant limitation to be considered in any attempt to study pan-regional data or reconstruct agricultural diffusion (for example, Zohary and Hopf 2000, p. 247). Potential problems include considerable variation in the sampling and recovery methods (for instance, collection by hand giving way

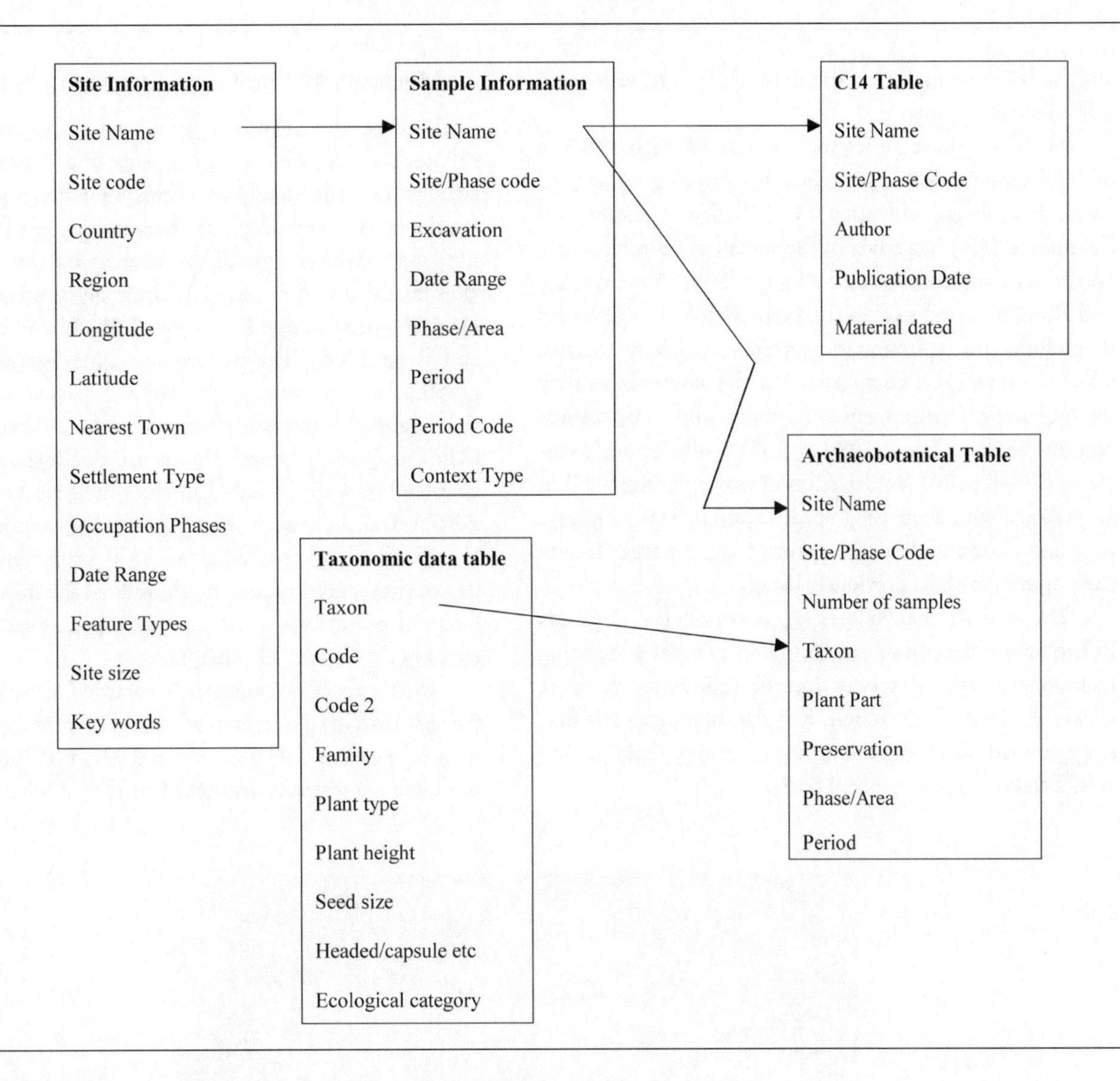

Figure 18.1 A model of the database of African Archaeobotany.

to systematic sampling, flotation, and sieving) and level of identification employed by different specialists, as well as different levels of experience and quality of reference material. For example, at El-Hibeh only material collected in a 4-mm sieve was examined, although small seeds adhering to the larger material were identified, thus providing at least a good range of taxa (Wetterstrom 1984).

Then there is potential variation between regions introduced not by floristic composition but by the nature of the archaeological site, with classical urban centres on the Mediterranean coast, inland Romanised farmsteads, mudbrick desert settlements, and large, often largely featureless sub-Saharan settlement mounds. In addition, the nature of preservation of the material itself varies according to the type of site and between different regions. For instance, desiccation tends to favour a greater range of plant parts and species, although grass caryopses tend to survive poorly, and preservation by charring is restricted to material that is more likely to be burned—for example, as fuel.

Finally are the biases introduced by cultural taphonomic processes; for instance, the various methods of processing applied to different crop types, particularly the Near Eastern cereal crops (wheat and barley) versus the sub-Saharan cereals (pearl millet and sorghum), are quite different owing to harvesting methods (for example, Reddy 1994, 1997). These harvesting methods subsequently affect the weed flora, harvested and brought to the settlement with the crop, with weeds seeds being more commonly retained with the grain for the Near Eastern cereal crops. Thus it might be expected that a far greater diversity of weed flora would be recovered from the northern wheat and barley growing areas than from the southern pearl millet/sorghum areas.

Many of these problems, associated with the use of archaeobotanical data covering large geographical areas, have been addressed by Colledge, Conolly, and Shennan (2004) in a study of the spread of farming across Mediterranean Europe and by Lange (1990) for Iron Age and Roman period sites in the Netherlands. Lange found that while the taphonomic processes are likely to have affected the original composition of the assemblages they did not appear to influence the more important trends seen in the data (Lange 1990, p. 135). Colledge and associates (2004, p. 46) similarly found that sites clustered in accordance with their geographic location, again suggesting that meaningful spatial patterns can emerge despite the potential pitfalls previously listed.

The analysis that follows explores only broad trends in the data rather than examining differential processing techniques, scale of production, or economic status of sites; therefore, on the whole, it is appropriate as the first stage toward observing the diffusion of crops and associated weeds on a pan-regional basis.

Multivariate Statistical Analysis of the Data

To explore the relationship between the archaeological flora at the Garamantian sites and those of other sites within the database, correspondence analysis was employed as a method of pattern searching (see Figure 18.2 for a map of the geographical and temporal distribution of sites included in the analysis). The analysis was conducted using the programme PAST ver. 1.38 (Hammer, Harper, and Ryan 2006). The pattern searching focused on geographical and chronological relationships and made use of the database's presence/absence data for all arable phases. (The Neolithic/pastoral phases are not included in this chapter but were included in the initial study in Pelling 2007.) The data were manipulated and standardised in Excel before being pasted into PAST for the analysis. The use of presence/absence should reduce the impact of differential preservation and context type as well as differences in the density of remains.

Correspondence analysis is regarded as well suited to pattern searching in terms of counts or presence/absence data (Greenacre 1984; Shennan 1997, pp. 308–52) and has been successfully employed to plot similar presence/

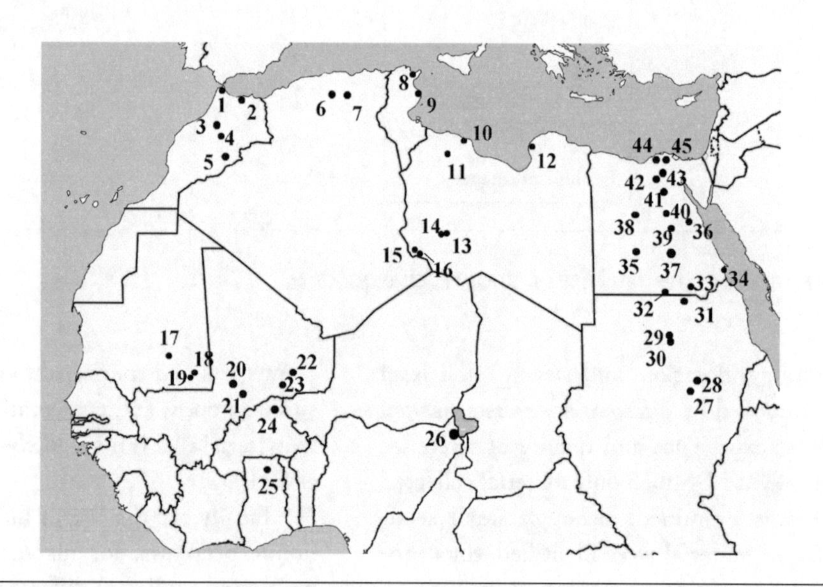

Figure 18.2 Sites included in the database (date ranges, publication references, and regions are given in Table 18.1): (1) Qasr es-Seghir; (2) North Moroccan coastal sites (Badis and Madinat en-Nakur); (3) Al Basra; (4) Volubilis; (5) Sijilmasa; (6) Setif; (7) Lambeisis; (8) Carthage; (9) Leptiminis; (10) Lepcis Magna; (11) ULVS sites; (12) Eusperides; (13) Central Wadi el-Ajal site (Jarma, Zinkekra); (14) Tinda B; (15) Akakus sites (Uan Tabu, Ti-N-Torha/Two Caves, Uan Afuda, Uan Muhuggiag); (16) Aghram Nadharif; (17) Dhar Tichitt; (18) Dhar Nema; (19) Oued Chebbi; (20) Inner Niger Delta sites (Dia, Jenné-Jeno); (21) Winde Koroji; (22) Karkarichinkat; (23) Gao; (24) Northern Burkina Faso sites (Saouga and Oursi); (25) Birimi; (26) Nigerian Chad Basin sites (Gajiganna, Kursakata, and Mege); (27) Soba East; (28) Toska West; (29) Kawa; (30) Nauri; (31) Semna; (32) Nabta Playa; (33) Qasr Ibrim; (34) Berenike; (35) Abu Ballas; (36) Mons Claudianus; (37) Southern Oases sites (Dakhla Oasis: Kellis, Kharga Oasis: Kharga Oasis North, Ayn- Manawir); (38) Farafra; (39) Epiphianus; (40) El Nana; el-'Amarna; (41) el-Hibeh; (42) Fayoum Basin sites (Karanis, Lahun); (43) El Omari; (44) Western Delta, Kom el-Hisn; (45) Eastern Delta Sites (Tell Ibrahim Awad and Minshat Abu Omar).

absence data to trace the spread of farming in the Eastern Mediterranean (Colledge, Conolly, and Shennan 2004). The data file on which the analysis was performed consisted of a matrix with sites/phases (samples) in rows and taxa in columns.

Initial Manipulation of the Data

A range of manipulations to the data was applied to produce plots that most reliably reflect differences in the data. In the first instance the presence/absence of all field crops and weeds represented by desiccation or charring was selected by site and phase. Fruits and tree/shrub species were omitted. Only taxa identified to the level of genus, species, or tribe were included, thus removing the categories of 'small' and 'large' grasses or 'Leguminous weeds', for example. The dataset was further reduced by omitting rare taxa (present in fewer than 5% of phases), and phases with fewer than 5 taxa (removal at the 10% level made no significant difference to the plots but reduced the dataset considerably, and the 5%/5 taxa cut-off was therefore adopted).

It is likely that by limiting the analysis to the level of absence/presence and by omitting rare taxa and small sites/phases the impact of variable taphonomic processes and recovery/analytical techniques, as discussed, will be minimised.

The Plots: Desiccated and Charred Material

The initial plot showing all arable sites/phases shows a clear geographical distribution (Figure 18.3). It is interesting to note that the Soba phases plotted more closely to

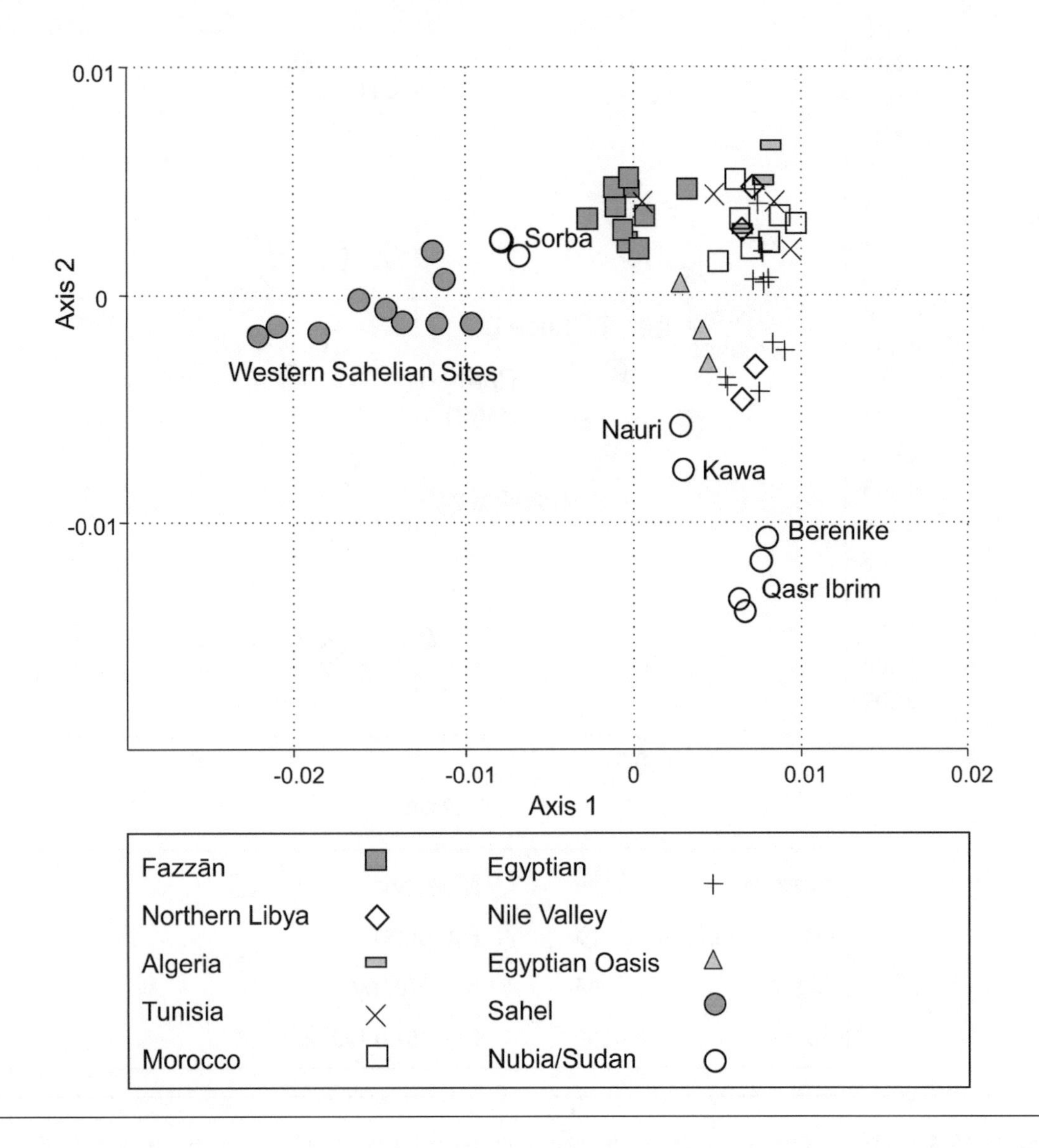

Figure 18.3 Correspondence Analysis (CA) plot of field crops and weeds from all agricultural period sites/phase (charred and desiccated).

the Western Sahelian sites to the left of axis 2, whereas the other Nubian sites plotted in the bottom right corner. This situation could reflect a greater Sahelian influence on the flora of Soba than Nilotic. The Fazzān, Egyptian Nile Valley, and Mediterranean sites form a tighter cluster. Qasr Ibrim and Berenike, both in southern Egypt, also plot away from the mass of Egyptian sites, with the other Nubian sites (Nauri and Kawa) being pulled in the same direction. Both sites produced remarkable assemblages with a very high density of desiccated crop remains displaying both Nilotic and Sahelian characteristics (for example, sorghum, cotton, and the near eastern cereals). It is unsurprising that these sites plot away from the mass of Mediterranean and Egyptian sites. The Sahelian sites, Soba, Qasr Ibrim, and Berenike, were excluded from further analysis to enable close examination of the data cloud and the relationship between the Fazzān sites/phases and those of the Mediterranean and Egyptian sites/phases.

Figure 18.4 shows the Egyptian Nile Valley, Mediterranean, Egyptian oasis, and Fazzān data comprising 39 sites and 54 taxa. In this plot there was clear separation along axis 1. The Fazzān sites, the Egyptian oases, and medieval Carthage were separated from the remaining sites by the 2nd axis, with Jarma and Tinda B plotting much farther to the left than Zinkekra and the oases sites, which clustered together. There was also some separation along axis 2; the Mediterranean phases, except medieval Carthage and the northern Libyan (UNESCO Libyan

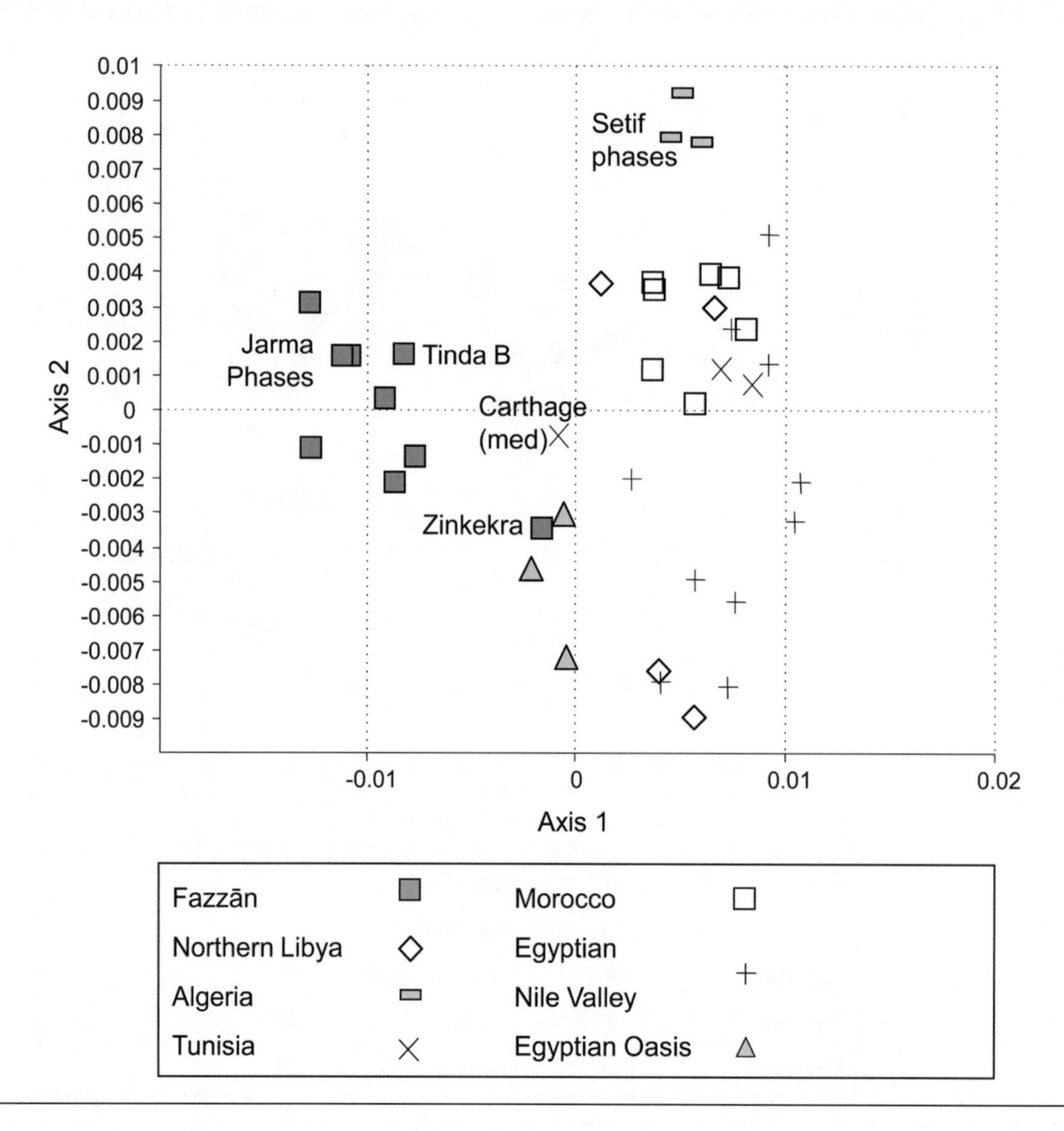

Figure 18.4 CA plot showing field crops and weeds from the Egyptian Nile Valley (excluding Qasr Ibrim), oases, and Maghreb/Fazzan sites/phases (charred and desiccated).

Valleys sites (ULVS)) sites/phases, plotted above axis 1, and the majority of the Egyptian sites/phases (except Tell Ibrahim Awad and Minshat Abu Omar plotted below axis 1. Setif (Algeria) plotted slightly away from the bulk of the Mediterranean sites/phases along axis 2.

CHARRED SAMPLES ONLY

To avoid bias due to differential preservation types, the analysis was repeated using charred data only (Figure 18.5). The matrix consisted of 46 phases and 60 taxa. Omitting all but the charred data automatically excluded some exceptional sites, such as Qasr Ibrim and Berenike, but also Zinkekra, a significant Fazzān site. As previously there is a clear separation of Mediterranean/Egyptian and Fazzān from Sahel sites along the 1st axis. Furthermore, the Fazzān, Sudanese, Inner Niger Delta, and Burkina Faso phases were separated from the rest by the 2nd axis, with the Kursakata phases plotting separately in the bottom-right quadrant. The Soba phases plotted more closely to the Fazzān than to the Sahel phases, suggesting preservation may have made some difference to the placing of these phases in the original plots.

The removal of the Sahelian sites (Figure 18.6, 35 phases) resulted in a strong separation of the Fazzān sites/phases from the main cloud by the 2nd axis and Soba from the rest in the top-left quadrant. The removal of the Soba data did not alter the distribution of the remaining samples significantly. The other Nubian site,

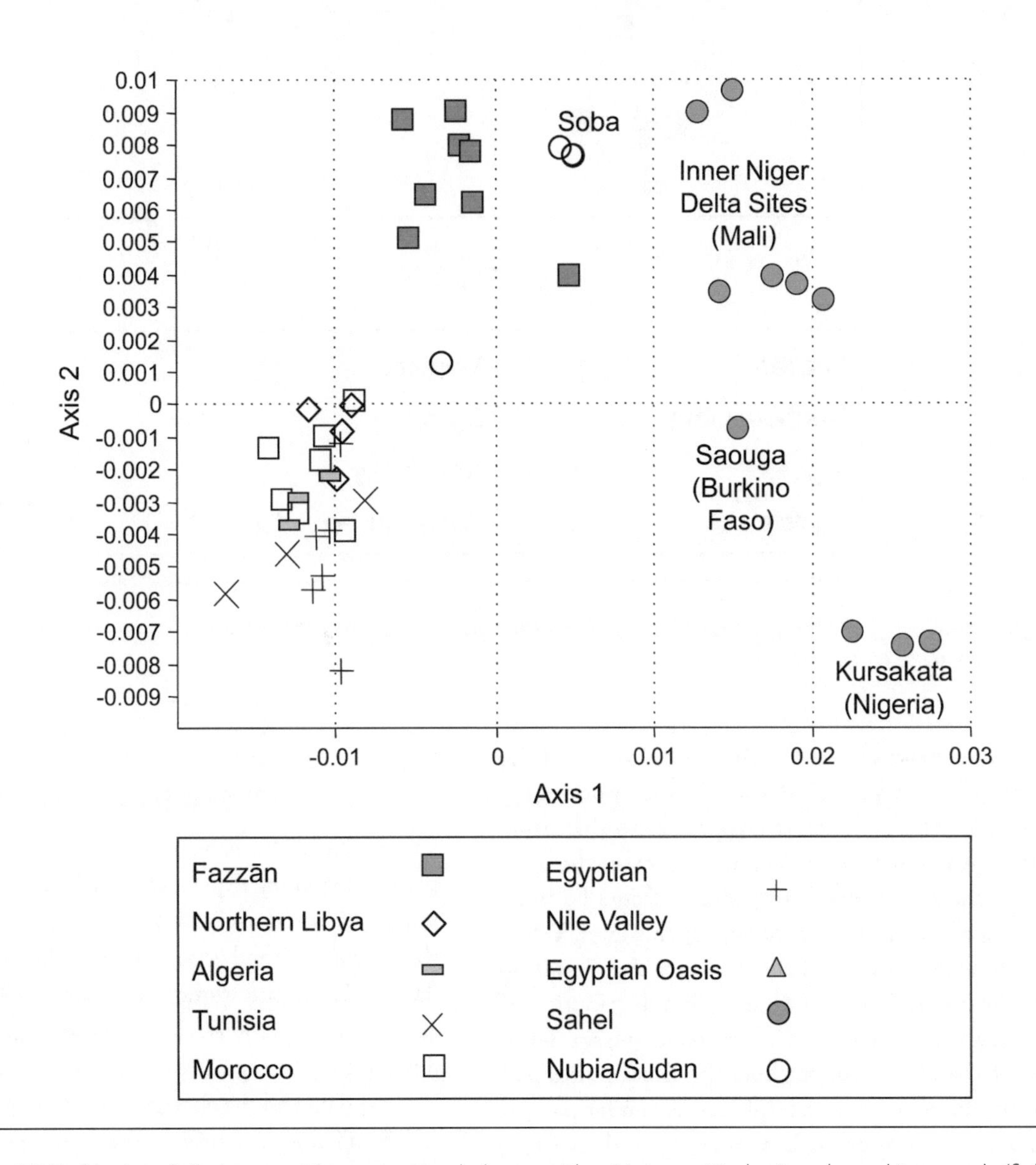

Figure 18.5 CA plot of all phases, with rare taxa and phases with <5 taxa omitted using charred items only (field crops and weeds).

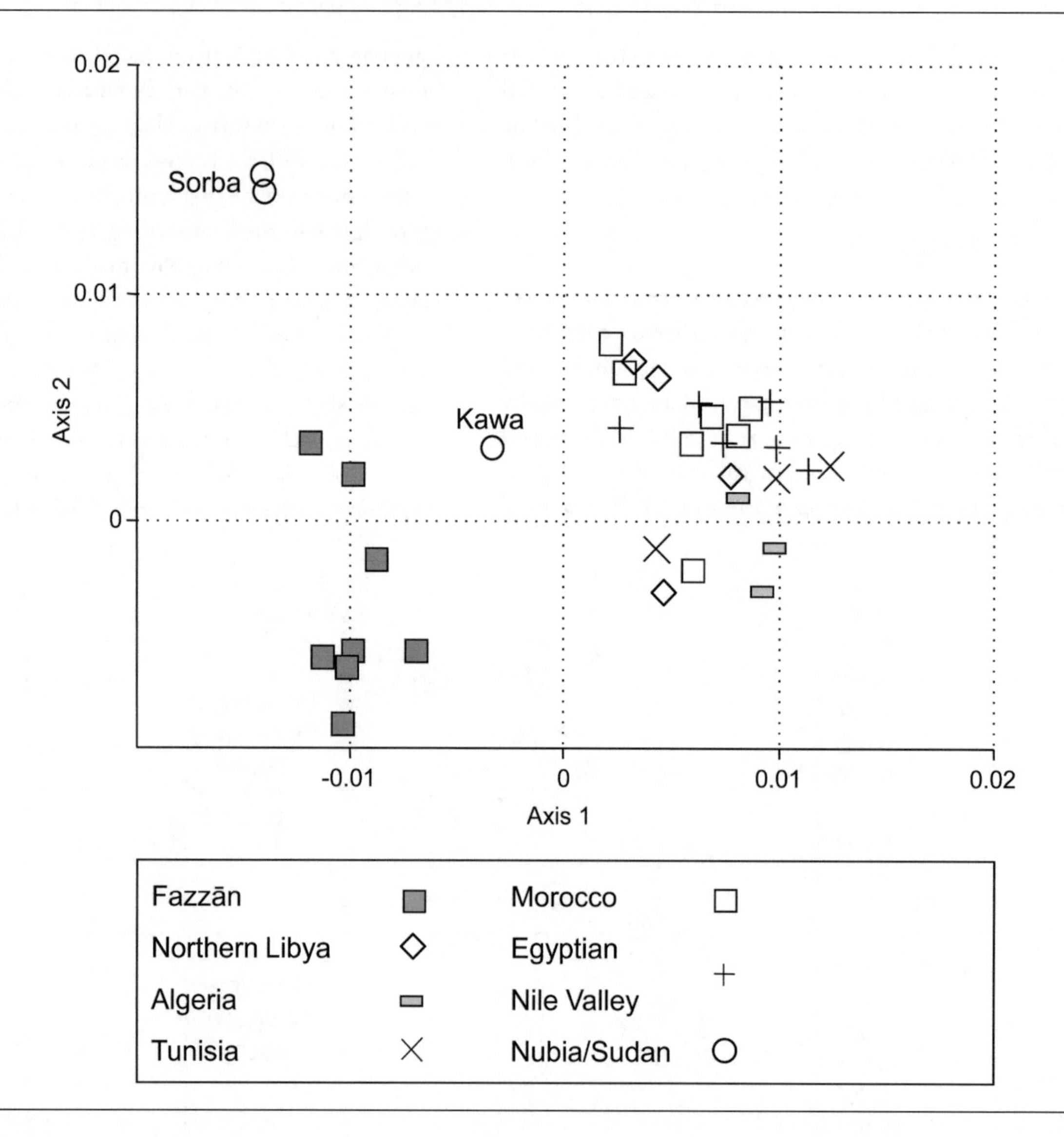

Figure 18.6 CA plot for all phases with Sahel sites removed (rare taxa and phases with <5 taxa removed) using charred items only (field crops and weeds).

Kawa, plotted between the Fazzān sites/phases and the Mediterranean/Nilotic sites/phases. The positioning of the Fazzān sites is therefore comparable to that using both desiccated and charred data. In most cases the positioning of the sites is not significantly affected by preservation when one is using taxa data at this level. When taxa codes are given broad habitat groups and flowering season (Figures 18.7 and 18.8), there is a distribution of summer crops and weeds and of desert/steppe species to the left of the plot, distinguishing the Fazzān sites and Soba from the remaining sites/phases. A similar pattern was obtained when applying habitat groups and flowering season to the combined desiccated and charred data plots (Pelling 2007).

Conclusion: The Evolution of an Cultivated Desert Flora

Correspondence analysis has been used as a means of pattern searching the pan-regional data to investigate the impact of introduced crops on the floristic composition of the Fazzān sites and in particular on the weed flora. In similar fashion to the studies by Lange (1990) and Colledge and associates (2004), clear geographical trends emerged during the plotting of both the combined desiccated and charred data and the charred data alone in this study.

In all the statistical analyses of the data presented, Jarma and Tinda for all phases plotted away from the Mediterranean/Egyptian sites, suggesting that there is a

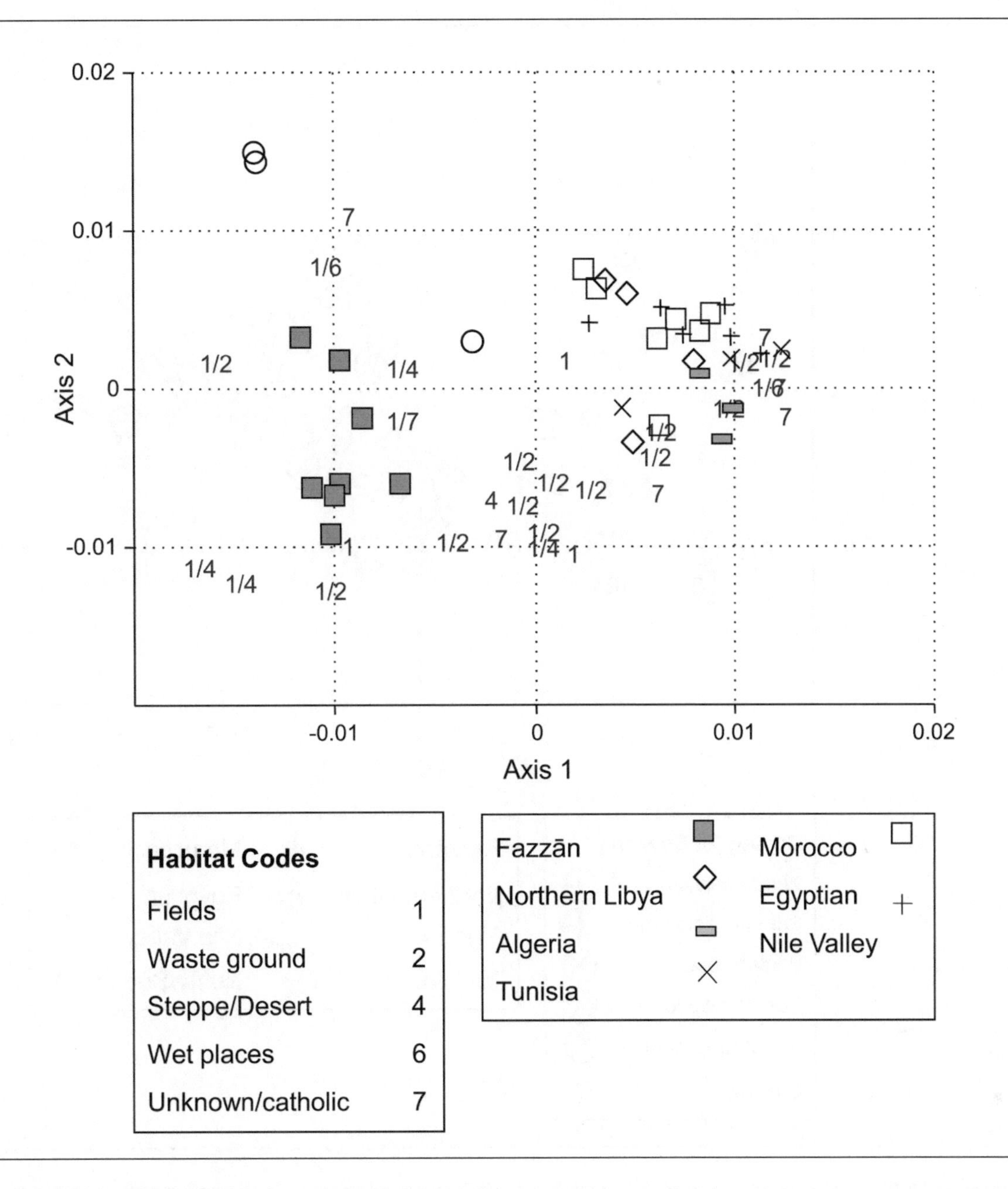

Figure 18.7 CA plot for all phases with Sahel/Nubian sites removed (rare taxa and phases with <5 taxa omitted), using charred field crop and weeds showing habitat codes.

real difference in their floristic composition. Although in part this separation must reflect the local vegetation, it possibly also reflects the adoption over time of new crops, notably those of Sahelian origin, and the formation of a flora that is uniquely central Saharan. Indeed, the flora may reflect the material culture and social customs of this region, particularly during the Garamantian period, in the sense that it is similarly composed of a range of indigenous and imported elements reflecting the Garamantes' geographical position at a major crossroads of cross-Saharan trade routes.

The apparent close association between Zinkekra and the Egyptian oasis sites is especially noteworthy, although it is unfortunate that weed seeds were rare in the oasis phases. The summer crops and weeds appear to cause some separation of Zinkekra from Jarma and Tinda, which may hint toward a chronological development of the flora, the Zinkekra material being earlier than that from Jarma and Tinda. The initial range of crops and their associated weed flora, as reflected at Zinkekra (van der Veen 1992; van der Veen and Westley 2010), may have been very similar to those of the Egyptian oases, composed of a mixture

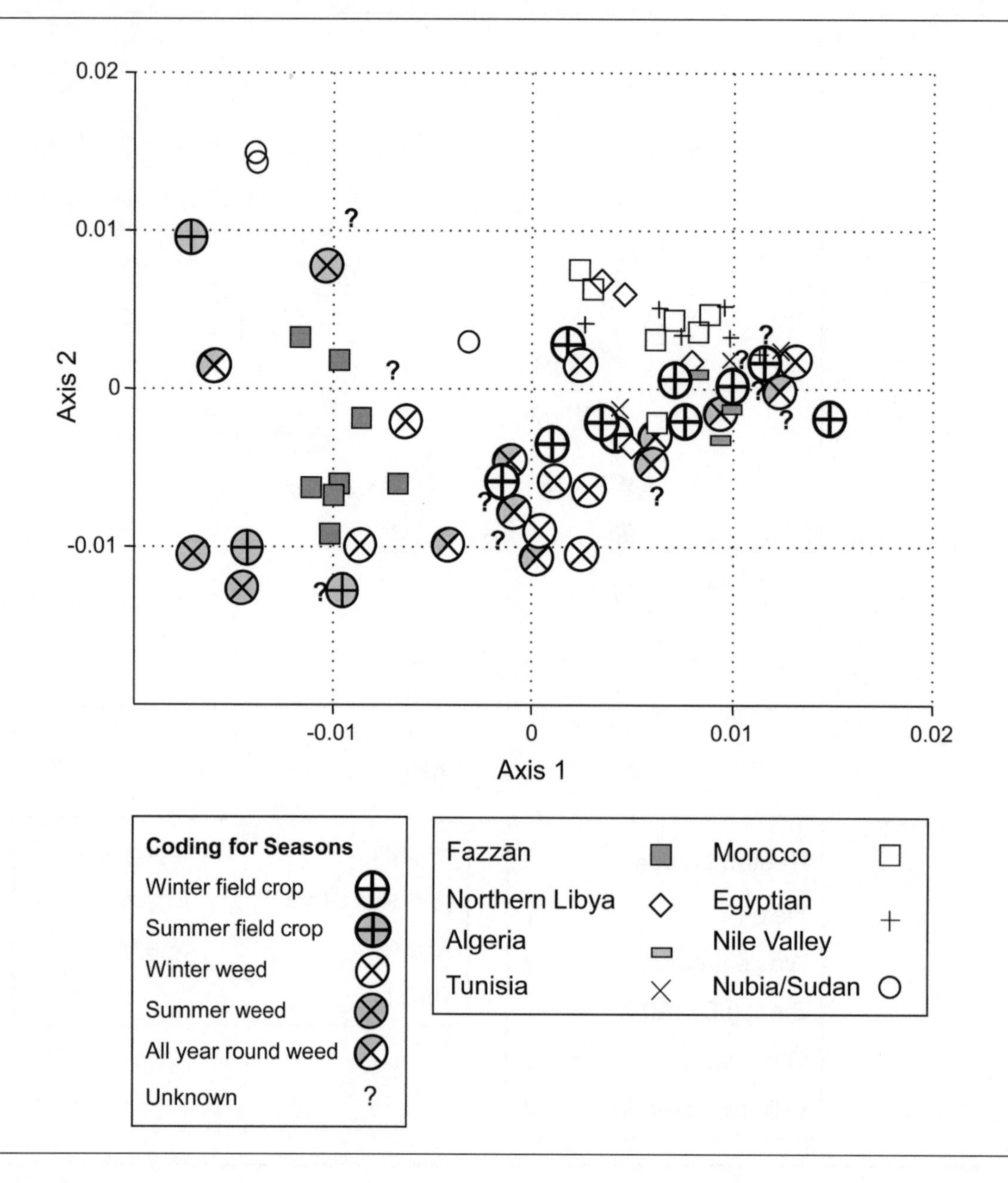

Figure 18.8 CA plot for all phases with Sahel/Nubian sites removed (rare taxa and phases with <5 taxa omitted), using charred field crops and weeds showing season codes.

of Near Eastern crop plants, imported weeds including a temperate northern flora, along with those species associated with oases. As new crops were adopted at Jarma and Tinda, in particular summer crops of Sahelian origin, a more localised flora appears to have evolved that is distinguished from the Egyptian oasis flora and that of Zinkekra by the presence of summer weeds. Although the presence of desert grasses in the weed signature may simply reflect the natural vegetation, it is significant that these grasses include some not recorded in the flora of the region today, notably *Brachiaria ramosa/deflexa* and *Brachiaria/Setaria* type grasses. *Brachiaria ramosa* and *B. deflexa*, although both edible grasses, also occur as weeds of cultivated crops in much of sub-Saharan Africa (Burkill 1994, pp. 191, 194). Although the presence of these grasses could suggest they were present as contaminants of imported pearl millet grain, the ubiquity of pearl millet does suggest this grain to be cultivated locally. It is possible therefore that *Brachiaria ramosa/deflexa*, as well as other summer weeds, were introduced with seed grain from the Sahel regions and persisted until recently as a weed.

The greatest difference in the data is between those sites in the northern Sahara (the Fazzān sites) and on the Mediterranean coast compared to those to the south (the Sahel sites), with the northern Saharan sites being more Mediterranean/Egyptian (or Near Eastern) in character than the Sahel sites. This association of the northern Saharan sites with the Mediterranean and Egyptian sites would suggest that despite the adoption of Sahelian crops the weed flora in the Fazzān continued to be dominated by more temperate zone weeds than the weeds of the Sahelian, summer rainfall belt. Because of the harvesting methods of pearl millet and sorghum, both usually cut high on the stem, it is possible that any associated weed flora would be unstable and variable through time; thus for weed species to fall out of the crop repertoire would not be unusual. Conversely, the weed flora of the temperate cereals is likely to be more stable and persistent over time given the greater chance of weeds being harvested and remaining with the seed grain.

Archaeobotany in Africa as a whole is still a relatively new subject, and there are considerable gaps in the data, both temporally and spatially; consequently, this initial study is by nature superficial. A more detailed study of the modern weed flora in Fazzān as well as detailed analysis of archaeobotanical assemblages from the area may prove fruitful in understanding the evolution of the weed floras' of summer crops outside their core area. In particular, it is hoped that the exercise has demonstrated the usefulness of a detailed database in terms of pan-regional analysis. The scope for more informative analysis can only increase as more data become available and the database is expanded.

ACKNOWLEDGEMENTS

This chapter was generated from work conducted as part of my Ph.D., funded by NERC. The Ph.D. was conducted at the Institute of Archaeology, UCL, under the supervision of Dorian Fuller and Lisa Fentress. Professor David Mattingly of Leicester University invited me to work on the material as part of the Fazzān Project, a multidisciplinary project funded by the Libyan Society. A number of specialists provided unpublished data as part of the study; these data are contained in the database. I am most grateful to Claire Newton, Alan Clapham, Dorian Fuller, Anna Maria Mercuri, and Shawn Murray. Thanks also to Chris Stevens for editing and commenting on the text and for redrawing Figures 18.3 to 18.8.

REFERENCES

Amblard, S., and J. Pernés (1989) The identification of cultivated pearl millet (Pennisetum) amongst plant impressions on pottery from Oued Chebbi (Dhar Oualata, Mauritania). *African Archaeological Review* 7, 117–26.

Bagnall, R. S. (1997) *The Kellis Agricultural Account Book (P. Kell. IV Gr. 96)* Dakhleh Oasis Project, Monograph 7, Oxbow Monograph 92. Oxford: Oxbow Books.

Barakat, H. (1990) Plant remains from El Omari (appendix IV). In F. Debono and B. Mortensen (Eds.), *El Omari, A Neolithic Settlement and Other Sites in the Vicinity of Wadi Hof, Helwan*, pp. 109–13. Mainz: Deutsches Archaeologisches Institut, Abteilung Kairo.

Barakat, H., and A. G. Fahmy (1999) Wild grasses as 'Neolithic' food resources in the Eastern Desert: A review of the evidence from Egypt. In M. van der Veen (Ed.), *The Exploitation of Plant Resources in Ancient Africa*, pp. 33–46. London: Kluwer Academic.

Bartlett, H. H. (1933) Fruits and other plants. In A. E. R. Boak (Ed.), *Karanis: The Temples, Coin Hoards, Botanical and Zoological Reports, Seasons 1924–31*, p. 88. Ann Arbor: University of Michigan Press.

Bedaux, R. M. A. (1991) The Tellem research project: The archaeological context. In R. Bolland (Ed.), *Tellem Textiles: Archaeological Finds from Burial Caves in Mali's Bandiagara Cliff*, pp. 14–36. Leiden: Mededelingen van het Rijksmuseum voor Volkenkunde.

Brett, M., and Fentress, E. (1996) *The Berbers*. Oxford: Blackwell Publishers.

Burkill, H. M. (1994) *The Useful Plants of West Tropical Africa*, Vol. 2 (Families E-I). London: Kew Publishing.

Camps, G. (1980) *Berbères: Aux marges de l'Histoire*. Toulouse: Hesperides.

Cappers, R. T. J. (1999) Trade and subsistence at the Roman port of Berenike, Red Sea coast, Egypt. In M. van der Veen (Ed.), *The Exploitation of Plant Resources in Ancient Africa*, pp. 185–97. London: Kluwer Academic/Plenum Publishers.

———. (2006) *Roman Foodprints at Berenike: Archaeobotanical Evidence of Subsistence and Trade in the Eastern Desert of Egypt*. Los Angeles: Cotsen Institute of Archaeology, UCLA.

———. (Ed.) (2007) *Fields of Change*. Proceedings of the 4th International Workshop for African Archaeobotany. Groningen: Barkhuis & Groningen University Library.

Cartwright, C. (1998) The wood, charcoal, plant remains and other organic material. In D. A. Welsby (Ed.), *Soba II: Renewed Excavations within the Metropolis of the Kingdom of Alwa in Central Sudan*, Memoirs of the British Institute in Eastern Africa: No. 15, pp. 255–68. London: British Museum Press.

Castelletti, L., E. Castiglioni, M. Cottini, and M. Rottoli (1999) Archaeobotanical analysis of charcoal, wood and seeds. In S. di-Lernia (Ed.), *The Uan Afuda Cave:*

Hunter-Gatherer Societies of Central Sahara, pp. 131–48. Firenze: All'Insegna del Giglio.

Chavane, B. A. (1985) *Villages de l'ancien Tekrour: Recherches archéologiques dans la moyenne vallée du fleuve Sénégal*. Paris: Éditions Karthala.

Clapham, A., and M. A. Dorri (Unpublished) *The Plant Remains from the North Kharga Oasis Survey 2004*. Unpublished report on the 2004 survey season.

Clapham, A., and P. Rowley-Conwy (2007) New discoveries at Qasr Ibrim, Lower Nubia. In R. Cappers (Ed.), *Fields of Change: Proceedings of the 4th International Workshop for African Archaeobotany*, pp. 157–64. Groningen: Barkhuis & Groningen University Library.

Colledge, S. M., J. Conolly, and S. Shennan (2004) Archaeobotanical evidence for the spread of farming in the eastern Mediterranean. *Current Anthropology 45* (supplement), 35–58.

Connah, G. (1981) *Three Thousand Years in Africa: Man and His Environment in the Lake Chad Region of Nigeria*. Cambridge: Cambridge University Press.

D'Andrea, A. C., M. Klee, and J. Casey (2001) Archaeobotanical evidence for pearl millet (*Pennisetum glaucum*) in sub-Saharan West Africa. *Antiquity 75*, 341–48.

Daniels, C. (1970) *The Garamantes of Southern Libya*. London: Oleander.

Di Lernia, S., G. Manzi, and F. Merighi (2002) Cultural variability and human trajectories in later prehistory of the Wadi Tanezzuft. In S. di Lernia and G. Manzi (Eds.), *Sands, Stones and Bones: The Archaeology of Death in the Wadi Tanezzuft Valley (5000–2000 BP)*, pp. 281–302. Firenze: All'Insegna del Giglio.

El-Hadidi, M. N. (1992) A historical flora of Egypt, a preliminary survey. In R. Friedman and A. Adams (Eds.), *The Followers of Horus: Studies Dedicated to M. A. Hoffman 1944–1990*, pp. 144–55. Oxford: Oxbow.

Fahmy, A. G. (1997) Evaluation of the weed flora of Egypt from Predynastic to Graeco-Roman times. *Journal of Vegetation History and Archaeobotany* 6, 241–47.

Ford, R. I., and N. Miller (1978) Paleoethnobotany I. In J. H. Humphrey (Ed.), *Excavations at Carthage 1976 Conducted by the University of Michigan*, Vol. 4, pp. 181–87. Ann Arbor: Kelsey Museum, University of Michigan.

Fuller, D. (2000) The botanical remains. In T. Insoll (Ed.), *Urbanism, Archaeology and Trade: Further Observations on the Gao Region (Mali). The 1996 Field season Results*. BAR International Series 829, pp. 28–35. Oxford: British Archaeological Reports.

———. (2004) Early Kushite agriculture: Archaeobotanical evidence from Kawa. *Sudan & Nubia 8*, 70–74.

Fuller, D., and D. N. Edwards (2001) Medieval plant economy in middle Nubia. Preliminary archaeobotanical evidence from Nauri. *Sudan and Nubia* 5, 97–103.

Fuller, D. Q, and R. Pelling (Forthcoming) Staple grains, cash crops and social differentiation: Archaeobotanical studies at Volubilis. in E. Fentress and H. Limane (Eds.), *Volubilis Excavation Report*.

Germer, R. (1998) The plant material found by Petrie at Lahun and some remarks on the problems of identifying Egyptian plant names. In S. Quirke (Ed.), *Lahun Studies*, pp. 84–91. Reigate: Sia Publishing.

Greenacre, M. J. (1984) *Theory and Applications of Correspondence Analysis*. London: Academic Press.

Haaland, R. (1995) Sedentism, cultivation, and plant domestication in the Holocene middle Nile region. *Journal of Field Archaeology* 22, 157–74.

———. (1999) The puzzle of the late emergence of domesticated sorghum in the Nile Valley. In C. Godsen and J. Hather (Eds.), *The Prehistory of Food*, pp. 397–418. New York: Routledge.

Hammer, Ø., D. A. T. Harper, and P. D. Ryan (2006) PAST: Palaeoentological statistics software package for education and data analysis. *Palaeoentological Electronical 4* (1), 9.

Harlan, J. R. (1992a) Indigenous African agriculture. In C. W. Cowan and P. J. Watson (Eds.), *The Origins of Agriculture*, pp. 53–60. Washington, D.C.: Smithsonian Institution Press.

———. (1992b) *Crops and Man*. Madison, WI: American Society of Agronomy and Crop Science Society of America.

Hoffman, E. S. (1981) Paleoethnobotany II: Plant remains from Vandal and Byzantine deposits in three cisterns. In J. H. Humphrey (Ed.), *Excavations at Carthage 1977 Conducted by the University of Michigan*, pp. 259–68. Ann Arbor: Kelsey Museum, University of Michigan.

Jezik, S. (2000) Flotation and soil sampling at Leptiminus in 1998–1999. In L. M. Stirling, D. L. Stone, and N. Ben Lazreg (Eds.), Roman kilns and rural settlement: Interim report of the 1999 season of the Leptiminus Archaeological Project. *Echos du Monde Classique/ Classical Views XLIV*, pp. 208–10.

Kahlheber, S. (1999) Indications for agroforestry: Archaeobotanical remains of crops and woody plants from medieval Sauga, Burkino Faso. In M. van der Veen (Ed.), *The Exploitation of Plant Resources in Ancient Africa*, pp. 89–100. London: Kluwer Academic/Plenum Publishers.

Kahlheber, S., K.-D. Albert, and A. Höhn (2001) A contribution to the palaeoenvironment of the archaeological site Oursi in northern Burkina Faso. In S. Kahlheber and K. Neumann (Eds.), *Man and Environment in the West*

African Sahel: An Interdisciplinary Approach, *Berichte des Sonderforschungsbereichs* 268 (17), pp. 145–59. Frankfurt am Main: SFB 268.

Klee, M., and B. Zach (1999) The exploitation of wild and domestic food plants at settlement mounds in north-east Nigeria (1800 cal BC to Today). In M. van der Veen (Ed.), *The Exploitation of Plant Resources in Ancient Africa*, pp. 81–88. London: Kluwer Academic/Plenum Publishers.

Klee, M., B. Zach, and K. Neumann (2000) Four thousand years of plant exploitation in the Chad Basin of northeast Nigeria 1: The archaeobotany of Kursakata. *Vegetation History and Archaeobotany* 9, 223–37.

Lange, A. G. (1990) *De Horden near Wijk bij Duurstede: Plant Remains from a Native Settlement at the Roman Frontier—A Numerical Approach*, Nederlandse Oudhenden, Kromme Rijn Projekt 3. Amersfoort: Rijksdienst voor het Oudheidkundig Bodemonderzoek.

Leighty, C. E. (1933) Cereals. In A. E. R. Boak (Ed.), *Karanis: The Temples, Coin Hoards, Botanical and Zoological Reports, Seasons 1924–1931*, pp. 87–88. Ann Arbor: University of Michigan Press.

Liverani, M. (2000a) The Garamantes: A fresh approach. *Libyan Studies 31*, 17–28.

———. (2000b) The Libyan caravan road in Herodotus IV, pp. 181–84. *Journal of the Economic and Social History of the Orient 43* (4), 496–520.

MacDonald, K. (1996) The Winde Koroji complex: Evidence for the peopling of the eastern Inland Niger Delta (2100–500 BC). *Préhistoire Anthropologie Méditerranée* 5, 147–65.

MacDonald, K., R. Vernet, D. Fuller, and J. Woodhouse (2003) New light on the Tichitt tradition: A preliminary report on survey and excavation at Dhar Nema, In P. Mitchell, A. Haour, and J. Hobart (Eds.), *Researching Africa's Past: New Contributions from British Archaeologists*, pp. 73–80. Oxford University School of Archaeology Monograph No. 57. Oxford: Oxford University School of Archaeology.

Mahoney, N. (1994) *Economy and Environment at Al-Basra: An Archaeobotanical Analysis of a Medieval Islamic Urban Centre*. M.A. thesis, George Washington University.

———. (Unpublished) *Archaeobotanical Remains from the 1993 Excavations at Sijilmasa, Morocco*. Unpublished excavation report, George Washington University.

Mattingly, D. J. (Ed.) (2003) *The Archaeology of Fazzān*, Vol. 1, *Synthesis*, Society for Libyan studies Monograph 5. London: Society of Libyan Studies and Department of Antiquities, Libya.

———. (Ed.) (2010) *The Archaeology of Fazzān*, Vol. 3, *Excavations of C. M. Daniels*, Society for Libyan studies Monograph 8. London: Society of Libyan Studies and Department of Antiquities, Libya.

Mattingly, D., T. Reynolds, and J. Dore (2003) Synthesis of human activities in Fazzān. In D. J. Mattingly (Ed.), *The Archaeology of Fazzān*, Vol. 1, *Synthesis*, Society for Libyan studies Monograph 5, pp. 327–73. London and Tripoli: Society of Libyan Studies and Department of Antiquities, Libya.

McIntosh, S. K. (1995) Paleobotanical and human osteological remains. In S. K. McIntosh (Ed), *Excavations at Jenné-Jeno, Hambarketolot, and Kaniana (Inland Niger Delta, Mali), the 1981 Season*, pp. 348–59. Berkeley and Los Angeles: University of California Press.

Mercuri, A. M. (2001) Preliminary analyses of fruits, seeds and other few plant macrofossils from the early Holocene sequence. In E. A. A. Garcea (Ed.), *Uan Tabu in the Settlement History of the Libyan Sahara*, Arid Zone Archaeology Monographs 2, pp. 189–210. Firenze: All'Insegna del Giglio.

Mercuri A. M., G. T. Grandi, G. Bosi, L. Forlani, and F. Buldrini (2005) The archaeobotanical remains (pollen, seeds/fruits, and charcoal). In M. Liverani (Ed.), *The Archaeology of Libyan Sahara*, Vol. II: *Aghram Nadharif: The Barkat Oasis (Sha'Abiya of Ghat, Libyan Sahara) in Garamantian Times*, Arid Zone Archaeology Monograph No. 5, pp. 335–448. Firenze: All'Insegna del Giglio.

Moens, M.-F., and W. Wetterstrom (1988) The agricultural economy of an Old Kingdom town in Egypt's Western Delta: Insights from the plant remains. *Journal of Near Eastern Studies* 47 (3), 159–73.

Munson, P. J. (1971) *The Tichitt Tradition: A Late Prehistoric Occupation of the Southwestern Sahara*. Ph.D. thesis, Urbana-Champaign, University of Illinois.

———. (1976) Archaeological data on the origins of cultivation in the southwestern Sahara and their implications for West Africa. In J. R. Harlan, J. M. J. D. Wet, and A. Stemler (Eds.), *Origins of African Plant Domestication*, pp. 187–209. The Hague: Mouton.

———. (1986) About 'Economie et societe neolithique de Dhar Tichitt (Mauritanie)', *Sahara 2*, 106–08.

Murray, S. S. (2005) Recherches archéobotanique. In R. M. A. Bedaux, J. Polet, K. Sanogo, and A. Schmidt (Eds.), *Recherches archéologiques à Dia dans le Delta intérieur du Niger (Mali): Bilan des saisons de fouilles 1998–2003*, Mededelingen van het Rijksmuseum voor Volkenkunde Series no. 33, pp. 386–400. Leiden: Research School CNWS, Leiden University.

Muzzolini, A. (1984) Les relations du Sahara néolithique avec les régions mediterranéennes ou nilotiques. *Mediterranea 14*, 26–33.

Neumann, K. (2003) The late emergence of agriculture in sub-Saharan Africa: Archaeobotanical evidence and ecological considerations. In K. Neumann, A. Butler, and S. Kahlheber (Eds.), *Food, Fuel and Fields: Progress in African Archaeobotany*, pp. 71–92. Köln: Heinrich-Barth-Institut.

Neumann, K., A. Butler, and S. Kahlheber (Eds.) (2003) *Food, Fuel and Fields: Progress in African Archaeobotany*. Köln: Heinrich-Barth-Institut.

Newton, C., T. Gonon, and M. Wuttmann (2006) Un jardin d'oasis d'é romaine a'Ayn-Manawir (Kharga, Egypte). *Bulletin Institut Français d'Archéologie Orientale (BIFAO) 105*, 167–96.

Nixon, S., M. A. Murray, and D. Q Fuller (2011) Plant use at an early Islamic merchant town in the West African Sahel: The archaeobotany of Essouk-Tadmakka (Mali). *Vegetation History and Archaeobotany 20* (3): 223–39.

Palmer, C. (1991) The botanical remains. In A. Mohamedi, A. Benmansour, A. A. Amamra, and E. Fentress (Eds.), *Fouilles de Setif (1977–1984)*, 5eme Supplement au Bulletin d'Archéologie Algerienne, pp. 260–67. Alger: Agence Nationale d'Archéologie et de Protection des Sites et Monuments Historiques.

Pelling, R. (2005) Garamantian agriculture and its significance in a wider north African context: The evidence of the plant remains from the Fazzan Project. *The Journal of North African Studies 10* (3-4), 397–411.

———. (2007) *Agriculture and Trade amongst the Garamantes and the Fezzanese: 3000 Years of Archaeobotanical Data from the Sahara and Its Margins*. Ph.D. thesis, University College London.

———. (2013) The archaeobotanical remains (Chapter 18) and the botanical data appendices (Chapter 28), in D. J. Mattingly, C. M. Daniels, J. N. Dore, D. Edwards, A. Leone, and D. C. Thomas (Eds.), *The Archaeology of Fazzān*, Vol. 4, *Survey and Excavations at Old Jarma (Ancient Garama) Carried Out by C. M. Daniels (1962–1969) and the Fazzān Project (1997–2001)*, pp. 473–94 and pp. 841–52. Society for Libyan Studies/Department of Antiquities: London.

Pelling, R., and S. al-Hassey (1998) New evidence for the agricultural economy of Eusperides, Cyrenaica: An interim report on the macroscopic plant remains. *Libya Antiqua New Series 4*, 27–34.

Reddy, S. N. (1994) *Plant Usage and Subsistence Modeling: An Ethnoarchaeological Approach to the Late Harappan of Northwest India*. Ph.D. dissertation, University of Wisconsin. Madison.

———. (1997) If the threshing floor could talk: Integration of agriculture and pastoralism during the Late Harappan in Gujarat, India. *Journal of Anthropological Archaeology 16*, 162–87.

Renfrew, J. (1985) Preliminary report on the botanical remains. In B. J. Kemp (Ed.), *Amarna Reports II: The Egypt Exploration Society, Occasional Publications 2*, pp. 175–90. London: Egypt Exploration Society.

Rowley-Conwy, P. (1991) The sorghum from Qasr Ibrim, Egyptian Nubia, c. 800 BC–AD 1811: A preliminary view. In J. Refrew (Ed.), *New Light on Early Farming*, pp. 191–212, Edinburgh: Edinburgh University Press.

Rowley-Conwy, P., W. Deakin, and C. H. Shaw (1999) Ancient DNA from Sorghum: The evidence from Qasr Ibrim, Egyptian Nubia. In M. van der Veen (Ed.), *The Exploitation of Plant Resources in Ancient Africa*, pp. 55–62. New York: Kluwer/Plenum.

Shennan, S. (1997) *Quantifying Archaeology*. Edinburgh: Edinburgh University Press.

Smith, A. B. (1992) *Pastoralism in Africa: Origins and Development Ecology*. London: Hurst & Company.

Smith, K. W. M. (2001) Environmental samples (1990–1994). In L. M. Stirling, D. J. Mattingly, and N. Ben Lazreg (Eds.), *Leptiminus (Lamta) Report No. 2: The East Baths, Cemeteries, Kilns, Venus Mosaic, Site Museum and Other Studies*, pp. 420–39, Journal of Roman Archaeology Supplementary series No. 41. Portsmouth, RI: Journal of Roman Archaeology

———. (2003) *Archaeobotanical Investigations of Agriculture at Late Antique Kom el-Nana (Tell el-Amarna)*, Seventieth Excavation Memoir, B. Kemp (Ed.). London: Egypt Exploration Society.

Stewart, R. (1984) Carbonized seeds. In H. R. Hurst and S. P. Roskams (Eds.), *Excavations at Carthage: The British Mission*, Vol. 1, *The Avenue du President Habib Bourguiba, Salammbo: The Site and Finds Other Than Pottery*. Institut National d'Archaéologie et d'Art de Tunisie and British Academy Carthage Committee. Sheffield: Published for the British Academy by the Department of Prehistory and Archaeology, University of Sheffield.

Thanheiser, U. (1992a) Plant food at Tell Ibrahim Awad: Preliminary report. In E. van der Brink (Ed.), *The Nile Delta in Transition: 4th–3rd Millennium B.C.*, Proceedings of the Seminar held in Cairo, October 21–24, 1990, at the Netherlands Institute of Archaeology and Arabic Studies, pp. 117–24. Jerusalem: The Israel Exploration Society.

———. (1992b) Plant remains from Minshat Abu Omar: First impressions. In E. van der Brink (Ed.), *The Nile Delta in Transition: 4th–3rd Millennium B.C.*, Proceedings of the Seminar held in Cairo, October 21–24, 1990, at the Netherlands Institute of Archaeology and Arabic Studies, pp. 167–70. Jerusalem: The Israel Exploration Society.

———. (1999) Plant remains from Kellis: First results. In C. A. Hope and A. J. Mills (Eds.), *Dakhleh Oasis Project:*

Preliminary Reports on the 1992–1993 and 1993–1994 Field Seasons, Dakhleh Oasis Project Monograph 8, pp. 89–93. Oxford: Oxbow Books.

van der Veen, M. (1991) The plant remains. In D. A. Welsby and C. M. Daniels (Eds.), *Soba: Archaeological Research at a Medieval Capital on the Blue Nile, Memoirs of the British Institute in Eastern Africa: No. 12*, 264–73. London: The British Institute in Eastern Africa.

———. (1992) Garamantian agriculture: The plant remains from Zinchecra, Fezzan. *Libyan Studies 23*, 7–39.

———. (Ed.) (1999) *The Exploitation of Plant Resources in Ancient Africa*. New York: Kluwer/Plenum.

———. (2001) The botanical evidence. In V. A. Maxfield and D. P. S. Peacock (Eds.), *Survey and Excavation at Mons Claudianus, 1987–1993*, Vol. 2, Excavations: Part 1, pp. 175–247. Cairo: Institut Français d'Archéologie Orientale.

van der Veen, M., A. Grant, and G. Barker (1996) Romano-Libyan agriculture: Crops and animals. In G. Barker, D. Gilbertson, B. Jones, and D. Mattingly (Eds.), *Farming the Desert: The UNESCO Libyan Valleys Archaeological Survey*, Vol. 1, *Synthesis*, Chapter 8. pp. 227–63. Paris: UNESCO.

van der Veen, M., and B. Westley (2010) Palaeoeconomic studies. In D. J. Mattingly (Ed.), *The Archaeology of Fazzān*, Vol. 3, *Excavation of M. Daniels*, Chapter 9. The Society for Libyan Studies Monograph 8, pp. 489–519. London: Society of Libyan Studies and Department of Antiquities, Libya.

van Zeist, W. (1983) Fruits in foundation deposits of two temples. *Journal of Archaeological Science 10*, 351–54.

———. (1994) Botanical remains. In H. R. Hurst (Ed.), *Excavations at Carthage: The British Mission*, Vol. 2, *The Circular Harbour, North Side: Excavations At Carthage, The British Mission*, p. 325. Oxford: Oxford University Press.

van Zeist, W., S. Bottema, and M. van der Veen (2001) *Diet and Vegetation at Ancient Carthage: The Archaeobotanical Evidence*. Unpublished internal report, University of Groningen.

Wasylikowa, K. (1992) Holocene flora of the Tadrart Acacus area, SW Libya, based on plant macrofossils from Uan Muhuggiag and Ti-N-Torha/Two Caves archaeological sites. *ORIGINI 16*, 125–52.

———. (1997) Flora of the 8000 years old archaeological site E-75-6 at Nabta Playa, Western Desert, southern Egypt. *Acta Palaeobotanica 37*, 99–205.

Wasylikowa, K., and J. Dahlberg (1999) Sorghum in the economy of the early Neolithic nomadic tribes at Nabta Playa, southern Egypt. In M. van der Veen (Ed.), *The Exploitation of Plant Resources in Ancient Africa*, pp. 11–31. London: Kluwer/Plenum.

Wasylikowa, K., J. H. Harlan, J. Evans, F. Wendorf, R. Schild, A. E. Close, H. Królik, and R. Housley (1993) Examination of botanical remains from early Neolithic houses at Nabta Playa, Western Desert, Egypt, with special reference to sorghum grains. In T. Show, P. Sinclair, B. Andah, and A. Okopoko (Eds.), *The Archaeology of Africa: Food, Metals and Towns*, pp. 154–64. London: Routledge.

Wasylikow, K., J. Mitka, F. Wendorf, and R. Schild (1997) Exploitation of wild plants by the early Neolithic hunter-gatherers in the Western Desert of Egypt: Nabta Playa as a case-study. *Antiquity 71*, 932–41.

Wasylikowa, K., R. Schild, F. Wendorf, H. Królik, L. Kubiak-Martens, and R. Harlan (1995) Archaeobotany of the early Neolithic site E-75-6 at Nabta Playa, Western Desert, south Egypt (preliminary results). *Acta Palaeobotanica 35*, 133–55.

Wetterstrom, W. (1984) The plant remains. In R. J. Wenke (Ed.), *Archaeological Investigations at el-Hibeh 1980: Preliminary Report*, American Research Center in Egypt Reports, Vol. 9, pp. 50–77. Malibu: Undena Publications.

———. (1998) The origins of agriculture in Africa: With particular reference to sorghum and pearl millet. *The Review of Archaeology 19* (2), 30–46.

Wheeler, M. (1954) *Beyond the Imperial Frontiers*. London: Bell.

Wigboldus, J. (1990) Disputable datings of early sorghum cultivation in the southern Old World: A case for tracing crop evolution and diffusion with help from history. In D. A. Posey, W. L. Overal, and M. P. E. Goeldi (Eds.), *Ethnobiology: Implications and Applications*, pp. 317–63. Belem: The Museum Press.

———. (1996) Early presence of African millets near the Indian Ocean. In J. Rease (Ed.), *The Indian Ocean in Antiquity*. London: Kegan Paul International/British Museum.

Winlock, H. E., and W. E. Cram (1926) *The Monastery of Epiphanius at Thebes, Part I: The Archaeological Material*, Egyptian expedition, Vol. 3. New York: Publications of the Metropolitan Museum of Arts.

Zohary, D., and M. Hopf (2000) *Domestication of Plants in the Old World: The origin and Spread of Cultivated Plants in West Asia, Europe, and the Nile Valley* (3rd ed.). Oxford: Clarendon Press.

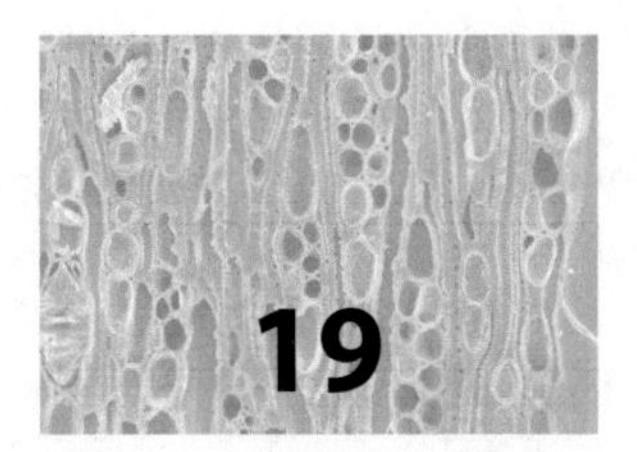

The Archaeobotany of Farming Communities in South Africa

A Review

Alexander Antonites and Annie Raath Antonites

Archer and Hastorf's (2000, p. 33) statement that 'there appears to be a conceptual gap between the role that plants and plant knowledge played in the past and the level of research interest and commitment to the study of archaeobotanical data within archaeology' rings especially true for the archaeology of South Africa's farming communities. Although archaeologists have a fair sense of the range of species cultivated by indigenous farmers, issues that relate to human-plant interaction remain understudied. The lack of archaeobotanical research is surprising, because ethnographic, historical, and archaeological information all suggest that plants formed an integral component of these societies.

In this chapter we examine the way in which archaeologists have approached the archaeobotany of farming communities since the 1980s and explore some of the reasons why they have commonly ignored its potential role in such studies.

South African Farming Communities

For the past 2,000 years Bantu-speaking people combining stock-keeping with agriculture have occupied most of southern Africa's summer rainfall zone. Generally, archaeologists distinguish between Early Farming Communities of the 1st millennium C.E. and Later Farming Communities of the 2nd millennium C.E. (see Huffman 2007; Mitchell 2002; Mitchell and Whitelaw 2005 for summaries). This division reflects changes in settlement location, grindstone shape, the location of smelting activities in relation to villages, and ceramic styles (Badenhorst 2009b; Maggs 1984a).

Ethnographic Records of Plant Use

Many of the cultures that make up South Africa's indigenous population have direct historical links to 2nd-millennium C.E. archaeological sites (Evers 1984, 1989; Huffman 2004). Because of this relatively recent history, archaeologists regularly use ethnographic material collected during the 19th and 20th centuries (for example, Boeyens 2003; Hall 1998; Huffman 1997; Schoeman 2006a, b) to understand prehistoric economy, subsistence, modes of production, and ritual. These ethnographies also provide substantial information on plant use. *Sorghum bicolor* (sorghum) and millet species, such as *Pennisetum glaucum* (pearl millet) and *Eleusine coracana* (finger millet), were the traditional staple crops (Krige and Krige 1943; Mönnig 1967; Quin 1959; Schapera and Goodwin 1937); however, by the time most ethnographies were written, *Zea mays* (maize) had replaced them as the principal cultigen. Subsidiary crops also played an

Archaeology of African Plant Use by Chris J. Stevens, Sam Nixon, Mary Anne Murray, and Dorian Q Fuller, Eds., 225–232

important role and include *Vigna unguiculata* (cowpea), *Vigna subterranea* (jugo bean), *Citrullus lanatus* (water melon), and *Lagenaria siceraria* (bottle gourd).

Cereals were also used in traditional beer—considered an essential food source and consumed in large quantities because of its low alcohol content. Beer also played an important role in social and ceremonial life (Krige and Krige 1943; Mönnig 1967; Schapera and Goodwin 1937). Stayt (1968, p. 48) notes of the Venda that beer 'is the first essential in all festivities, the one incentive to labour, the first thought in dispersing hospitality, the favourite tribute of subjects to their chief, and almost the only votive offering dedicated to their spirits'.

Wild plants fulfil a significant dietary role as sources of vitamin C and of amino and nicotinic acids (for example, niacin), nutrients mostly absent in cereal staples (Cunningham 1988, p. 444). Important wild resources include the leaves and roots of *Boscia albitrunca* (white stem), *Acacia senegal* (gum acacia), and fruits of *Englerophytum magalismontanum* (stamvrug), *Ficus* sp. (wild fig), *Vangueria infausta* (wild medlar), *Ximenia caffra* (sour plum), and *Sclerocarya birrea* (marula) (Coates-Palgrave 2002; Quin 1959; van Wyk and Gericke 2000). Wild plants further serve various economic, medicinal, and religious purposes (for example, Coates-Palgrave 2002; Plug 1987; Quin 1959; Sobiecki 2008; van Wyk and Gericke 2000; van Wyk, van Oudtshoorn, and Gericke 2005;). Beer made from the fruits of *Sclerocarya birrea* (marula), for example, carries specific social and ceremonial significance, especially during First Fruits festivals (Junod 1962, pp. 394–404).

The direct historical relationship between present-day and 2nd millennium C.E. archaeological populations in South Africa suggests that many of the historically recorded uses of plants possibly occurred much earlier. Archaeobotany, however, remains a relatively unexplored field.

ARCHAEOBOTANY IN SOUTH AFRICA

As a measure of archaeobotanical research, we analysed the frequency with which research articles containing botanical information have been published in three main archaeological journals since 1980 (cf. Archer and Hastorf 2000, pp. 33–35; Lepofsky, Moss, and Lyons 2001, p. 49); the *South African Archaeological Bulletin*, *Southern African Field Archaeology*, and the *Southern African Humanities*. These publications are the primary forums for local archaeologists and are relatively sensitive to local trends in the discipline; therefore, they provide the best indication of what South African archaeologists are doing in practise.

In these journals only 12 articles have had a specific archaeobotanical theme, and of these only three apply directly to farming communities (Cunningham 1988; Cunningham and Gwala 1986; Langejans 2006). Although many articles mention botanical information, the majority of them are associated with Stone Age material.

Southern African Humanities (formerly the *Natal Museum Journal of Humanities*, and previously the *Annals of the Natal Museum*) has by far the most articles containing botanical data, the majority being attributed to Tim Maggs and Aron Mazel, both affiliated with the Natal Museum during the 1980s. Of the 171 articles related to southern African archaeology that appeared between 1980 and 2012, 25% ($n = 42$) mention botanical data. However, only 10 (6%) of these relate directly to farming communities.

The *South African Archaeological Bulletin* (*SAAB*) and *Southern African Field Archaeology* (*SAFA*) contain fewer articles on botanical remains. SAAB published 434 research articles and technical/field reports from 1980 to 2012, of which 55 mention botanical remains. Similarly, in SAFA only 16 of the 118 articles published between 1992 and 2005 (when the publication was discontinued) refer to archaeobotanical material. In both journals, a mere 3% ($n = 11$ and $n = 3$, respectively) of these are associated with farming communities.

The botanical data contained in these journals fall into four broad themes: environmental reconstructions, economic systems, and, to a lesser extent, political complexity and social identity. Our discussion of these themes includes other published and unpublished material; we do, however, restrict it to South African research, even though these journals contain examples from neighbouring countries.

ECOLOGICAL PERSPECTIVES

Topics under this theme include the reconstruction and impact of human settlement on the environment (Feely and Bell-Cross 2011; Hall 1980, 1984; Prins 1993, 1994–1995), and its relationship with human action (Maggs 1982a; Prins 1993; Whitelaw 1991). Using the results of wood charcoal and phytolith analyses and the identification of macro remains, Prins (1993, 1994–1995) demonstrated that climatic fluctuations and the range of the summer rainfall zone probably determined the southern limit of 1st millennium C.E. settlement along the KwaZulu-Natal and Eastern Cape coast. Also investigated was the role of early farmers in the evolution of the modern vegetation mosaic of the region (Prins 1993; Prins and Granger 1993). The presence of *Urochloa* sp., a pioneer grass, combined with a decrease in the number of tree species represented in charcoal samples, from such archaeological sites as Ntistsana (Figure 19.1), suggested

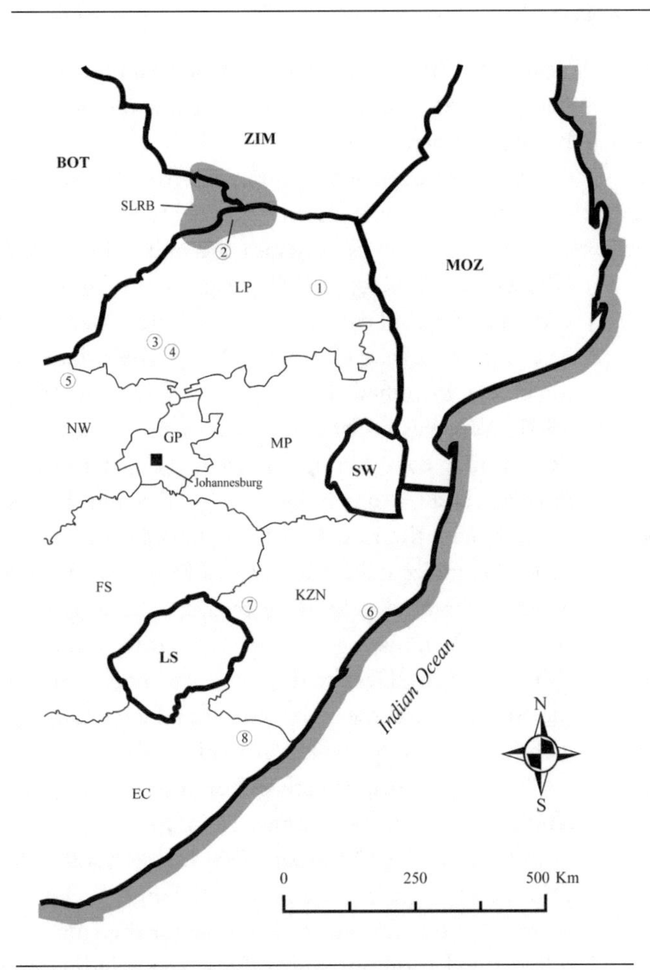

Figure 19.1 Location of archaeological sites mentioned in text. Neighbouring Countries: BOT = Botswana, LS = Lesotho, MOZ = Mozambique, SW = Swaziland, ZIM = Zimbabwe; South African Provinces: EC = Eastern Cape, FS = Free State, GP = Gauteng, KZN = KwaZulu-Natal, LP = Limpopo, MP = Mpumalanga, NW = North West; Archaeological sites/areas: SLRB = Shashe-Limpopo River Basin, 1 = Silver Leaves, 2 = Mapungubwe, 3 = Thabazimbi, 4 = Rooiberg, 5 = Marico/Dwarsberg area, 6 = Ndondondwane, 7 = Mgoduyanuka, 8 = Ntsitsana.

severe environmental degradation. Additionally, comparisons of the phytolith composition of archaeological dung samples with modern cattle byre deposits further implied a decrease in tree species. Prins (1993, pp. 117–20) concluded that woodland clearance for farming, the construction of settlements, and the collection of firewood had a noteworthy impact on vegetation, particularly around settlements.

Feely (1980) observed secondary effects on vegetation previously assumed to be in a relatively undisturbed natural state, and Hall (1984) also recorded the effect of human land use on the composition of woodland and grassland in KwaZulu-Natal. The analysis of the pollen content of cattle dung from farming community settlements substantiates a long-term and sustained human impact on the landscape. Carión and associates (2000) showed that, although the overall landscape was covered with trees, the immediate areas around settlements were cleared of woody vegetation.

For some areas, the selective process of clearing trees for charcoal production for iron smelting may have had enduring effects. Iron smelting requires comparatively large quantities of charcoal to process small volumes of ore, and smelting sites are often located in tree-rich areas substantial distances away from ore bodies (van der Merwe and Killick 1979). Hall (1984) considers that iron smelting actively contributed to the formation of *Euclea divinorum* woodland in areas of KwaZulu-Natal. Identification of charcoal from furnaces suggested the selective use of certain tree species for smelting, such as *Acacia* sp. (particularly harder species—for example, *A. karroo* and *A. caffra*) but also *Olea europaea* and *Spirostachys africana* (Maggs 1982a; Maggs and Ward 1984; Whitelaw 1991). The specific distribution of *A. caffra*, and its correlation with smelting sites, may have been a major factor in influencing their location in KwaZulu-Natal (Maggs 1982a, p. 140).

The archaeological evidence suggests that South Africa's farming communities had a conspicuous impact on vegetation in the areas where they settled. Although important, this impact was, however, less severe than initially suggested. Following Acocks's (1953) seminal work, ecologists believed that woodland clearance by farming communities had created the high-altitude grasslands of KwaZulu-Natal and the Eastern Cape. Feely (1985), however, demonstrated that the grasslands contain little evidence for farming settlements and that areas where agriculturalists did settle retained their woody vegetation. He concluded that the grasslands resulted from prevalent soil regimes rather than agricultural clearance.

Cunningham (1988, p. 445) in turn highlights the positive selective influence played by farming communities through (1) clearance for farmland while maintaining certain tree species for shade or fruit, (2) dispersal and germination of species at homesteads, (3) changing the sex ratios of species, (4) the increase of spinach varieties on disturbed soils, and (5) the deliberate burning of veld to stimulate the fruiting of favourable species. The effect of farming communities on vegetation mosaics, therefore, was probably a varied and intricate process.

Economy and Diet

Economic systems and dietary reconstructions are mainly based on the analyses of macro-remains, but research rarely moves beyond the identification of crop species.

However, there are some pertinent discussions that centre on the role of different crops in the domestic economy.

One topic that remains poorly understood is the relative importance of wild and domestic plant foods in the diet, a situation confounded by the general absence of wild plant remains in the archaeobotanical record. Although many wild species were probably consumed, their processing and storage rarely leave visible archaeological evidence. Ethnographic studies, however, indicate that wild plant species may constitute significant proportions of the overall diet (for example, Cunningham 1988).

Although isotopic analyses are widely used in Later Stone Age dietary reconstructions (for instance, Dewar 2010; Sealy 2006), studies of 1st and 2nd millennium C.E. farmer diets are largely based on faunal evidence. One isotope study of 2nd millennium agriculturalist skeletal material revealed a subsistence base that is mainly C_4, either through consuming C_4 crops such as sorghum and millet or grazing domestic stock. However, the proportions of cultivars (all C_4) versus wild plant foods (mainly C_3) in the diet varied significantly within and between different biomes, suggesting that farmers changed their subsistence strategies according to local conditions. This dietary flexibility calls into question the conventional uniformity in the diet as suggested by ethnographies (Lee-Thorp, Sealy, and Morris 1993, p. 116).

Agricultural systems and land use, although acknowledged, are rarely discussed in the archaeology of farming communities. Yet this is an area where archaeobotany could play an extremely important role, especially in the study of the terracing of the high-altitude grasslands of western Mpumalanga. Archaeologists have always assumed that these features were used in extensive agricultural production systems; however, ethnographic evidence is scanty, and questions regarding the range of crops used by these 2nd millennium C.E. farmers remain unanswered.

These terraces are distinctly different from the agricultural system of earlier farmers. First millennium C.E. settlements, for example, are located mostly on the deep alluvial soils of valley floors, implying that agriculture was an important consideration in determining their location (Maggs 1984a; Maggs and Ward 1984, p. 135). Activity areas on these settlements are often separated by large open spaces devoid of archaeological material, interpreted as the location of horticultural gardens (Greenfield, Fowler, and van Schalkwyk 2005; Maggs and Ward 1984). Excavations at Ndondondwane (Figure 19.1) showed that these open areas were created when settlement expanded beyond existing gardens. Accordingly, Greenfield and associates (2005, p. 325), suggest that 1st millennium gardens were located not only inside the settlement but also beyond the periphery.

Recovery of plant macro-remains from Ndondondwane and other 1st millennium settlements showed the predominant crops to be sorghum and millet (Hanisch 1980; Klapwijk 1974; Maggs 1984b; Maggs and Ward 1984). The earliest evidence of cultivation comes from Silver Leaves (ca. 3rd–5th centuries C.E.) (Figure 19.1), where domesticated millet was identified from pottery impressions (Klapwijk 1974). Whereas millet and sorghum arrived with the earliest farming communities, maize is thought to have been introduced at a much later stage through Portuguese contact along the East Coast. Predictably, the earliest evidence for maize cultivation comes from KwaZulu-Natal, where carbonised cobs were recovered at Mgoduyanuka (Figure 19.1), dating to the 17th/18th centuries C.E. (Maggs 1982b). Despite this evidence, there is no consensus on when and how maize spread to the interior, an argument that presently revolves around grindstones.

In the absence of archaeobotanical remains, archaeologists have used grindstones to inform on crop cultivation (for example, Huffman 1993; Maggs 1980). Narrow elliptical-type grindstones, typically found on 1st millennium C.E. sites, are seen as evidence for the cultivation of soft-grained sorghum and millets. The grinding method is one whereby an upper stone, held in one hand, is used to crush the grains against the sides of a long channel in the lower stone (Huffman 2004; Maggs 1984a). Because maize has a much harder kernel, it requires grinding with a larger and heavier two-handed upper stone, which is usually pitted to catch the kernels and used against a large, flat lower stone (Huffman 2004).

To Huffman (2006), the difference between the two types of grindstones signifies the development of separate technologies to accommodate different plant species and, combined with their historic distribution, proves maize cultivation without the need to obtain direct archaeobotanical information. A calibrated ^{14}C date of 1485–1615 C.E. (400 ± 10 BP, Pta-9543) from Thabazimbi, associated with the change of grindstone technology, predates the Mgoduyanuka evidence by 150 years. This early date suggests a rapid spread into the interior, probably by means of long-distance trade networks linking coastal sites with the Rooiberg tin mines near Thabazimbi (Figure 19.1) (Huffman 2006, p. 68).

Starch analysis of the Thabazimbi grindstones, however, failed to provide evidence that they were used to grind maize (Langejans 2006). Furthermore, Boeyens's (2003) review of 19th century traveller accounts from the Marico and Dwarsberg region in the interior (Figure 19.1) indicate that Sotho-Tswana communities were reliant on sorghum and that maize was largely unknown at the time. A more likely explanation, according to Boeyens (2003), is that maize was introduced from KwaZulu-Natal after the 1820s, during the period known as the *difaqane*, a time of large-scale political unrest throughout southern Africa. Conversely, Huffman (2006, p. 68) argues that, after its 16th century introduction, maize disappeared owing to a severe drought in the early 18th century, which would explain why historical accounts written after this period make no mention of maize.

Despite questions regarding the timing of maize's introduction, there is little argument as to its importance as a staple from the middle 19th century onward and its role in the cumulative processes that set off the *difaqane*.

Sociopolitical Issues

The third theme in archaeobotanical studies is sociopolitical complexity. Although archaeobotanical information typically does not form the focal point of these studies, information such as agricultural output and field systems is garnered in support of large-scale political developments, as well as issues of social and community organization.

As stated, maize played a significant role in political developments during the latter half of the 2nd millennium. Under optimal conditions maize yields greater returns than sorghum and millet and, along with the increased agricultural output, stimulates population growth (Huffman 2004). However, maize is less resistant to drought, and a dry spell in the early 19th century (Hall 1976) probably led to the widespread famine animated in oral histories and contributed to events leading to the *difaqane* (Huffman 2004, pp. 104–06).

Increased crop production also features in research directed at discerning the origins of the 13th century Mapungubwe state (Figure 19.1). Huffman (2000; cf. Schoeman 2006b) sees a dynamic interplay between increased floodplain agriculture, climate, ideology, and trade, leading to the formalisation of class structures. He links fluctuations in rainfall to a system of production centred on the flood cycles of the Shashe and Limpopo rivers and their tributaries, in which agriculturalists planted different sorghum and millet varieties to obtain multiple yields during the year. The increased agricultural output in turn enabled the population of the Shashe-Limpopo basin to flourish (Huffman 2000, pp. 25–26, 2007, pp. 382–84; cf. Smith, Lee-Thorp, and Hall 2007).

Phytolith analysis has been uniquely applied to identify the 1st millennium presence of a settlement model known as the Central Cattle Pattern (CCP). This ethnographically derived system sets out the spatial relationship between the cattle byre, or *kraal*, and other activity areas (for example, Huffman 1998; Prins 1993; Prins and Granger 1993; van Schalkwyk 1994). To demonstrate the CCP archaeologically, a central cattle *kraal* area has to be identified, with huts and other activity areas spaced around it (Huffman 1993). This distinct spatial pattern has resulted in a large body of research aimed at identifying cattle *kraals* vis-à-vis small stock enclosures on sites. Phytolith analysis has proven useful in distinguishing between grazer and browser dung deposits in this regard. (Huffman 1993, p. 221; van Schalkwyk 1994, p. 126; but see Badenhorst 2009a for critique). However, beyond verifying spatial layout the wider application of phytolith research remains underdeveloped.

Social Identity

Social identity is a developing theme in the archaeology of farming communities, and archaeobotany can play an important role in future research, especially in the spheres of gender and ritual.

Rain-control rituals form an important part of the agricultural cycle (Krige and Krige 1943), and excavations of 9th to 13th century hilltop sites in the Shashe-Limpopo basin uncovered a range of wild and domestic plant macroremains known ethnographically to be important in such rituals. These included *Grewia flava* and *Strychnos* sp., both associated with drought prevention and *Vigna subterranea*, which is linked to lightning protection. These findings in conjunction with other evidence led to the interpretation of these sites as the locations of rain-control rituals (Schoeman 2006a).

Historically, agriculture, as elsewhere in Africa, is characterised by the gendered division of labour. Women complete most activities related to planting, processing, cooking, and storing. Evidence such as granaries and grindstones, common on farming community sites, therefore opens the possibility to actively engage with gender issues and the division of labour. Hall (1998) addresses this topic in his analysis of Tswana homesteads through the application of cognitive structures from Southern Bantu ethnography. Spatial organisation is expressed through concepts of left/right, front/back, public/private, and secular/sacred (Kuper 1980), and these opposites materialise in the layout of 18th century Tswana homesteads. Here grain bin bases indicate that cereals were stored in the back courtyard behind the hut, with lower grindstones permanently in place under the eaves of the back veranda and a small separate enclosure to the side of the hut for cooking. The activities related to the storage and preparation of plant foods all take place in the private space of the household, which is the domain of women. This layout of segmented activities can be contrasted to earlier Tswana settlements of the 13th to 15th centuries, where remains of fireplaces, permanently fixed grindstones, pots, bone waste, and sorghum imply that cereals were stored and food prepared inside the hut near a sacred, male-dominated area at the back (Hall 1998). The later segmentation indicates an increase in the physical separation of gendered activity, and Hall (1998, p. 246) relates this structured complexity to a system controlled by men through ritual that sanctioned labour and the agricultural cycle.

Recovery and Reporting

The preceding discussion serves to illustrate some of the more pertinent ways in which archaeologists have engaged with archaeobotanical data, although overall, insufficient discussions of recovery and sampling methods are major pitfalls. Additionally, samples collected during excavations are often small or lack sufficient context.

Our discussions with colleagues and reading of published data revealed some general perceptions about archaeobotany: (1) plants did not play an important role; (2) archaeobotany is of limited importance, since plant use is already documented in ethnographic sources; and (3) plant remains do not preserve well in southern Africa. These misconceptions all play a large part in the current lack of sampling and reporting of botanical information. The last one in particular relates to problems of sampling strategies and research design. The preservation of macro-remains on open-air sites in South Africa continues to be a problem (Mason 1986). However, when specific sampling strategies (including flotation) were applied, recovery of archaeobotanical material increased significantly (Maggs 1984b; Maggs and Ward 1984; Prins 1993; Schoeman 2006a).

Conclusions

For the most part, the problems experienced with archaeobotany in South Africa are not unique (cf. Archer and Hastorf 2000; Lepofsky, Moss, and Lyons 2001), and, although the conundrum we present might seem somewhat negative, it is evident that there is a useful and ever-expanding body of accumulated data available. Recent research by Scott (2005) was the first South African dissertation that not only dealt exclusively with the in-depth analysis of seeds but also tackled methodological, taphonomical, and depositional issues. Likewise, the work of Langejans (2006) and Greenfield and associates (2005) suggest that there is indeed a move into more complex and useful approaches to botanical remains. The biggest challenge for South African farming community archaeology is to assimilate archaeobotany into mainstream practice—on a par with ceramic, lithic, and faunal analyses. However, such integration can be achieved only with more appropriate recovery techniques and when farming community archaeologists utilise the full potential of macro- as well as microbotanical techniques.

Acknowledgements

We presented an earlier version of this chapter at the 5th IWAA in 2006; the National Research Foundation (South Africa) provided funding for ARA to attend the workshop. We would like to thank Christine Sievers, Neels Kruger, Jan Boeyens, and Maria van der Ryst for comments on earlier drafts.

References

Acocks, J. P. H. (1953) *Veld Types of South Africa*. Pretoria: Government Printer.

Archer, S., and C. A. Hastorf (2000) Paleoethnobotany and archaeology 2000: The state of paleoethnobotany in the discipline. *Society for American Archaeology Bulletin 18*, 33–38.

Badenhorst, S. (2009a) Phytoliths and livestock dung at Early Iron Age sites in southern Africa. *South African Archaeological Bulletin 64*, 45–50.

———. (2009b) The Central Cattle Pattern during the Iron Age of southern Africa: A critique of its spatial features. *South African Archaeological Bulletin 64*, 148–55.

Boeyens, J. C. A. (2003) The Late Iron Age sequence in the Marico and early Tswana history. *South African Archaeological Bulletin 58*, 63–78.

Carión, J. S., L. Scott, T. N. Huffman, and K. Dreyer (2000) Pollen analysis of Iron Age cow dung in southern Africa. *Vegetation History and Archaeobotany 9*, 239–49.

Coates-Palgrave, M. (2002) *Keith Coates-Palgrave Trees of Southern Africa*. Cape Town: Struik.

Cunningham, A. B. (1988) Collection of wild plant foods in Tembe Thonga society: A guide to Iron Age gathering practices? *Annals of the Natal Museum 29*, 433–46.

Cunningham, A. B., and B. R. Gwala (1986) Building methods and plant species used in Tembe-Thonga hut construction. *Annals of the Natal Museum 27*, 491–511.

Dewar, G. (2010) Late Holocene burial cluster at Diaz Street Midden, Saldanha Bay, Western Cape, South Africa. *South African Archaeological Bulletin 67*, 26–34.

Evers, T. M. (1984) Sotho-Tswana and Moloko settlement patterns and the Bantu Cattle Pattern. In M. Hall, G. Avery, D. M. Avery, M. L. Wilson, and A. J. B. Humphreys (Eds.), *Frontiers: Southern African Archaeology Today*, British Archaeological Reports International Series 207, pp. 236–47. Oxford: British Archaeological Reports.

———. (1989) *The Recognition of Groups in the Iron Age of Southern Africa*, Ph.D. thesis, University of the Witwatersrand.

Feely, J. M. (1980) Did Iron Age man have a role in the history of Zululand's wilderness landscapes? *South African Journal of Science 76*, 150–52.

———. (1985) Smelting in the Iron Age of Transkei. *South African Journal of Science 81*, 10–11.

Feely, J. M., and S. M. Bell-Cross (2011) The distribution of Early Iron Age settlement in the Eastern Cape: Some historical and ecological implications. *South African Archaeological Bulletin 66*, 105–12.

Greenfield, H. J., K. D. Fowler, and L. O. van Schalkwyk (2005) Where are the gardens? Early Iron Age horticulture in the Thukela River Basin of South Africa. *World Archaeology 37*, 307–28.

Hall, M. (1976) Dendroclimatology, rainfall and human adaptation in the Later Iron Age of Natal and Zululand. *Annals of the Natal Museum 22*, 693–703.

———. (1980) *The Ecology of the Iron Age in Zululand*. Ph.D. thesis, University of Cambridge.

———. (1984) Prehistoric farming in the Mfolozi and Hluhluwe valleys of southeast Africa: An archaeobotanical survey. *Journal of Archaeological Science 11*, 223–35.

Hall, S. (1998) A consideration of gender relations in Late Iron Age 'Sotho' sequence of the Western Highveld, South Africa. In S. Kent (Ed.), *Gender in African Prehistory*, pp. 235–58. Walnut Creek, CA: AltaMira Press.

Hanisch, E. O. M. (1980) *An Archaeological Interpretation of Certain Iron Age Sites in the Limpopo/Shashe* Valley. M.A. dissertation, University of Pretoria.

Huffman, T. N. (1993) Broederstroom and the Central Cattle Pattern. *South African Journal of Science 89*, 220–26.

———. (1997) Snakes and crocodiles: Power and symbolism in ancient Zimbabwe. *South African Archaeological Bulletin 52*, 125–43.

———. (1998) The antiquity of lobola. *South African Archaeological Bulletin 53*, 57–62.

———. (2000) Mapungubwe and the origins of the Zimbabwe Culture. *South African Archaeological Society Goodwin Series 8*, 14–29.

Huffman, T. N. (2004) The archaeology of the Nguni past. *Southern African Humanities 16*, 79–111.

———. (2006) Maize grindstones, Madikwe pottery and ochre mining in precolonial South Africa. *Southern African Humanities 18*, 51–70.

———. (2007) *Handbook to the Iron Age: The Archaeology of Pre-Colonial Farming Societies in Southern Africa*. Scottsville: University of KwaZulu-Natal Press.

Junod, H. A. (1962) *The Life of A South African Tribe*. Vol. 1. New York: University Books.

Klapwijk, M. (1974) A preliminary report on pottery from the North-Eastern Transvaal. *South African Archaeological Bulletin 29*, 19–23.

Krige, E. J., and J. D. Krige (1943) *The Realm of a Rain Queen*. Cape Town: Juta.

Kuper, A. (1980) Symbolic dimensions of the southern Bantu homestead. *Africa 50*, 8–23.

Langejans, G. H. J. (2006) Starch grain analysis on Late Iron Age grindstones from South Africa. *Southern African Humanities 18*, 71–91.

Lee-Thorp, J. A., J. C. Sealy, and A. G. Morris (1993) Isotopic evidence for diets of prehistoric farmers in southern Africa. In J. B. Lambert and G. Grupe (Eds.), *Prehistoric Human Bone: Archaeology at the Molecular Level*, pp. 99–120. Berlin: Springer-Verlag.

Lepofsky, D., M. L. Moss, and N. Lyons (2001) The unrealized potential of paleoethnobotany in the archaeology of northwestern North America: Perspectives from Cape Addington, Alaska. *Arctic Archaeology 38*, 48–59.

Maggs, T. M. O'C. (1980) Msuluzi Confluence: A seventh century Early Iron Age site on the Tugela River. *Annals of the Natal Museum 24*, 111–45.

———. (1982a) Mabhija: Pre-colonial industrial development in the Tugela Basin. *Annals of the Natal Museum 25*, 123–41.

———. (1982b) Mgoduyanuka: Terminal Iron Age settlement in the Natal grasslands. *Annals of the Natal Museum 25*, 83–113.

———. (1984a) The Iron Age South of the Zambezi. In R. Klein (Ed.), *Southern African Prehistory and Paleoenvironments*. Boston: A.A. Alkema.

———. (1984b) Ndondonwane: A preliminary report on an Early Iron Age site in the lower Tugela River. *Annals of the Natal Museum 26*, 71–94.

Maggs, T. M. O'C., and V. Ward (1984) Early Iron Age sites in the Muden area of Natal. *Annals of the Natal Museum 26*, 105–40.

Mason, R. J. (1986) *Origins of the Black People of Johannesburg and the Southern, Western and Central Transvaal AD 350–1880*, University of the Witwatersrand Archaeological Research

Unit Occasional Paper 16. Johannesburg: University of the Witwatersrand Archaeological Research Unit.

Mitchell, P. (2002) *The Archaeology of Southern Africa*. Cambridge: Cambridge University Press.

Mitchell, P., and G. Whitelaw (2005) The archaeology of southernmost Africa from c.2000 BP to the early 1800s: A review of recent research. *Journal of African History* 46, 209–41.

Mönnig, H. O. (1967) *The Pedi*. Pretoria: J. L. van Schaik.

Plug, I. (1987) *An Analysis of Witchdoctor Divining Sets*. Pretoria: National Culture History Museum.

Prins, F. E. (1993) *Aspects of Iron Age Ecology in Transkei*. M.A. dissertation, University of Stellenbosch.

———. (1994–1995) Climate, vegetation and early agriculturist communities in Transkei and KwaZulu-Natal. *Azania 29/30*, 179–86.

Prins, F. E., and J. E. Granger (1993) Early farming communities in northern Transkei: The evidence from Ntsitsana and adjacent areas. *Natal Museum Journal of Humanities* 5, 153–74.

Quin, J. P. (1959) *Food and Feeding Habits of the Pedi*. Johannesburg: Witwatersrand University Press.

Schapera, I., and A. J. H. Goodwin (1937). Work and wealth. In I. Schapera (Ed.), *The Bantu Speaking Tribes of South Africa*. London: Routledge.

Schoeman, M. H. (2006a) *Clouding Power? Rain-Control, Space, Landscapes and Ideology in Shahse-Limpopo State Formation*. Ph.D. thesis, University of the Witwatersrand.

———. (2006b) Imagining rain-places: Rain-control and changing ritual landscapes in the Shashe-Limpopo Confluence Area, South Africa. *South African Archaeological Bulletin 61*, 152–65.

Scott, C. (2005) *Analysis and Interpretation of Botanical Remains from Sibudu Cave, KwaZulu-Natal*. M.A. dissertation, University of the Witwatersrand.

Sealy, J. C. (2006) Diet, mobility and settlement pattern among Holocene hunter-gatherers in southernmost Africa. *Current Anthropology 47*, 569–95.

Smith, J., J. Lee-Thorp, and S. Hall (2007) Climate change and agropastoralist settlement in the Shashe-Limpopo River Basin, southern Africa: AD 880 to 1700. *South African Archaeological Bulletin 62*, 115–25.

Sobiecki, J. F. (2008) A review of plants used in divination in southern Africa and their psychoactive effects. *Southern African Humanities 20*, 333–51.

Stayt, H. A. (1968) *The Bavenda*. London: Frank Cass and Co.

van der Merwe, N. J., and D. Killick (1979) Square: An iron smelting site near Phalaborwa. *South African Archaeological Society Goodwin Series* 3, 86–93.

van Schalkwyk, L. O. (1994) Wosi: An Early Iron Age village in the lower Tugela Basin, Natal. *Natal Museum Journal of Humanities* 6, 65–117.

van Wyk, B. E., and N. Gericke (2000) *People's Plants*. Pretoria: Briza.

van Wyk, B. E., B. van Oudtshoorn, and N. Gericke (2005) *Medicinal Plants of South Africa*. Pretoria: Briza.

Whitelaw, G. (1991) Precolonial iron production around Durban and in southern Natal. *Natal Museum Journal of Humanities* 3, 29–39.

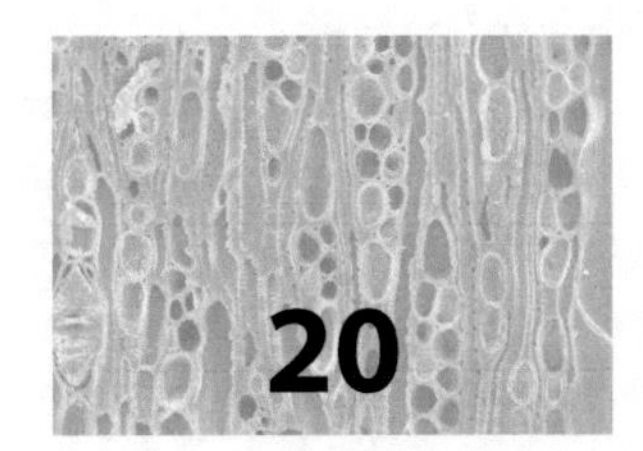

Linguistic Evidence and the Origins of Food Production in Africa

Where Are We Now?

Christopher Ehret

Diagnosing Food Production from Lexicon

The historical implications of the lexicons of food production across Africa have been surveyed on a succession of occasions over the past half century (Blench 2006; Ehret 1984; Greenberg 1964). The most recent continent-wide survey (Blench 2006) is extensive in its coverage but relies for the most part on evidence that, in itself, is nondiagnostic of food production, as its author recognizes. So what lexical evidence is diagnostic of food production? There are two requirements for tracing food production back into the past via the evidence of word histories.

The Requirement of Systematic Historical Linguistic Reconstruction

First and fundamental to the process is the existence of a historical linguistic reconstruction of the language family, or the branch of a family, in which the relevant lexicon occurs. Because sound change histories in languages proceed according to regular rules, the systematic formulation of such changes in a language family constitutes the essential analytical apparatus for determining whether the surface similarities between two words of like meaning are due to chance, to borrowing, or to actual common derivation from a root word in the proto-language. To somewhat oversimplify a complex process of analysis, words in different languages of a particular family can be considered reflexes of an ancient root word *only* if their modern-day forms show regular sound correspondences throughout the stem portion of the word. (By *stem* we mean the root element of the word, with any prefixes or suffixes disregarded; English 'oxen', for example, consists of the stem *ox-* plus an old plural suffix, *-en*.) The lack of regular correspondence at any point in a word stem normally indicates that we are dealing with a later word borrowing or with a chance resemblance. (For an introduction to this kind of analysis see Ehret 2000).

In a majority of cases, the root identifications in Blench (2006) do not meet this requirement. They are not anchored in systematic historical linguistic reconstructions. Instead, many rest on surface phonetic similarities. Blench takes explicit account of this problem by preceding his proposed roots with the sign #, in place of the asterisk used by linguists, to mark systematically reconstructed roots. Often the surface similarities in his proposed roots actually do derive from underlying systematic phonological patterns; in those cases his inferences and conclusions work. But he also groups words that, once we take into account the available published treatments of regular sound correspondences, can be shown to come from different old roots. In addition, without a systematic accounting of sound change

Archaeology of African Plant Use by Chris J. Stevens, Sam Nixon, Mary Anne Murray, and Dorian Q Fuller, Eds., 233–242

histories, one cannot confidently distinguish word borrowings from shared inheritance of a root word.

An especially arresting illustration of the degree of confusion that can arise when the evidence is not critically evaluated against a systematic phonological reconstruction is Blench's identification of a '#k-r root for goat' (Blench 2006, Table 9.14), proposed to be present in both Nilo-Saharan and Afrasian languages. This proposed item conflates the reflexes of up to seven distinct root words, including the Nilo-Saharan roots #31, 33, and 34 (Ehret 2001, pp. 496, 588, 589) presented in Table 20.1; proto-Eastern Cushitic *xolaḣ 'goat wether' (Ehret 1991); a Kanuri word *k´la/wo* for female kid, in which the actual underlying root is *-lag- and *k´- is a prefix; and two separate roots, one in Omotic and the other in the Eastern Cushitic language Burji, both of which begin with ejective /k'/. This consonant corresponds with full regularity to /k'/ in most Eastern Cushitic languages (although to /q/ in Soomaali and /q'/ in Yaaku) but never to /k/ or /x/. (See Ehret 1991, 2001; Sasse 1979 for the regular sound correspondences and morphology of these instances.)

Now a person can argue that the existing systematic language family reconstruction relevant to a particular case is wrong, or that it is contentious and that therefore one is not going to make use of it. But to do only that is to abdicate one's responsibilities in this kind of study. If a person doubts an existing systematic language family or branch reconstruction but still means to reconstruct the past from language evidence, there are just two possible courses to take:

1. immerse oneself in the evidence, test the reconstructions and the regular sound correspondence rules on which they are based against the evidence, and determine where they work and where they do not; or
2. undertake one's own full and systematic reconstruction of the relevant language family or branch of the family.

There is no third alternative. If one does not undertake rigorous linguistic historical reconstruction first, or does not make use of existing reconstructions, it does not matter how sophisticated one's understanding is of the possibilities and pitfalls of historical linguistics. One can indeed make educated guesses and, often, guesses with a high probability of being correct, but guesses they remain nevertheless.

Semantic Requirements

The second requirement is a semantic one: a root word that systematic linguistic reconstruction allows us to trace back to an earlier period in time must also specifically and always have a food-production-connected meaning.

For livestock-raising this second criterion is best filled by verbs and certain categories of nouns. Verbs that mean specifically and always 'to herd (animals)' directly confirm the tending of animals in some manner. Even more tellingly, a verb 'to milk' indicates that the process of domestication is underway. One key noun, meaning 'livestock pen', has a similar diagnostic strength for inferring livestock-raising. A still stronger criterion of livestock-raising is the presence of breeding taxonomies. In the case of cattle, to cite one example, the reconstruction of ancient terms for either of two particular breeding states—'heifer' (a female about to be of breeding age) or 'steer' (a castrated bull unable to beget young)—is a sure indicator of ancient cattle-keeping. In contrast, the presence of an old root word for 'bull' is not nearly as significant, because hunter-gatherer peoples often have specific terms differentiating the males of large, wild subsistence animals.

The reconstructed presence of an old generic term for a particular domestic animal—and this is an important point—is not diagnostic by itself of the raising of the animal. The existence of an ancient root word in the Afrasian (Afroasiatic) language family for 'donkey' (Blench 2006; Ehret 2000; Orel & Stolbova 1995) indicates no more than that the original Afrasian lands included areas where donkeys lived. Even though Afrasian peoples were certainly the domesticators of the donkey, the ancient word for the animal does not tell us when or where its domestication took place. The exceptions to this rule are cases in which we can demonstrate with certainty that the animal in question was domesticated in some other distant area. In most parts of Africa, for example, our ability to reconstruct a particular root word for the Asian-domesticated goat or sheep back to a particular point in language history would solidly indicate the arrival of domestic goats or sheep by or before that stage in time.

As for the tending and growing of plant foods, similarly, plant names are normally not diagnostic by themselves of food production, because the names could well originally have referred to their wild progenitors. Reconstructed old verbs referring to the manipulation of plant productivity usually provide the key diagnostic evidence for incipient cultivation and for cultivation. Ancient verb roots the reflexes of which in modern-day languages everywhere connote activities associated with cultivation, such as field clearing for planting, weeding, and so on, fulfill this criterion. The reconstruction of verb roots meaning specifically either 'to sow' or 'to plant' (seed or part of plant)' suggest a further inference. If one deliberately sows or plants seed or shoots, then the technology of plant manipulation has shifted toward activities that eventually bring about crop domestication. Only a single category of noun, words that consistently denote a cultivated field of some kind, can be taken as surely diagnostic of cultivation. Words for 'digging stick', for example, go just as well with plant-collecting activities as with farming. Of course, cultivated crops such as animals can and do spread to areas far away from those in which they occur wild, and when that happens, the spread of a term for a plant can be an indicator of agricultural spread.

Illustrating the Criteria: The Nilo-Saharan Case

The evolution of food-production lexicon in the history of the Nilo-Saharan family offers a long series of examples of how we can determine what is diagnostic of food production and what is not. For this family we have a full-scale, systematic reconstruction, including morphological aspects as well as phonology (Ehret 2001). For the well-known languages of the family, the sound correspondences are strongly attested. Many uncertainties remain, though, about the numerous still poorly known languages of the family. The data in the lexicon relating to food production, shown in Table 20.1, have mostly been previously published. But reconsideration of some of the earlier published materials, as well as the addition of new lexical data, provides an updated reconstruction of these terminologies. These data include several newly discovered root words relating to livestock-raising that have not yet been separately published.

The earliest creation of vocabulary diagnostic of the deliberate manipulation and tending of food sources belongs to the proto-Northern Sudanic stratum in Nilo-Saharan history. The next two subsequent periods, the proto-Saharo-Sahelian and the proto-Sahelian eras saw successively more extensive creation of new vocabulary diagnostic of plant and animal tending as well as new developments in the scale of settlement. In Table 20.1, the sets of new terms reconstructed to each of the three successive historical stages have been grouped by semantic category. The subcategories provide multiple illustrations of what constitutes diagnostic evidence and what does not. Included in the table are the names of the Nilo-Saharan subgroups from which reflexes of the particular root words are so far known. For these various groups and a family tree of their relationships, see Ehret (2006).

Leaving aside the addition of goats and sheep at the third stage in this sequence of periods, we can infer that the rest of the developments reflected in the creation of these new lexicons originated indigenously among these particular societies. The reason is that most of the newly created words consistently derive from existing Nilo-Saharan roots that previously had no reference to food production (Ehret 1999, 2000, 2001). They are, in other words, older indigenous terms readapted in meaning to specify a new set of knowledge and activities.

Furthermore, whole new semantic categories appear in the innovated lexicon of each period, localizing the particular new developments depicted by those words specifically to the periods in which the words first appear. At the proto-Northern Sudanic period, livestock tending terms and a word for 'cow' came into use; at the proto-Saharo-Sahelian stage, the first lexicons of plant tending and sedentary settlements appeared; and at the proto-Sahelian era, a lexicon relating to ovicaprids was added.

If we reexamine the time distributions of the specific kinds of diagnostic root words, we can add a new degree of nuance to the historical analysis not present in my previous treatments (Ehret 1993, 1999, 2000).

For the proto-Northern Sudanic (PNSud) period, the existence of a term for 'cow', but for no other subsequently domesticated animal, allows us to specify the cow as the animal that was herded. The existence of a PNSud root word for a thornbush livestock pen tells us that the Northern Sudanic people penned and protected the cows from predators at night—they were, at the least, beginning the process of domestication. On the other hand, our inability so far to reconstruct a breeding taxonomy for the PNSud stratum in Nilo-Saharan history may mean that the cattle tended by the PNSud society were still relatively few in number and not yet central to their subsistence. The Northern Sudanians, in other words, may have been 'pastro-foragers' (to use a term from Kuper and Kröpelin 2006)—hunter-gatherers who practised some cattle-raising and, it appears, may even have utilized cows for milk as well as meat. In addition, a reconstructed verb for 'pot-making' reveals that they possessed ceramic technology.

In contrast, elaborations of the breeding taxonomy of cattle in the subsequent proto-Saharo-Sahelian (PSS) and proto-Sahelian (PSah) periods, as well as the development of a new verb for 'milking', make the case that full domestication of cattle may have come about only in these two eras. The most striking new addition at the PSS period was a verb lexicon of plant-tending. At the least, the deriving of the new verbs, along with a noun for 'cultivated plot', in PSS imply that the proto-Saharo-Sahelian society engaged in the preparing of plots in which to encourage the growth of favored plants and in the weeding and protecting of those plants. In this period as well, a new set of terms for a more sedentary residence pattern, with enclosed homestead areas, granaries, and round houses, came into use—just the kind of settlement attested at Nabta from the

Table 20.1 Lexicon of Early Nilo-Saharan Subsistence and Settlement

A. Proto-Northern Sudanic Subsistence Lexicon

Lexicon diagnostic of livestock-raising

1. ***nd^{y}ɔw** 'to milk': Kunama; Eastern Sahelian (Taman; Jebel; Rub)
2. ***s^{y}uuk** 'to drive' (domestic animals): Kunama; Saharan; Eastern Sahelian (Nubian)
3. ***yaat 1** 'to water' (livestock): Kunama; Saharan
4. ***ɔroh** 'thornbush livestock pen': Kunama 'thornbush pen'; Saharan (Zaghawa 'livestock, herd'); Eastern Sahelian (Kir: Didinga 'village [enclosed in thornbush fence]')

Subsistence lexicon, nondiagnostic of food production

5. ***yaayr** 'cow (generic)': Kunama; Songay; Eastern Sahelian (Nara; Kir)
6. ***way** *or* ***'way** 'grain': Kunama; For; Eastern Sahelian (Astaboran, Kir)
7. ***keen** 'ear of grain': Kunama; Songay
8. ***p'ɛl** 'grindstone': Kunama; Eastern Sahelian (Kir)

Lexicon of other material culture

9. ***saap**, ***saab**, *or* ***saaƃ** 'temporary shelter': Kunama; Songay; Eastern Sahelian (Kir)
10. ***ted** 'to make pot': Kunama: Maban; Eastern Sahelian (Kir)

B. Proto-Saharo-Sahelian Stage

Lexicon diagnostic of the tending of food plants

11. ***d^{y}ip^{h}** 'to cultivate': Saharan; Songay; Eastern Sahelian (Jebel)
12. ***tɔɔk(ɔɔp)** 'to cultivate": Saharan; Songay; Eastern Sahelian (Rub)
13. ***t^{h}ayp^{h}** 'to clear plot for cultivation': Saharan; Songay; Eastern Sahelian (Jebel; Rub)
14. ***k^{h}ay** 'to clear (weeds, stubble)': Saharan; Songay; Eastern Sahelian (Kir)
15. ***ḍomp** 'cultivated field': Saharan: E'rn Sahelian (Kir)

Grain preparation lexicon, nondiagnostic of cultivation

16. ***ŋak** *or* ***ŋag** *or* ***ŋaɠ** 'to grind (grain) coarsely': Saharan; Eastern Sahelian (Kir)
17. ***p^{h}eθ** 'to winnow': Saharan; Songay; Eastern Sahelian (Kir)

Other material culture lexicon: residential

18. ***ƃoreh** 'thornbush cattle pen': Saharan; Eastern Sahelian (Rub)
19. ***k^{h}al** 'fence': Saharan; Songay; Eastern Sahelian (Kir)
20. ***dɔŋ** *or* ***ɗɔŋ** *or* **ɗ^{y}ɔŋ** 'yard, enclosure of homestead': Saharan; For; Eastern Sahelian (Kir)
21. ***ɗ^{y}or** 'open area of settlement': Saharan; Eastern Sahelian (Nubian; Jebel; Kir)
22. ***piḓah** 'granary': Saharan; For
23. ***ɗonk'ol** 'circular roll of reeds/grass that supports roof of round house': Saharan; Eastern Sahelian (Kir)

Additional livestock terminology, diagnostic of livestock-raising

(19. * **ƃoreh** 'thornbush cattle pen')

24. ***yokw** 'to herd': Saharan; Songay; Eastern Sahelian (Kir; Rub)
25. ***wer** *or* ***'wer** 'heifer': Saharan; Eastern Sahelian (Kir; Rub)
26. ***opu** 'bull': Saharan; Eastern Sahelian (Kir)
27. ***ŋgɛt̪'** 'to milk': Saharan; Eastern Sahelian (Kir)

C. Proto-Sahelian Stage

Lexicon diagnostic of sheep and goat-raising

28. ***ay** 'goat': For; Eastern Sahelian (Kir)
29. ***ɗent** 'he-goat': Songay; Eastern Sahelian (Rub)

Table 20.1 Continued

30. ***wɛd** *or* *'**w ɛd** 'sheep': For; Eastern Sahelian (Nubian; Kir-Abbaian)
31. ***meŋk** 'ram' *or* 'sheep': Maba; Eastern Sahelian
32. ***wel̪** *or* *'**wel̪** 'ram': For; Eastern Sahelian (Kir; Rub)
33. ***k'er** 'ewe-lamb, female kid': Songay; Eastern Sahelian (Kir-Abbaian)
34. ***θagw** 'young male goat or sheep (?)': Songay; Eastern Sahelian (Kir)

Additional cattle terminology, nondiagnostic

35. ***hɛw** 'cow (generic)': For; Songay
36. ***t̪ɛ** *or* ***t̪ɛh** 'cow (generic)': Maban; Eastern Sahelian (Astaboran, Kir)

Additional cattle terminology, diagnostic of cattle-raising

37. ***owiŋ** *or* ***o'wiŋ** 'bull': For; Eastern Sahelian (Kir)
38. ***maawr** 'ox': Maban; Eastern Sahelian (KA: Kir)
39. ***yagw** *or* ***yaɠw** 'young cow': Songay 'young bull'; Eastern Sahelian (Nubian 'small stock'; Kir 'heifer')
40. ***oŋa** 'heifer': For; Eastern Sahelian (Kir)
41. ***kaa** *or* ***kaah** 'cattle camp (?), cattle pen': Songay 'nomad encampment; parc à betail'; Eastern Sahelian (Kir 'homestead [enclosed in thornbush fence]')

Additional crop lexicon

42. ***ɗuT** 'a kind of gourd': Songay; Eastern Sahelian (Rub)
43. ***k^{h}ul** 'a kind of gourd': Songay; Eastern Sahelian (Kir)
44. ***Kɛdɛh** 'bottle gourd': For; Eastern Sahelian (Kir)
45. ***bud̪** 'edible gourd': For; Eastern Sahelian (Kir)

Additional lexicon diagnostic of cultivation

46. ***p^{h}ad** 'to cultivate': Songay; For; Eastern Sahelian (Kir)
47. ***t'um** 'to sow, plant': Songay; Eastern Sahelian (Kir)
48. ***p^{h}aal̪** 'bush, uncultivated land': For; Eastern Sahelian (Nubian, Kir)

Food preparation lexicon, nondiagnostic of food production

49. ***p'ent'uh** 'winnowing tray': Songay; Eastern Sahelian (Kir)

late 8th millennium B.C.E. onward (Wendorf et al., 1998, 2001). The proto-Saharo-Sahelians, the lexical evidence implies, developed a fuller-fledged cattle pastoralism, coupled with the tending and protection of at least a few of their main subsistence plants. Sorghum, its morphology not yet indicative of domestication, occurs at Nabta Playa. Was this a crop protected and tended, although not domesticated by the proto-Saharo-Sahelians?

For the subsequent proto-Sahelian (PSah) stage, the linguistic evidence supports the argument that by then further developments toward cultivation were underway. Two terms, a new verb for 'to plant, sow' and a noun distinguishing uncultivated from cultivated land, carry this implication. An interesting idea, suggested by the traceability of four separate terms for different kinds of gourds back to the PSah language, is that the selective breeding of different gourd types may already have been underway at this time and that therefore African cucurbits may conceivably have been the earliest actually domesticated species of Sudanic agriculture (see Fuller 2005, Wasylikowa & Van der Veen 2004).

The developments of the successive Northern Sudanic, Saharo-Sahelian, and Sahelian periods, it can be argued, played out in a relatively restricted set of regions. But from the end of the proto-Sahelian period onward, the linguistic evidence requires a rapid spreading out of the cattle-, sheep-, and goat-raising descendants of the proto-Sahelian society over large areas of the southern half of the Sahara.

I have previously argued in a variety of places (Ehret 1993, 1999, 2000, 2011a) that the sequence of cultural changes evident in the lexical record of the Northern Sudanic, Saharo-Sahelian, and proto-Sahelian periods of Nilo-Saharan history—in subsistence, in technology (very early ceramics), in residence, and in the timing of the appearance of ovicaprids—matches up at each significant point with the sequence of changes evident from 8500 to 6000 B.C.E. in the archaeology of portions of the eastern Sahara. The subsequent linguistically attested spreading out of the descendants of the proto-Sahelian society across large portions of the southern half of the Sahara is matched in the archaeology by the wide dispersal after 6000 B.C.E. of pastoralism across those same regions.

FOOD PRODUCTION IN THE HORN OF AFRICA

Along with the eastern Sahara, the West African savannas and the Horn of Africa seem also to have been independent major centers of agricultural innovation. The complexities of the evidence and the as-yet weak comparative linguistic apparatus for the Niger-Congo family means that the West African center cannot be adequately considered here. For the Horn, however, the linguistic tools are well developed, and extensive evidence is available. The language evidence in fact identifies two separate seminal regions of agricultural development in and around the Horn of Africa. One is attested in the linguistic evidence of the Cushitic branch of the Afrasian (Afroasiatic) language family; the other, in the evidence of the Omotic branch. For the sake of nonspecialists, we must first deal with a number of mistaken ideas that have been propagated recently concerning the origin periods of food production among peoples of the Afrasian language family.

The most serious mistaken claim, because two non-linguists of major standing in their own fields have, unfortunately, propagated it to a wider audience (Diamond and Bellwood 2003), is Alexandr Militarev's (2003) proposal that vocabulary of cultivation can be reconstructed back to the proto-Afrasian language. Leaving aside the problem that a majority of Militarev's proposed root words appear to be composites of two or more distinct proto-Afrasian roots (Ehret 2007), his items fail the very first test for being diagnostic of cultivation. None of his 30-plus putative root words consistently have solely agriculture-connected meanings. In some instances their attributed meaning is a plant name or, more often, a category of plants. In other cases Militarev attributes a cultivation meaning to a noun root, when the actual reflexes he cites may denote a cultivated field in one branch of Afrasian but a meadow or grassland in another branch. His proposed verb roots similarly are not diagnostic of cultivation, because their reflexes in different branches of the family sometimes have agricultural connotations and sometimes have reference to food collecting activities or to activities such as digging, equally applicable in pre-agricultural subsistence.

Uncritical reading of certain other sources might also convey the impression that proto-Afrasians kept livestock. In particular, Orel and Stolbova's *Hamito-Semitic Etymological Dictionary: Materials for a Reconstruction* (1995) lists in its index 29 proposed cognate sets with meanings relating to goats and sheep. But a majority of the reconstructions in the monograph cannot be taken at face value. Over 1,600 of the 2,600-plus putative roots in this book fall back on ad hoc claims of irregular sound correspondences (Ehret 2007). In keeping with these tendencies of the volume overall, 16 of the 29 proposed cognate sets relating to ovicaprids invoke irregular sound changes. Another 10 of the 29 sets are nondiagnostic even if they turn out to be valid reconstructions, because their original semantic application could have been to any of several animals, at least some of which existed wild in Africa. Just three of the root words for ovicaprids in Orel and Stolbova appear to be validly diagnostic of the presence of one or both of these animals fairly early on.

Of these three root words—*bag- 'sheep', *bok- 'he-goat', and *kVrr- 'lamb' (Orel and Stolbova items 173, 309, and 1432)—the last two have pan-regional distributions that fit with their having spread long ago by diffusion into and across parts of the Sahara, presumably accompanying the original spread of ovicaprids into those regions. The remaining root, *bag-, occurs not only in Berber and Chadic languages but also in Ethiopia in the Agaw sub-branch of Cushitic, in the Ethiopian Semitic languages, and in a few languages of the Omotic primary branch of the family. Before the 16th-century Oromo expansions, however, the Ethiopian occurrences of *bag- formed a tight, continuous, unbroken distribution. This pattern argues for diffusion by borrowing to Ethiopian Semitic and Omotic from Agaw (see Ehret 1999).

An old word for 'cow' also reconstructs to an early period in Afrasan. The regular sound correspondences between the Cushitic and Chadic reflexes of this root clearly establishes its original shape as *ɬ(the sign ɬ is pronounced like the double letter, *ll*, of Welsh). Most writers (Blench 2006, p. 161; Orel and Stolbova 1995) have presumed that the resemblant proto-Semitic term *ɬaʔ 'cow belongs to this root. But the regular Semitic sound correspondent of proto-Cushitic and proto-Chadic *ɬ is not *l but proto-Semitic *ś (see Ehret 1995; Steiner 1977). So proto-Semitic *ɬaʔ 'cow is either a chance resemblance to old Afrasian *ɬo 'cow' or a borrowing. Since cows were wild animals of the northern Sahara and, at the beginning of the Holocene wet phase, of the central and southeastern Sahara as well, the Cushitic and Chadic sharing of a common term for 'cow' cannot be considered diagnostic of cattle-keeping.

To sum up, the evidence overall does not support the postulation of proto-Afrasian livestock-raising, although a very early Afrasian acquaintance with wild donkeys is clear (Ehret 2000). Afrasian peoples gained knowledge of wild cattle at a later period in their history. In addition, the great majority of scholars who have studied this issue do not doubt that the proto-Afrasian origin areas lay in Africa and, within the continent, most probably in or adjacent to the Horn of Africa (for example,, Blench 2006; Diakonoff 1981, 1998; Ehret 1980b; Ehret, Keita, and Newman 2004; Fleming 1983 place the homeland farther north toward Egypt but still in Africa).

Having dealt with the issues of Afrasian origins and early subsistence, we can again return to the primary focus of this chapter, the lexical evidence that *is* diagnostic of food production.

Table 20.2 Early Cushitic Subsistence Lexicon

A. Proto-Cushitic Stage (pre-6000 b.c.e.)

Lexicon of livestock, nondiagnostic of herding

1. *šaʕ- 'cow'
2. *ɬo- 'cow' (earlier Afrasian root word)
3. *ħarle '(wild?) donkey'

Lexicon of livestock, diagnostic of herding

4. *galaal- 'to herd, take to pasture'
5. *der- 'to herd'
6. *dall- 'livestock pen'
7. *yaw- or *ʔaw- 'bull'
8. *ʔayz- 'sheep'
9. *ʔanaaʕ- 'goat, sheep'
10. *ʔaff- 'kid, lamb'"
11. *rangan- 'ewe-lamb, young female kid'
12. *ʔogr- *or* *ʔorg- 'he-goat'

Grain lexicon, nondiagnostic of cultivation

13. *ʕag- 'sorghum (bicolor?)';
14. *ʕayl- 'grain species'
15. *bil- *or* *bal- 'grain plant'
16. *puzn- 'flat bread'

B. Proto-Agaw-Southern Stage (5000 B.C.E.??)

Livestock lexicon, diagnostic of herding

17. *mawr- 'cattle pen'
18. *mal- 'to watch livestock'
19. *dak^{w}'- '(domestic?) donkey'

Grain lexicon, diagnostic of cultivation

20. *ʔabr- *or* *ʔirb- 'to cultivate'
21. *baayr- *or* *paayr- 'cultivated field'

Grain lexicon, nondiagnostic of cultivation

22. *dingawc- *or* *dangawc- 'finger millet'
23. *tl'eff- 'grain species'
24. *muuɬɬ- 'legume sp. (pigeon pea?)'

CUSHITIC INNOVATIONS IN FOOD PRODUCTION

The proto-Cushitic language dates back to the Early Holocene and was spoken in or near the southern Red Sea hills. Inferences from the linguistic geography of the deep branches of Cushitic best locate the proto-Cushitic society in the still archaeologically almost unknown areas north of Eritrea (argued *in extenso* in Ehret 1999, 2000, 2006). Its chronological placement follows from the timing of the proto-Sahelian borrowing of a particularly salient word, the generic term *ay for 'goat' (see Table 20.1 and also Table 20.2), from an early language of the Northern branch of Cushitic. The borrowing of this word has two implications:

1. that goat-raising diffused to the proto-Sahelians via the Northern Cushites;
2. that the proto-Cushitic society had already diverged into its primary daughter branches, one of which was Northern Cushitic, *before* the late 7th millennium B.C.E., because that is the period of the first appearance of goats in the archaeology of the probable eastern Saharan homeland areas of the proto-Sahelians (Kuper and Kröpelin 2006).

Contacts between the early Northern Cushites and the proto-Sahelians require locating the Northern Cushites of ca. 6000 B.C.E. also in the eastern Sahara, presumably east of the Nile (that is, northward from Eritrea).

After the initial breakup of the proto-Cushitic society, Cushitic-speaking populations spread successively over a period of several thousand years, first probably around the northern edges of the Ethiopian highlands and then into the areas in and to the east of the Ethiopian Rift Valley (Ehret 1976, 1999). The last major southward extension of Cushites, a settlement for which we do have an archaeological dating correlation (Ambrose 1982), was that of the Southern Cushites, who first moved southward into northern East Africa around the late 3000s B.C.E.

Linguistic reconstruction of proto-Cushitic subsistence lexicon reveals the early Cushitic societies to have been above all pastoralists, with cattle, sheep, and goats, and with the addition of the donkey at an early period, although not necessarily as early as the proto-Cushitic era (Table 4a). Along with livestock-raising, an important source of food in their diet came from the collection of wild grains (Ehret 1999, 2006). The areas extending from the middle Red Sea Hills south to Eritrea, as the likely early Cushitic lands, are crucial places for future archaeozoological and archaeobotanical research.

After the initial breakup of proto-Cushitic into two daughter languages, proto-Northern and proto-Agaw-East-South Cushitic, a second kind of development in food production began (Table 4b). This is attested in new words developed in the proto-Agaw-East-South Cushitic language, which was most likely spoken along the northern edges of the Ethiopian highlands. To that language we can reconstruct a verb "to cultivate" and a noun for a cultivated plot (Table 20.2A). Finger millet (*Eleusine coracana*) and *Eragrostis tef*, both domesticated in the Ethiopian highlands, are possible candidates for the early important plants cultivated by these communities (see Appleyard 2006; Ehret 1980a, 1987, 1991; Ehret and Ali 1985; Sasse 1979). A much more extensive listing of the relevant reconstructed roots for each stage in early Cushitic linguistic history, along with supporting data and sound correspondence charts for all the Cushitic languages, appears in Ehret 2011b.

The Origins of Omotic Agriculture

The Omotic peoples of the southern Ethiopian highlands are almost universally associated today with an agriculture of entirely different inspiration, in which the enset plant is the staple. The enset, which resembles and is related to the banana, is valued not for its fruit but for its large corm, the esculent portion of which extends up into the lower stem of the plant. It has long been thought that the cultivation of this plant may have initiated a separate invention of agriculture in southern Ethiopia some thousands of years ago (Simoons 1964).

The Omotic branch of the Afrasian family certainly goes back thousands of years in southern Ethiopia; a time frame of the 6th millennium B.C.E. for the proto-Omotic society would not be out of the question. The median minimal cognation in the 100-word basic-meaning (Swadesh) list between languages of the two primary branches of the family runs at just below 10%, significantly lower than the middle-teens percentages typical of the Indo-European family. Proto-Indo-European, from the evidence of its technological lexicon, must date to just around 4000–3500 B.C.E. (contra Colin Renfrew 1987; for a penetrating statement of the evidence see Anthony 1995; see also Comrie 2003). Proto-Omotic, with its significantly lower percentages of cognation, must be placed earlier than proto-Indo-European.

Shiferaw Assefa (2011) brings a large body of new lexical evidence to bear on the sequence of developments in this separate emergence of agriculture. The early Omotic societies took shape, it is clear, in the farthest southwest corner of the Ethiopian highlands, a region of tropical mountain rainforest environments broken by deep, much drier river valleys. From the proto-Omotic era of perhaps the 6th millennium B.C.E. down through the proto-North Omotic era of around the 4th millennium, Omotic communities remained restricted to those regions.

Assefa's evidence indicates that enset was already an important food source at the proto-Omotic period. The proto-Omotic language possessed at least two terms for parts of the enset plant, including a term that most probably connoted the edible corm and inner stalk, and, in addition, terms for tools used uniquely in the processing of enset for food. At the least, enset would seem to have been a major foraged food source for the original proto-Omotic society, although whether cultivated or not is unknown.

The earliest root word certainly diagnostic of cultivation traces back to the second major era in Omotic history, the proto-North Omotic period, of roughly the 4th millennium B.C.E., 2,000 years or more after the proto-Omotic era. This particular word, a verb *tokk-, specifically connoted the planting, as opposed to the sowing, of a crop, exactly the kind of activity that enset cultivation involves. The proto-North Omotic language also possessed terms for each of the four major stages in the growth of enset to productive maturity. Enset cultivation, it seems, had become a complex, fully established productive system sometime before the close the proto-North Omotic era.

Assefa shows that several additional elements began to be blended into this subsistence system during the same era. Terms for cows and milking of cows and for sorghum both trace back to proto-North Omotic. The ethnographic evidence indicates that Ethiopian yam species also had a place in the subsistence system by this time, but the lexical evidence in this case requires further study.

The knowledge of sheep and goats may possibly go back even earlier. The earliest word for sheep in Omotic appears to be a borrowing, specifically from a Nilo-Saharan language spoken in South Sudan not far from the early Omotic lands, and sorghum is likely also to have been an introduction from Nilo-Saharan peoples living to the immediate west of the southern Ethiopian highlands. Assefa's interpretation is that the early Omotic farmers may particularly have exploited the altitudinal transition zones of their region, cultivating sorghum in fields lower down their mountains, and enset higher up.

Beginning around 1000 B.C.E., a new eastward spread of Omotic peoples into the highlands around Ethiopian Rift Valley transferred the knowledge and practices of enset-raising to the Eastern Cushitic peoples of those areas and set in motion a long era of extensive agricultural interchange and intensification across the southern Ethiopian highlands (Assefa 2011).

CONCLUDING THOUGHTS

Overall, the various bodies of lexical evidence—strong in the case of Nilo-Saharan; very strong for Cushitic; and strong for Omotic—especially direct our attention toward the urgency of discovering much more about the plant-based aspects of subsistence in the archaeobotany of the African Early Holocene. We can say that the evidence of early cultivation is simply not there in the material record. The linguistic record, however, avers that evidence once did exist even if it does not any more. Is the evidence simply gone in some regions? Or are the difficulties in finding it more a reflection of the vastness of the landscape and the fewness of the workers. For the periods before domestication changed the genetics and phenotype of a crop, how do we distinguish the cultivation and gathering of a plant from just gathering it? How, especially, can we uncover such distinctions in higher rainfall regions of West Africa or southern Ethiopia, where decay of vegetative material is rapid and, moreover, where yams are likely to have been the early staple? What new kinds of indirect indicators of cultivation in Africa might we learn to recognize? The possibilities beckon.

REFERENCES

Ambrose, S. H. (1982) Archaeology and linguistic reconstructions of history in East Africa. In C. Ehret and M. Posnansky, *The Archaeological and Linguistic Reconstruction of African History*, pp. 104–57. Berkeley and Los Angeles: University of California Press.

Anthony, D. W. (1995) Horse, wagon and chariot: Indo-European languages and archaeology. *Antiquity* 59: 554–65.

Appleyard, D. (2006) *A Comparative Dictionary of the Agaw Languages*. Köln: Rüdiger Köppe Verlag.

Assefa, S. (2011) *Omotic Peoples and the Early History of Agriculture in Southern Ethiopia: A Linguistic Approach*. Ph.D dissertation, University of California at Los Angeles.

Blench, R. (2006) *Archaeology, Language, and the African Past*. Lanham, MD: AltaMira Press.

Comrie, B. 2003. Farming dispersal in Europe and the spread of the Indo-European family. In P. Bellwood and C. Renfrew (Eds.), *Examining the Farming/Language Dispersal Hypothesis*, pp. 409–19. Cambridge: McDonald Institute for Archaeological Research.

Diakonoff, I. 1981. Earliest Semites in Asia: Agriculture and animal husbandry, according to the linguistic data. *Altorientalische Forschungen 8*: 23–74.

———. 1998. The earliest Semitic society. *Journal of Semitic Studies 43*: 209–19.

Diamond, J., and P. Bellwood (2003) Farmers and their languages: The first expansions. *Science 300* (5619): 597–603.

Ehret, C. (1976) Cushitic prehistory. In M. L. Bender (Ed.), *The Non-Semitic Languages of Ethiopia*, pp. 85–96. East Lansing: Michigan State University.

———. (1980a) *The Historical Reconstruction of Southern Cushitic Phonology and Vocabulary*. Berlin: Reimer.

———. (1980b) Omotic and the subclassification of the Afroasiatic language family. In R. Hess (Ed.), *Proceedings of the Fifth International Conference on Ethiopian Studies, Session B*, pp. 51–62. Chicago: University of Illinois.

———. (1984) Historical/linguistic evidence for early African food production. In J. D. Clark and S. Brandt (Ed.), *From Hunters to Farmers*, pp. 26–35. Berkeley and Los Angeles: University of California Press.

———. (1987) Proto-Cushitic reconstruction. *Sprache und Geschichte in Afrika* 8: 7–180.

———. (1991) Revising the consonant inventory of proto-Eastern Cushitic. *Studies in African Linguistics 22* (3): 211–75.

———. (1993) Nilo-Saharans and the Saharo-Sudanese Neolithic. In T. Shaw, P. Sinclair, B. Andah, and A. Okpoko (Eds.), *The Archaeology of Africa: Food, Metals and Towns*, pp. 104–25. London: Routledge.

———. (1995) *Reconstructing Proto-Afroasiatic (Proto-Afrasian): Vowels, Tone, Consonants, and Vocabulary*. Berkeley and Los Angeles: University of California Press.

———. (1999) Who were the rock artists? Linguistic evidence for the Holocene populations of the Sahara. In A. Muzzolini and J.-L. Le Quellec (Eds.), *Symposium 13d: Rock Art and the Sahara*. In *Proceedings of the International Rock Art and Cognitive Archaeology Congress*. Turin: Centro Studie Museo d'Arte Prehistorica.

———. (2000) Language and history. In B. Heine and D. Nurse (Eds.), *African Languages: An Introduction*, pp. 272–97. Cambridge: Cambridge University Press.

———. (2001) *A Comparative Historical Reconstruction of Proto-Nilo-Saharan*. Köln: Rüdiger Köppe Verlag.

Ehret, C. (2006) Linguistic stratigraphies and Holocene history in northeastern Africa. In M. Chlodnicki and K. Kroeper (Eds.), *Archaeology of Early Northeastern Africa*, pp. 1019–55. Posnan: Posnan Archaeological Museum.

———. (2007) Applying the comparative method in Afroasiatic (Afrasan, Afrasisch). In R. Voigt (ed.), *From Beyond the Mediterranean: Akten des 7. internationalen Semitohamitistenkongresses* (VII. ISHaK), Berlin, September 13–15, 2004, pp. 43–70. Aachen: Shaker Verlag.

———. (2011a) *History and the Testimony of Language*. Berkeley and Los Angeles: University of California Press.

———. (2011b) A linguistic history of cultivation and herding in northeastern Africa. In A. G. Fahmy, S. Kahlheber, and A. C. D'Andrea (Eds.), *Windows on the African Past: Current Approaches to African Archaeobotany*, Vol. 3. Frankfurt am Main: Africa Magna Verlag.

Ehret, C., and M. N. Ali (1984) Soomaali classification. In T. Labahn (Ed.), *Proceedings of the Second International Congress of Somali Studies*, Hamburg, August 1983, Vol. 1, pp. 201–69. Hamburg: Buske Verlag.

Ehret, C., P. Newman, and S. Y. O. Keita (2004) The origins of Afroasiatic. *Science 306* (3): 1680–81.

Fleming, H. C. (1983) Chadic external relations. In E. Wolff and H. Meyer-Bahlburg, *Studies in Chadic and Afroasiatic Linguistics*. Hamburg: Buske.

Fuller, D. Q (2005) Crop cultivation: The evidence. In K. Shillington (Ed.), *Encyclopedia of African History*, Vol. 1 A-G, pp. 326–28.

Greenberg, J. H. (1964) Historical inferences from linguistic research in sub-Saharan Africa. In J. Butler (Ed.), *Boston University Papers in African History 1*: 1–15.

Kuper, R., and S. Kröpelin (2006) Climate-controlled Holocene occupation in the Sahara: Motor of Africa's evolution. *Science 313*: 803–07.

Militarev, A. (2003) The prehistory of a dispersal: The proto-Afrasian (Afroasiatic) farming lexicon. In P. Bellwood and C. Renfrew (Eds.), *Examining the Farming/Language Dispersal Hypothesis*, pp. 135–50. Cambridge: The MacDonald Institute.

Orel, V. E., and O. Stolbova (1995) *Hamito-Semitic Etymological Dictionary: Materials for a Reconstruction*. Leiden: E. J. Brill.

Renfrew, C. (1987) *Archaeology and Language: The Puzzle of Indo-European Origins*. Cambridge: Cambridge University Press.

Sasse, H.-J. (1979) The consonant phonemes of proto-East Cushitic: A first approximation. *Afroasiatic Linguistics 7* (1): 1–66.

Simoons, F. J. (1964) Some questions on the economic prehistory of Ethiopia. *Journal of African History* 6 (1): 1–13.

Steiner, R. C. (1977) *The Case for Fricative-Laterals in Proto-Semitic*. New Haven, CT: American Oriental Society.

Wasylikowa, K., and M. van der Veen (2004) An archaeobotanical contribution to the history of watermelon, Citrullus lanatus (Thunb.) Mats. & Nakai (syn. C. vulgaris Schrad.). *Vegetation History and Archaeobotany 13* (4): 213–17

Wendorf, F., and R. Schild (1998) Nabta Playa and its role in the northeastern African history. *Anthropological Archaeology* 20: 97–123.

Wendorf, F., R. Schild, et al. (2001) *Holocene Settlement of the Egyptian Sahara*, Vol. 1. *The Archaeology of Nabta Playa*. New York: Kluwer Academic/Plenum Publishers.

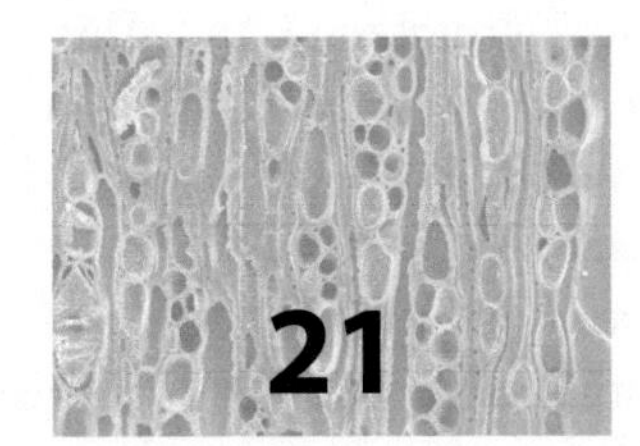

African Agricultural Tools

Implications of Synchronic Ethnography for Agrarian History

Roger Blench

Although the last few years have seen considerable advances in African archaeobotany, so that we now have a broader picture of the evolution of African agriculture from the point of view of crops, our understanding of the techniques informing that agriculture remains poor. Although Africa has a rich diversity of agricultural tools, they are known principally from synchronic descriptions rather than excavation.

A certain urgency is suggested by the rapid erosion of traditional tool production and use. Animal traction and tractors are replacing hand tools in some areas, but probably more significant is the replacement of implements made locally by blacksmiths with standardised factory-made tools. This replacement can follow from aid projects but is also often a consequence of social disruption and war. Once NGOs and international agencies get into the business of resupplying communities following civil society they do not often enquire closely into traditional implements but supply those easily available from industrial sources. There is also a noticeable difference between the ethnographic tools illustrated in early monographs or taken from 19th- and early-20th century collections and those in use today, even if they have been made by 'traditional' blacksmiths. The growth of the nation-state, with improved long-distance trade, agricultural schools, and development projects, has tended to make the tools more uniform over much greater areas. The availability of scrap iron and improved blacksmithing techniques have made possible greater specialisation and economies of scale, and this is affecting tool repertoires.

Despite these changes, the majority of African farmers probably still use some traditional tools and have them repaired by village blacksmiths, so that evidence can be recovered for their names, construction, and use. However, this information has remained a poorly exploited source of data for archaeologists and prehistorians. This chapter is a preliminary survey of the tools in use, their classification, and the hypotheses that can be suggested concerning their evolution and development. In assessing whether African tools are essentially indigenous or have spread from outside the region, it is useful to have comprehensive comparative materials. Unfortunately, these are also sparse. Roman implements have been well covered in White (1967, 1975), and Stuhlmann (1912) is a valuable guide to the Maghreb, but many questions posed by speculations in this chapter have no immediate answers.

The definition of African tools can also be rather fluid; almost anything can be developed into a hoe or an earth-shaping tool. Dupré (2000) illustrates this point with the agricultural knife of the Congo, which at one extreme resembles the bush-cutting knife but in some examples develops a wide blade that can also be used for planting and uprooting. She calls this agricultural knife an *outil polyvalent*, which seems appropriate. A similar case is digging sticks; those on the edge of the Sahara have gradually developed extra wide blades and now often resemble long-handled hoes. Thus it is always helpful to investigate tools in use, rather than to assume that their function can be deduced from their morphology.

Archaeology of African Plant Use by Chris J. Stevens, Sam Nixon, Mary Anne Murray, and Dorian Q Fuller, Eds., 243–258

A perplexing problem in describing agricultural tools is that many African types have no standardised name. French scholars, who have been more active in this area of research, have devised a number of terms, often by reexcavating old rural names, but these have yet to be adapted into English. Wigboldus (2000), attempting to describe the wooden spade-like tool used in the Sahel, proposed the term 'long-handed scuffle' but later admitted defeat and returned to *iler*, a regional term. Nonetheless, this seems unsatisfactory; this chapter makes some further efforts to introduce descriptive terms for African tools.

Three Phases of the Availability of Iron

Although agricultural tools can clearly also be made of stone and wood, there is little doubt that African agriculture was transformed by the production of iron and that this development led to a major diversification of tool morphology. Unlike the Mediterranean world, in Africa copper was widely smelted contemporaneously with iron, but copper seems never to have been used for tools and only occasionally for weapons (Cornevin 1993). The only exception is in the subdesertic regions, especially Mauritania, where a widespread industry existed from ca. 2000 B.C.E., using both native copper and later copper alloys. Lambert (1983, pp. 79–80) illustrates a number of copper implements (with arsenic) that she describes as *haches*. The illustrations are not clear enough to allow us to be sure, but some of the triangular blades with truncated tops could be hoe blades. The method of fixing these blades to a handle is uncertain. The great majority of tools in these Mauritanian finds are weapons, arrow and spear tips, so if there was an excursion from copper into agricultural tools it made almost no impact on sub-Saharan Africa.

Iron became available in three significantly different phases, and this scale of availability has had different effects on the types and numbers of tools created by blacksmiths. These phases can be delineated as follows: first, the introduction of iron smelting on a small scale (ca. 500 B.C.E.); second, the increase in the availability of iron with the import of pig-iron from the 16th century onward; third, the availability of iron on a large scale (ca.1950 C.E. onward) through access to scrap iron from discarded industrial products. Needless to say, the spread of iron was geographically determined, and the more remote an area, the longer it took for the effect of cheaper iron to take effect. For example, iron began to be traded from the coast in the 16th century, but it was not until the 1930s that it competed effectively with locally smelted iron in the interior, causing smelting industries to decline and eventually disappear. This process of elimination is now virtually complete, but in an extremely remote area in southwest Ethiopia, smelting still competes with imported scrap, because the nearest road is still a week away on horseback, making transport costs very high (Haaland 2004). A fourth phase could also be suggested—the purchase of finished iron products from European industries. It seems that the first item in this category to be imported was the cutlass, or *panga*. Other imported iron tools were the ploughshare, the harrow, and in more recent times, spades, shovels, rakes, and various types of hoe blade. The significance of these has been highly variable according to how useful farmers perceive them to be. Imports appear to have been much more influential in eastern and southern Africa than in West Africa.

Ethnographic and Archaeological Evidence for Tools

The use of material culture from the recent past is still a store of information largely unexploited by archaeologists. It is sometimes thought that ethnoarchaeology covers this entire field, but, in fact, the emphasis on pottery and house-forms has been pursued to the near exclusion of all other types of evidence (Blench 2006). Indeed, the mapping of existing African agricultural tools and their associated terminology is still in its infancy. There are, however, a variety of ethnological descriptions and overviews that are useful background material. The German ethnologists took considerable interest in this topic: Baumann (1944) published a very detailed description of the morphology and distribution of African farmers' tools. Two edited volumes in French provide rich material as yet unmined by archaeologists: Seignobos (1984); Seignobos, Marzouk, and Sigaut (2000). The latter material has the classic problem of Francophone publications—their tendency to halt at linguistic boundaries. Papers use 'Northern Cameroun' as a unit of analysis, even though there is no evidence that such a boundary is relevant to the tools under discussion. Like so much in the field of material culture, documentation is urgently required, because there is a significant process of homogenisation at work, even where blacksmiths are still making tools. Descriptions of tools are scattered in hard-to-obtain monographs, such as Coulibaly (1978) for the Senufo in Cote d'Ivoire and Mudindaambi (1976) for the Mbala in DRC. Seignobos (2000) documents this process in Northern Cameroun, where a relatively few tool types are becoming dominant, and the variety that he illustrates is gradually disappearing. Throughout the continent, factory-made tools and tractors are replacing traditional cultivation techniques.

A striking feature of African agricultural tools is the comparative rarity of preexisting models in other materials. Although agriculture clearly preceded iron, we have only a sketchy idea of what tools were in use before the introduction of iron. It is possible to make a wooden hoe blade for use in light, sandy soils, but whether wooden hoes preceded iron ones is doubtful. Although stone sickles for cutting grass existed as far back as 10,000 B.C.E. in West Africa (Shaw and Daniels 1984), these were not the precursors of the iron sickle of the present Sahel, which is a late trans-Saharan introduction. But it seems doubtful that many of the techniques characteristic of African agriculture could be pursued without iron tools—for example, the raising of large furrows and yam mounds. A case where it is possible to see something of this limited pre-iron repertoire is Fernando Po. This island was settled by a Stone Age Bantu group, the Bubi, some 3,000–4,000 years ago. Although the Gabonese Fang people reached the island prior to European contact (supposedly 800 years ago), they brought little iron, with the result that most Bubi were still using lithic technology when Europeans first made contact (Tessmann 1922).

THE EARLY RARITY OF IRON AND ITS VALUE

Iron produced by smelting is a lengthy process, and, particularly when smelting was first introduced, iron was presumably rare and costly. One of the consequences of this was that the metal was constantly reused. Hoe blades would have been forged over and over again, and when the blade became too fragile, its pieces would have been made into ornaments and other items unrelated to tools. As a consequence, remnants of agricultural tools are rarely found in early sites, even where furnace remains show that iron must have been produced in quantity. As skills developed, iron production was gradually on a larger and larger scale, leading to subindustrial sites. One of the more well-known sites is Meröe in Upper Nubia, from the last centuries B.C.E. and early centuries C.E. (Cornevin 1993, p. 141; see Fuller, Chapter 14 this volume). Somewhat later is the Igbo site of Leja, which although it has dates as early as 200 cal B.C.E., probably began high volume production in the 15th and 16th centuries, where most dates are clustered (Okafor 1993, p. 438). Sukur, in the Mandara mountains of Northern Nigeria, probably became a major producer in the 17th century (David 1996, p. 598). Nonetheless, until the immediate precolonial period, iron remained a rarity in many remote societies and is difficult to trace in the history of tools.

Figure 21.1 'Marriage' hoe from the Mandara mountains (drawn from author's photo).

The value of iron in the era of smelting and its use in agriculture soon became related to iron's use in currency systems. Although there is some evidence for the use of copper in trade, notably the Katanga crosses, iron was probably more important as a local currency. The most well-known example of this use is the 'Kissi penny' or *guinzé*, a long, thin strip of iron with a flattened end used in exchanges in a zone between the Liberian coast and Southern Guinea at the end of the 19th century (Béavogui 2000). The exact antiquity of this device is hard to gauge, since it probably reflects the abundance of iron following post-European imports. Hoe blades were frequently used in currency-like contexts, such as bride-price payments, throughout much of West Africa. However, as the culture of ritual exchange developed, and as the total amount of iron in circulation gradually increased, ritual blades became morphologically transformed until they were no longer useful as hoes but functioned only within the context of exchange. The Mandara mountains in Northern Cameroun are particularly notable for the wide range of hoe-like objects (see Seignobos 2000) manufactured and circulated (Figure 21.1).

HOE CULTURE AND THE DIVISION OF LABOUR

Agricultural tools do not exist in a sociological vacuum; farm labour in Africa has always been strongly divided along gender lines, and this situation is frequently reflected in the tools themselves. The discussion about the division of labour was taken up in detail by Baumann (1928) and later Goody (1976). Broadly speaking, men tend to do the 'heavy lifting', clearing the bush and raising large furrows and yam heaps, while women tend kitchen gardens and carry out secondary tasks such as weeding. In early versions of this argument, hoe culture was connected to 'matriarchy' [matriliny in modern terms], but whether such correlations are useful can be debated. There are many variations on this pattern, from some of the extreme societies in the Senegambian region where women carry out virtually all agricultural tasks to societies where women are not allowed on the farm. The advent of a puritanical Islam has also effected a significant transformation in parts of West Africa, where men have increasingly taken over all farming tasks, considering that to expose women outdoors contravenes Islam. The consequence of this division is that many societies have 'men's' and 'women's' tools. The focus of this chapter is on the gross morphology of tools, which often does not reflect this difference, but typically men's hoes and axes have different shapes and weights and may be adapted to slightly different tasks.

LINGUISTIC EVIDENCE

Despite the importance of African agricultural tools, historical linguists have so far ventured very few reconstructions (see also Ehret, Chapter 21 this volume). Table 21.1 (see also Figure 21.2) shows all the proto-forms in the Bantu language groups relating to agricultural practices and tools. The complex of terms around farming and cultivation, attested in A and B groups/zones close to the Bantu homeland, argues fairly convincingly that the proto-Bantu had some form of agriculture (see also Bostoen, Ehret this volume). There is an overlap of words for 'hoe', 'axe', and 'razor', especially partway through the Bantu expansion (group/zone C onward). This overlap probably corresponds to the period of the introduction of iron tools, some 2,500 years ago. Initially, they would have been rare and expensive, and there would have been a tendency to call them by the name of their material, leading to a polysemy that is uncommon in the present.

Another important study is that by Tourneux (1984) of the names of agricultural implements in Northern Cameroun. Although the languages he considers are quite closely related, the vernacular terms are very diverse, making it difficult to extract useful historical information. Two lessons can be learned from this: linguistic sources are often not very accurate in terms of descriptions of material culture, and there is considerable shifting of terms from one implement to another.

Table 21.1 Bantu Reconstructions Indicating Agriculture (zones are shown in Figure 21.2)

Root	Gloss	*Form	Zones	Regions
I	hoe, axe	bàgò	A J P	NW NE SE
		bògà	A B	NW
II	hoe	cúkà	C F G J L M S	NW C NE SE
		kácù	D K L M	NC
		púkà	A J	NW, NC
III	cultivate (especially with hoe)	dɪ̀m	B C E F G J K L M N P R S	Throughout
	cultivated field	dɪ̀mɪ̀	J L M	NC
	field sp.	dɪ̀mɪ̀dò	J	NC
	cultivated field	dɪ̀ma	J S	NC
	field sp.	dɪ̀mé	J L M	NC
	farmer	dɪ̀mì	J L	NC
	work	dɪ̀mò	C F G H J K L M N S	Throughout
IV	hoe; axe; spear-head; knife	gèmbè	C D E F G J M P	NW C NE SE
	shave; cut hair	gèmb	J	NC
	razor	gèmbè	D F J L	NE
	axe; hoe	dèmbè	S	SE
	axe; hoe	jèmbè	E G L M N S	NE C SE

Source: Bantu Lexical Reconstructions, 3rd ed. (BLR3) maintained by the Musée Royale de l'Afrique Centrale (MRAC), http://linguistics.africamuseum.be/BLR3.html.

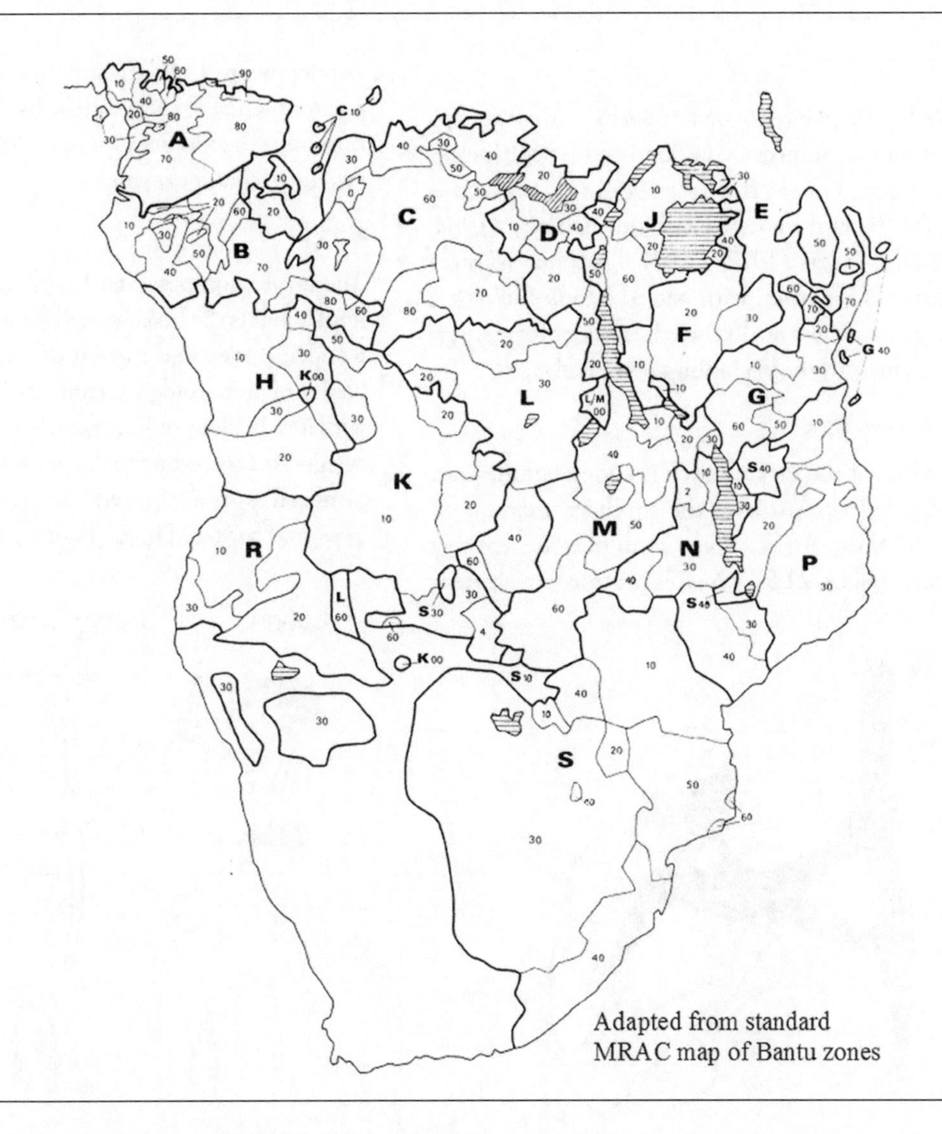

Figure 21.2 Map of Bantu zones defined by Guthrie. **Bold** letters mark Guthrie/Tervuren zones (see also Table 21.1), and numbers mark subgroups within those zones.

DIGGING STICKS AND PLANTERS

DIGGING STICK

The simplest tool still in use in agriculture is the digging stick. In its most rudimentary form, it is a stick with a pointed end, used for making holes in which seeds can be dropped. More complex sticks have iron tips, and these can gradually widen and flatten out until they resemble long-handled hoes, used for more sophisticated earth manipulation. Some digging sticks still in use in the 20th century had perforated stone weights on the top of the stick to increase the penetration of the pointed end. Such implements are recorded from Ethiopia (Gascon 1977). Perforated ring-stones have been recorded from a number of archaeological sites in Africa. Digging-ticks are pre-agricultural, used, for example, by foragers to dig out yams. Vincent (1985) records the Hadza of Central Tanzania digging for tubers with sharp, pointed sticks. However, the Hadza use decidedly modern cutlasses to sharpen the stick as they dig, so this practice cannot be a model for the pre-Neolithic. The Bubi of Fernando Po still used an all-wooden digging stick in the 1920s when Tessmann (1922) visited them (Figure 21.3).

Figure 21.3 Wooden digging stick from Fernando Po (after Tessmann 1922).

Planter

Across the Sahel, pointed wooden sticks with iron tips are used as planters, especially in flood-retreat cultivation (Figure 21.4; for example, Raynaut 1984). This system is described for Senegal by Lericollais and Schmitz (1984, p. 440). Wente-Lukas (1977, p. 92) illustrates planters from Northern Cameroun with angled handles like that of a walking cane. These are likely to be ancient forms, perhaps coincident with the beginnings of agriculture.

Arrowhead Digging Stick

A development of the digging stick is a long handle with an arrow-headed tip, used on the southern margins of the desert in West Africa. Some examples are entirely made of iron (Figure 21.5), which is presumably a recent development. They are suitable for turning soil in sandy environments. It seems most likely they are an introduction from the medieval period, although there is no direct evidence to support this.

Hoes

The most widespread and significant African agricultural implement is the hoe. A small number of African hoes have wooden blades, and there is debatable evidence for copper blades in archaeological contexts in West Africa, but hoes with iron blades predominate. Iron-bladed hoes can be subcategorised according to the method of fixing the blade. The principal types are bound hoes, transpierced hoes, gripped hoes, and socketed hoes (Figure 21.6; Figure 21.7). A fifth

Figure 21.4 Hausa planter in Maradi, Niger (drawn from Raynaut 1984).

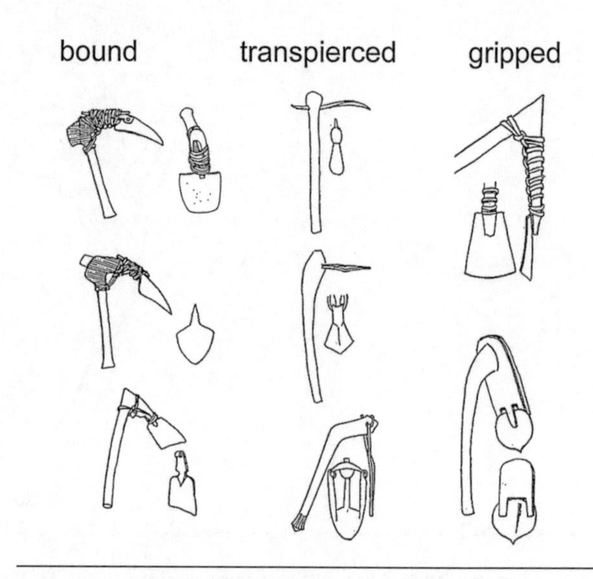

Figure 21.6 Comparison of hoes with iron blades: bound (after Seignobos 2000), transpierced, and gripped types (after Baumann 1944).

Figure 21.5 Head of Tamashek arrowhead digging stick, Mali (courtesy of the author).

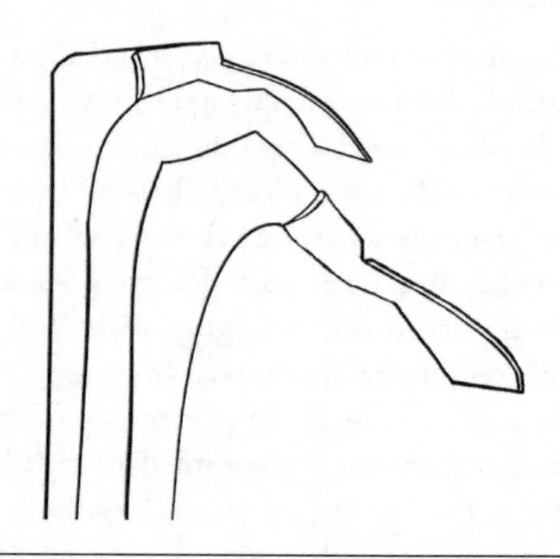

Figure 21.7 Socketed hoes, Burkina Faso (drawn from author's photo).

type of hoe, where a straight handle passes through a ring soldered to the top end of the hoe blade, is characteristic of most European hoe designs and is the common type imported into Africa. Such hoes have been recorded traditionally from Morocco and down the Nile as far as Sennar. Recent industrial hoes are of this type and can be seen in recent descriptions of African tools (for example, FIDA 1999) but are never illustrated in older ethnographic texts.

Wooden Hoes

Baumann (1944, p.207) illustrates a number of wooden hoe types from across Africa (Figure 21.8; see also Arkell 1937a). Baumann's map (p. 208) shows a strip across the continent from Senegambia to Southern Ethiopia. A wooden hoe will function only in soil that is relatively light and without too many stones. Otherwise the blade will break extremely quickly. Morphologically, there seems to be little unity between these implements, and it is not unlikely that they are not precursors of the iron hoe but rather back-formations—that is, copies in wood of iron implements. Some indeed have iron tips and may simply be designs to save on iron. Some also fall within the next category 'bound hoes'.

Bound Hoes

Earlier sources for African agricultural implements illustrate a wide variety of bound hoes (Fr. *houe à surliure*) (Baumann 1944; Seignobos 2000; Figure 21.6), where the blade is simply attached to the handle by a cord. Meek's account of Mambila hoes makes this point: "They had no hoes and carried out their operations by means of digging sticks. When they first obtained the iron hoe head they used it without affixing a handle. At the present time the hoe head is fixed to the iron handle by the primitive method of binding with palm-fibre" (Meek 1931, Vol. 1, p. 562).

Personal observation in the 1980s suggests that this type of hoe had been completely replaced by 'modern' hoes from Nigeria, where the tang transpierces the handle. Technically, this method is highly inefficient as the impact of the hoe against the ground will loosen the binding very rapidly. The fact that so many survived into the ethnographic era underlines the point that iron was expensive until very recently and widespread access to iron hoes relatively new.

Transpierced Hoes

Transpierced hoes have a metal tang that projects from the blade and passes through the wooden handle (Figure 21.5; Figure 21.9). This is probably the simplest method of fixing a metal blade, but it seems to have no wooden analogue, since the first blow against the earth would probably split the wood. Archaeological evidence for this type of hoe is quite abundant—for example, from Iron Age Zambia (Figure 21.10). Lancaster (1975) reported similar hoes traded long distances in recent times.

Gripped Hoes

The gripped hoe is widespread but like the bound hoe seems to have a highly diverse morphology. Baumann (1944) illustrates a number of types distributed from Ghana to Chad and parts of the Congo (see Figure 21.5), and Seignobos (2000) shows that these are particularly

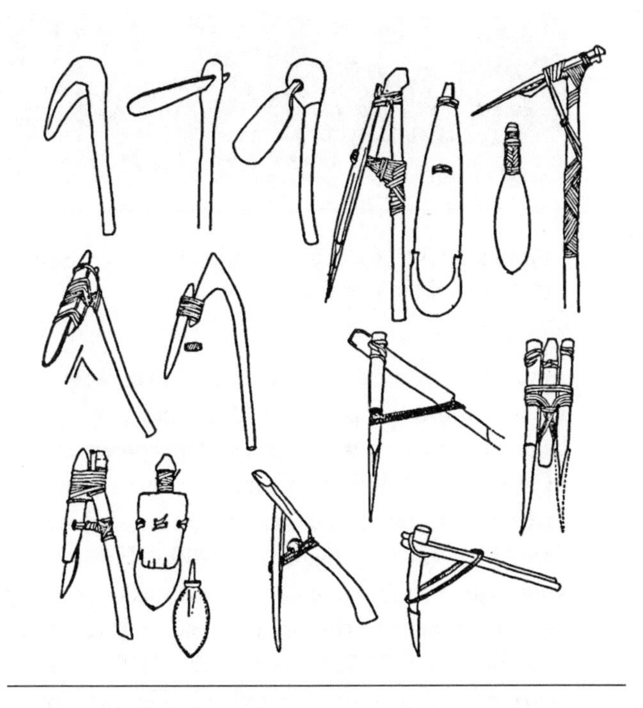

Figure 21.8 African hoes with wooden blades (after Baumann 1944).

Figure 21.9 Transpierced hoe, Supyire, Southeast Mali (courtesy of the author).

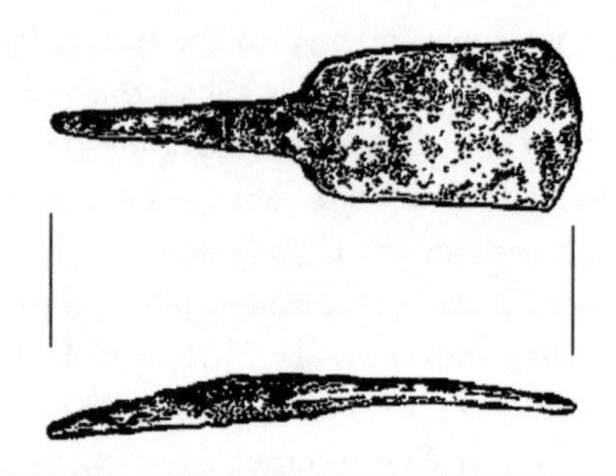

Figure 21.10 Archaeological hoe blade from Kumadzulo, Zambia, 6th-century site C.E. (after Vogel 1973).

diverse in Northern Cameroun. In some cases gripping can be combined with binding, but the principle is that the blade is gripped between wooden projections on the handle. The advantage of this construction method is that the hoe need use only a very small iron tip, with most of the blade made of wood. The disadvantage is that the iron piece probably comes loose with some regularity. Like the bound hoe, the gripped hoe may not represent a stable morphological type but rather a transition to the transpierced and socketed types.

Socketed Hoes

Another very characteristic type is the socketed hoe, found almost throughout the continent. Instead of piercing the handle with a tang, the blade is folded into a tube and usually fitted to a naturally bent handle (Figure 21.7). Socketed hoes require more iron than the other types do but are probably more stable in use. Eggert and associates (2006, Figure 4) illustrate a decorated hoe blade from the site of Akonétyé in Southern Cameroun dated to 130–420 C.E. On this site, see also Kahlheber, Höhn, and Neumann, Chapter 10 this volume; they remark that the socketed design is unusual for Central Africa, but socketed hoes are common in West Africa. They also suggest that the thin blade may imply that it was manufactured for ritual purposes rather than everyday use. However, it may also simply reflect the scarcity of iron at this early date.

Knives, Cutlasses

Knives

Knives appear to be very old in African culture, and cane knives probably preceded iron types. An old root for 'knife' can be identified in Niger-Congo languages (Table 21.2). Cognates between Benue-Congo and Kru suggest an ancestral node that must be 6,000–7,000 years old at a minimum, older than the introduction of

Table 21.2 A Root for 'Knife' in Niger-Congo Languages

#-gbeN	knife		
Group	**Language**	**Attestation**	**Comment**
Kru	Aizi	bɛ	
	Bete	gblɛ̀	
Yoruboid	Igala	obe	
Edoid	Bini	ábɛ́è	
Igboid	Ekpeye	ògè	? loan from Ogoni. cf. Kana gɛə̀
Akpes	Daja	oyùŋ̀gbà	
Nupoid	Nupe	ebì	
Oko	Magongo	igbegbẽ	
Idomoid	Idoma	àgbàgá	
Plateau	Tyap	a̱baai	
	Shall	nbaa	
	Jijili	obã	
Cross River	E. Ogbia	ɔ̀-gyɛ̀	pl. ɛ̀-
Dakoid	Dɔ̃	gbaa	
Mambiloid	Camba	bu	
	Somyev	bi	
	Len	mbɛ̀tɛ́	
Bantu	PEG	*-bé`	

Figure 21.11 Agricultural knife from the Dogon, Mali (courtesy of the author).

iron. There is no proof that such knives were used only for agriculture proper; it is likely that they would have been adapted to a wide variety of purposes. Excavations in Akonétyé, Southern Cameroun, have recovered the earliest iron knife so far recorded in sub-Saharan Africa, dated 130–420 C.E. (Eggert et al. 2006). However, African knives quite closely resemble curved sickles and may have evolved in form as the sickle spread in West Africa (for example, Figure 21.11). Wente-Lukas (1977, p. 100) illustrates two types of harvest knife used by the Bana people in Northern Cameroun, which are straight-bladed knives rather than true sickles.

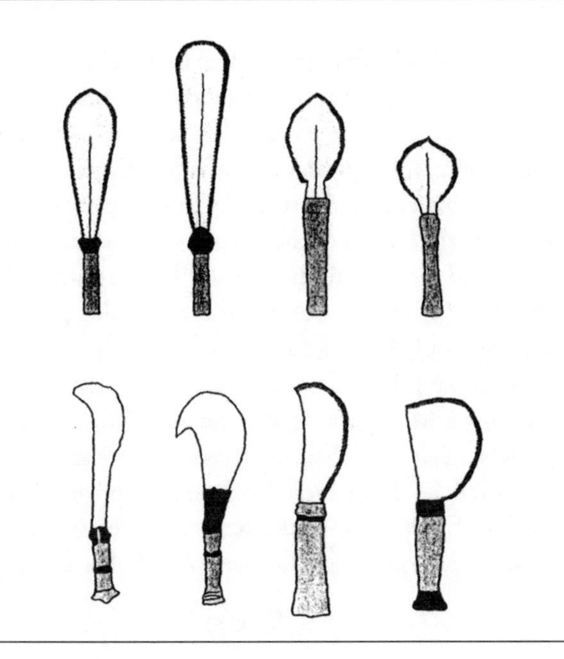

Figure 21.12 Selected variations of the cultivating knife (after Dupré 2000).

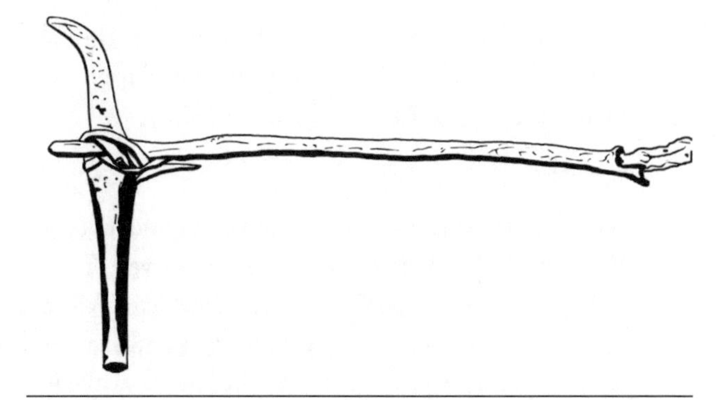

Figure 21.13 Banana-cutter, Uganda (redrawn from photo from FIDA 1999).

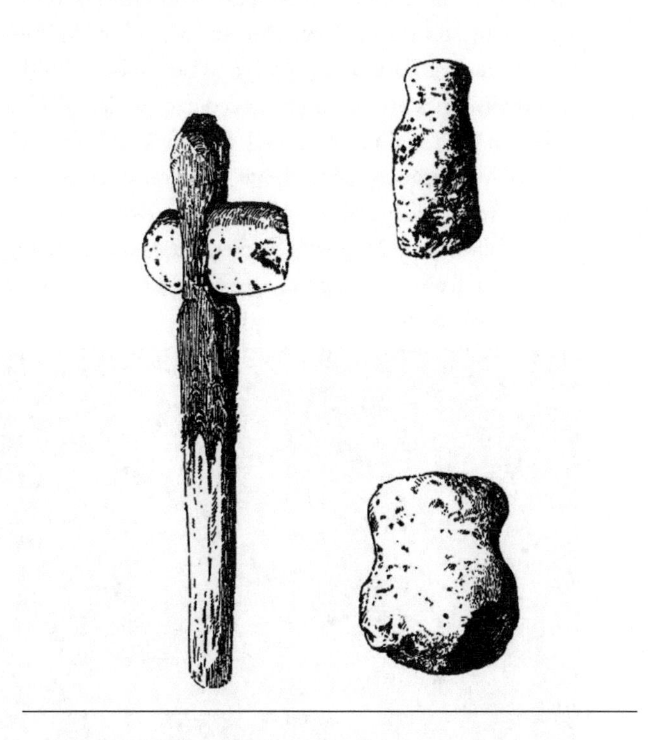

Figure 21.14 Hafted stone axes of the Bubi (Fernando Po) from the 1920s (from Tessmann 1922).

Cultivating Knife

An intriguing and little-known implement from West-Central Africa is the cultivating knife (*couteau de culture*) described and illustrated by Dupré (2000); a variety are reproduced in Figure 21.12. These are knives with either asymmetrical blades adapted for cutting or symmetrical, leaf-shaped blades, similar to a trowel, also used in planting operations. Dupré (2000) reports a distribution in eastern Gabon, Congo, and upper Sanga, but to judge by an illustration in Seignobos (2000, Figure 6) these knives are found as far north as the Cameroun Grassfields.

Cutlass

The African cutlass, matchet or *panga*, is essentially a large knife used for cutting undergrowth and woody stems as part of general ground clearance. This is not a typically European tool, so it was presumably designed as an improved version of an existing African tool, most likely the cultivating knife. Perhaps it was remodelled in Europe in the 19th century and reexported to Africa. Since it has a 'new' name in most places, it is likely that the category was unfamiliar to most buyers. It is employed virtually continent-wide as an implement and is often used to symbolise African culture.

Banana-Cutter

In regions where bananas are a staple, such as Uganda, a variant of the knife is used to cut down bunches of bananas (FIDA 1999). This knife has a curved blade attached to a forked stick (Figure 21.13). It is not clear whether this implement is widespread.

Axes

Hafted Stone Axes

Stone axes are a common lithic artefact described in pre-iron age archaeology, but the evidence for their contexts of use remains slight. Widstrand (1958, p. 88) points out that the axe features in many 'civilising hero' myths, often to the exclusion of other tools. Whether these narratives date from the pre-iron period is unclear, but they point to the early role of axes in chopping away vegetation for farming. Waisted stones axes are found widely across much of

Africa and survived into historic times on the island of Fernando Po because of the island's isolation from the mainland (Figure 21.14; Tessmann 1922).

IRON AXES

Axes with iron blades occur almost throughout the continent. Widstrand (1958) is a comprehensive survey of axe types, including tools, weapons, and ceremonial axes. Most societies recognise a distinction between adzes and axes; and adze has the cutting edge transverse to the handle, while in an axe the blade is parallel with the shaft. Adzes are used principally for woodcarving and are not treated here. Axes are used in many societies as much for warfare as for cutting wood, and for this reason they are not usually covered in synthesising sources such as Baumann (1944) and Seignobos (2000). The two main types are the transpierced axe (which Widstrand calls the 'slot-shafted' axe) and the socketed axe (Figure 21.15). Widstrand (1958) shows that transpierced axes are found from Senegambia to Zululand, with some records on the Nile in the Maghreb. Socketed axes are more limited and occur from Senegambia across to the Horn of Africa, with a small island of occurrences in Angola, but are otherwise absent from eastern and southern Africa. Akonétyé in Southern Cameroun produced an axe blade (ca. 130–420 C.E.), but it is unclear how it was hafted (Eggert et al. 2006).

A third axe-form is recorded from northeastern Africa, in which the handle of the axe passes through a ring, either at the end opposite to the cutting blade or in the middle of the blade (in which case it starts to resemble a pickaxe morphologically). These axes are recorded only from the East African coast, Uganda, Ethiopia, and a strip up the Nile, as well as on Madagascar (Widstrand 1958). There is some evidence that even these sub-Saharan African occurrences may well be from local blacksmiths copying European models. Similar axe-blades have also been imported into other regions of Africa (for example, Mudindaambi 1976).

Figure 21.15 Iron axes, transpierced type (top) and socketed type (bottom), both from the Dogon in Mali (courtesy of the author).

SICKLES

Two types of sickles for harvesting cereals occur across Sahelian Africa, the curved sickle, with a hooked blade in a cylindrical wooden handle (Figure 21.16), and a lateral sickle (Figure 21.17).

Figure 21.16 Curved sickle (drawn from Raynaut 1984, p. 531).

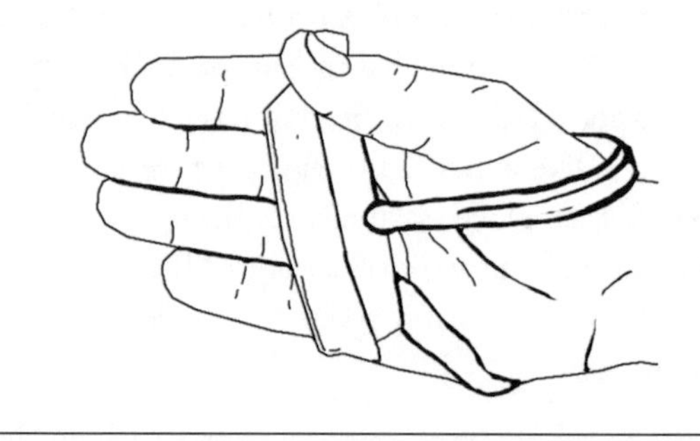

Figure 21.17 Lateral sickle (drawn from Raynaut 1984).

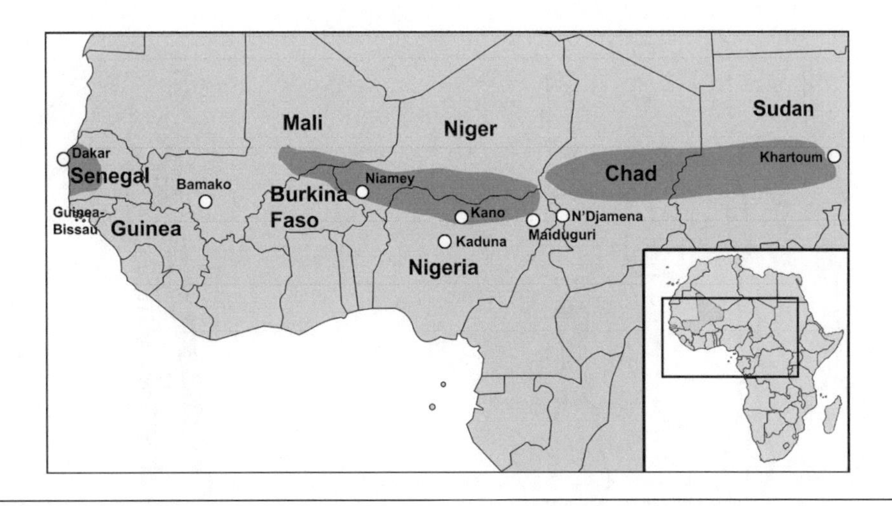

Figure 21.18 Distribution of the *iler* spade (after Raynaut 1984).

Curved Sickle

The curved sickle closely resembles small Mediterranean sickles. A Moroccan parallel can be seen in Stuhlmann (1912, p. 72). Raynaut (1984, pp. 530–31) points out that both inserted tang and socketed types of sickle exist in West Africa. In many Nigerian languages, the term is borrowed from the Hausa *lauje*, and it seems likely to have been spread southward from Hausaland. There are thus reasons to infer this is a relatively recent introduction, perhaps from the medieval period (see also Arkell 1937b), although a recent Iron Age excavation in Cameroun produced a fragment of a possible sickle from 1700–2000 B.C.E. (Meister 2010).

Lateral Sickle

The lateral sickle is an iron blade with a leather or wooden holder, attached to the hand by a loop of cord found across Sahelian West Africa (Figure 21.17). It allows the harvester to cut off the head of grain with considerable precision.

Spades, Shovels, Trowels

Compared with the hoe, the spade principle, where the blade is in line with the handle instead of perpendicular to it, is very rare in Africa. Long-handled spades, sometimes with wooden blades, occur across the Sahel, and a variety of implements similar to trowels occur in Cameroun and adjacent regions.

The Iler

Across the West African Sahel, a long-handled spade (the *iler*) is used to move earth, especially in flood plains. Daniel (1931) may have been the first to draw attention to this implement, which he records being used in the area of Sokoto in northwestern Nigeria. Pelissier (1966, *Planche* 43) includes a comprehensive series of photographs of the use of the *iler* among the Diola of the Casamance in Senegambia. This is an old North African tool and is also recorded ethnographically from Morocco. Figure 21.18 shows its approximate distribution across Sahelian Africa.

There are two discussions in print of this tool, Raynaut (1984) and Wigboldus (2000), both of whom conclude the *iler* is a relatively recent trans-Saharan migrant, although they differ on the date of its transmission. To judge by its geography, it may well have diffused across the Sahara at different times along a variety of routes, so there may be no final solution.

Trowels

The trowel is a rare agricultural implement in Africa (Figure 21.19). The literature states that it occurs in a restricted area in Cameroun and adjacent Congo and Gabon (Baumann 1944; Seignobos 2000). However, it is apparently more widespread, evidenced by a trowel collected among the Nigerian Igbo (Figure 21.20). Blade shapes vary, but at least some examples have the classic diamond shape typical of European trowels. Wente-Lukas (1977, p. 92) and Seignobos (2000) show trowels from the Mandara mountains that have a distinctive T-shaped wooden handle and socketed blades rather than the inserted blades illustrated by Baumann. Nonetheless, usage appears to be identical, to judge by Seignobos (2000, Figure 4).

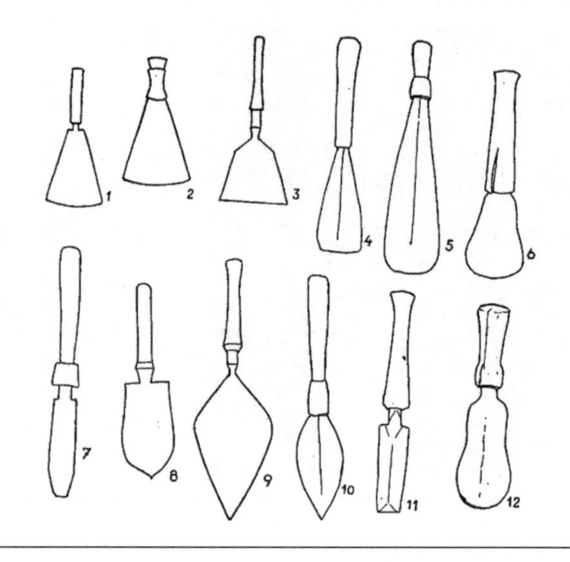

Figure 21.19 African trowel types (after Baumann 1944).

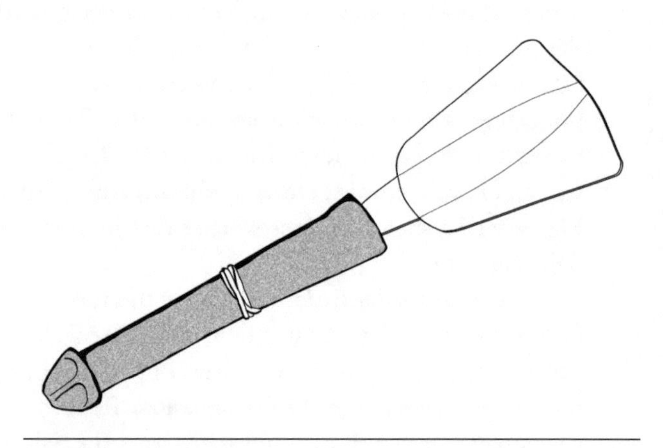

Figure 21.20 Igbo trowel, Nigeria (drawn from author's collection).

MISCELLANEA

FRUIT-HOOK

An implement of unknown date is the fruit-hook, an angled knife on a long pole, used to cut off stalks and bring down fruits from high trees (Figure 21.21). Fruit-hooks are made by the Dogon peoples of Northern Mali, who depend on a wide variety of economic trees for their subsistence. Rather more temporary implements are made widely throughout Africa, usually long bamboo canes with a bent piece of wire inserted into one end. Rather charmingly, these are known in Nigerian vernacular English as a 'go-to-hell', apparently from their resemblance to a bishop's crozier.

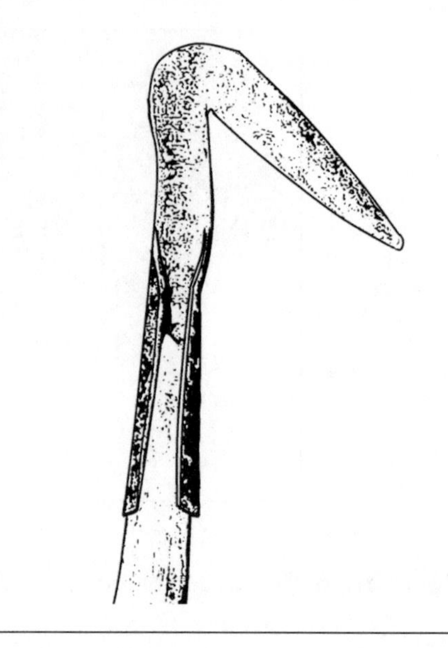

Figure 21.21 Fruit-hook from the Dogon in Mali (drawn from author's photograph).

LANGALANGA

The langalanga, or *coupe-coupe*, is a simple flat strip of metal, bent at one end, which can be used to slash at rampant grass. It is not a European tool or one of any great antiquity in Africa, and it seems to have developed in the colonial period in West Africa, based on scrap metal. Even today, it is not uncommon to see lines of schoolchildren in Nigeria disconsolately advancing across a school ground, using this tool to slash away at the grass. Although the langalanga is also known in Ghana, little information is available about its origin or distribution. 'Langalanga' was adopted as a pen name by a colonial officer for his memoirs in 1927, so it may be dated to at least this period. Moñino (1984) illustrates a type of langalanga with a cylindrical wooden handle in use among the Gbaya in CAR. FIDA (1999) pictures a very similar implement from Uganda, so it is probably widespread across Africa.

YAM EXTRACTOR

Bahuchet (2000) draws attention to a quite idiosyncratic tool, a *tarière*, used by the Aka and Baka pygmies of the Central African rain forest. It is a stick with the far end split into five strips and the free strips bent outward to form a sort of cradle. It is used for extracting a particular species of yam, *Dioscorea semperflorens*. Once the ground has been pierced by a digging stick, the yam extractor is used to dig down and pull out the yam tuber (Figure 21.22). The Aka call it *dìsó* and the Baka *bòndùngà*. Seignobos (2000)

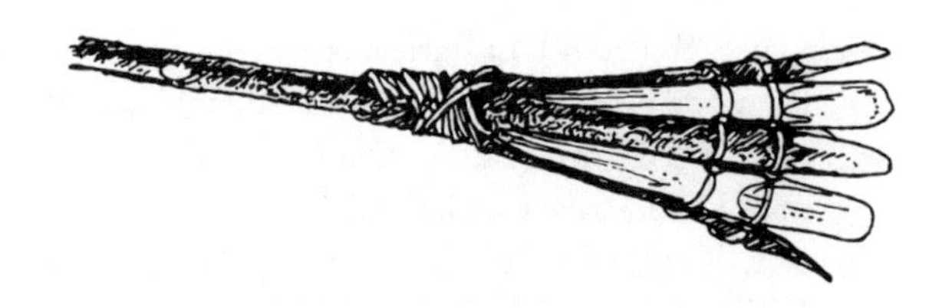

Figure 21.22 Yam extractor, Aka pygmies (after Bahuchet 2000).

Figure 21.23 Palm-wine tapping knife, Igbo, Nigeria (courtesy of the author).

also mentions a similar tool among the Vute of the Grassfields, so this implement may not be confined to the pygmies.

PALM-WINE TAPPING KNIFE

Figure 21.23 shows a typical palm-wine tapping knife used by the Igbo in Nigeria to pierce the trunk of the palm tree to drain off the sap to make alcoholic drinks. Such knives occur all along the forest zone of West Africa, although their exact distribution seems not to have been mapped.

PITCHFORK AND RAKE

Raynaut (1984, p. 528) illustrates a pitchfork and rake from the Maradi area of Niger, known in Hausa as *mashaarii* (pitchfork) and *mayayaa* (rake). The pitchfork is made from a naturally forked wooden stick and is used to lift straw. Baumann (1944, p. 298) shows a similar implement from the Oromo in Southern Ethiopia. The rake is made from a stick with one end split into tongues, which are kept spread out by transverse sticks. This tool may be adopted from North Africa, although Baumann (1944, p. 298) illustrates a rake from the Sandawe in Tanzania. Copies of modern rakes of European design (that is, where the handle ends in a transverse bar along which are fixed a series of lateral strakes perpendicular to it) are now made by West African blacksmiths, but this is a recent development.

THE ROAD NOT TAKEN: MEDITERRANEAN TOOLS THAT FAILED TO SPREAD

Although there is a case for the diffusion of some implements across the Sahara, to a large extent sub-Saharan Africa seems to have followed its own path with respect to agricultural tools. The most notable example of an implement that failed to spread is the plough; ploughs were unknown in sub-Saharan Africa until introduced by missionaries and the colonial authorities in the 1920s. Ethiopia, as so often, seems to have quite a different history from elsewhere. The plough characteristic of Ethiopia, an ard that fractures and disturbs the soil, seems to be have been introduced following the migrations of Ethiosemitic speakers across from Yemen. The Amharic term for plough, *maräša*, ማረሻ, has been borrowed into all the main languages of Ethiopia. Even where this term is not used, the local terms for plough turn out to be constructs ('hoe of cow', and so on), which indicate its recent adoption. Barnett (1999, p. 24) canvasses ideas of introductions from Arabia or Egypt around 2000–1000 B.C.E., but the linguistic evidence suggests a more recent date. Neither the design of the Ethiopian plough nor its name points to external origin, and it is quite likely that this tool was constructed locally through stimulus diffusion; that is, a plough seen elsewhere was redesigned for local conditions.

All forms of animal traction have an ancient history in North Africa and, in theory at least, the plough could have been transmitted across the Sahara with the caravan trade along with food crops and irrigation techniques (Bulliet 1975). Indeed, as Bernus (1981, p. 286) points out, simple camel-drawn ground-preparation tools (*ashek n egdri*) are used in Saharan oases by the Tamasheq. This raises the interesting question as to why the introduction of the plough in the 1920s was so successful, if it had previously been rejected. The answer may lie in the challenge from the trypanosomoses. Until recently, cattle could survive the challenge of tsetse in sub-Saharan Africa through careful management by herders; subjected to work stress and kept in a single location they often died. Once better nutrition and simple trypanocides were introduced, traction cattle could stay alive and were thus an economic option.

There are other North African implements that failed to cross the desert. One of these is the pick-axe, widely used around the Mediterranean for breaking rocks and hard earth but not recorded south of the desert. The rake and the pitchfork, although with a couple of records discussed here are probably very recent introductions.

CONCLUSIONS

African agricultural tools remain remarkably little studied by archaeologists and ethnographers, and much of what *has* been published relates to Francophone countries, giving a skewed image of the continent as a whole. Yet the introduction of iron tools introduced a revolution in the agriculture of the continent, and the need to produce iron by smelting effected a major economic transformation. This chapter has not covered the relationship between specific tool categories and agriculture practices, but there is no doubt that without iron the exploitation of the equatorial forest for subsistence other than by foraging was virtually impossible. Some broad conclusions can be drawn from our present understanding of the data.

1. The archaeological evidence suggests that African agriculture takes off relatively late, although before the introduction of iron. However, iron made it possible to exploit a range of new environments inaccessible with stone tools.
2. The relative diversity of African agricultural implements is probably strongly related to the availability of iron. For much of the period following the introduction of smelting iron, tools were expensive and designs were intended to save iron.
3. Wooden hoes are probably not precursors of iron hoes but subsequent copies.
4. Hoes with bound and gripped blades do not reflect a single design but the gradual introduction of traded blades to individual societies.
5. A significant number of new implements have spread across the Sahara in the last thousand years. Hence there is a relative diversity on the southern edge of the desert compared with the continent as a whole.
6. The abundant iron now available from scrap has led to a second phase of diversification of tools. However, at the same time, increased long-distance trade has tended to replace highly local tools with common designs.

REFERENCES

Arkell, A. J. (1937a) An extinct Darfur hoe. *Sudan Notes and Records XX*, 146–50.

———. (1937b) The tigda, or reaping knife in Darfur. *Sudan Notes and Records XX*, 306–07.

Bahuchet, S. (2000) La tarière à igname des Pygmées de l'ouest de bassin congolais. In C. Seignobos, Y. Marzouk, and F. Sigaut, (Eds.), *Outils Aratoires en Afrique*, pp. 237–6. Paris: Karthala/IRD.

Barnett, T. (1999) *The Emergence of Food Production in Ethiopia*, British Archaeological Reports: International Series 763. Oxford: Archaeopress.

Baumann, H. (1928) The division of work according to sex in African hoe culture. *Africa: Journal of the International African Institute 1* (3), 289–319.

———. (1944) Zur Morphologie des afrikanischen Ackergerätes. In H. Baumann (Ed.), *Koloniale Völkerkunde*, pp. 192–322. Horn, Austria: Verlag Ferdinand Berger.

Béavogui, F. (2000) Circulation monétaire en Afrique de l'Ouest: Le cas du *guinzé* (Guinée, Liberia). In C. Seignobos, Y. Marzouk, and F. Sigaut (Eds.), *Outils aratoires en Afrique*. pp. 175–90. Paris: Karthala/IRD.

Bernus, E. (1981) *Touaregs nigériens*. Paris: ORSTOM.

Blench, R. M. (2006) *Archaeology, Language, and the African Past*. Lanham, MD AltaMira.

Bulliet, R. W. (1975) *The Camel and the Wheel*. Cambridge, MA: Harvard University Press.

Cornevin, M. (1993) *Archéologie africaine*. Paris: Maisonneuve and Larose.

Coulibaly, S. (1978) *Le paysan senoufo*. Abidjan-Dakar: Les Nouvelles Éditions Africaines.

Daniel, F. (1931) An agricultural implement from Sokoto, Nigeria. *Man XXXI*, 48.

David, N. (1996) A new political form? The classless industrial society of Sukur (Nigeria). In G. Pwiti and R. Soper (Eds), *Aspects of African Archaeology*, Proceedings of the 10th Pan-African Congress on Prehistory and Related Studies (Harare, June 1995), pp. 593–00. Harare: University of Zimbabwe Press.

Dupré, M.-C. (2000) Un outil agricole polyvalent: Le couteau de culture dans les monts du Chaillu (Congo, Gabon). In C. Seignobos, Y. Marzouk, and F. Sigaut (Eds.), *Outils aratoires en Afrique*, pp. 191–25. Paris: Karthala/IRD.

Eggert, M. K. H., A. Höhn, S. Kahlheber, C. Meister, K. Neumann, and A. Schweize. (2006) Archaeological and archaeobotanical research in Southern Cameroun. *Journal of African Archaeology 4* (2), 273–98.

FIDA (1999) *Outils agricoles utilisées par les femmes africaines*. Rome: Fonds International de Développement Agricole.

Gascon, A. (1977) Le dangwara, pieu à labourer d'Éthiopie. *Journal d'agriculture tropicale et botanique appliquée XXIV*, 111–26.

Goody, J. R. (1976) *Production and Reproduction*. Cambridge: Cambridge University Press.

Haaland, R. (2004) Iron smelting a vanishing tradition: Ethnographic study of this craft in south-west Ethiopia. *Journal of African Archaeology* 2 (1), 65–79.

Lambert, N. (1983) Nouvelle contribution à l'étude du Chalcolithique de Mauritanie. In N. Echard (Ed.), *Métallurgies africaines. Nouvelles contributions*. Mémoires de la Société des Africanistes 9, pp. 63–87. Paris: Société des Africanistes.

Lancaster, C. S. (1975) Later Iron Age hoes from Southern Zambia. *Current Anthropology* 16 (2), 283.

Lericollais, A., and J. Schmitz. 1984 La calebasse et la houe : Techniques et outils des cultures de décrue dans la vallée du Sénégal. *Cahiers ORSTOM, séries Sciences Humaines XX* (3-4), 427–52.

Meek, C. K. (1931) *Tribal Studies in Northern Nigeria* [2 vols.]. London: Kegan Paul, Trench and Trubner.

Meister, C. (2010) Remarks on Early Iron Age Burial Sites from Southern Cameroon. *African Archaeological Review* 27, 237–49.

Moñino, Y. (1984) Histoires d'houes: Instruments aratoires centrafricains. *Cahiers ORSTOM, séries Sciences Humaines XX* (3-4), 585–95.

Mudindaambi, L. (1976) *Objets et techniques de la vie quotidienne Mbala*. Bandundu: CEEBA publications.

Okafor, E. E. (1993) New evidence on early iron-smelting from southeastern Nigeria. In T. Shaw, P. Sinclair, B. Andah, and A. Okpoko (Eds.), *The Archaeology of Africa*, pp. 432:48. London: Routledge.

Pelissier, P. (1966) *Les paysans du Sénégal*. Saint-Yrieix: Imprimerie Fabrègue.

Raynaut, C. (1984) Outils agricoles de la région de Maradi (Niger). *Cahiers ORSTOM, séries Sciences Humaines XX* (3-4), 505–36.

Seignobos, C. (Ed.) (1984) Les instruments aratoires en Afrique tropicale. *Cahiers ORSTOM, séries Sciences Humaines XX* (3-4).

Seignobos, C. (Ed.) (2000) Nomenclature commentée des instruments aratoires du Cameroun. In C. Seignobos, Y. Marzouk, and F. Sigaut (Eds.), *Outils aratoires en Afrique*, pp. 297–337. Paris: Karthala/IRD.

Seignobos, C., Y. Marzouk, and F. Sigaut (Eds.) (2000) *Outils aratoires en Afrique*. Paris: Karthala/IRD.

Shaw, T., and S. G. H. Daniels (1984) Excavations at Iwo Eleru, Ondo State, Nigeria. *West African Journal of Archaeology* 14, 1–269.

Stuhlmann, F. (1912) *Ein kulturgeschichtlicher Ausflug in den Aures (Atlas von Süd-Algerien)*. Hamburg: L. Friederichsen and Co.

Tessmann, G. (1922) *Die Bubi auf Fernando Po*. Berlin: Dietrich Reimer.

Tourneux, H. (1984) Vocabulaires comparés des instruments aratoires dans le Nord-Cameroun. *Cahiers ORSTOM, séries Sciences Humaines XX* (3-4), 597–612.

Vincent, A. S. (1985) Plant foods in savanna environments: A preliminary report of tubers eaten by the Hadza in Northern Tanzania. *World Archaeology* 17 (2), 131–48.

Vogel, J. O. (1973) Early Iron Age hoe blades from Southern Zambia. *Current Anthropology* 14 (5), 529–32.

Wente-Lukas, R. (1977) *Die materielle Kultur der Nicht-Islamischen Ethnien von Nordkamerun und Nordostnigeria*. Wiesbaden: Steiner Verlag.

Widstrand, C. G. (1958) *African Axes*. Studia Ethnographica Upsaliensia XV. Uppsala: Almqvist and Wiksells Boktryckeri.

Wigboldus, J. 2000 The early history of the *iler*: Raulin's hypothesis revisited. In C. Seignobos, Y. Marzouk, and F. Sigaut, (Eds.), *Outils aratoires en Afrique*, pp. 149–72. Paris: Karthala/IRD.

White, K. D. (1967) *Agricultural Implements of the Roman World*. Cambridge: Cambridge University Press.

———. (1975) *Farming Equipment of the Roman World*. Cambridge: Cambridge University Press.

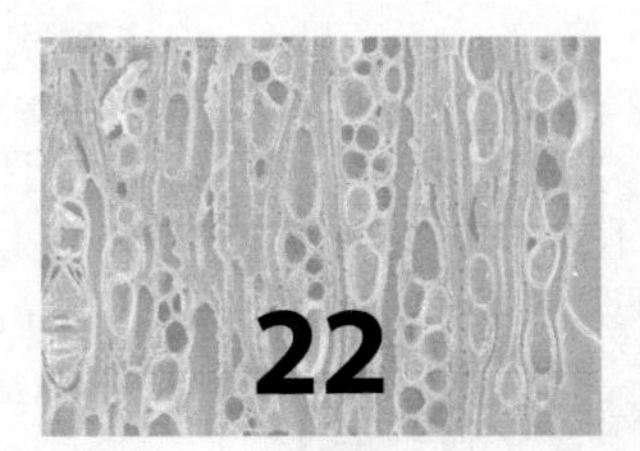

22

Leaving a Lasting Impression

Arable Economies and Cereal Impressions in Africa and Europe

Meriel McClatchie and Dorian Q Fuller

Studies of arable agricultural systems throughout the world have frequently utilised data from cereal grain and chaff impressions on ceramic vessels in the reconstruction of past economies (for example, Costantini 1983; Helbaek 1952, 1959; Jessen and Helbaek 1944; Klee and Zach 1999; Munson 1976; Stemler 1990; Vishnu-Mittre 1969). Cereal components can become incorporated into ceramic vessels during manufacture and may be preserved through charring or destroyed during the firing of a pot, leaving morphologically identifiable impressions of the material in the fabric of vessels (Figure 22.1; see also Manning and Fuller this volume). It has often been suggested that the frequency of various cereal types recorded from ceramic vessels represents the relative economic importance of each cereal type (for instance, Costantini 1983; Godwin 1975, p. 405; Helbaek 1952; Jessen and Helbaek 1944, p. 10; Possehl 1999, p. 459). A range of processes and behavioural patterns may, however, have affected the ways in which cereals were incorporated into ceramic vessels, and the predominance of certain cereal types at various times is unlikely to be related directly to their economic importance (Hubbard 1975, p. 200).

In this chapter we argue for an alternative, crop-processing framework. We develop this from discussion of a European example, with apparent contradictions between impressions and charred macro-remains assemblages in Bronze Age Ireland. We then reconsider some of the African record in light of a crop-processing framework, including the contrasts between earlier and later Late Stone Age and Iron Age impressions from West Africa, and the contrasts between Mesolithic and Neolithic impressions in Sudan. Evidence suggests that some of the temporal shifts in impressions data reveal more about changes in the organisation of potting vis-à-vis plant processing, rather than simply changes in diet. Nevertheless, there may a recurrent signal in chaff-tempering linked to the initial intensification of reliance on early cultivated pearl millet across West Africa.

Rethinking Impressions: A Case Study from Bronze Age Ireland

For many years, plant impressions on ceramic vessels constituted the main macro-remains evidence for arable agriculture in Bronze Age Ireland (2300–600 B.C.E.). A number of studies indicated that barley (*Hordeum* spp.), particularly naked barley, was the focus of arable activity at this time, with wheat (*Triticum* spp.) playing a very minor role in Bronze Age agriculture (Hartnett 1957; Jessen and Helbaek 1944; Monk 1986; Ó Ríordáin and Waddell 1993). The recent collation of published and unpublished charred macro-remains assemblages from Bronze Age sites in Ireland (McClatchie 2009) has, however, provided a very different picture with regard to the economic status

Archaeology of African Plant Use by Chris J. Stevens, Sam Nixon, Mary Anne Murray, and Dorian Q Fuller, Eds., 259–265

Figure 22.1 Bronze Age vessel from Graney West, Co. Kildare, Ireland. Barley grain impressions highlighted in white circles (image courtesy of the National Museum of Ireland).

of various cereal types (Figures 22.2 and 22.3). Wheat appears to have been much more significant than previously considered, particularly during the Middle Bronze Age, and hulled barley also played a prominent role. It is clear that this new study does not correlate well with previous findings from analyses of cereal impressions on ceramic vessels.

The significance of cereal impressions in Irish and British ceramics has been interpreted in various ways. It has often been proposed that the incorporation of cereals into the fabric of prehistoric ceramic vessels is a result of the presence of crops in manufacturing areas, whereby components are inadvertently incorporated (Cleary 1987, p. 35; Godwin 1975, p. 405; Jessen and Helbaek 1944, p. 10). The intentional inclusion of cereals may also have occurred owing to technical requirements of potters—for example, in the use of chaff as a tempering agent (Boreland 1996, p. 22; Gibson 2002, p. 35, 2003, p. vi; Gibson and Woods 1997, p. 33) and as a result of symbolic, social, or stylistic reasons (Darvill 2004, p. 204, n.2; Gibson 2003, p. vi; Schiffer and Skibo 1987, p. 596). As containers and also as crucibles for the transformation of foodstuffs from one state to another (raw to cooked; sour to sweet; milk to yoghurt; mash to ale; and so on), ceramic vessels may have been perceived as dynamic agents, with attached meanings perceptible to their creators and users (Darvill 2004, p. 196).

The recent availability of systematically collected flotation samples makes it clear that data of impressions differ from typical charred macro-remains. Charred macro-remains are inferred to mainly represent the waste from routine processing activities (Fuller and Stevens 2009; Stevens 2003; van der Veen 2007) and are thus useful for indicating the relative importance of various crops

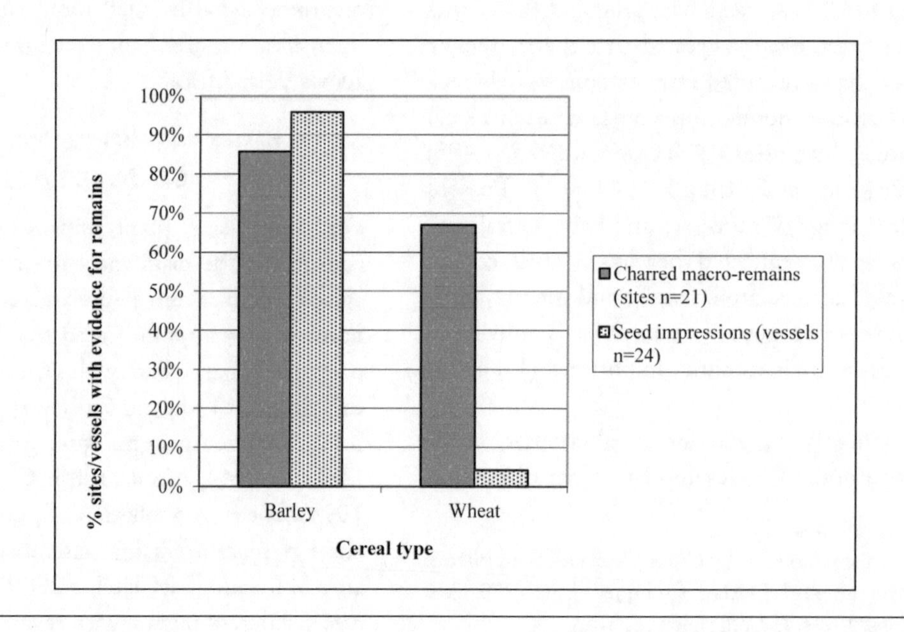

Figure 22.2 Relative occurrence of cereal types in Bronze Age Ireland, comparing data from seed impressions on ceramic vessels with plant macro-remains from archaeological deposits.

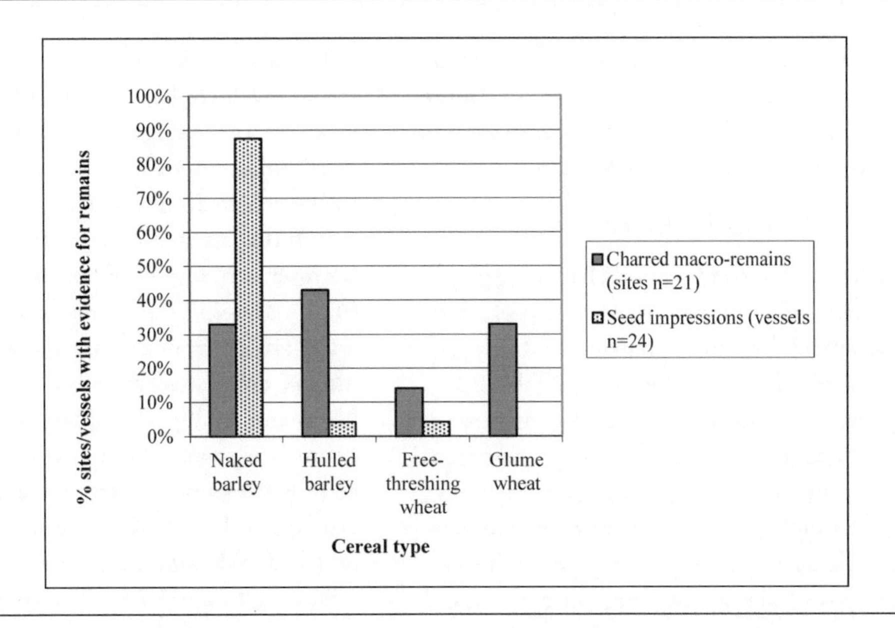

Figure 22.3 Relative occurrence of naked/free-threshing and hulled/glume cereal types in Bronze Age Ireland, comparing data from seed impressions on ceramic vessels with plant macro-remains from archaeological deposits.

that were regularly processed; however, the discrepancy between these data and Jessen and Helbaek's findings from cereal impressions is striking.

The strong association between naked barley and ceramic vessels in Bronze Age Ireland may be better viewed as representing a relationship between the processing of naked barley and the locations of ceramic vessel production. It may be significant that naked cereals, mainly barley, and in one case free-threshing wheat, are predominant in the cereal types identified from seed impressions, whereas hulled barley and glume wheats are less well represented. It is possible that crops requiring a greater amount of processing (dehusking), such as hulled barley and glume wheats, were not fully processed before storage, being stored in spikelet or sheaf form. Whole spikelets would have been less attractive as temper and less likely to become incorporated into the fabric of vessels owing to their mass. Indeed, the only incidence of hulled barley in the seed impression record is of a floret rather than a seed alone (Jessen and Helbaek 1944, p. 21), perhaps the accidental inclusion of threshing waste. Routine crop-processing may have followed a different pathway when compared with less frequent crop-processing associated with certain occasions—the former into the charred record, the latter into the ceramic record.

Harvest is likely to have occurred in late summer or early autumn (McClatchie 2009, p. 191). Harvest thereby coincides with a time when the production of ceramic vessels would have been advantageous, coming at the end of the driest season and thus facilitating the preparation and firing of ceramic vessels, which need to be slowly dried to the 'leatherhard' state, as well as the preparation of fuel (Arnold 1985, pp. 61–77). Rather than occupying all available labour and thereby creating a scheduling conflict (Arnold 1985, p. 99; Kramer 1985, p. 80), harvest time in temperate areas, such as Ireland, may have roughly coincided with ceramic production, the latter being to some extent a seasonal activity with a proportion of it being scheduled to occur around harvest time. It has previously been suggested that pottery production was a seasonal endeavour, since cereals were available for incorporation into fabrics at harvest time (for example, Howard 1981, p. 25; Jones 1999). This approach fails to recognise that cereals can be stored, and therefore utilised, over relatively long periods. A ready supply of cereal components would, however, become available for intentional/unintentional incorporation, but significantly only of those species that are more easily fully processed into clean grain and light chaff components (such as naked barley) rather than the hulled cereals (including hulled barley and glume wheats), which are likely to have been stored as semiprocessed spikelets or sheafs. Large numbers of people may have been mobilised for harvest and selective processing, and some proportion of them could have been involved in the production and distribution of vessels.

The exclusive use of evidence from cereal impressions in ceramic vessels does not, therefore, seem appropriate in determining the economic roles of various cereal types in arable agricultural economies. Instead, a crop-processing perspective allows ceramic impression data to contribute to the integration of archaeobotanical evidence in a more complete reconstruction of economic scheduling.

PATTERNS IN AFRICAN PLANT IMPRESSIONS

Casts of plant parts preserved in ceramics have long played a role in the study of plant use in African prehistory, sometimes supplemental to macro-remains and in many cases as the only available source of evidence (Table 22.1). A reliance on plant impressions has been particularly true at either end of the Sahel zone, in the Khartoum region from Sudan and in Mauritania. In some cases, this evidence has provided important indications for the presence of domesticated crops, such as the impressions of domesticated pearl millet involucres and spikelets from Dhar Nema and Oualata in southeast Mauritania (Amblard and Pernes 1989; Fuller, Macdonald, and Vernet 2007; MacDonald et al. 2003) and early domesticated pearl millet in northern Mali (Manning and Fuller this volume; Manning et al. 2011). In sherds from this region, starting before 2000 B.C.E. in the Tilemsi valley and by ca. 1600 B.C.E. in southeast Mauritania, the threshing and winnowing waste of processing domesticated pearl millet (*Pennisetum glaucum*) was added to some pottery as temper. Pearl millet is known to be domesticated by the occasional presence of nonshattering involucre base types and the recurrence of paired spikelets (Manning and Fuller this volume). Similarly, pearl millet impressions occur in eastern Nigeria during the later Late Stone Age (2000–1000 B.C.E.).

In the latter case, contrasts with charred remains are, however, striking: pearl millet grains dominate Iron Age (450 B.C.E.–450 C.E.) samples at Kursukata but are generally absent from impressions of this period, which are instead dominated by culm fragments (Klee, Zach, and Neumann 2000). In earlier phases at the site, charred grains are present, and many impressions of chaff are recorded. A similar pattern is recorded at Ganjigana, which overlaps with early Kursukata and extends back to early-mid 2nd millennium B.C.E. Later periods have many pearl millet chaff impressions, like at early Kursukata but with a diverse mix of other wild grass spikelets, wild rice, and a small number of culms (Klee, Zach, and Stika 2004). The earliest periods have much less vegetable tempering of any kind. These patterns of change suggest a shift in the use of plant-processing by-products in potting. Perhaps as pearl millet cultivation became a more routine part of each household's economy in West Africa, the seasonal availability of bulk threshing waste increased, making it a more attractive additive to ceramics.

Table 22.1 Synopsis of Patterns of Impression Occurrence across Sherds (XX = dominant/frequent; x = regularly present-rare; + = very rare)

Country	Period	Number sites	Culms	*Oryza*	Paniceae	*Pennisetum* chaff	*Setaria*	*Echinochloa*	*Sorghum*	Leaves/twigs	Other	References
Sudan	Mesolithic (Early Khartoum)	3	XX	–	+?	–	–	–	–	XX	x	Abdel-Magid 1989, 2003; Fuller and Smith 2004
Sudan	Neolithic	9	X	–	x	–	XX	–	XX	x	x	Abdel-Magid 1989, 2003; Stemler 1990
Mali: Tilemsi valley	Neolithic	4	–	–	–	XX	–	–	–	–	–	Manning et al. 2011; Manning and Fuller this volume
Mauritania	Late Stone Age and Iron Age	3	–	–	–	XX	–	–	–	–	+	Amblard and Pernes 1989; MacDonald et al. 2003
Nigeria	Early Late Stone Age (2nd mill. B.C.E.)	1	XX	x	XX	+	–	x	–	–	–	Klee and Zach 1999
Nigeria	Later Late Stone Age 1200–400 B.C.E.	2	–	x	x	XX	–	x	–	–	–	Klee and Zach 1999; Klee et al. 2000
Nigeria	Iron Age	1	XX	x	x	–	–	–	–	–	–	Klee et al. 2000

A different pattern can be drawn from early Sudan. Here in the Early Khartoum tradition (8000–4500 B.C.E.), some rocker-stamp decorated pottery is heavily tempered with culm and grass leaves. Examination of such material from the Bayuda survey failed to record identifiable spikelets or chaff of domesticates or wild crop progenitors, although inflorescence parts (perhaps immature?) are sometimes present (Fuller 1998; Fuller and Smith 2004). Abdel-Magid (1989, 2003), however, has reported a few *Sorghum* grain/spikelet impressions, and even a *Setaria* sp. spikelet, but compared to the vast majority of vegetable-tempered sherds of the Early Khartoum tradition these are very rare.

By contrast, in pottery of the later Neolithic period (Shaheinab tradition), after ca. 4000 B.C.E., spikelets of millet-grasses (including prominently wild *Sorghum* and *Setaria* spp.) occur in sherds together with culms and leaves (Abdel-Magid 1989, 2003; Stemler 1990). This mixture suggests that material from mature (or near mature) grass panicles is occurring together with straw. While the presence of the millets testifies to their exploitation by people, it is not clear how their use in potting can be related to use in diet, nor is their absence from the earlier pottery indicative of their absence from diet. Rather we may need to consider this change in terms of the scheduling of grass collection activities vis-à-vis ceramic production. The Early Khartoum potters may have been gathering grasses as temper when these were greens, in contrast to the use of drier waste from cereal gathering in later periods, or the use of crop-processing waste that is recorded for parts of Neolithic West Africa (see Manning and Fuller this volume).

CONCLUSIONS

This study highlights the importance of considering formation processes in the creation of plant impressions. Social organisation of potting and activities that produced the appropriate plant parts are significant factors. It is also important to consider how such activities are organised in the settlement systems and scheduled through the seasons. Evidence from Bronze Age Ireland is interpreted as indicating a link between autumn crop harvest and the production of ceramic vessels. In this case, the cereal types that were more fully processed at an early stage are more likely to become incorporated into pots. Evidence from Neolithic Sudan suggests a convergence in time of the scheduling of pottery making and grass harvesting for food (the start of dry season), perhaps associated with new scheduling regimes accompanying the adoption of cattle in the Neolithic. In West Africa, the early phase has pottery with less vegetable temper. As millet cultivation became prominent, it provided a convenient source for large quantities of tempering material at the start of the dry season. Its use implies that potting and harvesting/threshing were being carried out in the same community, perhaps fairly close together in seasonal time. In the later period, a shift away from pearl millet chaff may signal the separation of potting (in time/community) from cultivation/harvesting, perhaps indicating craft specialisation. In short, the examples discussed here highlight how vegetable impressions in pottery may relate to changing patterns in plant exploitation and potting. Although this category of evidence may capture traits that relate to domestication, the presence or absence of such evidence is biased by other practices that relate to the seasonal and social organization of potting.

ACKNOWLEDGEMENTS

McClatchie's analysis of Bronze Age data from Ireland formed part of Ph.D. research funded by the National University of Ireland Travelling Studentship in Archaeology. For access to ceramic vessels, McClatchie would like to thank the National Museum of Ireland, and for access to plant remains data, she thanks: Archive Unit, National Monuments Service; CFA Archaeology Ltd.; Discovery Programme; Eachtra Archaeological Projects Ltd.; Martin Doody; Margaret Gowen & Co. Ltd.; Northern Archaeological Consultancy; QUB School of Geography, Archaeology and Palaeoecology; Stafford McLoughlin Archaeology; UCD School of Archaeology; Valerie J. Keeley Ltd.

REFERENCES

Abdel-Magid, A. (1989) *Plant Domestication in the Middle Nile Basin: An Archaeoethnobotanical Case Study*. British Archaeological Series: International Series 523. Oxford: British Archaeological Series.

———. (2003) Exploitation of food-plants in the Early and Middle Holocene Blue Nile area, Sudan and neighbouring areas. *Complutum 14*, 345–72.

Amblard, S., and J. Pernes (1989) The identification of cultivated pearl millet (*Pennisetum*) amongst plant impressions on pottery from Oued Chebbi (Dhar Oualata, Mauritania). *The African Archaeological Review 7*, 117–26.

Arnold, D. E. (1985) *Ceramic Theory and Cultural Process*. Cambridge: Cambridge University Press.

Boreland, D. (1996) Late Bronze Age pottery from Haughey's Fort. *Emania 14*, 21–28.

Cleary, R. M. (1987) Appendix VI: The pottery from Ballyveelish 2 and 3. In R. M. Cleary, M. F. Hurley, and E. A. Twohig (Eds.), *Archaeological Excavations on the Cork-Dublin Gas Pipeline (1981–1982)*, pp. 31–35. Cork: Department of Archaeology, University College, Cork.

Costantini, L. (1983) The beginning of agriculture in the Kachi Plain: The evidence of Mehrgarh. In B. Allchin (Ed.), *South Asian Archaeology 1981*, pp. 29–33. Cambridge: Cambridge University Press.

Darvill, T. (2004) Soft-rock and organic tempering in British Neolithic pottery. In R. Cleal and J. Pollard (Eds.), *Monuments and Material Culture, Papers in Honour of an Avebury Archaeologist: Isobel Smith*, pp. 193–206. Salisbury: Hobnob Press.

Fuller, D. Q (1998) Palaeoecology of the Wadi Muqaddam: A preliminary report on plant and animal remains from the Omdurman-Gabolab Survey 1997. *Sudan and Nubia 2*, 52–60.

Fuller, D. Q, K. Macdonald, and R. Vernet (2007) Early domesticated pearl millet in Dhar Nema (Mauritania): Evidence of crop-processing waste as ceramic temper. In R. Cappers (Ed.), *Fields of Change: Progress in African Archaeobotany*, Groningen Archaeological Studies 5, pp. 71–76. Groningen: Barkhuis publishing.

Fuller, D. Q, and L. Smith (2004).The prehistory of the Bayuda: New evidence from the Wadi Muqaddam. In T. Kendall (Ed.), *Nubian Studies 1998. Proceedings of the Ninth Conference of the International Society of Nubian Studies, August 21–26, 1998, Boston, Massachusetts*, 265–281. Boston: Department of African-American Studies, Northeastern University.

Fuller, D. Q, and C. J. Stevens (2009) Agriculture and the development of complex societies: An archaeobotanical agenda. In A. Fairbairn and E. Weiss (Eds.), *From Foragers to Farmers: Papers in Honour of Gordon C. Hillman*, pp. 37–57. Oxford: Oxbow.

Gibson, A. (2002) Aspects of manufacture and ceramic technology. In A. Woodward and J. D. Hill (Eds.), *Prehistoric Britain: The Ceramic Basis*, pp. 34–37. Oxford: Oxbow.

———. (2003) Prehistoric pottery: People, pattern and purpose—Some observations, questions and speculations. In A. Gibson (Ed.), *Prehistoric Pottery: People, Pattern and Purpose*, v–xi. British Archaeological Reports, International Series 1156. Oxford: Archaeopress.

Gibson, A., and A. Woods (1997) *Prehistoric Pottery for the Archaeologist*. London: Leicester University Press.

Godwin, H. (1975) *History of the British Flora*. Cambridge: Cambridge University Press.

Hartnett, P. J. (1957) Excavation of a passage grave at Fourknocks, Co. Meath. *Proceedings of the Royal Irish Academy 58C*, 197–277.

Helbaek, H. (1952) Early crops in southern England. *Proceedings of the Prehistoric Society 18*, 194–233.

———. (1959) The domestication of food plants in the old world. *Science 130*, 365–72.

Howard, H. (1981) In the wake of distribution: Towards an integrated approach to ceramic studies in prehistoric Britain. In H. Howard and E. Morris (Eds.), *Production and Distribution: A Ceramic Viewpoint*, British Archaeological Reports International Series 120, pp. 1–30. Oxford: British Archaeological Reports.

Hubbard, R. N. L. B. (1975) Assessing the botanical component of human palaeo-economies. *Bulletin of the Institute of Archaeology: University College London 12*, 197–205.

Jessen, K., and H. Helbaek (1944) Cereals in Great Britain and Ireland in prehistoric and early historic times. *Det Kongelige Danske Videnskabernes Selskab: Biologiske Skrifter 3* (2), 1–68.

Jones, A. (1999) The world on a plate: Ceramics, food technology and cosmology in Neolithic Orkney. *World Archaeology 31* (1), 55–77.

Klee, M., and B. Zach (1999) The exploitation of wild and domesticated food plants at settlement mounds in North-East Nigeria (1800 cal. BC to today). In M. van der Veen (Ed.), *The Exploitation of Plant Resources in Ancient Africa*, pp. 81–88. New York: Kluwer/Plenum.

Klee, M., B. Zach, and K. Neumann (2000) Four thousand years of plant exploitation in the Chad Basin of northeast Nigeria I: The archaeobotany of Kursakata. *Vegetation History and Archaeobotany 9*, 223–37.

Klee, M., B. Zach, and H.-P. Stika (2004) Four thousand years of plant exploitation in the Lake Chad Basin (Nigeria), Part III: Plant impressions in potsherds from the Final Stone Age Gajiganna Culture. *Vegetation History and Archaeobotany 13*, 131–42.

Kramer, C. (1985) Ceramic ethnoarchaeology. *Annual Review of Anthropology 14*, 77–102.

MacDonald, K., R. Vernet, D. Q Fuller, and J. Woodhouse (2003) New light on the Tichitt Tradition: A preliminary report on survey and excavation at Dhar Nema. In P. Mitchell, A. Haour, and J. Hobart (Eds.), *Researching Africa's Past: New Contributions from British Archaeologists*, Monograph No. 5, pp. 773–80. Oxford: Oxford University School of Archaeology.

Manning, K., R. Pelling, T. Higham, J.-L. Schwenniger, and D. Q Fuller (2011) 4500-year old domesticated pearl millet (*Pennisetum glaucum*) from the Tilemsi Valley, Mali: New insights into an alternative cereal domestication pathway. *Journal of Archaeological Science 38* (2), 312–332.

McClatchie, M. (2009) *Arable Agriculture and Social Organisation: A Study of Crops and Farming Systems in Bronze Age Ireland*. Ph.D. thesis, Institute of Archaeology, University College London.

Monk, M. A. (1986) Evidence from macroscopic plant remains for crop husbandry in prehistoric and early historic Ireland: A review. *Journal of Irish Archaeology* 3, 31–36.

Munson, P. J. (1976) Archaeological data on the origins of cultivation in the southwestern Sahara and their implications for West Africa. In J. R. Harland, J. M. J. de Wet, and A. B. L. Stemler (Eds.), *Origins of African Plant Domestication*, pp. 187–209. The Hague: Mouton.

Ó Ríordáin, B., and J. Waddell (1993) *The Funerary Bowls and Vases of the Irish Bronze Age*. Galway: Galway University Press.

Possehl, G. L. (1999) *Indus Age: The Beginnings*. Philadelphia: University of Pennsylvania Press.

Schiffer, M. B., and J. M. Skibo (1987) Theory and experiment in the study of technological change. *Current Anthropology* 28 (5), 595–622.

Stemler, A. B. (1990) A scanning electron microscopic analysis of plant impressions in pottery from sites of Kadero, El Zakiab, Um Direiwa and El Kadada. *Archeologie du Nil Moyen* 4, 87–106.

Stevens, C. (2003) An investigation of agricultural consumption and production models for prehistoric and Roman Britain. *Environmental Archaeology* 8 (1), 61–76.

van der Veen, M. (2007) Formation processes of desiccated and carbonized plant remains: The identification of routine practice. *Journal of Archaeological Science* 34, 968–90.

Vishnu-Mittre (1969) Remains of rice and millet. In H. D. Sankalia, S. B. Deo, and Z. D. Ansari (Eds.), *Excavations at Ahar (Tambavati)*, pp. 229–36. Pune: Deccan College Postgraduate and Research Institute.

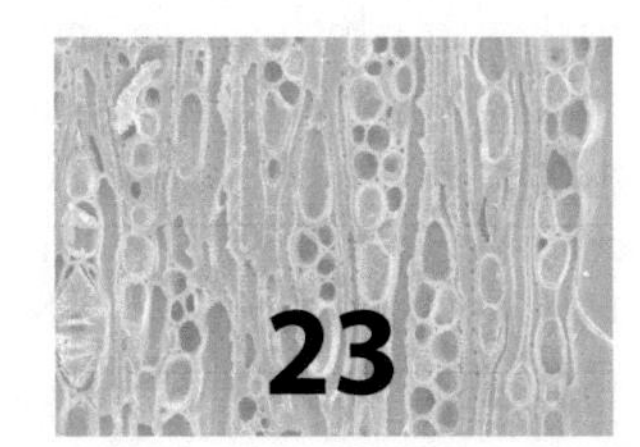

The Use of Plants in Iron Production

Insights from Smelting Remains from Buganda

Louise Iles

Successful iron production requires the complex amalgamation of several inter-related specialist technologies (including charcoal production, tuyère production, furnace construction and so on), many of which incorporate a use of plant materials. However, despite the fact that plant use is an integral and necessary component of iron production, it is an aspect of smelting that has been rarely touched upon in the existing archaeological and archaeometallurgical literature. Dedicated studies of plant use within archaeometallurgy are even fewer in number (although notable exceptions include Lyaya 2008; Mikkelsen 1997, 2003; Thompson and Young 1999).

In an effort to investigate this under-explored topic further, research was carried out into the use of plant material in some of the precolonial iron production industries of the former kingdom of Buganda, situated in what is now modern Uganda (Kiwanuka 1971; Reid 2003). This chapter summarises the results of that research. The datasets for this study were derived from blocks of slag (which are generated as one of the waste products of iron production) that had been observed to bear numerous well-preserved, macroscopic plant impressions. Such slag blocks were recovered from a number of iron smelting sites from two regions within the former kingdom, as part of wider fieldwork run by Dr. Andrew Reid of UCL Institute of Archaeology. Ultimately, this study aimed to utilise the available plant data obtained from these sources in order to identify patterns in the selection and use of plants within the smelting technologies of these two regions, interpreting these patterns within the social context of the kingdom.

To obtain the necessary data from the slag remains, a novel methodology was specifically developed for this research. To gain optimal information on the plant impressions observed in the slag, non-destructive casts were taken of the plant impressions on-site in Uganda, using a polyvinylsiloxane dental gel. These casts could then easily be transported to London for further, more detailed microscopic examination. During the subsequent analysis, where possible the casts were used to identify the impressions to the level of plant family, and this data was then employed to reveal variations and uniformities in plant selection between the distinct areas of smelting. Additional information, drawn from local informants and ethnographic sources, as well as data concerning known vegetation patterns in the region, was also considered in order to generate hypotheses concerning raw material selection criteria, thereby facilitating a discussion of the social, ecological and technical factors involved in these iron production technologies.

African Iron Production and Plant Use

The spread of iron production and working throughout Africa in the 1st millennium B.C.E. and 1st millennium C.E. had a powerful transformative effect on the continent.

Archaeology of African Plant Use by Chris J. Stevens, Sam Nixon, Mary Anne Murray, and Dorian Q Fuller, Eds., 267–274

It provided a new strong and tough material with which to create weapons and tools, thereby facilitating the clearance of forest and the intensification of agriculture as well as providing a new medium by which to accrue power and status. Partly because of this, iron production in sub-Saharan Africa has long been an area of archaeological interest (Childs and Killick 1993). Early studies tended to be heavily influenced by preconceptions of the pre-industrial nature of African technologies, yet it has developed into a discipline that carries great academic importance. Today, there is a heightened appreciation of the sophistication and complexity of indigenous, pre-colonial iron production and working techniques, which are recognised to contribute valuable new perspectives to the more general field of world archaeometallurgy.

Undoubtedly, one of its most significant contributions to the wider body of archaeometallurgical knowledge has arisen from the opportunity for ethnographers working within Africa to record living iron production practices (for example Childs 2000; Haaland 1985; Reid and MacLean 1995; van der Merwe and Avery 1987 among many others). This has contributed towards the development of arguably a more complete understanding of both the physical, social and knowledge processes that make up these technologies, and have provided an opportunity to appreciate the full *chaîne opératoires* of many examples of iron production technologies within African contexts. Such ethnographic examples have demonstrated the relevance of the social contexts of iron production, and of the importance of giving more generous consideration to the supporting technologies, such as charcoal production and ore extraction, which are essential to the iron smelting process, and yet which are often overlooked.

Through these sources, the roles played by plant resources in the iron production process have been highlighted. Most obviously this is in the form of charcoal as fuel. However, ethnographic data has shown that plants also fulfil other vital roles, for example as part of ritual or symbolic activities prior to or during smelting (for example, van der Merwe and Avery 1987), or through their incorporation into the furnace structure itself (for instance, Kagwa 1934; Roscoe 1911). Similarly, ethnographic examples have also indicated that the motivations that drive plant exploitation and utilisation are highly complex and vary widely, especially with regards to iron production (for example, Childs 2000; Thompson and Young 1999; Tabuti, Dhillion, and Lye 2003). Traditions, such as restrictions of access to sacred groves or the necessity of negotiating with forest spirits before exploiting forest resources, plus a multitude of taboos concerning the procurement and use of certain plant species all are known to act as cultural regulators of plant exploitation in the Great Lakes region (Schmidt 1997; Tabuti, Dhillion, and Lye 2003).

Nevertheless, the means by which to examine these aspects of smelting technologies are somewhat limited. The most commonly encountered archaeological remains of smelting episodes are the blocks of slag that are left behind as waste products. Fortunately, these durable, rock-like remains provide valuable clues as to the environments, physical and chemical, in which they formed.

The solid-state 'bloomery' process of iron production was the method commonly employed across pre-colonial sub-Saharan Africa (Bachmann 1982; Miller and van der Merwe 1994). During this process, iron ore and charcoal are loaded into a lit furnace, with air introduced in a controlled manner through tuyères, or air-pipes. At a temperature of around 1150°C (below the melting point of iron) the gangue (unwanted rock minerals that occur in an ore deposit) begins to melt and physically separate from the iron oxides within the ore, which remain solid throughout. The reducing atmosphere within the furnace further chemically reduces these iron oxides to iron, which come together to form a spongy iron bloom, whereas the unwanted gangue materials cool and solidify to form slag (Figure 23.1). A successful smelt requires not only a high temperature and a controlled, reducing atmosphere, but also a means within the furnace structure to provide a way of *physically* separating the liquid slag waste from the solid metal bloom.

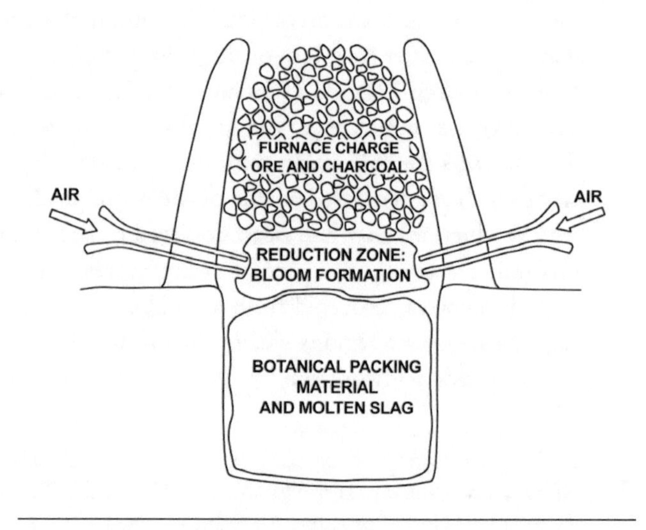

Figure 23.1 Schematic diagram showing the working of a pit furnace.

There are two major ways in which to achieve this separation. In the case of a slag-tapping furnace, the molten slag is drained away from the furnace periodically throughout the smelt, generally into a shallow pit to the side of the furnace structure. With a pit furnace, a pit is dug beneath the furnace shaft, which is packed with a rigid plant material. Straw, twigs, small branches and heather have all been documented as being used for this purpose (Mikkelsen 2003). This plant packing provides initial support for the furnace charge of ore and charcoal, and as the smelt progresses it becomes a receptacle for the molten slag, which runs through the fill structure and cools around the pit filling, leaving impressions of the packing material both on the surface of the slag and throughout it. In this way, tangible remains of the original packing material are preserved in the slag, presenting a unique opportunity to access some of the past plant use strategies employed in these technologies.

ARCHAEOLOGICAL CONTEXT OF THE STUDY

The Buganda kingdom is situated in the Great Lakes region of eastern Africa, and falls within the borders of present-day Uganda, lying on the north and northwest shores of Lake Victoria (Figure 23.2). It had become one of the most powerful and influential kingdoms in the region by the 19th century, growing in significance through a combination of banana plantation agriculture and military expansion from the 17th century. Iron was an important commodity needed both for weapons and for agricultural implements, such as the billhooks that were used to harvest bananas, Buganda's main food staple

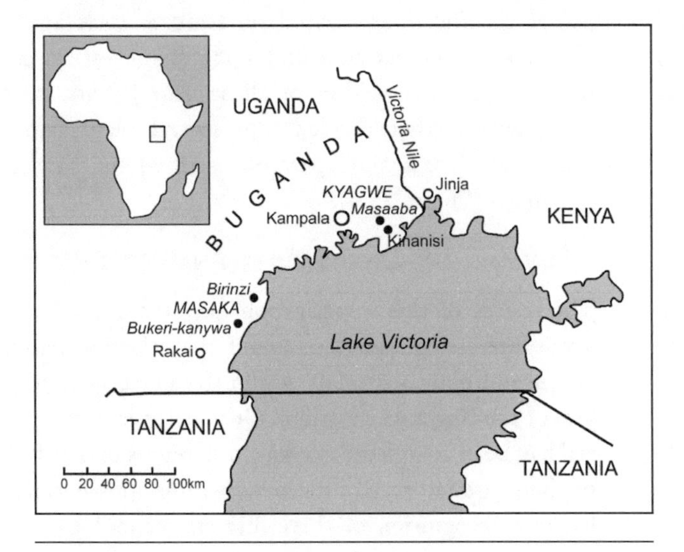

Figure 23.2 The northwest shore of Lake Victoria, showing sites mentioned in the text.

(Reid 2001). However, in the kingdom's formative years, a major strategic impediment was a lack of this crucial iron and the skilled labour capable of working this iron. Buganda responded to this need with a policy of expansion into neighbouring kingdoms, eventually encompassing the territories of Kyagwe to the east and Masaka to the west, areas which had both plentiful natural iron resources and populations that were adept in the production and working of iron (Reid 2002).

Archaeological survey between 2000 and 2003 pinpointed two main Late Iron Age iron production locations in both of these areas: Kinanisi and Masaaba in Kyagwe district, and Bukeri-Kanywa and Birinzi 100km to the west in Masaka district (cf. Reid 2003). At the site of Kinanisi in Kyagwe, numerous slag clusters were found consisting of over a hundred individual slag blocks, each of which represents the waste materials from a single smelting episode. Two furnace bases were also excavated. Only 10 km away at Masaaba, more slag clusters were found and a similarly high density of slag was recovered, although the slag was more brittle and tended to be more fragmentary. Further around the lakeshore in Masaka district, slag was recovered from two furnace bases that were excavated at Bukeri-Kanywa, and additional slag clusters were also encountered at Birinzi (Iles 2004; see Figure 23.2 for site locations). Archaeological and associated ceramic evidence suggests that all but one of these sites date to the eighteenth and nineteenth centuries; Bukeri-Kanywa alone is suggested to date to the 16th century (D. A. M. Reid, pers. comm.).

On first inspection, the slag encountered at each of these sites appeared markedly different from each other in several ways. Not only were there differences in terms of slag flows, shape, size, density and brittleness, but variation was also noted in terms of the nature of the plant impressions visible on the slag surfaces. These plant impressions were often of such good condition that they showed detailed morphological features of the plant material—structures such as culms, inflorescences and leaves, as well as venation patterns and nodes (Figure 23.3). It was felt that these features might lead to the identification of the plants that the past smelters had chosen to use in these smelting episodes.

APPLIED METHODOLOGY

The initial challenge was to transform the plant impressions in the slag (individual blocks of which could weigh up to 200 kg) into manageable and comparable samples for analysis. This was accomplished using a method adapted

Figure 23.3 Plant impressions in slag: (A) papyrus inflorescence; (B) Musaceae pseudostem impression; (C) dicotyledonous leaf impression; (D) reed impression (photographs courtesy of D. A. M. Reid).

from one previously used to identify impressions of plant material in pottery (for example, Abdel-Magid 1989; Stemler 1990; Fuller, MacDonald, and Vernet 2007; Klee, Zach, and Neumann 2000), which entailed casting the remains using a plastic impression material. One major advantage of this technique is that it allows the casts to be taken off-site to be studied using reflected light microscopy, whilst leaving the original impression undamaged.

Several casting options were considered, including latex- and silicone-based casting materials, but after much experimentation undertaken within the archaeobotanical laboratory at the UCL Institute of Archaeology prior to the 2003 field season, an addition-type polyvinylsiloxane dental gel was considered most appropriate (see also Fuller, MacDonald, and Vernet 2007). Several factors rendered the other materials inappropriate in this instance: liquid latex rubber was found to shrink and distort whilst drying, resulting in non-diagnostic casts; silicone rubber was found to require a high level of accuracy in weighing the two addition parts, an important practical consideration when working in the field. Polyvinylsiloxane was considered the most suitable as it dries quickly and without major distortion, it requires no mould and it is simple to mix and use.

Several stages were required for successful casting in the field. First, all the slag blocks selected for sampling were cleaned with water and brushes, and any plant impressions were inspected for their suitability for casting in terms of the preservation of characteristic features and morphology, and the extent of erosional damage. A paraloid consolidant (5% paraloid, 95% acetone) was then applied to the dry slag, and the two-part polyvinylsiloxane was mixed at a ratio of 1:1 and applied immediately to the selected impressions. After only a few minutes (although this was dependent on the ambient temperature and humidity), the cast was dry and able to be gently removed and washed of any residual dirt. If necessary, a second cast was taken from an impression if embedded dirt had resulted in a cast lacking in sufficient anatomical detail.

In total, nearly 500 samples were cast from selected slag blocks across the slag clusters. Samples were chosen on the basis of the clarity of impressions, and whether these impressions displayed clear morphological features. Importantly, they were also chosen to represent the proportions of different plant types present on a slag block, and detailed notes were taken of each slag block that was sampled from. The slag blocks that had been selected for additional metallurgical study were prioritised, in order to facilitate inter-disciplinary comparisons. The samples were then taken back to UCL Institute of Archaeology for analysis. In addition, several examples of local known plant species were collected, in order to be examined microscopically in conjunction with a small comparative collection. Although it was hoped that plant species would be classified through a detailed system based upon the identification of venation patterns specific to each species, unfortunately due to time restrictions and the lack of a fully comprehensive comparative collection, it was only possible to make identifications to the level of plant family. Initial classifications—based on features such as culm shape, venation patterns and node shape—were made to class (monocotyledon or dicotyledon), with monocotyledons further divided into several plant families: Gramineae (grasses), Cyperaceae (sedges) and Musaceae (bananas) (Figure 23.4).

SUMMARY AND DISCUSSION OF ANALYTICAL RESULTS

The results of this investigation (see Table 23.1) posed some interesting questions about the selection patterns and use of plant materials within the smelting technologies of the Buganda kingdom. Once analysis of the material had been completed, striking differences in the choices of plants used in each of the areas became apparent, which led to a recognition of three discrete technologies, each using distinct plant-selection or plant-use strategies.

Undoubtedly, the largest body of data was collected from Kinanisi in the Kyagwe region. Here, grasses were by far the most dominant plant material encountered, generally constituting about 90% of the samples that were positively identified, with sedges present at less than 10%. The prevalence of grasses in these samples appears to correspond with the modern vegetation landscape around Kinanisi, which is dominated by grass/savannah/forest mosaic, with limited patches of *Cyperus papyrus* swamp also accessible within a 7 km radius. In addition to the identification of this overall trend, the sampling strategy from the individual slag-blocks at Kinanisi (which each represented a single smelting episode) allowed for the plant selection strategies to be considered on a smelt-by-smelt basis. By comparing the frequencies of plant species between smelts, it was possible to see that there were high levels of continuity in plant choice from smelt to smelt. This uniformity in technological procedure was also reflected in the metallurgical analyses (cf. Humphris 2004; see also Humphris and Iles 2013).

The technological uniformity observed at Kinanisi was not unexpected. Because of iron's high material and cultural value, and the high cost of a failed smelt, many recent iron production technologies in the Great Lakes region have been seen to be steeped in ritual and tradition, which may serve to ensure technical repeatability and control over the process (de Barros 2000). Certain symbolic aspects of smelting that have been documented across sub-Saharan Africa also appear to be present in several examples of recent iron production in Buganda, as conveyed by two written accounts that document Bugandan iron smelting in the early 20th century. These accounts record nontechnical features of these technologies—for example, sexual prohibition and the attribution of gender to the different types of ore used that were integral to the success of the smelts (Roscoe 1911; Kagwa 1934). Thus it may be suggested that past smelters also had a variety of mechanisms in place that acted in keeping their materials and actions the same in every smelt they performed, in order to increase their probability of success. This technical conservatism is what may be seen reflected in the uniformity of the plant materials in use at Kinanisi.

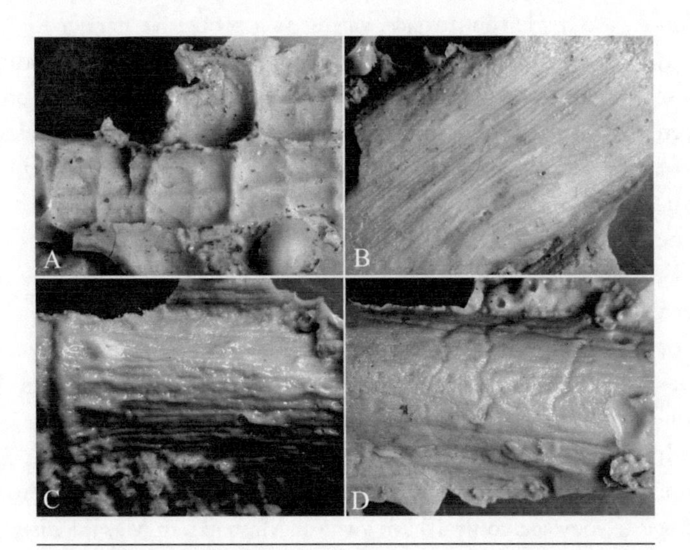

Figure 23.4 Casts of plant impressions: (A) Musaceae pseudostem; (B) sedge; (C) grass; (D) dicotyledon.

The second body of data, from the sites in Masaka, presents a slightly different scenario of smelting technology and plant procurement to that seen in Kinanisi. The metallurgical analyses showed that the smelting technologies of Birinzi and Bukeri-Kanywa were technically very similar, with both sites utilising deep pit furnaces and using kaolinitic clays to produce highly refractory, greyish tuyères (Humphris 2004). The furnace design and technical ceramics are in contrast to the shallow pit furnaces and reddish brown tuyères in evidence at Kinanisi, and also in contrast to the technology in Masaaba, which is discussed in more detail shortly. Significantly less smelting waste was found at the two sites in Masaka district, and the small volume of slag is reflected in the number of casts taken from these sites. Unfortunately, the sample size from Bukeri-Kanywa was extremely limited, with only six casts taken from a single block of slag, which represented a single smelting episode. Many of the plant impressions on this slag were small and unsuitable for casting owing to surface

Table 23.1 Number of Positively Identified Plants at Each Site

District	Site	Grasses (large)	Grasses (small)	Sedges (large)	Sedges (small)	Dicotyledons	Leaf	Palm	Musa	Total
Kyagwe	Kinanisi	227	21	3	22	6	0	0	0	279
	Masaaba	48	18	10	3	0	19	0	3	101
Masaka	Birinzi	1	0	72	0	0	0	3	0	76
	Bukeri-Kanywa	2	3	1	0	0	0	0	0	6

erosion of the slag block. However, it was still possible to see that both grass and sedge impressions were present. At Birinzi, in contrast to the evidence from Kinanisi, large sedges were the most dominant plant impression encountered, constituting about 75% of the sample set, generally stems, although numerous clear impressions of papyrus inflorescences were also present. The prevalence of sedges appears also to correspond well with modern vegetation patterns at both Birinzi and Bukeri-Kanywa; although there are surrounding grasslands, papyrus swamps are closer and more accessible than at Kinanisi.

Looking at the Kinanisi and Masaka datasets side by side, we see that although plant selection is likely to have been significantly influenced by local vegetation patterns, it is possible to suggest that selection criteria concerning the choice of plants used within the furnaces were not dictated by the availability of plant materials alone. Assuming that local vegetation patterns have remained relatively unchanged over the past few hundred years or so, both sedges and grasses are likely to have been available to the smelters at Kinanisi and in Masaka, yet in Kinanisi grasses were clearly preferred over sedges, whereas in Masaka, sedges dominated. Other influencing factors may have had an effect, although their nature, whether related to functional or cultural prerequisites, is difficult to ascertain. Certain types of plant may have been considered stronger to use or were easier to collect and carry. Alternative uses of some plants, such as grasses for thatching or papyrus for making mats, may also have shaped the value and application of certain plants in these regions. As mentioned previously, plant selection may also have been regulated by the need to avoid sacred groves, or taboos or restrictions concerning certain plants or other social issues of access. However, despite the differences in the proportions of different plant types, smelting practices in both of these areas, Kinanisi and Bukeri-Kanywa/Birinzi, were utilising the plant material in very similar ways, as packing material for pit furnaces.

The final set of botanical data returns us to Kyagwe district and the site of Masaaba, only 10 km from Kinanisi. Here, the botanical results appeared markedly different from the other sites discussed. In addition to a large percentage of grass impressions (roughly 65%), both Musaceae pseudostem impressions (ca. 3%) and dicotyledenous leaf impressions (ca. 20%) were also found on the surfaces of some of the slag fragments (see also Iles 2009). These findings were not initially understood. However, the metallurgical analysis of this material was able to discern that the slag from Masaaba had been part of a very different smelting process than had been seen at any of the other sites (Humphris 2004). At Masaaba, the macro- and micro-structures of the slag samples indicated that the slag had been tapped from the furnace and had not formed within a furnace pit as at other sites. It appeared that this distinctive slag-tapping technology in use at Masaaba might also be linked to a significantly different approach to the use of plants.

At Masaaba plant matter was not required to fill or pack furnace pits, but plants clearly continued to play a valuable role. Roscoe (1911) describes the use of green leaves to quench the fire and protect the iron bloom as it cooled, in the iron smelting that he witnessed in Buganda at the turn of the last century. It is possible that the leaf and pseudostem impressions that are apparent at Masaaba may be a result of a similar use of plants. Whereas grasses may have been used to a certain extent as a packing and supporting material within the furnace body, it is possible that banana pseudostems were used to form bunds around the tapping pits in order to control the dangerous molten slag (D. A. M. Reid pers. comm.); green, leafy branches might have been used to cover this liquid slag when it was outside the furnace, acting as a protective barrier for the smelters. Although the various suggestions as to why these plants were used are interesting, clearly a much larger body of archaeobotanical data is required before the technological associations of such plant use are fully illuminated.

Conclusion

The botanical evidence presented here, supported by the complementary archaeometallurgical data, suggests that several distinct smelting technologies were active within the Buganda kingdom at the height of its influence in the 18th and 19th centuries. The method of production and the plant use seen in the pit furnaces and the resultant slag at Kinanisi contrast greatly with the slag-tapping technology seen only 10 km away at Masaaba. In Masaka district, at Birinzi and Bukeri-Kanywa, the types of plants being used as packing material for the furnaces were markedly different from those at Kinanisi in the Kyagwe region, yet they fulfilled a similar role. So, although differences in plant availability may have had some effect on different plant utilisation strategies, it is clear that the Buganda kingdom had several iron production traditions. This diversity is potentially related to the territorial expansion of the Buganda state at this time, which resulted in the assimilation of various groups into the kingdom, each using distinct, perhaps clan-based, smelting procedures to produce iron (Humphris et al. 2009; see also Iles 2011, 2013). In this way, the sociopolitical setting of the kingdom may have given rise to the existence of such diverse iron-producing industries, as seen reflected both in the plant remains and in the metallurgical data.

The development and implementation of the methodology used here have been able to effectively highlight differences in plant use within iron production in Buganda. In conjunction with a second investigative technique, in this case archaeometallurgy, this methodology has provided insights into an aspect of iron production that is often overlooked. Unfortunately, the lack of an extensive comparative collection for this region, coupled with a restrictive time limit, meant that this project could not reach its full potential. To be of maximum use, the methodology needs to be refined further, to encompass a greater sample range and to facilitate the identification of plants to genus or species level. Then it will be possible to comprehensively address questions of technological choice and plant utilisation, which will help to confirm and highlight further details of the precolonial metal-producing technologies of this area of Great Lakes Africa. Nevertheless, I hope that this study demonstrates that even a basic identification of plant type is a worthwhile and valuable strand of archaeological investigation into iron technologies if a more complete understanding of the potential variation in these production processes is to be achieved.

ACKNOWLEDGEMENTS

Many thanks must go to those who took part in the fieldwork in Uganda in 2003, and especially to Meriel McClatchie and Dorian Fuller for their assistance with the botanical elements of this research, and to Andrew Reid for his continual support. This chapter is based on the dissertation that formed part of my Bachelor of Science degree in Archaeology, awarded by UCL Institute of Archaeology in 2004.

REFERENCES

Abdel-Magid, A. (1989) Plant domestication in the Middle Nile Basin: An archaeological case study, British Archaeological Reports International Series, 523. Oxford: British Archaeological Reports.

Bachmann, H.-G. (1982) *The Identification of Slags from Archaeological Sites*. London: Institute of Archaeology.

de Barros, P. (2000) Iron metallurgy: Sociocultural context. In M. Bisson, S. T. Childs, P. de Barros, and A. Holl, *Ancient African Metallurgy: The Socio-cultural Context*, pp. 147–98. Walnut Creek, CA: AltaMira.

Childs, S. T. (2000) Traditional iron working: A narrated ethnoarchaeological example. In M. Bisson, S. T. Childs, P. de Barros, and A. Holl (Eds.) *Ancient African Metallurgy: The Socio-Cultural Context*, pp. 199–253. Walnut Creek, CA: AltaMira.

Childs, S. T., and D. Killick (1993) Indigenous African metallurgy: Nature and culture. *Annual Review of Anthropology* 22, 317–37.

Fuller, D. Q, K. MacDonald, and R. Vernet (2007) Early domesticated millet in Dhar Nema (Mauritania): Evidence of crop processing waste as ceramic temper. In R. Cappers (Ed.), *Fields of Change: Progress in African Archaeobotany*, pp. 71–76. Groningen: Barkhuis and Groningen University Library.

Haaland, R. (1985) Iron production: Its socio-cultural context and ecological implication. In R. Haaland and P. Shinnie (Eds.), *Ancient African Iron-Working: Ancient and Traditional*, pp. 50–72. Oslo: Norwegian University Press.

Humphris, J. (2004) *Reconstructing a Forgotten Industry: An Investigation of Iron Smelting in Buganda*. M.A. thesis, University College London, Institute of Archaeology.

Humphris, J., and L. Iles (2013) Pre-colonial iron production in Great Lakes Africa: Recent research at UCL Institute of Archaeology. In J. Humphris and T. Rehren (Eds.), *The World of Iron*, pp. 56–65. London: Archetype.

Humphris, J., M. Martinón-Torres, T. Rehren, and A. Reid (2009) Variability in single smelting episodes: A pilot study using iron slag from Uganda. *Journal of Archaeological Science* 36, 359–69.

Iles, L. (2004) *Supporting the Smelt: An Archaeological Investigation into the Selection and Use of Plants within the Buganda Iron Smelting Tradition*. B.Sc. dissertation, University College London, Institute of Archaeology.

———. (2009) Impressions of banana pseudostem in iron slag from eastern Africa. *Ethnobotany Research and Applications* 7, 283–91.

———. (2011) *Reconstructing the Iron Production Technologies of Western Uganda: Reconciling Archaeometallurgical and Ethnoarchaeological Approaches*. Ph.D. thesis, University College London, Institute of Archaeology.

———. (2013) The development of iron technology in precolonial western Uganda. *Azania: Archaeological Research in Africa* 48, 65–90.

Kagwa, A. (1934) *The Customs of the Baganda*. New York: Columbia University Press.

Kiwanuka, M. (1971) *A History of Buganda from the Foundation of the Kingdom to the Present Day*. Longman: London.

Klee, M., B. Zach, and K. Neumann (2000) Four thousand years of plant exploitation in the Chad basin of northeast Nigeria 1: The archaeobotany of Kursaka. *Vegetation History and Archaeobotany* 9, 223–37.

Lyaya, E. (2008) Archaeological field research in Njombe, Tanzania. *Nyame Akuma 70*, 21–29.

Mikkelsen, P. (1997) Straw in slag pit furnaces. In L. Nørbach (Ed.), *Early Iron Production: Archaeology, Technology and Experiments*. Technical report no. 3, pp. 63–66. Lejre: Historical-Archaeological Experimental Centre.

———. (2003) Slag—with an impression of agricultural practices. In L. Nørbach (Ed.), *Prehistoric and Medieval Direct Iron Smelting in Scandinavia and Europe: Aspects of Technology and Science*, pp. 43–48. Aarhus: Aarhus University Press.

Miller, D., and N. van der Merwe (1994) Early metal working in sub-Saharan Africa: A review of recent research. *Journal of African History* 35, 1–36.

Reid, D. A. M. (2001) Bananas and the archaeology of Buganda. *Antiquity* 75, 811–12.

———. (2003) Recent research on the archaeology of Buganda. In P. Mitchell, A. Haour, and J. Hobert (Eds.), *Researching Africa's Past: New Contributions from British Archaeologists*, pp. 110–17. Oxford: Oxford University School of Archaeology.

Reid, D. A. M., and R. MacLean (1995) Symbolism and the social contexts of iron production in Karagwe. *World Archaeology* 27, 144–61.

Reid, R. (2002) *Political Power in Pre-colonial Buganda: Economy, Society and Welfare in the Nineteenth Century*. London: James Currey.

Roscoe, J. (1911) *The Baganda*. London: Macmillan and Co.

Schmidt, P. (1997) Archaeological views on a history of landscape change in east Africa. *Journal of African History 38*, 393–421.

Stemler, A. (1990) A scanning electron microscope analysis of plant impressions in pottery from the sites of Kadero, El Zakiab, Um Direiwa and El Kadada. *Archeologie du Nil Moyen 4*, 87–105.

Tabuti, J., S. Dhillion, and K. Lye (2003) Firewood use in Bulamogi County, Uganda: Species selection, harvesting and consumption patterns. *Biomass and Bioenergy 3*, 1–16.

Thompson, G., and R. Young (1999) Fuels for the furnace: Recent and prehistoric ironworking in Uganda and beyond. In M. van der Merwe (Ed.), *The Exploitation of Plant Resources in Ancient Africa*, pp. 221–40. New York: Plenum Press.

van der Merwe, N., and D. Avery (1987) Science and magic in African technology: Traditional smelting in Malawi. *Africa* 57, 143–72.

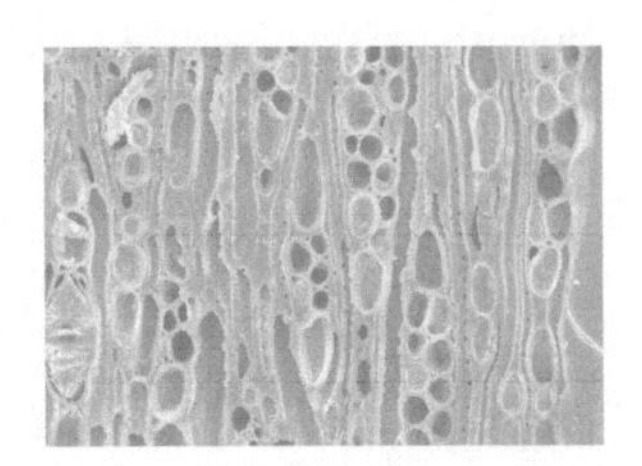

Index

Subject Index

SPECIES INDEX

D

E

F

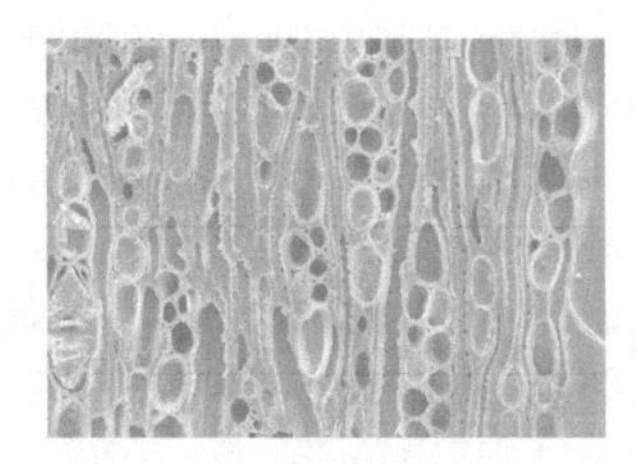

About the Contributors

Alexander Antonites is a senior lecturer at the University of Pretoria. He has an M.A. in Archaeology from University of Pretoria (2005) and completed a Ph.D. at Yale University in 2012. He has worked on salt production and consumption of farming communities in South Africa. His research interests include 1st millennium farmers in South Africa, ceramic sociology, and systems of production and consumption.

Annie Raath Antonites is currently completing a Ph.D. at Yale University. Her thesis research focuses on early farming community foodways in the Limpopo valley. She is a member of the Association of Southern African Professional Archaeologists (ASAPA) and the International Council for Archaeozoology (ICAZ) and is currently affiliated with the University of Pretoria, Archaeology and Anthropology Department. Research interests include archaeozoology and foodways.

Ceri Ashley is a senior lecturer at the University of Pretoria. She completed her Ph.D. on Iron Age ceramics and societies around Lake Victoria (Kenya, Uganda) in 2006 at University College London. She was the Cotsen Visiting Scholar at UCLA 2006–2007 and a British Academy Postdoctoral Fellow at the Institute of Archaeology, UCL (2007–2010).

Roger Blench is the chief research officer for the Kay Williamson Educational Foundation, Cambridge. He is a linguist, ethnomusicologist, and development anthropologist. He has written extensively on the relationship between linguistics and archaeology in Africa, and more recently East Asia. He is currently working on a long-term project to document the languages of Central Nigeria.

Koen Bostoen is Professor at the Department of Languages and Cultures, Ghent University. His main field of research is the historical, comparative, and descriptive study of Africa's largest language group, the Bantu languages. He has been working on the integration of language data into wider interdisciplinary approaches to studying the African past.

Caroline R. Cartwright is a research scientist at the Department of Conservation and Scientific Research, British Museum, working on the identification and interpretation of many types of organic remains. She has worked and led environmental teams on a wide range of sites covering the Near East, Africa, the Caribbean, and Europe.

Alan Clapham is a free-lance archaeobotanist who has worked extensively on material from Egypt, Europe, Japan, and the Middle East. He specialises in plant macroremains, with extensive experience of charred, desiccated, and waterlogged material. His Ph.D. undertaken at John Moores University, Liverpool, was on British submerged forests. He is currently working and publishing on the botanical material from the Egyptian sites at Amarna, Hisn al-Bab, and Qasr Ibrim.

Richard Cowling is a professor at the Department of Botany, Nelson Mandela Metropolitan University. He is a plant ecologist and has worked extensively on the conservation and management of plant biodiversity in the species-rich Cape Floral Kingdom of South Africa. He is also researching plant evolution, in particular biome evolution in Mediterranean climate ecosystems and the identification of the palaeoscapes that sustained the evolution of modern humans along the South African Cape coast.

Alioune Déme received his Ph.D. in Anthropology from Rice University. He currently teaches archaeology in the Department of History at the Cheikh Anta Diop University, Dakar, Senegal. His area of interest is West Africa, specifically the Middle Senegal Valley, where he researches state formation, subsistence, and technological evolution, alongside palaeoclimate change.

Christopher Ehret is the Distinguished Research Professor at UCLA in African history and historical linguistics. His main field of research has been within the correlation of linguistic taxonomy and reconstruction with the archaeological record. Most recently he has been working on the history and evolution of early human kinship systems.

Barbara Eichhorn is a research fellow at the Goethe University, Institute of Archaeological Sciences, Frankfurt am Main. She has researched Holocene vegetation history, in particular in West and southwest Africa with respect to human and climatic influence using phytoliths and charcoal. She currently works in Burkino Faso, Mali, and Namibia.

Ahmed Gamal-El-Din Fahmy is Professor of Botany at Helwan University, Egypt. Ahmed did his Ph.D. at Cairo University and has since worked as an archaeobotanist (and also a botanist) on many sites throughout Egypt, among them Helwan (Australian Centre for Egyptology, Macquarie University), Hierakonpolis (British Museum), and the Valley of the Kings (Tauser Temple Project, University of Arizona). Ahmed hosted the 6th International Workgroup for African Archaeology (2009) and co-edited the subsequent proceedings *Windows on the African Past: Current Approaches to African Archaeobotany* (2011).

Dorian Q Fuller is Professor of Archaeobotany at the Institute of Archaeology, University College London. He completed his Ph.D. at Cambridge (2000) on the origins of agriculture in South India and has subsequently worked on archaeobotanical material and plant domestication studies in India, China, Sudan, Ethiopia, West Africa, and the Near East. He is co-author with Eleni Asouti of *Trees and Woodlands of South India: Archaeological Perspectives* (2008, Left Coast Press) and co-editor of *Climates, Landscapes, and Civilizations* (American Geophysical Union Monograph Series, Vol. 198). He jointly edits the journal *Archaeological and Anthropological Sciences.*

Michael Haslam is an ERC Research Fellow at the Research Laboratory for Archaeology and the History of Art, University of Oxford. His research has focused on human evolution and primate tool use, as well as stone tool function and technology using microscopic use-wear and residues. He has worked in South and Southeast Asia, Australia, Oceania, Tropical Africa, and Brazil. He currently leads a major project looking at the archaeology of nonhuman primates as an evolutionary context for the emergence of human technology.

Ingrid Heijen studied archaeology at the University of Leiden. She specialized in archaeobotany under C. C. Bakels, working on Dutch medieval material from several excavations. For her master's thesis she undertook the archaeobotanical analysis of material from the Mbewe ya Mitengo excavations in Malawi. Additionally, she took part in several other projects in Greece, Egypt, and Jordan during and after her studies. She has worked in commercial archaeobotany at BIAX*consult*. She is currently based at the Ministry of Infrastructure and the Environment

Gordon C. Hillman is retired, Honorary Professor, the Institute of Archaeology, University College London, where he taught archaeobotany from 1981 to 1997, and has influenced a generation of archaeobotanists. He stimulated new directions in archaeobotany, most notably collaborations between UCL and King's College on the role of plant processing in ancient human diet and health. He is also an accomplished field botanist and plant taxonomist. He pioneered research on the domestication of cereals through now legendary, innovative ethnographic and experimental crop-processing fieldwork in Turkey in the 1970s and archaeobotanical work on the site of Abu Hureyra. He co-edited with David R. Harris *Foraging and Farming: The Evolution of Plant Exploitation* (1989) and co-authored with A. M. T. Moore and A. J. Legge *Village on the Euphrates: From Foraging to Farming at Abu Hureyra*

(Oxford University Press, 2000). In 2007 the BBC broadcast a series that Gordon worked on together with Ray Mears on the use of wild foods in Britain.

Alexa Höhn is a research associate specialising in the analyses of archaeological charcoal. She is currently working in a project on charcoals from an archaeological site close to Douala, Cameroon, at the African archaeobotany section of the Institute of Archaeological Sciences, University of Frankfurt, Germany. Main research interests are vegetation changes caused by human or climatic influence and the interrelations between vegetation changes and cultural development.

Louise Iles is currently a Marie Curie Research Fellow at the Department of Archaeology, University of York. She obtained her Ph.D. from the Institute of Archaeology, University College London (2011). She specialises in archaeometallurgy and has worked on sites in Uganda, Kenya, and Tanzania. Her current research examines the link between iron production and environmental change.

Stefanie Kahlheber is a research associate in the African archaeobotany section of the Institute of Archaeological Sciences at the University of Frankfurt, Germany. Currently she is working in a multidisciplinary project on the Nok culture in Central Nigeria. Her main research interests are prehistoric and modern plant use and the domestication history of African crops.

Nozomu Kawai is Associate Professor at the Waseda Institute for Advanced Study, Waseda University, Japan. He completed his Ph.D. in Egyptology at Johns Hopkins University (2006). He has been the field director of the Abusir-Saqqara Project as part of the Waseda University Egyptian Expedition.

Katie Manning is a Research Associate at the Institute of Archaeology, University College London. She completed her D.Phil. at Oxford University (2008). Her research centres on livestock intensification and the ecology of early farming practices, and the evolution of dietary specialisation in the context of Neolithic Europe and Africa. She is the co-editor of *African Pottery Roulettes Past and Present: Techniques, Identification and Distribution* and *The Origins and Spread of Domestic Animals in Southwest Asia and Europe.*

Águedo Marrero is a biologist at the Jardín Botánico Canario Viera y Clavijo, Las Palmas de Gran Canaria. A research centre associated to the CSIC (Spanish National Research Council). He has researched and published on taxonomy, chorology, and endangered species of the Canary Islands. He has participated at several projects in Azores, Madeira, Cape Verde, Mauritania, Morocco, and Mediterranean islands. He has been central to the discovery and documentation of more than 20 new species within the Canary Islands, including a new species of Dragon tree (*Dracaena tamaranae*).

Meriel McClatchie is currently a National University of Ireland Postdoctoral Research Fellow in the Humanities at the School of Archaeology, University College Dublin. She completed her Ph.D. in archaeobotany at University College London (2009) and has also been a postdoctoral researcher at Queen's University, Belfast. Her research has focused on the archaeobotany of Ireland from Neolithic, Bronze Age, and later periods. She has participated in archaeological fieldwork in Uganda.

Jacob Morales-Mateos is a postdoctoral research fellow of the CSIC (Spanish National Research Council). He completed his Ph.D. on archaeobotany at the University of Las Palmas de Gran Canaria (2006), after which he held postdoctoral positions in Las Palmas, Cambridge University, and University of Leicester. He is author of *El uso de las plantas en la Prehistoria de Gran Canaria: Alimentación, Agricultura y Ecología* (2010, Cabildo Insular de Gran Canaria), which provides synthesis of archaeobotany and agricultural history of the Canary Islands. His recent research has included work on historical Egypt, prehistoric Libya, and the north African Neolithic.

Mary Anne Murray is currently an archaeobotanist/environmental archaeologist working in Qatar for the University of Copenhagen and University College London (UCL-Q). She is also an Honorary Research Associate at the Institute of Archaeology, University College London, where she received her Ph.D. In Africa, she has worked extensively on ancient Egyptian plant remains from several sites, as well as those from Mali and Senegal. For many years she has also worked as an archaeologist and/or archaeobotanist in the Middle East, Europe, North and South America, and the Caribbean.

Shawn S. Murray received her Ph.D. in Anthropology from the University of Wisconsin-Madison. Dr. Murray is currently a Research Associate in Anthropology working with Rice University in Houston, TX. She specializes in past plant use, subsistence, and the rise of agriculture in West Africa.

Katharina Neumann is in charge of the African archaeobotany section in the Institute of Archaeological Sciences, University of Frankfurt, Germany. Her research focuses on prehistoric plant use and the Holocene vegetation history of West and Central Africa, with special emphasis on woody plants, phytolith analysis, and the domestication history of African crops. Currently she runs three research projects in Nigeria, Cameroon, and Mali.

Sam Nixon completed his Ph.D. at the Institute of Archaeology, University College London (2008), an investigation of the archaeology of early Islamic trans-Saharan trading towns in West Africa. His principal areas of archaeological fieldwork have been in Mali, Benin, and Morocco, and he is currently completing a monograph on his excavations at the ruins of the early Islamic-era Saharan trading town of Tadmekka in northern Mali.

Emuobosa Akpo Orijemie is researcher and lecturer in the palynology laboratory of the Department of Archaeology and Anthropology, University of Ibadan, Nigeria. He completed his Ph.D. (2013) at the same department, palynological and archaeological investigation of the late Holocene human-landscape interactions in the rainforest of southwestern Nigeria. His main area of research has been on the changing dynamics of tropical rainforest and human-landscape interactions in the Holocene, mainly through studying pollen and spore assemblages within sediment cores and excavated units. He has also been working and studying the Owu people of western Nigeria.

John Parkington is an Emeritus Professor in the Department of Archaeology, University of Cape Town. He is a well-established authority on the stone age of South Africa and has published on rock art and the coastal communities of this area. His research interests continue to focus on seasonality, mobility, and the intertwining of ethnographic, ecological, historical, and archaeological evidence.

Ruth Pelling is a senior archaeobotanist with English Heritage. She undertook her Ph.D. at University College London (2007), working on the site at Jarma (Libya), compiling an archaeobotanical database for sites in North Africa. She has worked extensively on British sites from the Neolithic to the medieval period. She has also worked on sites in Libya and Morocco.

Andrew Reid is a senior lecturer in East African archaeology at the Institute of Archaeology, University College London. He completed a Ph.D. in Archaeology at Cambridge University (1991). His research interests include livestock management and butchery practices, as well as the archaeology of African and European contact. He is involved in the Archaeology of Bananas in Buganda project with the Uganda Museum and British Institute in Eastern Africa, as well as the Dikalate project and Archaeology of the Sowa Pan in Botswana with the Kalahari Conservation Society and the University of Botswana.

Amelia Rodríguez is a lecturer in prehistory at the Department of Historical Sciences, University of Las Palmas de Gran Canaria. She specialises in lithic technology and use-wear analysis, and her research interests include the prehistoric settlement of the Canary Islands and the process of Neolithisation in the Western Mediterranean basin (Iberian Peninsula and North Africa).

Christine Sievers finished her Ph.D. at University of the Witwatersrand, studying the archaeobotanical remains from Middle Stone Age deposits at Sibudu Cave. She lectures in the School of Geography, Archaeology and Environmental Studies of the University of the Witwatersrand. She has also undertaken archaeobotanical work from excavations in the Limpopo and North West provinces of South Africa.

M. Adebisi Sowunmi is a retired professor from Department of Archaeology and Anthropology, University of Ibadan, Ibadan, Nigeria. She completed her Ph.D. in 1967 on pollen morphology of the Palmae. She pioneered palynological studies in Nigeria and worked on recognition of anthropogenic impacts in pollen evidence from tropical Africa. She has also written on the social and economic context of the practice of archaeology in Africa and was a visiting professor at the University of Uppsala and the UCL Institute of Archaeology in the 1990s.

Chris J. Stevens is a Post-Doctoral Research Associate in Archaeobotany at the Institute of Archaeology, University College London, working on a European Research Council project on Comparative Pathways to Agriculture

(ComPAg). He completed his Ph.D. at Cambridge (1996) and was a postdoctoral fellow at McDonald Institute for Archaeological Research. He has worked extensively as an archaeobotanist in British developer-funded archaeology and as archaeobotanist on field projects in Italy and at Amarna, Egypt. He is co-author of *Environmental Archaeology: Approaches, Techniques and Applications* (Tempus, 2003).

Michèle Wollstonecroft is a Teaching Fellow in Environmental Archaeology at the Institute of Archaeology, University College London. She completed her Ph.D. at the Institute of Archaeology (2007) on food-processing intensification in the Epipalaeolithic Near East, followed by a British Academy Research Fellowship (2008–2011) to investigate the role of plant food processing in prehistoric economies, particularly how innovations in technology or techniques may have influenced species selection and dietary change. She has carried out fieldwork in western Canada, Portugal, Turkey, Ghana, and Ethiopia.

Sakuji Yoshimura is Professor Emeritus at Waseda University, Japan. He is the General Director of the Waseda University Egyptian Expedition. He has been directing excavations at Saqqara since 1991 and more recently also on the intact tombs at Dahshur. He is currently working on the restoration of the Second Boat of Khufu at Giza.